# Graph Algorithms and Applications 4

# Graph Algorithms and Applications 4

**EDITORS**

**Giuseppe Liotta**
*University of Perugia, Italy*

**Roberto Tamassia**
*Brown University, USA*

**Ioannis G Tollis**
*University of Crete, ICS-FORTH, Greece and*
*The University of Texas at Dallas, USA*

World Scientific

NEW JERSEY · LONDON · SINGAPORE · BEIJING · SHANGHAI · HONG KONG · TAIPEI · CHENNAI

*Published by*

World Scientific Publishing Co. Pte. Ltd.

5 Toh Tuck Link, Singapore 596224

*USA office:* 27 Warren Street, Suite 401-402, Hackensack, NJ 07601

*UK office:* 57 Shelton Street, Covent Garden, London WC2H 9HE

**British Library Cataloguing-in-Publication Data**
A catalogue record for this book is available from the British Library.

**GRAPH ALGORITHMS AND APPLICATIONS 4**



ISBN 981-256-844-1 (pbk)

Printed in Singapore.

# Preface

This book contains volume 7 of the *Journal of Graph Algorithms and Applications* (*JGAA*). Among other papers, the book contains two special issues.

Topics of interest for *JGAA* include:

*Design and analysis of graph algorithms:* exact and approximation graph algorithms; centralized and distributed graph algorithms; static and dynamic graph algorithms; internal- and external-memory graph algorithms; sequential and parallel graph algorithms; deterministic and randomized graph algorithms.

*Experiences with graph algorithms:* animations; experimentations; implementations.

*Applications of graph algorithms:* computational biology; computational geometry; computer graphics; computer-aided design; computer and interconnection networks; constraint systems; databases; graph drawing; graph embedding and layout; knowledge representation; multimedia; software engineering; telecommunication networks; user interfaces and visualization; VLSI circuits.

*JGAA* is supported by distinguished advisory and editorial boards, has high scientific standards, and takes advantage of current electronic document technology. The electronic version of *JGAA* is available on the Web at

http://jgaa.info/

We would like to express our gratitude to the members of the advisory board for their encouragement and support of the journal, to the members of the editorial board and guest editors for their invaluable service and to the many anonymous referees for their essential role in the selection process. Finally, we would like to thank all the authors who have submitted papers to *JGAA*.

Giuseppe Liotta
Roberto Tamassia
Ioannis G. Tollis

# Journal of Graph Algorithms and Applications

**Managing Editor:**

*Giuseppe Liotta*, University of Perugia

**Publication Editor:**

*Emilio Di Giacomo*, University of Perugia

**Editors-in-Chief:**

*Roberto Tamassia*, Brown University
*Ioannis G. Tollis*, University of Crete and ICS-FORTH

**Advisory Board:**

*I. Chlamtac*, CREATE-NET
*S. Even*, Technion
*G. N. Frederickson*, Purdue University
*T. C. Hu*, University of California at San Diego
*D. E. Knuth*, Stanford University
*C. L. Liu*, University of Illinois
*K. Mehlhorn*, Max-Planck-Institut für Informatik
*T. Nishizeki*, Tohoku University
*F. P. Preparata*, Brown University
*I. H. Sudborough*, University of Texas at Dallas
*R. E. Tarjan*, Princeton University
*M. Yannakakis*, Columbia University

**Editorial Board:**

*S. Albers*, Universität Freiburg
*L. Arge*, University of Aarhus
*U. Brandes*, Universität Konstanz
*A. L. Buchsbaum*, AT&T Labs – Research
*G. Di Battista*, University of Roma Tre
*P. Eades*, University of Sydney
*D. Eppstein*, University of California at Irvine
*M. Fürer*, Pennsylvania State University
*A. Gibbons*, King's College
*M. T. Goodrich*, University of California at Irvine

*X. He,* State University of New York at Buffalo
*A. Itai,* Technion
*Y. Kajitani,* University of Kitakyushu
*M. Kaufmann,* Universität Tübingen
*S. Khuller,* University of Maryland
*S. G. Kobourov,* University of Arizona
*E. W. Mayr,* Technischen Universität München
*H. Meijer,* Queen's University
*J. S. B. Mitchell,* State University of New York at Stony Brook
*B. M. E. Moret,* University of New Mexico
*P. Mutzel,* Universität Dortmund
*B. Raghavachari,* University of Texas at Dallas
*D. Wagner,* University of Karlsruhe
*T. Warnow,* University of Texas at Austin
*S. Whitesides,* McGill University

# Contents

## Volume 7:4 (2003) 305

**Advances in Graph Drawing. Special Issue on Selected Papers from the Ninth International Symposium on Graph Drawing, GD 2001.** *Guest Editor(s): Petra Mutzel and Michael Jünger.*

Volume 7:1 (2003)

# Journal of Graph Algorithms and Applications

http://jgaa.info/

vol. 7, no. 1, pp. 3–31 (2003)


# Statistical Analysis of Algorithms: A Case Study of Market-Clearing Mechanisms in the Power Industry


*Chris Barrett*  *Achla Marathe*  *Madhav Marathe*
Los Alamos National Laboratory

*Doug Cook*
Colorado School of Mines

*Gregory Hicks*
North Carolina State University

*Vance Faber*
Mapping Sciences Inc.

*Aravind Srinivasan*
University of Maryland, College Park

*Yoram Sussmann*
State University of West Georgia

*Heidi Thornquist*
Rice University



## Abstract

We carry out a detailed empirical analysis of simple heuristics and provable algorithms for bilateral contract-satisfaction problems. Such problems arise due to the proposed deregulation of the electric utility industry in the USA. Given a network and a (multi)set of pairs of vertices (contracts) with associated demands, the goal is to find the maximum number of simultaneously satisfiable contracts. Four different algorithms (three heuristics and a provable approximation algorithm) are considered and their performance is studied empirically in fairly realistic settings using rigorous statistical analysis. For this purpose, we use an approximate electrical transmission network in the state of Colorado. Our experiments are based on the statistical technique *Analysis of Variance* (ANOVA), and show that the three heuristics outperform a theoretically better algorithm. We also test the algorithms on four types of *scenarios* that are likely to occur in a deregulated marketplace. Our results show that the networks that are adequate in a regulated marketplace might be inadequate for satisfying all the bilateral contracts in a deregulated industry.



Communicated by Dorothea Wagner: submitted April 2002;
revised December 2002.

C. Barrett et al., *Statistical Analysis of Algorithms*, JGAA, 7(1) 3–31 (2003)  4

# 1  Introduction

The U.S. electric utility industry is undergoing major structural changes in an effort to make it more competitive [21, 17, 19, 11]. One major consequence of the deregulation will be to decouple the controllers of the network from the power producers, making it difficult to regulate the levels of power on the network; consumers as well as producers will eventually be able to negotiate prices to buy and sell electricity [18]. In practice, deregulation is complicated by the facts that all power companies will have to share the same power network in the short term, with the network's capacity being just about sufficient to meet the current demand. To overcome these problems, most U.S. states have set up an ISO (independent system operator): a non-profit governing body to arbitrate the use of the network. The basic questions facing ISOs are how to decide which contracts to deny (due to capacity constraints), and who is to bear the costs accrued when contracts are denied. Several criteria/policies have been proposed and/or are being legislated by the states as possible guidelines for the ISO to select a maximum-sized subset of contracts that can be cleared simultaneously [18]. These include: (a) Minimum Flow Denied: The ISO selects the subset of contracts that denies the *least* amount of proposed power flow. This proposal favors clearing bigger contracts first. (b) First-in First-out: The contract that comes first gets cleared first; this is the least discriminating to the contractors. (c) Maximum Consumers Served: This clears the smallest contracts first and favors the small buyers whose interests normally tend to go unheard.

There are three key issues in deciding policies that entail specific mechanisms for selecting a subset of contracts: *fairness* of a given policy to producers and consumers; the *computational complexity* of implementing a policy, and how *sound* a given policy is from an economic standpoint. (For instance, does the policy promote the optimal clearing price/network utilization etc.) Here we focus on evaluating the efficacy of a given policy with regard to its computational resource requirement and network resource utilization. It is intuitively clear that the underlying network, its capacity and topology, and the spatial locations of the bilateral contracts on the network, will play an important role in determining the efficacy of these policies. We do not discuss here the fairness and economics aspects of these policies: these are subjects of a companion paper. The work reported here is done as part of a simulation based analytical tool for deregulated electrical power industry being developed at the Los Alamos National Laboratory.

We experimentally analyze several algorithms for simultaneously clearing a maximal number of bilateral contracts. The qualitative insights obtained in this paper can be useful to policy makers who carry the ultimate responsibility of deploying the best clearing mechanism in the real world. The algorithms were chosen according to provable performance, ability to serve as a proxy for some of the above-stated policies, and computational requirement. The algorithms are as follows; see § 3 for their specification. The ILP-RANDOMIZED ROUNDING (RR) algorithm has a provable performance guarantee under certain conditions. The computational resource requirement is quite high, but the approach also

C. Barrett et al., *Statistical Analysis of Algorithms*, JGAA, 7(1) 3–31 (2003) 5

provides us with an upper bound on any optimal solution and proves useful in comparing the performance of the algorithms. The LARGEST-FIRST HEURISTIC (LF) is a proxy for the *Minimum Flow Denied* policy. The SMALLEST-FIRST HEURISTIC (SF) serves as a proxy for the *Maximum Contracts Served* policy. The RANDOM-ORDER HEURISTIC (RO) clears the contracts in the random order. This algorithm was chosen as a proxy for the *First-in First-out* policy. Such a policy is probably the most natural clearing mechanism and is currently in place at many exchanges.

To compare the algorithms in a quantitative and (semi-)rigorous way, we employ statistical tools and experimental designs. Many of the basic tools are standard in statistics and their use is common in other fields. But to the best of our knowledge, the use of formal statistical methods in experimental algorithmics for analyzing/comparing the performance of algorithms has not been investigated. *Analysis of Variance* (ANOVA) is one such technique that can help identify which algorithms and scenarios are superior in performance. We believe that such statistical methods should be investigated further by the experimental algorithmics community for deriving more (semi)-quantitative conclusions when theoretical proofs are hard or not very insightful. For instance, consider a given approximation algorithm that has a worst-case performance guarantee of $\rho$. First, the algorithm may perform much better on realistic instances that are of interest. Quantifying the special structure of such instances is often hard; this often makes it difficult to develop further theoretical improvements on the performance of the algorithm. Second, many heuristics that have poor worst-case performance perform very well on such instances. Statistical methods such as ANOVA can facilitate the comparison of such heuristics and provable algorithms in settings that are of interest to the users of such algorithms.

We used a coarse representation of the Colorado electrical power network (see § 4) to qualitatively compare the four algorithms discussed above in fairly realistic settings. The realistic networks differ from random networks and structured networks in the following ways: (i) Realistic networks typically have a very low average degree. In fact, in our case the average degree of the network is no more than 3. (ii) Realistic networks are not very uniform. One typically sees one or two large clusters (downtown and neighboring areas) and small clusters spread out throughout. (iii) For most empirical studies with random networks, the edge weights are chosen independently and uniformly at random from a given interval. However, realistic networks typically have very specific kinds of capacities since they are constructed with particular design goal.

From our preliminary analysis, it appears that although the simple heuristic algorithms do not have worst-case performance guarantees, they outperform the theoretically better randomized rounding algorithm. We tested the algorithms on four carefully chosen scenarios. Each scenario was designed to test the algorithms and the resulting solutions in a deregulated setting. The empirical results show that networks that are capable of satisfying all demand in a regulated marketplace can often be inadequate for satisfying all (or even a acceptable fraction) of the bilateral contracts in a deregulated market. Our results also confirm intuitive observations: e.g., the number of contracts satisfied

C. Barrett et al., *Statistical Analysis of Algorithms*, JGAA, 7(1) 3–31 (2003)  6

crucially depends on the scenario and the algorithm.

As far as we are aware, this is the first study to investigate the efficacy of various policies for contract satisfaction in a deregulated power industry. Since it was done in fairly realistic settings, the qualitative results obtained here have implications for policy makers. Our results can also be applied in other settings, such as bandwidth-trading on the Internet. See, e.g., [2]. Also, to our knowledge, previous researchers have not considered the effect of the underlying network on the problems; this is an important parameter especially in a free-market scenario.

The rest of this paper is organized as follows. The problem definitions and algorithms considered are described in Sections 2 and 3 respectively. Our experimental setup is discussed in Section 4. Section 5 presents our experimental results and analyzes them and Section 6 concludes the paper. In the appendix, we discuss interesting optimization issues that arise from deregulation, and also show problem instances on which our algorithms do not perform well.

## 2  Problem Definitions

We briefly define the optimization problems studied here. We are given an undirected network (the power network) $G = (V, E)$ with capacities $c_e$ for each edge $e$ and a set of source-sink node pairs $(s_i, t_i)$, $1 \leq i \leq k$. Each pair $(s_i, t_i)$ has: (i) an integral *demand* reflecting the amount of power that $s_i$ agrees to supply to $t_i$ and (ii) a negotiated *cost* of sending unit commodity from $s_i$ to $t_i$. As is traditional in the power literature, we will refer to the source-sink pairs along with the associated demands as *a set of contracts*. In general, a source or sink may have multiple associated contracts. We find the following notation convenient to describe the problems. Denote the set of nodes by $N$. The contracts are defined by a relation $R \subseteq (N \times N \times \Re \times \Re)$ so that tuple $(v, w, \alpha, \beta) \in R$ denotes a contract between source $v$ and sink $w$ for $\alpha$ units of commodity at a cost of $\beta$ per unit of the commodity. For $A = (v, w, \alpha, \beta) \in R$ we denote $source(A) = v$, $sink(A) = w$, $flow(A) = \alpha$ and $cost(A) = \beta$. Corresponding to the power network, we construct a digraph $H = (V \cup S \cup T \cup \{s, t\}, E')$ with source $s$, sink node $t$, capacities $u : E' \to \Re$ and costs $c' : E' \to \Re$ as follows. For all $A \in R$, define new vertices $v_A$ and $w_A$. Let $S = \{v_A \mid A \in R\}$ and $T = \{w_A \mid A \in R\}$. Each edge $\{x, y\}$ from $G$ is present in $H$ as the two arcs $(x, y)$ and $(y, x)$ that have the same capacity as $\{x, y\}$ has in $G$, and with cost 0. In addition, for all $A = (v, w, \alpha, \beta) \in R$, we introduce: (i) arcs $(v_A, v)$ and $(w, w_A)$ with infinite capacity and zero cost; (ii) arc $(s, v_A)$ with capacity $flow(A) = \alpha$ and cost 0; and (iii) arc $(w_A, t)$ with capacity $flow(A) = \alpha$ and cost equaling $cost(A)$. By this construction, we can assume without loss of generality that each node can participate in exactly one contract. A *flow* is simply an assignment of values to the edges in a graph, where the value of an edge is the amount of flow traveling on that edge. The value of the flow is defined as the amount of flow coming out of $s$ (or equivalently the amount of flow coming in to $t$). A generic *feasible flow* $f = (f_{x,y} \geq 0 : (x, y) \in E')$ in

C. Barrett et al., *Statistical Analysis of Algorithms*, JGAA, 7(1) 3–31 (2003)  7

$H$ is any non-negative flow that: (a) respects the arc capacities, (b) has $s$ as the only source of flow and $t$ as the only sink. Note that for a given $A \in R$, in general it is not necessary that $f_{s,v_A} = f_{w_A,t}$. For a given contract $A \in R$, $A$ is said to be *satisfied* if the feasible flow $f$ in $H$ has the additional property that for $A = (v, w, \alpha, \beta)$, $f_{s,v_A} = f_{w_A,t} = \alpha$; i.e., the contractual obligation of $\alpha$ units of commodity is shipped out of $v$ and the same amount is received at $w$. Given a power network $G(V, E)$, a contract set $R$ is *feasible* (or *satisfied*) if there exists a feasible flow $f$ in the digraph $H$ that satisfies every contract $A \in R$. Let $B = supply(s) = demand(t) = \sum_{A \in R} flow(A)$.

In practice, it is typically the case that $R$ does not form a feasible set. As a result we have two possible alternative methods of relaxing the constraints: (i) relax the notion of feasibility of a contract and (ii) try and find a subset of contracts that are feasible. Combining these two alternatives we define the following types of "relaxed feasible" subsets of $R$. We will concern ourselves with only one variant here. A contract set $R' \subseteq R$ is a *0/1-contract satisfaction feasible set* if, $\forall A = (v, w, \alpha, \beta) \in R'$, $f_{s,v_A} = f_{w_A,t} = \alpha$.

**Definition 2.1** *Given a graph $G(V, E)$ and a contract set $R$, the (0/1-VERSION, MAX-FEASIBLE FLOW) problem is to find a feasible flow $f$ in $H$ such that $\sum_{A \in R'} f(A)$ is maximized where $R'$ forms a 0/1-contract satisfaction feasible set of contracts. In the related (0/1-VERSION, MAX-#CONTRACTS) problem, we aim to find a feasible flow $f$ in $H$ such that $|R'|$ is maximized, where $R'$ forms a 0/1-contract satisfaction feasible set of contracts.*

Though such electric flow problems have some similarities with those from other practical situations, there are many basic differences such as reliability, indistinguishability between the power produced by different generators, short life-time due to inadequate storage, line effects etc. [22]. The variants of flow problems related to power transmission studied here are intuitively harder than traditional multi-commodity flow problems, since we *cannot distinguish between* the flow "commodities" (power produced by different generators). As a result, current solution techniques used to solve single/multi-commodity flow problems are not directly applicable to the problems considered here.

## 3  Description and Discussion of Algorithms

We work on the (0/1-VERSION, MAX-#CONTRACTS) problem here. Let $n$ and $m$ respectively denote the number of vertices and edges in the network $G$. In [5], it was shown that (0/1-VERSION, MAX-#CONTRACTS) is $NP$-hard; also, unless $NP \subseteq ZPP$, it cannot be approximated to within a factor of $m^{1/2-\epsilon}$ for any fixed $\epsilon > 0$, in polynomial time. Thus, we need to consider good heuristics/approximation algorithms. First, there are three simple heuristics. The SMALLEST-FIRST HEURISTIC considers the contracts in non-decreasing order of their demands. When a contract is considered, we accept it if it can be feasibly added to the current set of chosen contracts, and reject it otherwise. The LARGEST-FIRST HEURISTIC is the same, except that the contracts are ordered

C. Barrett et al., *Statistical Analysis of Algorithms*, JGAA, 7(1) 3–31 (2003)  8

in non-increasing order of demands. In the RANDOM-ORDER HEURISTIC, the contracts are considered in a random order.

We next briefly discuss an approximation algorithm of [5]. This has proven performance only when all source vertices $s_i$ are the same; however, this algorithm extends naturally to the multi-source case which we work on. An integer linear programming (ILP) formulation for the problem is considered in [5]. We have indicator variables $x_i$ for the contract between $s_i$ and $t_i$, and variables $z_{i,e}$ for each $(s_i, t_i)$ pair and each edge $e$. The intended meaning of $x_i$ is that the total flow for $(s_i, t_i)$ is $d_i x_i$; the meaning of $z_{i,e}$ is that the flow due to the contract between $(s_i, t_i)$ on edge $e$ is $z_{i,e}$. We write the obvious flow and capacity constraints. Crucially, we also add the valid constraint $z_{i,e} \leq c_e x_i$ for all $i$ and $e$. In the integral version of the problem, we will have $x_i \in \{0, 1\}$, and the $z_{i,e}$ as non-negative reals. We relax the condition "$x_i \in \{0, 1\}$" to "$x_i \in [0, 1]$" and solve the resultant LP; let $y^*$ be the LP's optimal objective function value. We perform the following rounding steps using a carefully chosen parameter $\lambda > 1$. (a) Independently for each $i$, set a random variable $Y_i$ to 1 with probability $x_i/\lambda$, and $Y_i := 0$ with probability $1 - x_i/\lambda$. (b) If $Y_i = 1$, we will choose to satisfy $(1 - \epsilon)$ of $(s_i, t_i)$'s contract: for all $e \in E$, set $z_{i,e} := z_{i,e}(1 - \epsilon)/x_i$. (c) If $Y_i = 0$, we choose to have no flow for $(s_i, t_i)$: i.e., we will reset all the $z_{i,e}$ to 0. A deterministic version of this result based on *pessimistic estimators*, is also provided in [5]; see [5] for further details.

**Theorem 3.1** *([5]) Given a network $G$ and a contract set $R$, we can find an approximation algorithm for (0/1-VERSION, MAX-#CONTRACTS) when all source vertices are the same. Let $OPT$ be the optimum value of the problem, and $m$ be the number of edges in $G$. Then, for any given $\epsilon > 0$, we can in polynomial time find a subset of contracts $R'$ with total weight $\Omega(OPT \cdot \min\{(OPT/m)^{(1-\epsilon)/\epsilon}, 1\})$ such that for all $i \in R'$, the flow is at least $(1 - \epsilon)d_i$.*

## 4   Experimental Setup and Methodology

To test our algorithms experimentally, we used a network corresponding to a subset of a real power network along with contracts that we generated using different scenarios. The network we used is based on the power grid in Colorado and was derived from data obtained from PSCo's (Public Service Company of Colorado) Draft Integrated Resources Plan listing of power stations and major sub stations. The network is shown in Figure 1. We restricted our attention to major trunks only.

**Sources:** The location and capacities of the sources was roughly based upon data obtained from PSCo's Draft Integrated Resources Plan listing of power stations and major sub stations.

**Sinks:** The location and capacity of the sinks were roughly based upon the demographics of the state of Colorado. In order to determine the location and capacity of the sinks we used the number of households per county obtained from

C. Barrett et al., *Statistical Analysis of Algorithms*, JGAA, 7(1) 3–31 (2003)  9

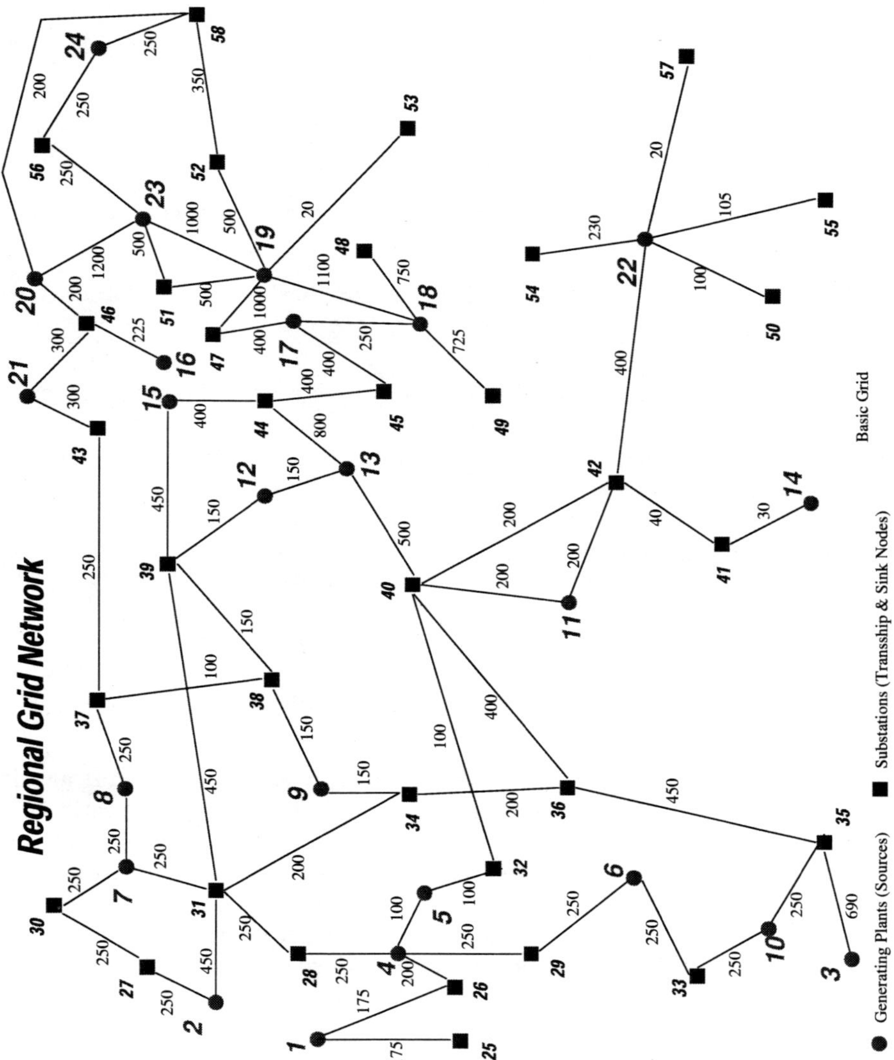



Figure 1: This shows the network with node numbered as they are referenced in all scenarios and edge capacities labeled at values used for **Scenarios 1 & 2**. The placement of the nodes and edges are what is probably the final form. The least number of edges cross and the nodes in the upper right are spread out a little bit maintaining the general feel of the distribution while allowing easier reading.

C. Barrett et al., *Statistical Analysis of Algorithms*, JGAA, 7(1) 3–31 (2003) 10

the US Census bureau. By assigning counties (load) to specific sub stations (sink nodes) the data for the sinks were derived.

The following three websites can be accessed to obtain the necessary information:

- `http://www.census.gov/population/estimates/county/co-99-1/99C1_08.txt` gives the population per county as of 1995.

- `http://www.census.gov/datamap/www/08.html` contains a map of Colorado counties.

- Finally, `http://ccpg.basinelectric.com/` is the PSCo Colorado Website.

**Edges:** The edge capacities were derived from test data obtained by running the network through a max-flow program with the source and sink capacities at maximum and no capacity limits placed upon the connecting edges. The total sink capacity equaled the total source capacity. The sink capacity was distributed to the various sink nodes in correspondence with population percentages assigned to each sink node. The edge capacities were then roughly assigned and the model was rerun through the max-flow program until all edge limits were defined. The criteria used for defining all of edge limits was that the network must be feasible under the condition of maximum source/sink capacity. Once the feasibility criteria was satisfied, some edge limits were set at capacity, while others were set higher than capacity in order to provide flexibility in contract development for the later problems.

**Software and Data Format.** DIMACS (http://dimacs.rutgers.edu) has developed a standard format for storing network data for input into existing network solvers. For the network being examined the need exists to include two directed arcs for each edge since the network is undirected. Addition of a master source and a master sink node with edges to the individual source and sink nodes was needed in order to conform to the format requirement of a single source and a single sink node. The edge capacities of the edges from the master source and sink nodes were set to be the capacities of the respective individual source or sink node.

## 4.1   Creation and Description of Test Cases

All the test cases were generated from the basic model. The general approach we used was to fix the edge capacities and generate source-sink contract combinations, using the capacities and aggregate demands in the basic model as upper bounds. To ensure that the test cases we generated corresponded to (1) difficult problems, i.e. infeasible sets of contracts, and (2) problems that might reasonably arise in reality, we developed several scenarios that included an element of randomness (described in § 4.2).

C. Barrett et al., *Statistical Analysis of Algorithms*, JGAA, 7(1) 3–31 (2003) 11

## 4.2 Description of Scenarios

The current implementation is still based upon a network which should be feasible only if the total source capacity is greater than the total sink capacity and the only requirement is that the total sink capacity be satisfied regardless of which source provides the power. **Scenarios 1, 2, 3** and **4** are based around the network with total generating capacity **6249 MW**, and reduced sink capacities near **4400MW** combined. See Figures 2–4.

1. **Scenario 1:**  This scenario is based upon the network with a total sink capacity (i.e. customer demand) of 4400MW. The source capacity (supplier's maximum production capacity) was reduced by a constant proportion from the total generating capacity based upon population density of Colorado counties. The source capacity of the network was reduced until the running the MAXFLOW code indicated that the maximum flow in the network to be slightly less than the demand. This reduction in the sources total production capacity increased the chances of refusing customers (contracts).

2. **Scenario 2:**  For this scenario, we took the basic network and increased the sink capacity while the source capacity remained fixed.

3. **Scenario 3:**  For generating instances for this scenario, the edge capacities were adjusted, reduced in most cases, to limit the network to a maximum flow of slightly more than 4400MW given its source and sink distribution. Here, if the load is allowed to be fulfilled from any source (as is normally done with centralized control), the network and the edge capacities are enough to handle a total of 4400MW. However, if we insist that a particular source needs to serve a particular sink (as is done in bilateral contract satisfaction), then the capacities may not be enough to handle the same load of 4400MW.

4. **Scenario 4:**  For this scenario, we took the network of Scenario 3 and biased the selection of source nodes towards the lower valued source units.

## 4.3 Methodology

We worked with the four scenarios and ran all four algorithms for each. For the three greedy heuristics the implementations are fairly straightforward, and we used public-domain network flow codes. Implementing the randomized rounding procedure requires extra care. The pessimistic estimator approach of [5] works with very low probabilities, and requires significant, repeated re-scaling in practice. Thus we focus on the randomized version of the algorithm of [5]; five representative values of $\epsilon$ varying from .1 to .5 were chosen. We believe that satisfying a contract partially so that a contract is assigned less than .5 of the required demand is not very realistic. For each scenario, and for each of the 5 values of $\epsilon$, the programs implementing the algorithms under inspection produced 30 files from which the following information could be extracted:

C. Barrett et al., *Statistical Analysis of Algorithms*, JGAA, 7(1) 3–31 (2003) 12

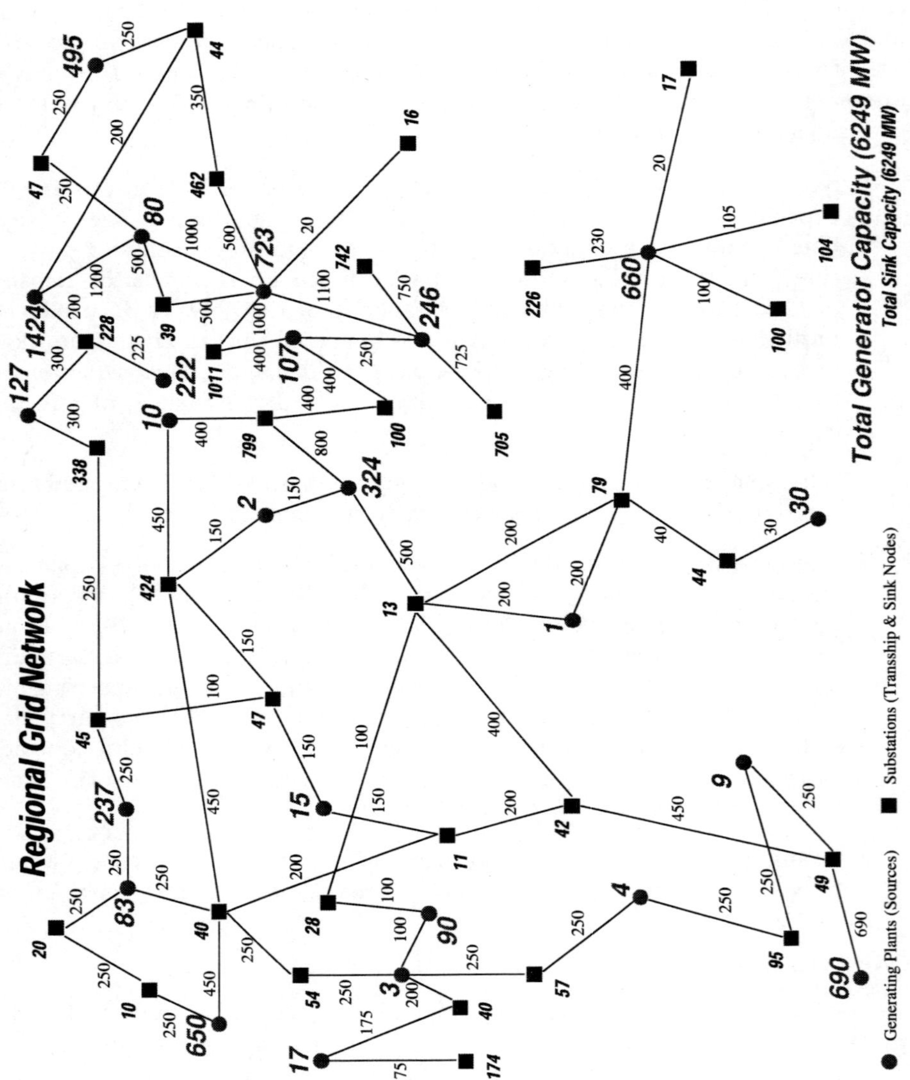


Figure 2: Shows the maximum capacities of the nodes and edges at the values used in **Scenario 2**. The positioning of the nodes and edges have not been changed to the same as the previous figure.

C. Barrett et al., *Statistical Analysis of Algorithms*, JGAA, 7(1) 3–31 (2003) 13

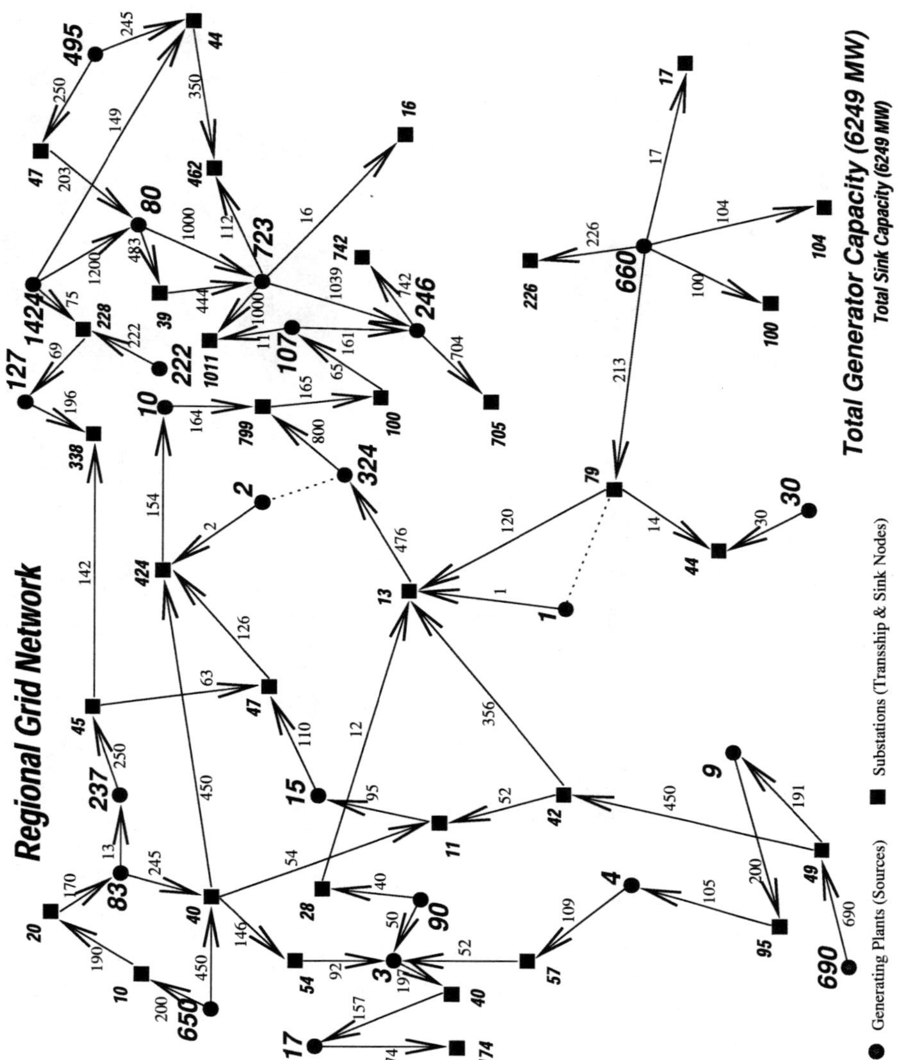


Figure 3: Shows the same network as the maximum capacities except the edges have been modified with arrows indicating direction of flow and the numbers associated with the edges are the flow values not the capacities. The edges with no flow have been changed to dotted lines although one or two of the dotted lines may look solid.

C. Barrett et al., *Statistical Analysis of Algorithms*, JGAA, 7(1) 3–31 (2003) 14

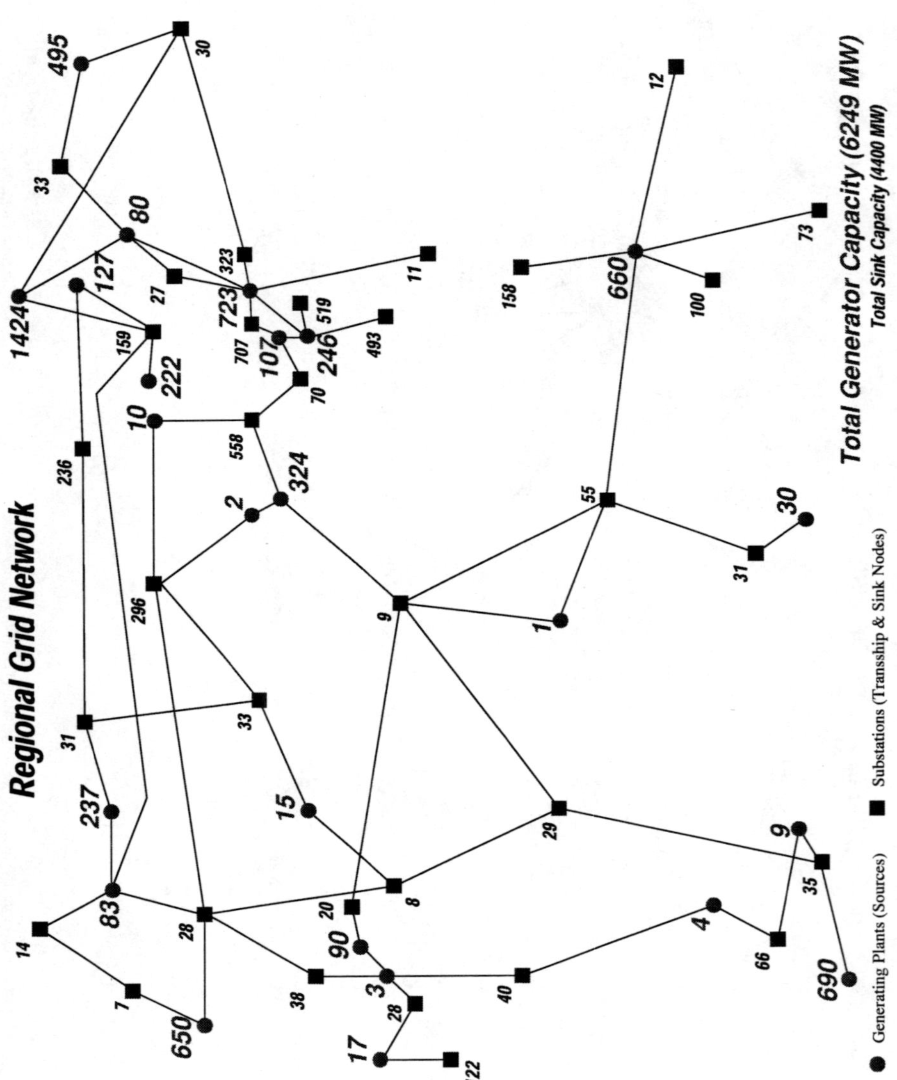


Figure 4: Shows the general network with the node capacities labeled with the sink capacities reduced to a total of 4400 MW. These are the basic capacities used in the creation of **Scenarios 1, 2, 3, & 4.**

C. Barrett et al., *Statistical Analysis of Algorithms*, JGAA, 7(1) 3–31 (2003) 15

1. The running time of each algorithm.

2. Total number of satisfied contracts by each algorithm.

3. The LP upper bound on the IP and thus an upper bound on the approximations given by the algorithms.

4. The IP approximation and objective function value.

The number 30 was chosen to ensure that a statistically "large" sample of each measure would be provided in order to make valid statistical inference. We consider two parameters to measure the performance of our algorithms – (i) the running time and (ii) the quality of the solution obtained. More attention is given to the quality of solution measure since from a social standpoint contract satisfaction may leave little room for finding solutions that are far from optimal.

We now describe how these measures are used to make inferences about the qualitative performance of these algorithms with respect to one another and independently. Since the intent is to make inferences concerning solution quality, a measure of this sort must be derived from the data generated. To do this, the data provided by the LP relaxation is examined. The $\lfloor y^*_{\mathcal{AS}} \rfloor = \lfloor y^* \rfloor$ provides the best-case number or upper bound on the objective function value our algorithms can produce for a scenario. Hence, if an algorithm produces an objective function value of $\lfloor y^*_{\mathcal{AS}} \rfloor$, it has produced an optimal solution for a given scenario. For a given algorithm $\mathcal{A}$ and scenario $\mathcal{S}$, let $Value_{\mathcal{AS}}$ denote the number of contracts that are satisfied by $\mathcal{A}$ under $\mathcal{S}$. The fraction

$$p_{\mathcal{AS}} = \frac{Value_{\mathcal{AS}}}{\lfloor y^*_{\mathcal{AS}} \rfloor}$$

provides a measure of the quality of the algorithm's solution.

## 4.4   Experimental Objective

The objective of our experiments was to find out which, if any, of the algorithms discussed here performs better than the others, in terms of quality of solution and running time for different contract scenarios. The design of the experiment was developed keeping this objective in mind. Since the performance depends on the type of algorithm used and the contract scenario, these are our factors of interest. As mentioned in the section 4.3, for a given $\epsilon$, 30 runs were performed for each algorithm-scenario pair. We perform two separate sets of experiments, one for the quality of solution as measured by $p_{\mathcal{A}}$ and the other for running time. This was done because the quality of solution and running time were independent of each other. The number of contracts satisfied do not depend upon the length of the time it takes to run the algorithm.

C. Barrett et al., *Statistical Analysis of Algorithms*, JGAA, 7(1) 3–31 (2003) 16

# 5 Results and Analysis

## 5.1 General Conclusions

We first present general conclusions obtained from our results and experimental analysis. These will be elaborated on subsequently.

1. Although there exist instances where the three heuristics produce solutions as large as $\Omega(n)$ times the optimal fractional solution, most of our tests show that we could find integral solutions fairly close to optimal.

2. Our experiments show that different scenarios make a *significant* difference in the type of solutions obtained. For example, the quality of solution obtained using the fourth scenario is significantly worse than the first three scenarios. The sensitivity to the scenarios poses interesting questions for infrastructure investment. The market will have to decide the cost that needs to be paid for expecting the necessary quality of service. It also brings forth the equity-benefit question: namely, who should pay for the infrastructure improvements?

3. It is possible that for certain scenarios, the underlying network is incapable of supporting even a minimum acceptable fraction of the bilateral contracts. This observation – although fairly intuitive – provides an extremely important message, namely, networks that were adequate to service customers in a completely regulated power market *might* not be adequate in deregulated markets. This makes the question of evicting the bilateral contracts more important.

4. One expects a trade-off between the number of contracts satisfied and the value of $\epsilon$, for the randomized rounding algorithm: as $\epsilon$ increases, and the demand condition is more relaxed, a higher number of contracts should get satisfied. But our experiments show that the change in the number of contracts satisfied for different values of $\epsilon$ is insignificant. Also, $\lambda = 2$ gave the best solutions in our experiments.

5. In practical situations, the Random-Order heuristic would be the best to use since it performs very close to the optimal in terms of quality of solution and has very low running time. Furthermore, though the Smallest-First heuristic does even better on many of our experiments, Random-Order is a natural proxy to model contracts arriving in an unforeseen way. Also, since the heuristics deliver solutions very close to the LP upper bound, we see that this LP bound is tight and useful. To further evaluate the randomized rounding algorithm, we need to implement its deterministic version [5], which is a non-trivial task.

## 5.2 Statistical Background

We use a statistical technique known as *analysis of variance* (ANOVA) to test whether differences in the sample means of algorithms and scenarios reflect differences in the means of the statistical populations that they came from or are just sampling fluctuations. This will help us identify which algorithm and

C. Barrett et al., *Statistical Analysis of Algorithms*, JGAA, 7(1) 3–31 (2003) 17

scenarios perform the best.[1]

ANOVA has the following three advantages over individual $t$-tests[2] when the number of groups being compared is greater than two. See [9] for more details. In our case, we have four algorithms and four scenarios. Standard statistics terminology for a hypothesis that we wish to test, is *null hypothesis*.

- It gives accurate and known type-I error probability.[3]

- It is more powerful i.e. if null hypothesis is false, it is more likely to be rejected.

- It can assess the effects of two or more independent variables simultaneously.

## 5.3   Mathematical Model

**Quality of Solution:** We first describe the experiment for the quality of solution i.e. $p_{AS}$. We use a two-factor ANOVA model since our experiment involves two factors which are:

1. The algorithms: $\mathcal{A}_i, i = 1, 2, 3$ and $4$.

2. The scenario: $\mathcal{S}_j, j = 1, 2, 3$ and $4$.

Following classical statistics terminology, we will sometimes refer to algorithms as *treatments* and the scenarios as *blocks*. We will use $\mathcal{A}$ to denote the set of algorithms and $\mathcal{S}$ to denote the set of scenarios. For each algorithm-scenario pair we have *30 observations (or replicates)*. When testing the efficacy of the algorithms, we use 4 algorithms, each having 120 observations (30 for each scenario) from the corresponding population. The design of experiment used here is a *fixed-effect complete randomized block*. *Fixed-effect* because the factors are fixed as opposed to randomly drawn from a class of algorithms or scenarios; the conclusions drawn from this model will hold only for these particular algorithms and scenarios. *Complete* implies that the number of observations are the same for each block. *Randomized* refers to the 30 replicates being drawn randomly. We wish to test the hypothesis:

---

[1] The populations in each of the groups are assumed to be normally distributed and have equal variances. The effect of violation of ANOVA assumptions of normality and homogeneity of variances have been tested in the literature ([10]) and the results show:

- Non-normality has negligible consequences on type-I and II error probabilities unless the populations are highly skewed or the sample is very small.

- When the design is balanced, i.e. the number of observations are the same for each group, violation of homogeneity of variance assumption has negligible consequences on the accuracy of type-I error probabilities.

[2] $t$-test checks for the significance of the difference in the means of *two* samples. It can assess whether the difference in sample means is just due to sampling error or they really are from two populations with different means.

[3] The probability of rejecting a null hypothesis when it is actually true.

C. Barrett et al., *Statistical Analysis of Algorithms*, JGAA, 7(1) 3–31 (2003) 18

Is the mean quality of solution provided by different algorithms the same, against the alternative hypothesis that some or all of these means are unequal?

The model for randomized block design includes constants for measuring the scenario effect (block effect), the algorithm effect (treatment effect) and a possible interaction between the scenarios and the algorithms. The appropriate mathematical model is as follows:

$$X_{ijk} = \mu + \tau_i + \beta_j + (\tau\beta)_{ij} + \varepsilon_{ijk},$$

where $X_{ijk}$ is the measurement $(p_{AS})$ for the $kth$ sample within the $ith$ algorithm and the $jth$ scenario. $\tau_i$ is the algorithm effect. $\beta_j$ is the scenario effect. $(\tau\beta)_{ij}$ captures the interaction present between the algorithms and the scenarios. $\varepsilon_{ijk}$ is the random error. See [8, 9] for further details on ANOVA.

We use S-Plus [15] software to run two-factor ANOVA to test the following three different null hypotheses.

1. Are the means given by the 4 different algorithms equal? The null hypothesis here is, $H_0 : \tau_1 = \tau_2 = \tau_3 = \tau_4$.

2. Are the means given by the 4 different scenarios equal? The null hypothesis here is, $H_0 : \beta_1 = \beta_2 = \beta_3 = \beta_4$.

3. Is there any interaction between the two factors? The null hypothesis here is, $H_0 : (\tau\beta)_{ij} = 0$.

The results of two-factor ANOVA are shown in Table 1 and Table 2. In the following discussion, we explain the meaning of each column in Table 1. $DF$ refers to the degrees of freedom, $SS$ refers to the sum of squared deviations from the mean. $MS$ refers to the mean square error, which is the sum of squares divided by the degrees of freedom.[4]

---

[4]The sum of squares for the algorithm factor can be calculated as:

$$SS_A = nJ\Sigma_i(\overline{X}_{i..} - \overline{X}...)^2$$

where $n$ is the number of replicates, $J$ is the number of scenarios, $\overline{X}_{i..}$ is the mean of algorithm $i$ across all scenarios and $\overline{X}...$ is the grand mean across all algorithms and scenarios. Recall that in our case $n = 30$ and $J = 4$ yielding a total sample size of 120.

The sum of squares for scenario factor can be calculated as:

$$SS_S = nI\Sigma_j(\overline{X}_{.j.} - \overline{X}...)^2$$

where as before $n$ is the number of replicates, $I$ is the number of algorithms and $\overline{X}_{.j.}$ is the mean of scenario $j$ across all algorithms. Again, in our case $n = 30$ and $I = 4$.

The sum of squares for algorithms and scenario interaction is:

$$SS_{AS} = n\Sigma j\Sigma_i[\overline{X}_{ij.} - (\overline{X}... + \hat{\tau}_i + \hat{\beta}_j)]^2$$

Here $\overline{X}_{ij.}$ is the mean of observations for the algorithm $i$ scenario $j$ pair. $\hat{\tau}_i$ and $\hat{\beta}_j$ are respectively the estimated least square values of $\tau_i$ and $\beta_j$. The sum of squares "within" refers to the squared difference between each observation and the mean of the scenario and algorithm of which it is a member. It is also referred as the residual sum of squares. This can be calculated as:

$$SS_W = n\Sigma j\Sigma_i\Sigma_k(X_{ijk} - \overline{X}_{ij.})^2$$

C. Barrett et al., *Statistical Analysis of Algorithms*, JGAA, 7(1) 3–31 (2003) 19

The $p$-value gives the smallest level of significance at which the null hypothesis can be rejected.[5] The lower the $p$-value, the lesser the agreement between the data and the null hypothesis. Finally the $F$-**test** is as follows. To test the null hypothesis, i.e., whether the population means are equal, ANOVA compares two estimates of $\sigma^2$. The first estimate is based on the variability of each population mean around the grand mean. The second is based on the variability of the observations in each population around the mean of that population. If the null hypothesis is true, the two estimates of $\sigma^2$ should be essentially the same. Otherwise, if the populations have different means, the variability of the population mean around the grand mean will be much higher than the variability within the population. The null hypothesis in the $F$-test will be accepted if the two estimates of $\sigma^2$ are almost equal.

In a two-factor fixed-effect ANOVA, three separate $F$-**tests** are performed: two tests for the factors, and the third for the interaction term. The null hypothesis for the first factor can be written as:

$$H_0^A : \mu_{1..} = \mu_{2..} = \cdots = \mu_{j..}$$

which is equivalent to writing: $H_0 : \tau_1 = \tau_2 = \tau_3 = \tau_4$. The $F$-test is:

$$F_A = \frac{SS_A/(I-1)}{SS_W/IJ(n-1)}$$

and the null hypothesis for the second factor can be written as:

$$H_0^S : \mu_{.1.} = \mu_{.2.} = \cdots = \mu_{.j.}$$

which is equivalent to writing: $H_0 : \beta_1 = \beta_2 = \beta_3 = \beta_4$. The $F$-test is:

$$F_S = \frac{SS_S/(J-1)}{SS_W/IJ(n-1)}$$

and the null hypothesis for the interaction term can be written as:

$$H_0^{AS} : (\tau\beta)_{ij} = 0.$$

The $F$-test is:

$$F_{AS} = \frac{SS_{AS}/(I-1)(J-1)}{SS_W/IJ(n-1)}$$

If this $F$-ratio is close to 1, the null hypothesis is true. If it is considerably larger – implying that the variance between means is larger than the variance

---

The total sum of squares is

$$SS_T = SS_A + SS_S + SS_{AS} + SS_W$$

[5]To obtain a $p$-value for say $F_A$, the algorithm effect, we would look across the row associated with 3 degree of freedom in the numerator and 464 degrees of freedom in the denominator in the $F$-distribution table and find the largest value that is still less than the one obtained experimentally. From this value, we obtain a $p$-value of 0 for $F_A$.

C. Barrett et al., *Statistical Analysis of Algorithms*, JGAA, 7(1) 3–31 (2003) 20

| Source | $DF$ | $SS$ | $MS$ | $F$-test | $p$-value |
|---|---|---|---|---|---|
| Scenario (Block) | 3 | 0.14 | 0.05 | 43.38 | 0 |
| Algorithm (Treatment) | 3 | 22.78 | 7.59 | 6792.60 | 0 |
| Scenario:Algorithm | 9 | 0.12 | 0.01 | 15.90 | 0 |
| Residuals | 464 | 0.40 | .0008 | | |
| Total | 479 | 23.45 | | | |

Table 1: **Results of Two-Factor ANOVA:** This table shows results of two-factor ANOVA where the factors are algorithms and scenarios. The measurement is the quality of solution, given by $p_{AS}$. The $p$-values show that the algorithm effect, scenario effect and the interaction between the algorithms and scenarios are all significant at any level of confidence.

within a population – the null hypothesis is rejected. The $F$ distribution table should be checked to see if the $F$-ratio is significantly large.

The results in Table 1 show that all the above three null hypothesis are rejected at any significance level. This implies that the performance (measured by $p_{AS}$) of at least one of the algorithms is significantly different from the other algorithms. Also, different scenarios make a difference in the performance. Finally, the scenarios and the algorithms interact in a significant way. The interaction implies that the performance of the algorithms are different for different scenarios.

### 5.3.1 Contrasts

The next question of interest is what really caused the rejection of the null hypothesis; just knowing that at least one of the algorithms is different does not help us identify which algorithm is significantly different. To answer this we use a procedure called *contrast*. A contrast $C$ among $I$ population means ($\mu_i$) is a linear combination of the form

$$C = \Sigma_i \alpha_i \mu_i = \alpha_1 \mu_1 + \alpha_2 \mu_2 + \cdots + \alpha_I \mu_I$$

such that the sum of contrast coefficients $\Sigma_i \alpha_i$ is zero. In the absence of true population means, we use the unbiased sample means which gives the estimated contrast as:

$$\hat{C} = \Sigma_i \alpha_i \overline{X}_i = \alpha_1 \overline{X}_1 + \alpha_2 \overline{X}_2 + \cdots + \alpha_I \overline{X}_I.$$

The contrast coefficients $\alpha_1$, $\alpha_2$, $\cdots$, $\alpha_I$ are just positive and negative numbers that define the particular hypothesis to be tested. The null hypothesis states that the value of a parameter of interest for every contrast is zero, i.e., $H_0 : C =$

C. Barrett et al., *Statistical Analysis of Algorithms*, JGAA, 7(1) 3–31 (2003) 21

| Performance Measure: Quality of Solution (in %) | | | | | |
|---|---|---|---|---|---|
| | RR | SF | LF | RO | Scn. Means |
| Scn. 1 | $\overline{X}_{11.}{=}48.68$ | $\overline{X}_{21.}{=}99.73$ | $\overline{X}_{31.}{=}97.97$ | $\overline{X}_{41.}{=}97.78$ | $\overline{X}_{.1.}{=}86.02$ |
| Scn. 2 | $\overline{X}_{12.}{=}46.91$ | $\overline{X}_{22.}{=}99.56$ | $\overline{X}_{32.}{=}98.38$ | $\overline{X}_{42.}{=}98.93$ | $\overline{X}_{.2.}{=}85.94$ |
| Scn. 3 | $\overline{X}_{13.}{=}45.69$ | $\overline{X}_{23.}{=}99.25$ | $\overline{X}_{33.}{=}97.10$ | $\overline{X}_{43.}{=}98.82$ | $\overline{X}_{.3.}{=}85.22$ |
| Scn. 4 | $\overline{X}_{14.}{=}46.99$ | $\overline{X}_{24.}{=}98.03$ | $\overline{X}_{34.}{=}88.65$ | $\overline{X}_{44.}{=}93.41$ | $\overline{X}_{.4.}{=}81.77$ |
| Algo. Means | $\overline{X}_{1..}{=}47.07$ | $\overline{X}_{2..}{=}99.14$ | $\overline{X}_{3..}{=}95.51$ | $\overline{X}_{4..}{=}97.24$ | $\overline{X}_{...} = 84.74$ |

Table 2: **The Mean Values of the Quality of Solution:** This table shows the mean values of the quality of solution for each algorithm and each scenario.

0. The value of the contrast is tested by an $F$-test to see if the observed value of the contrast is significantly different from the hypothesized value of zero.

Table 2 shows the average value of the quality of solution for each algorithm-scenario pair. e.g. $\overline{X}_{11.}$ means that we fix $i = 1$ and $j = 1$ and take the average of $X_{ijk}$ over all $k$. From Table 2, it is clear that the randomized rounding algorithm (RR) is different from all the other algorithms for all four scenarios. On an average, RR algorithm satisfies 49% less contracts than the Largest-First (LF) heuristic and 50% less than the Random-Order (RO) heuristic and 52% less contracts than the Smallest-First (SF) heuristic. The difference between SF, RO and LF heuristics appears only marginal. Based on this observation, we constructed the following contrast that tests if RR is statistically significantly different from the other three algorithms:

$$\mathcal{C}_{1Q} = \frac{1}{3}(\overline{X}_{2..}) + \frac{1}{3}(\overline{X}_{3..}) + \frac{1}{3}(\overline{X}_{4..}) - \overline{X}_{1..}$$

Using the value of algorithm means from Table 2 we can calculate the value of $\mathcal{C}_{1Q}$ ($Q$ stands for the quality of solution) to be equal to 0.50.[6] The sum of squares of a contrast is expressed as:

$$SS(\mathcal{C}_{1Q}) = \frac{(\mathcal{C}_{1Q})^2}{\Sigma_i \alpha_i^2 / N_i}$$

Here $\alpha_i$ are the coefficients of the contrast and $N_i = 120$ is the number of observations (i.e. sample points for each algorithm across all scenarios). This results in $SS(\mathcal{C}_{1Q}) = 22.68$. Now we can use the following F-test to see the significance of the contrast:

$$SS(\mathcal{C}_{1Q})/MSE \sim F(1, 464)$$

---

[6]The table values are shown in percentages, but here we use actual values.

C. Barrett et al., *Statistical Analysis of Algorithms*, JGAA, 7(1) 3–31 (2003) 22

MSE stands for the mean square error of the residuals. The contrast has one degree of freedom and residuals have 464 degrees of freedom (see table 1). F = 22.68/.0008 = 28350, since the observed value of F-test is greater than the critical $F$-value given in the $F$-distribution table, for any significance level, the null hypothesis is rejected. This confirms our earlier observation that the RR algorithm is significantly inferior in performance compared to the other three algorithms. The sum of squares of $SS(\mathcal{C}_{1Q})$ = 22.68 shows that 98% of the variation in factors sum of squares (total factors sum of squares being 23.05 i.e. total SS - residual SS, see table 2) is due to the difference in RR algorithm versus the other three algorithms.

Table 2 shows that the first three scenarios clear about 86% of the optimal number of contracts while under the fourth scenario, the number of contracts cleared is less than 82% of the optimal. Even though the difference in the number of cleared contracts is not very big, one would be curious to find out if the difference in performance under the first three scenarios versus the fourth scenario is significant or not. To answer this we created the following contrast which is orthogonal[7] to the first contrast $(\mathcal{C}_{1Q})$:

$$\mathcal{C}_{2Q} = \frac{1}{3}(\overline{X}_{.1.}) + \frac{1}{3}(\overline{X}_{.2.}) + \frac{1}{3}(\overline{X}_{.3.}) - \overline{X}_{.4.}$$

Just like $\mathcal{C}_{1Q}$, we can calculate the value of $\mathcal{C}_{2Q}$ using table 2:

$$SS(\mathcal{C}_{2Q})/MSE \sim F(1,464) = 0.14/.0008 = 175$$

Again, the null hypothesis is rejected implying that the fourth scenario is indeed significantly different from the other three scenarios.

Now we look at two more contrasts to check if SF and LF are significantly different $(\mathcal{C}_{3Q})$ and LF and RO are significantly different $(\mathcal{C}_{4Q})$.

$$\mathcal{C}_{3Q} = \overline{X}_{2..} - \overline{X}_{3..}$$

$$\mathcal{C}_{4Q} = \overline{X}_{3..} - \overline{X}_{4..}$$

$$SS(\mathcal{C}_{3Q})/MSE \sim F(1,464) = 2.178/.0008 = 2722.5$$

$$SS(\mathcal{C}_{4Q})/MSE \sim F(1,464) = 1.038/.0008 = 1297.5$$

For both $\mathcal{C}_{3Q}$ and $\mathcal{C}_{4Q}$, the observed value of the F-test is greater than the critical $F$-value given in the $F$-distribution table, the null hypothesis in both cases are rejected, implying that SF provides a better solution than LF and also that RO performs significantly better than LF.

---

[7]Two contrasts $\mathcal{C}_1$ and $\mathcal{C}_2$ are said to be orthogonal if the sum of the products of their corresponding coefficients is zero. It is desirable to have independent or orthogonal contrasts because independent tests of hypotheses can be made by comparing the mean square of each such contrast with the mean square of the residuals in the experiment. Each contrast has only one degree of freedom.

C. Barrett et al., *Statistical Analysis of Algorithms*, JGAA, 7(1) 3–31 (2003) 23

| Source | $DF$ | $SS$ | $MS$ | $F$-test | $p$-value |
|---|---|---|---|---|---|
| Scenario (Block) | 3 | 21152 | 7050.8 | 56.97 | 0 |
| Algorithm (Treatment) | 3 | 2161199 | 720399.8 | 5821.07 | 0 |
| Scenario:treatment | 9 | 28156 | 3128.5 | 47.78 | 0 |
| Residuals | 464 | 30381 | 65.5 | | |
| Total | 479 | 2240888 | | | |

Table 3: **Results of Two-Factor ANOVA:** This table shows results of two-factor ANOVA where the factors are algorithms and scenarios. The measurement is the running time of the algorithm-scenario pair. The $p$-values show that the algorithm effect, scenario effect and the interaction between the algorithms and scenarios are all significant at any level of confidence.

In summary, all algorithms show significantly different performance when measured in terms of quality of solution. On an average, the best solution is given by the SF heuristic and the worst by the RR.

**Running Time:** Tables 3 and 4 show results of the same experiment when performance is measured by the *running time* of the algorithm. The factors, number of observations, kinds of tests, etc. remain the same as before, except the performance measure. Table 3's results clearly demonstrate that different algorithms take significantly different time to run and that different scenarios have significantly different running time. The interaction term is significant at any level of confidence implying that the running time of an algorithm is different for different scenarios.

Table 4 shows that the RR algorithm takes noticeably more time to run as compared to the other three heuristics. Among the three heuristics, LF and RO take about the same time but SF takes about 19 megaticks more than the LF and RO. Similarly, scenario 3 and 4 take about the same time but scenario 1 and 2 look different. To test all the above mentioned observations, we create the following different contrasts:

$$C_{1t} = \frac{1}{3}(\overline{X}_{2..}) + \frac{1}{3}(\overline{X}_{3..}) + \frac{1}{3}(\overline{X}_{4..}) - \overline{X}_{1..}$$

$$C_{2t} = \frac{1}{2}(\overline{X}_{3..}) + \frac{1}{2}(\overline{X}_{4..}) - \overline{X}_{2..}$$

$$C_{3t} = \overline{X}_{3..} - \overline{X}_{4..}$$

$$C_{4t} = \overline{X}_{.1.} - \overline{X}_{.2.}$$

C. Barrett et al., *Statistical Analysis of Algorithms*, JGAA, 7(1) 3–31 (2003) 24

| Performance Measure: Running Time (in Megaticks) | | | | |
|---|---|---|---|---|
| | RR | SF | LF | RO | Scn. Means |
| Scn. 1 | $\overline{X}_{11.}=163.33$ | $\overline{X}_{21.}=41.23$ | $\overline{X}_{31.}=24.57$ | $\overline{X}_{41.}=25.50$ | $\overline{X}_{.1.}=63.66$ |
| Scn. 2 | $\overline{X}_{12.}=218.23$ | $\overline{X}_{22.}=49.63$ | $\overline{X}_{32.}=29.73$ | $\overline{X}_{42.}=30.23$ | $\overline{X}_{.2.}=81.96$ |
| Scn. 3 | $\overline{X}_{13.}=181.70$ | $\overline{X}_{23.}=45.70$ | $\overline{X}_{33.}=23.30$ | $\overline{X}_{43.}=26.43$ | $\overline{X}_{.3.}=69.28$ |
| Scn. 4 | $\overline{X}_{14.}=184.33$ | $\overline{X}_{24.}=44.53$ | $\overline{X}_{34.}=27.00$ | $\overline{X}_{44.}=27.27$ | $\overline{X}_{.4.}=70.78$ |
| Algo. Means | $\overline{X}_{1..}=186.90$ | $\overline{X}_{2..}=45.27$ | $\overline{X}_{3..}=26.15$ | $\overline{X}_{4..}=27.36$ | $\overline{X}_{...}=71.42$ |

Table 4: **The Mean Values of the Running Time:** This table shows the mean values of the running time for each algorithm and each scenario.

All the above contrasts are orthogonal to each other. The first contrast, $C_{1t}$ (here $t$ stands for running time), checks if the RR algorithm takes more time to run than the other three heuristics. The second contrast, $C_{2t}$, will find if the SF heuristic is significantly different from the LF and RO heuristic. The third contrast, $C_{3t}$, checks if the LF and RO heuristics take about the same time to run. Finally, contrast $C_{4t}$, check if the first scenario takes less time to run than the second scenario. The results of all the contrasts are shown below.

$$SS(C_{1t})/MSE \sim F(1, 464) = 2133700.9/65.5 = 32575.5$$

$$SS(C_{2t})/MSE \sim F(1, 464) = 27424.4/65.5 = 418.69$$

$$SS(C_{3t})/MSE \sim F(1, 464) = 72.6/65.5 = 1.11$$

$$SS(C_{4t})/MSE \sim F(1, 464) = 20093.4/65.5 = 306.76$$

As can be seen by looking at the F-distribution table, all the above contrasts except $C_{3t}$ show that the observed value of the $F$-test is greater than the critical $F$-value. Hence the null hypothesis i.e. $H_0 : C_{it} = 0$ where $i = 1, 2, 4$ can be rejected at any level of significance. This confirms our earlier hypothesis that RR indeed takes longer to run than the other three heuristics. SF takes more time to run than the LF and RO heuristics and the second scenario takes significantly more time to run than the first scenario.

The mean difference in running time across different algorithms shows that all algorithms are significantly different in terms of running time except for the *Largest-First* and the *Random-Order* heuristics. These two heuristics take about the same time to run and indeed a contrast done i.e. $C_{3t}$ on LF and RO proves that and the null hypothesis, $H_0 : C_{3t} = 0$, is accepted.

The randomized rounding algorithm takes significantly more time to run than any of the other heuristics. The gap in running time narrows when RR is compared against SF. RR takes 141 megaticks more time than the SF heuristic,

C. Barrett et al., *Statistical Analysis of Algorithms*, JGAA, 7(1) 3–31 (2003) 25

160 megaticks more than the LF and RO. SF takes more time to run than LF and RO but it clears more contracts than LF and RO.

All the above analysis was performed while keeping the value of $\epsilon$ constant at 0.1. The performance of the randomized rounding algorithm[8] does not change in any significant way, both in terms of $p_{AS}$ and running time when $\epsilon$ varied from 0.1 to 0.5. So all the above results hold for $\epsilon = 0.1, 0.2, 0.3, 0.4$ and $0.5$.[9]

**Summary:** It is clear that SF heuristic clears the most contracts, almost as good as the optimal but takes more running time as compared to LF and RO. However, it takes only a quarter of the time as compared to RR. As far as scenarios go, the first scenario clears most contracts in the least amount of time. From a practical standpoint, the RO heuristic seems to be the best since it performs very well both in terms of running time and quality of solution and is trivial to implement. It performs very close to optimal in terms of clearing contracts and yet takes minimal time to do it as compared to the other algorithms. The RR algorithm, although it gives good theoretical lower bounds, is not very appropriate for real-life situations where both time and a high level of contract satisfaction have a very high priority.

# 6 Discussion and Concluding Remarks

We carried out an empirical study to evaluate the quality and running time performance of four different market clearing mechanisms. A novel aspect of our work is the use of statistical technique, *Analysis of Variance*, to compare the performance of different market clearing mechanisms. This technique allows us to formally test which algorithm performs better in terms of each of the performance measures.

One heuristic was based on using a relaxation of integer linear program followed by randomized rounding of the fractional solution to yield an approximate integral solution. Although the algorithm had a provable performance guarantee, experiments suggest that the algorithm is not likely to be practically useful given the running time and the quality of solution produced. The result is not entirely unexpected; it has been observed that many approximation algorithms that are designed to work in the worst case typically do not have a very good average case behavior.

We also studied three different simple heuristics: experimental results suggest that each is likely to perform better than the theoretically provable approximation algorithm. This is in spite of the fact that it is very easy to construct instances where the heuristics have unboundedly poor performance guarantee.

One of the heuristics: the *random-order heuristic* was studied to emulate a simple "first-come first-serve" type clearing mechanism that is currently employed by many ISO. The heuristic performs surprisingly well even compared to a bound on an optimal solution obtained via linear programming. The results

---

[8]Other heuristics do not depend on the value of $\epsilon$.

[9]The results are available from the authors upon request.

C. Barrett et al., *Statistical Analysis of Algorithms*, JGAA, 7(1) 3–31 (2003) 26

suggest that this simple clearing mechanism currently employed might result in near-optimal utilization of the network resources.

Our overall assessment is that for the purposes of developing large-scale microscopic-simulations of the deregulated power industry, the three heuristic methods give sufficiently good performance in terms of the quality of solution and the computational time requirement.

An interesting direction for future research is to study (both theoretically and experimentally) the performance of these algorithms when we have flow constraints modeling resistive networks. The additional constraints imposed on the system could conceivably make the problem easier to approximate. See [1] for further discussions on this topic.

## Acknowledgments

A preliminary version of this work appeared as the paper "Experimental Analysis of Algorithms for Bilateral-Contract Clearing Mechanisms Arising in Deregulated Power Industry" in *Proc. Workshop on Algorithm Engineering*, 2001. Work by Doug Cook, Vance Faber, Gregory Hicks, Yoram Sussmann and Heidi Thornquist was done when they were visiting Los Alamos National Laboratory. All authors except Aravind Srinivasan were supported by the Department of Energy under Contract W-7405-ENG-36. Aravind Srinivasan was supported in part by NSF Award CCR-0208005.

We thank Dr. Martin Wildberger (EPRI) for introducing us to the problems considered herein, and for the extensive discussions. We thank Mike Fugate, Dean Risenger (Los Alamos National Laboratory) and Sean Hallgren (Cal Tech) for helping us with the experimental evaluation. Finally, we thank the referees of Workshop on Algorithmic Engineering (WAE) 2001 for their helpful comments on the preliminary version of this paper, and the JGAA referees for very helpful suggestions on the final version.

Authors affiliations: Chris Barrett, Achla Marathe and Madhav Marathe are at the National Infrastructure Simulation and Analysis Center and CCS3, P.O. Box 1663, MS M997, Los Alamos National Laboratory, Los Alamos NM 87545. Email: `barrett, achla, marathe@lanl.gov`. Doug Cook is at the Department of Engineering, Colorado School of Mines, Golden CO 80401. Gregory Hicks is at the Department of Mathematics, North Carolina State University, Raleigh. Email: `gphicks@unity.ncsu.edu`. Vance Faber's email address is `vance@mappingscience.com`. Aravind Srinivasan is at the Department of Computer Science and Institute for Advanced Computer Studies, University of Maryland, College Park, MD 20742. Parts of his work were done while at Lucent Technologies, Murray Hill, NJ 07974-0636, and while at the National University of Singapore. Email: `srin@cs.umd.edu`. Yoram Sussmann is at the Department of Computer Science, State University of West Georgia, Carrollton, GA 30118. Email: `yoram@westga.edu`. Heidi Thornquist is at the Department of Computational and Applied Mathematics, Rice University, Houston, TX. Email: `heidi@caam.rice.edu`.

C. Barrett et al., *Statistical Analysis of Algorithms*, JGAA, 7(1) 3–31 (2003) 27

# References


[1] Ahuja, R.K., Magnanti, T.L., Orlin, J.B. Network flows: theory, algorithms, and applications. Prentice Hall, Englewood Cliffs, New Jersey. (1993)

[2] Example sites for information on bandwidth trading: `http://www.lightrade.com/dowjones2.html` and `http://www.williamscommunications.com/prodserv/network/webcasts/bandwidth.html`

[3] Cardell, J.B., Hitt, C.C., Hogan, W.W. Market Power and Strategic Interaction in Electricity Networks. Resource and Energy Economics Vol. 19 (1997) 109-137.

[4] California Public Utilities Commission. Order Instituting Rulemaking on the Commission's Proposed Policies Governing Restructuring California's Electric Services Industry and Reforming Regulation, Decision 95-12-063 (1995)

[5] Cook, D., Faber, V., Marathe, M., Srinivasan, A., Sussmann, Y.J. Low Bandwidth Routing and Electrical Power Networks. Proc. 25th International Colloquium on Automata, Languages and Programming (ICALP) Aalborg, Denmark, LNCS 1443, Springer Verlag, (1998) 604-615.

[6] Barrett, C.L., Marathe A. , Marathe, M. V. Parametrized Models for Large Scale Simulation and Analysis of Deregulated Power Markets. Internal Technical Report, Los Alamos National Laboratory (2002).

[7] The Changing Structure of the Electric Power Industry: Selected Issues. DOE 0562(98), Energy Information Administration, US Department of Energy, Washington, D.C. (1998)

[8] Fryer, H.C. Concepts and Methods of Experimental Statistics. Allyn and Bacon Inc. (1968)

[9] G. Glass and K. D. Hopkins. Statistical Methods in Education and Psychology, 3rd ed. Allyn and Bacon, 1996.

[10] Glass G.V., Peckham P.D. and J.R. Sanders. "Consequences of failure to meet assumptions underlying the fixed-effects analysis of variance and covariance" Review of Educational Research, 42, (1972) 237-288.

[11] The U.S. Federal Energy Regulatory Commission. Notice of Proposed Rulemaking ("NOPRA"). Dockets No. RM95-8-000, RM94-7-001, RM95-9-000, RM96-11-000, April 24 (1996)

C. Barrett et al., *Statistical Analysis of Algorithms*, JGAA, 7(1) 3–31 (2003) 28


[12] Garey, M.R., Johnson, D.S. Computers and Intractability. A Guide to the Theory of NP-Completeness. Freeman, San Francisco CA (1979)

[13] Hogan, W. Contract networks for electric power transmission. J. Regulatory Economics. (1992) 211-242.

[14] Kolliopoulos, S.G., Stein, C. Improved approximation algorithms for unsplittable flow problems. In Proc. IEEE Symposium on Foundations of Computer Science. (1997) 426-435.

[15] S-Plus5 "Guide to Statistics" MathSoft Inc. September (1988)

[16] Srinivasan, A. Approximation Algorithms via Randomized Rounding: A Survey. Lectures on Approximation and Randomized Algorithms (M. Karoński and H. J. Prömel, eds.), Series in Advanced Topics in Mathematics. Polish Scientific Publishers PWN, Warsaw.(1999) 9-71.

[17] Thomas, R. J. and T. R. Schneider. "Underlying Technical Issues in Electricity Deregulation" Power Systems Engineering Research Center, (1997). http://www.pserc.wisc.edu/ecow/get/publicatio/1997public/

[18] Wildberger, A. M. Issues associated with real time pricing. Unpublished Technical report Electric Power Research Institute (EPRI) (1997)

[19] Websites: http://www.magicnet.net/~metzler/page2d.html and http://www.eia.doe.gov/ (Energy Information Administration)

[20] Wolak, F. An Empirical Analysis of the Impact of Hedge Contracts on Bidding Behavior in a Competitive Electricity Market http://www.stanford.edu/~wolak. (2000)

[21] The Wall Street Journal, August 04 2000, A2, August 14 2000, A6, and August 21 2000, November 27 2000, December 13, 14 2000.

[22] Wood, A.J., Wollenberg, B.F. Power Generation, Operation and Control. John Wiley and Sons. (1996)

C. Barrett et al., *Statistical Analysis of Algorithms*, JGAA, 7(1) 3–31 (2003) 29

# Appendix

## A   Illustrative Examples

**Example 1.**   This example illustrates the issues encountered as a result of deregulation. Figure 5(a) shows an example in which there are two power plants $A$ and $B$, and two consumers. Let us assume that each consumer has a demand of 1 unit. Before deregulation, say both $A$ and $B$ are owned by the same company. If we assume that the plants have identical operating and production costs, then the demands can be satisfied by producing 1 unit of power at each plant. Now assume that due to deregulation, $A$ and $B$ are owned by separate companies. Further assume that $A$ provides power at a much cheaper rate and thus both the consumers sign contracts with $A$. It is clear that both the consumers now cannot get power by $A$ alone. Although the total production capacity available is more than total demand and it is possible to route that demand through the network under centralized control, it is not possible to route these demands in a deregulated scenario.

**Example 2.**   Here, the graph consists of a simple line as shown in Figure 5(b). We have three contracts each with a demand of 1. The capacity of each edge is also 1. A feasible solution is $f(s_1, t_3) = f(s_2, t_1) = f(s_3, t_2) = 1$. The crucial point here is that *the flow originating at $s_i$ may not go to $t_i$ at all* — since power produced at the sources are indistinguishable, the flow from $s_i$ joins a stream of other flows. If we look at the connected components induced by the edges with positive flow, we may have $s_i$ and $t_i$ in a different component. Thus we do not have a path or set of paths to round for the $(s_i, t_i)$-flow. This shows a basic difference between our problem and standard multi-commodity flow problems, and indicates that traditional rounding methods *may not be directly applicable.*

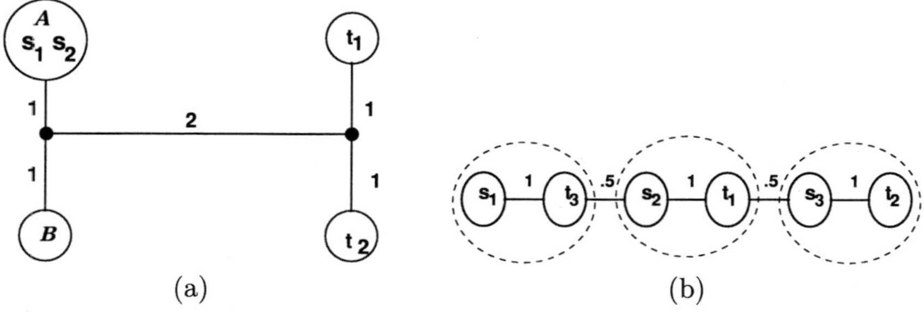


Figure 5: Figures for Examples 1 and 2

**Example 3:**   In this example, we illustrate how different policies can yield different solutions. The graph is shown in Figure 6 with edge capacities as listed. Again, we have three contracts, whose details are given as follows:

C. Barrett et al., *Statistical Analysis of Algorithms*, JGAA, 7(1) 3–31 (2003) 30

1. Contract 1 – $(s_1, t_1)$ demand $d_1 = 2$ and cost/unit $c_1 = .5$

2. Contract 2 – $(s_2, t_2)$ demand $d_1 = 1$ and cost/unit $c_1 = 1$

3. Contract 3 – $(s_3, t_3)$ demand $d_1 = 1$ and cost/unit $c_1 = 2$

The various solutions obtained under different policies are given below:

1. (0/1-VERSION, MAX-FEASIBLE FLOW): Two possible solutions: (i) $f(s_1, t_1) = 2$, (ii) $f(s_2, t_2) = f(s_3, t_3) = 1$. Both solution route 2 units of flow in the network.

2. (0/1-VERSION, MAX-#CONTRACTS): In contrast to the previous case only one solution is possible: $f(s_2, t_2) = f(s_3, t_3) = 1$. This also routes 2 units of flow.

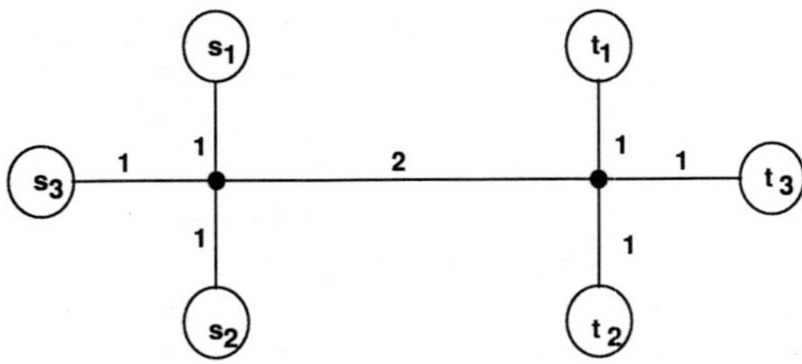


Figure 6: Example illustrating the various solutions under different contracts.

# B    Worst-Case Examples

The three heuristic methods of § 3 can be shown to have worst case performance guarantee that is $\Omega(n)$. (Recall that the performance guarantee of an approximation algorithm for a maximization problem $\Pi$ is the supremum of the ratio of the optimal solution to the heuristic solution over all instances $I$ of $\Pi$.) **Example 4** shows that all the heuristics can perform poorly w.r.t. an optimal solution. This is not too surprising given that the optimal solution gets to look at all of the input before clearing the contracts.

**Example 4:** Consider a network with two nodes $A$ and $B$ joined by an edge $(A, B)$. The capacity of the edge $(A, B)$ is $C < 1$. There is an even number $n$ of contracts $(s_1, t_1), \ldots, (s_n, t_n)$. Odd-numbered contracts have demand of 1 unit and the sources and sinks of these contracts are distributed as follows: source-nodes $s_1, s_3, \ldots s_{n-1}$ are located at node $A$ and their corresponding consumers $t_1, t_3, \ldots t_{n-1}$ are located at $B$. Let us call this set *Odd-Set*. For the even numbered contracts (denoted *Even-set*) we have a demand of $1 + \frac{C}{2n}$ per contract and the source sink locations are reversed: the sources are located at $B$ and the sinks at $A$. Note that

C. Barrett et al., *Statistical Analysis of Algorithms*, JGAA, 7(1) 3–31 (2003) 31

1. All *Odd-set* contracts have demand that is less than every contract in *Even-Set*.

2. In the absence of any other contracts, only one *Odd-set* contract can be cleared; similarly, exactly one *Even-set* contract can be cleared.

Now consider how many contracts can be satisfied by the each of three heuristic methods.

1. SMALLEST-FIRST HEURISTIC will clear only one contract $(s_1, t_1)$.

2. LARGEST-FIRST HEURISTIC will also clear exactly one contract $(s_2, t_2)$.

3. RANDOM-ORDER HEURISTIC will also perform poorly with high probability. This is because there is are a total of $n!$ ways to arrange the contracts and roughly only $O((\frac{n}{2})!(\frac{n}{2})!)$ good ways to do it.

4. An optimal solution can clear all the contracts simultaneously, since the flows from *Odd-set* appropriately cancel the flows from *Even-Set*. Thus the performance guarantee of the SMALLEST-FIRST HEURISTIC and LARGEST-FIRST HEURISTIC is $\Omega(n)$. The performance guarantee of RANDOM-ORDER HEURISTIC is also $\Omega(n)$ with high probability.

**Example 5:** Again, we have a single edge as the network. Denote the edge by $(A, B)$ as before, with the endpoints being $A$ and $B$ respectively and the edge capacity being 1. We have $n$ contracts. As before we divide them into *Even-Set* and *Odd-set* of contracts. The contracts' demands are strictly increasing: the $i^{th}$ contract has demand $1 + (i - 1)\epsilon$. The value $\epsilon$ is chosen so that $0 < \epsilon < 1$ and $(n - 1)\epsilon > 1$. It is clear that SMALLEST-FIRST HEURISTIC can clear all the contracts, while LARGEST-FIRST HEURISTIC can clear exactly one contract. Again, a simple calculation shows that RANDOM-ORDER HEURISTIC will perform poorly with high probability.

Journal of Graph Algorithms and Applications

http://jgaa.info/

vol. 7, no. 1, pp. 33–77 (2003)

# Lower Bounds for the Number of Bends in Three-Dimensional Orthogonal Graph Drawings

*David R. Wood*

School of Computer Science
Carleton University
Ottawa, Canada

davidw@scs.carleton.ca
http://www.scs.carleton.ca/~davidw

## Abstract

This paper presents the first non-trivial lower bounds for the total number of bends in 3-D orthogonal graph drawings with vertices represented by points. In particular, we prove lower bounds for the number of bends in 3-D orthogonal drawings of complete simple graphs and multigraphs, which are tight in most cases. These result are used as the basis for the construction of infinite classes of $c$-connected simple graphs, multigraphs, and pseudographs ($2 \leq c \leq 6$) of maximum degree $\Delta$ ($3 \leq \Delta \leq 6$), with lower bounds on the total number of bends for all members of the class. We also present lower bounds for the number of bends in general position 3-D orthogonal graph drawings. These results have significant ramifications for the '2-bends problem', which is one of the most important open problems in the field.

Communicated by Dorothea Wagner: submitted January 2001;
revised December 2002.

A preliminary version of this paper appeared in *Proc. 8th International Symp. on Graph Drawing* (GD 2000). Volume 1984 of Lecture Notes in Comput. Sci., pages 259-271, Springer, 2001.

Supported by NSERC. Partially completed while a Ph.D. student in the School of Computer Science and Software Engineering, Monash University (Melbourne, Australia), and at the School of Information Technologies, The University of Sydney (Sydney, Australia), where supported by the ARC.

D. Wood, *Lower Bounds for 3-D Drawings*, JGAA, 7(1) 33–77 (2003)     34

# 1   Introduction

The *3-D orthogonal grid* consists of *grid-points* in 3-space with integer coordinates, together with the axis-parallel *grid-lines* determined by these points. Two grid-points are said to be *collinear* if they are contained in a single grid-line, and are *coplanar* if they are contained in a single grid-plane. A *3-D orthogonal drawing* of a graph positions each vertex at a distinct grid-point, and routes each edge as a polygonal chain composed of contiguous sequences of axis-parallel segments contained in grid-lines, such that (a) the end-points of an edge route are the grid-points representing the end-vertices of the edge, and (b) distinct edge routes only intersect at a common end-vertex.

For brevity we say a 3-D orthogonal graph drawing is a *drawing*. A drawing with no more than $b$ bends per edge is called a *b-bend drawing*. The graph-theoretic terms 'vertex' and 'edge' also refer to their representation in a drawing. The *ports* at a vertex $v$ are the six directions, denoted by $X_v^+$, $X_v^-$, $Y_v^+$, $Y_v^-$, $Z_v^+$ and $Z_v^-$, which the edges incident with $v$ can use. For each dimension $I \in \{X, Y, Z\}$, the $I_v^+$ (respectively, $I_v^-$) port at a vertex $v$ is said to be *extremal* if $v$ has maximum (minimum) $I$-coordinate taken over all vertices.

Clearly, 3-D orthogonal drawings can only exist for graphs with maximum degree at most six. 3-D orthogonal drawings of maximum degree six graphs have been studied in [3, 4, 6, 10–13, 17, 19, 21, 22, 33, 35–37]. By representing a vertex by a grid-box, 3-D orthogonal drawings of arbitrary degree graphs have also been considered; see for example [5, 8, 21]. 3-D graph drawing has applications in VLSI circuit design [1, 2, 18, 23, 26] and software engineering [15, 16, 24, 25] for example. Note that there is some experimental evidence suggesting that displaying a graph in three dimensions is better than in two [28, 29].

Drawings with many bends appear cluttered and are difficult to visualise. In VLSI layouts, bends in the wires increase the cost of production and the chance of circuit failure. Therefore minimising the number of bends, along with minimising the bounding box volume, have been the most commonly proposed aesthetic criteria for measuring the quality of a drawing. Using straightforward extensions of the corresponding 2-D NP-hardness results, optimising each of these criteria is NP-hard [12]. Kolmogorov and Barzdin [17] established a lower bound of $\Omega(n^{3/2})$ on the bounding box volume of drawings of $n$-vertex graphs. In this paper we establish the first non-trivial lower bounds for the number of bends in 3-D orthogonal drawings. Lower bounds for the number of bends in 2-D orthogonal drawings have been established by Tamassia *et al.* [27] and Biedl [7].

A graph with no parallel edges and no loops is *simple*; a *multigraph* may have parallel edges but no loops; and a *pseudograph* may have parallel edges and loops. We consider $n$-vertex $m$-edge graphs $G$ with maximum degree at most six, whose vertex set and edge set are denoted by $V(G)$ and $E(G)$, respectively.

A $j$-edge matching is denoted by $M_j$; that is, $M_j$ consists of $j$ edges with no end-vertex in common. $K_p \setminus M_j$ is the graph obtained from the complete graph $K_p$ by deleting a $j$-edge matching $M_j$ (where $2j \leq p$). The 2-vertex multigraph

D. Wood, *Lower Bounds for 3-D Drawings*, JGAA, 7(1) 33–77 (2003)      35

with $j$ edges is denoted by $j \cdot K_2$, and $L_j$ is the 1-vertex pseudograph with $j$ loops. An *m-path* is a path with $m$ edges. By $C_m$ we denote the cycle with $m$ edges, which is also called an *m*-cycle. A *chord* of a cycle $C$ is an edge not in $C$ whose end-vertices are both in $C$. We say two cycles are *chord-disjoint* if they do not have a chord in common. Note that chord-disjoint cycles may share a vertex or edge. A *chordal path* of a cycle $C$ is a path $P$ whose end-vertices are in $C$, but the internal vertices of $P$ and the edges of $P$ are not in $C$.

**Lower bounds for the maximum number of bends per edge:**

Obviously every drawing of $K_3$ has at least one bend. It follows from results in multi-dimensional orthogonal graph drawing by Wood [32], Wood [35] that every drawing of $K_5$ has an edge with at least two bends. It is well known that every drawing of $6 \cdot K_2$ has an edge with at least three bends, and it is easily seen that $2 \cdot K_2$ and $3 \cdot K_2$ have at least one edge with at least one and two bends, respectively.

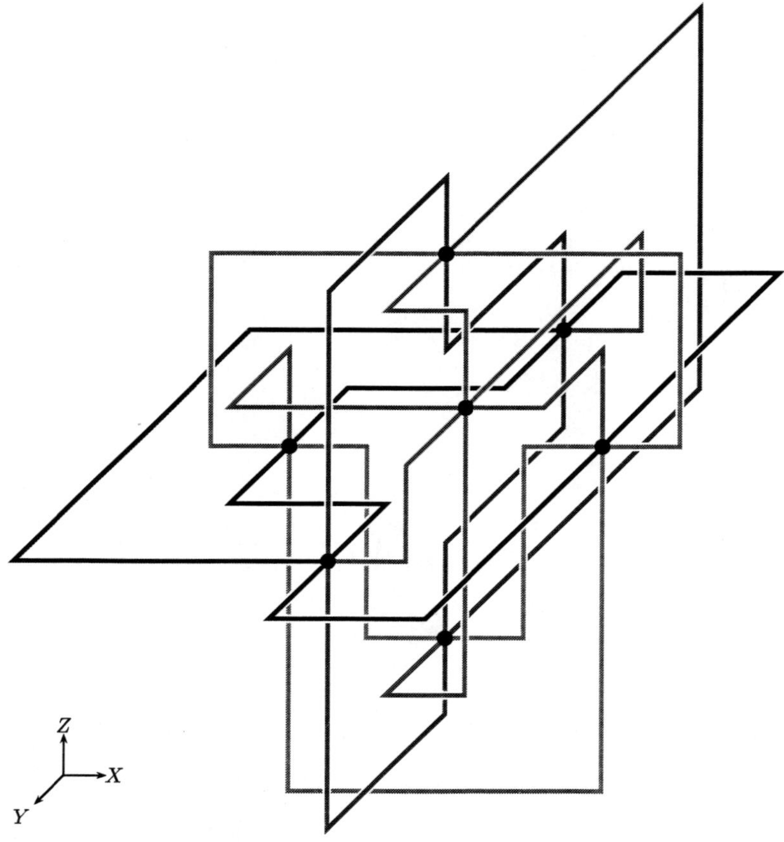


Figure 1: A 2-bend drawing of $K_7$.

D. Wood, *Lower Bounds for 3-D Drawings*, JGAA, 7(1) 33–77 (2003)     36

Eades *et al.* [13] originally conjectured that every drawing of $K_7$ has an edge with at least three bends. A counterexample to this conjecture, namely a drawing of $K_7$ with at most two bends per edge, was first exhibited by Wood [32]. A more symmetric drawing of $K_7$ with at most two bends per edge is illustrated in Figures 1 and 2. This drawing[1] has the interesting feature of rotational symmetry about the line $X = Y = Z$.

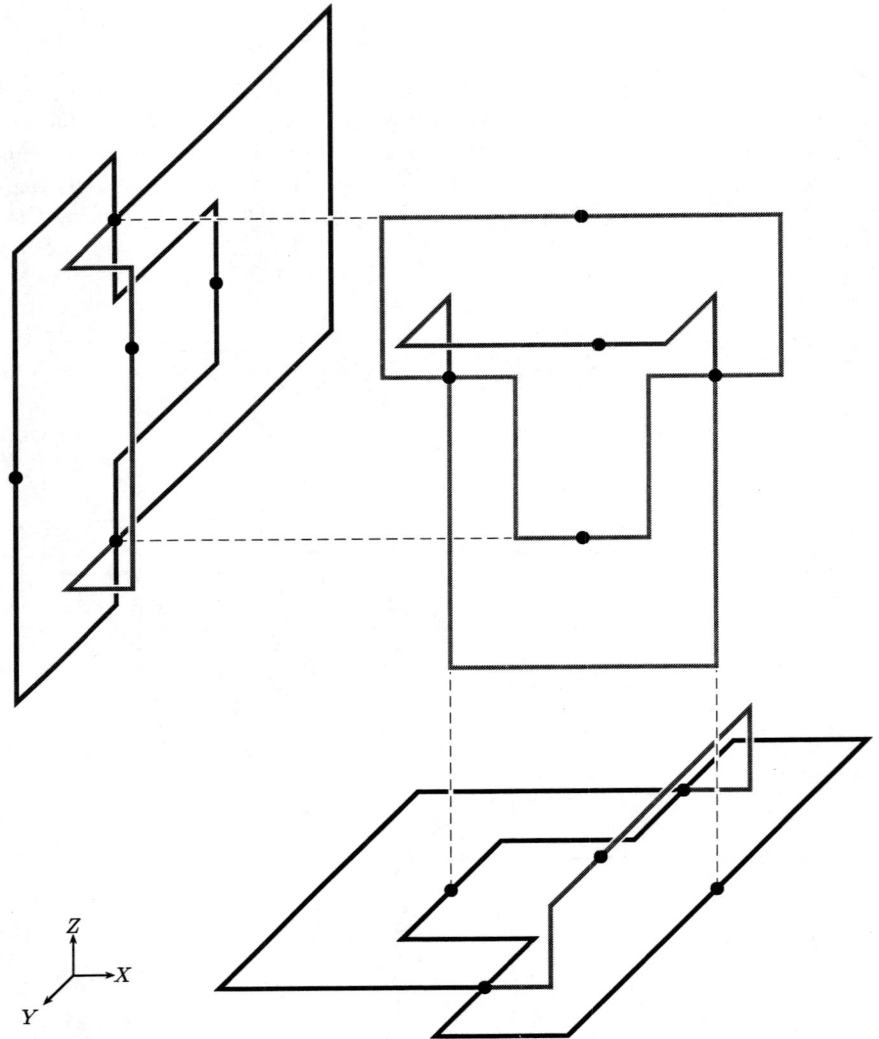



Figure 2: Components of a 2-bend drawing of $K_7$.

[1] A physical model of this drawing is on display at the School of Computer Science and Software Engineering, Monash University, Clayton, Melbourne, Australia.

D. Wood, *Lower Bounds for 3-D Drawings*, JGAA, 7(1) 33–77 (2003)    37

One may consider the other 6-regular complete multi-partite graphs $K_{6,6}$, $K_{3,3,3}$ and $K_{2,2,2,2}$ to be potential examples of simple graphs requiring an edge with at least three bends. However, 2-bend drawings of these graphs were discovered by Wood [35].

**Lower bounds for the total number of bends:**

The main result in this paper is the construction of infinite families of graphs of given connectivity and maximum degree, with a lower bound on the average number of bends in a drawing of each graph in the class. As a first step toward this goal we establish lower bounds on the minimum number of bends in drawings of small complete graphs, and the graphs obtained from small complete graphs by deleting a matching; see Table 1. For many of these graphs the obtained lower bound is tight; that is, there is a drawing with this many bends. The main exception being $K_7$ and the graphs derived from $K_7$ by deleting a matching. In particular, we prove a lower bound of $20 - 3j$ for the number of bends in drawings of $K_7 \setminus M_j$, whereas the best known drawings have $24 - 4j$ edges; see Figure 19. We conjecture that there is no drawing of $K_7 \setminus M_j$ with fewer than $24 - 4j$ edges for each $j \in \{0, 1, 2, 3\}$. There is also a gap in our bounds in the case of $K_6 \setminus M_3$. Here we have a lower bound of seven bends, whereas the best known drawing of $K_6 \setminus M_3$ has eight bends, which we conjecture is bend-minimum.

Table 1: Bounds for the minimum number of bends in drawings of complete graphs minus a matching $K_p \setminus M_j$.

|         | $p = 3$ | $p = 4$ | $p = 5$ | $p = 6$ | $p = 7$ |
|---------|---------|---------|---------|---------|---------|
| $j = 0$ | 1       | 3       | 7       | 12      | $20 \ldots 24$ |
| $j = 1$ | 0       | 2       | 5       | 10      | $17 \ldots 20$ |
| $j = 2$ | -       | 1       | 4       | 8       | $14 \ldots 16$ |
| $j = 3$ | -       | -       | -       | $7 \ldots 8$ | $11 \ldots 12$ |

Furthermore we show that drawings of the multigraphs $j \cdot K_2$ with $2 \leq j \leq 6$ have at least 2, 4, 6, 8 and 12 bends, respectively. Since a loop has at least three bends in every drawing, the pseudograph $L_j$ with $1 \leq j \leq 3$ has at least $3j$ bends.

We use the above lower bounds as the basis for the construction of infinite families of $c$-connected graphs of maximum degree $\Delta$ with lower bounds on the number of bends for each member of the class, as summarised in Table 2.

**Upper bounds:**

A number of algorithms have been proposed for 3-D orthogonal graph drawing [3, 6, 9–13, 17, 21, 22, 33, 35–37]. We now summarise the best known upper

Table 2: Lower bounds on the average number of bends in drawings of an infinite family of $c$-connected graphs with maximum degree at most $\Delta$.

| type | simple graphs | | | | | multigraphs | | | | | pseudographs | | | | |
|---|---|---|---|---|---|---|---|---|---|---|---|---|---|---|---|
| $\Delta$ | 6 | 5 | 4 | 3 | 2 | 6 | 5 | 4 | 3 | 2 | 6 | 5 | 4 | 3 | 2 |
| $c=0$ | $\frac{20}{21}$ | $\frac{4}{5}$ | $\frac{7}{10}$ | $\frac{1}{2}$ | $\frac{1}{3}$ | 2 | $\frac{8}{5}$ | $\frac{3}{2}$ | $\frac{4}{3}$ | 1 | 3 | 3 | 3 | 3 | 3 |
| $c=2$ | $\frac{17}{21}$ | $\frac{2}{3}$ | $\frac{1}{2}$ | $\frac{1}{3}$ | - | $\frac{4}{3}$ | $\frac{6}{5}$ | 1 | $\frac{2}{3}$ | - | 2 | $\frac{3}{2}$ | $\frac{3}{2}$ | - | - |
| $c=3$ | $\frac{8}{11}$ | $\frac{14}{23}$ | $\frac{8}{19}$ | $\frac{2}{9}$ | - | 1 | $\frac{4}{5}$ | $\frac{1}{2}$ | - | - | $\frac{4}{3}$ | $\frac{6}{5}$ | - | - | - |
| $c=4$ | $\frac{12}{17}$ | $\frac{7}{12}$ | $\frac{2}{5}$ | - | - | $\frac{2}{3}$ | $\frac{2}{5}$ | - | - | - | 1 | - | - | - | - |
| $c=5$ | $\frac{24}{35}$ | $\frac{14}{25}$ | - | - | - | $\frac{1}{3}$ | - | - | - | - | - | - | - | - | - |
| $c=6$ | $\frac{2}{3}$ | - | - | - | - | - | - | - | - | - | - | - | - | - | - |

bounds for the number of bends in 3-D orthogonal drawings. The 3-BENDS algorithm of Eades *et al.* [13] and the INCREMENTAL algorithm of Papakostas and Tollis [21] both produce 3-bend drawings[2] of multigraphs[3] with maximum degree six. As discussed above there exist simple graphs with at least one edge having at least two bends in every drawing. The following open problem is therefore of interest:

*2-Bends Problem:* Does every (simple) graph with maximum degree at most six admit a 2-bend drawing? [13]

The DIAGONAL LAYOUT & MOVEMENT algorithm of Wood [37] (also see [33]) solves the 2-bends problem in the affirmative for simple graphs with maximum degree five. For maximum degree six simple graphs, the same algorithm uses a total of at most $\frac{16}{7}m$ bends, which is the best known upper bound for the total number of bends in 3-D orthogonal drawings.

In this paper we provide a negative result related to the 2-bends problem. A 3-D orthogonal graph drawing is said to be in *general position* if no two vertices lie in a common grid-plane. The general position model is used in the 3-BENDS and DIAGONAL LAYOUT & MOVEMENT algorithms. In this paper we show that the general position model, and the natural variation of this model where pairs of vertices share a common plane, cannot be used to solve the 2-bends problem, at least for 2-connected graphs.

The remainder of this paper is organised as follows. In Section 2 we establish

---

[2]The 3-BENDS algorithm [13] produces drawings with $27n^3$ volume. By deleting grid-planes not containing a vertex or a bend the volume is reduced to $8n^3$. The INCREMENTAL algorithm [21] produces drawings with $4.63n^3$ volume. A modification of the 3-BENDS algorithm by Wood [36] produces drawings with $n^3 + o(n^3)$ volume.

[3]The 3-BENDS algorithm [13] explicitly works for multigraphs. The INCREMENTAL algorithm, as stated in [21], only works for simple graphs, however with a suitable modification it also works for multigraphs [A. Papakostas, private communication, 1998].

D. Wood, *Lower Bounds for 3-D Drawings*, JGAA, 7(1) 33–77 (2003)     39

a number of introductory results concerning 0-bend drawings of cycles. These results are used to prove our lower bounds on the total number of bends in drawings of complete graphs and graphs obtained from complete graphs by deleting a matching, which are established in Section 3. In Section 4 we use these lower bounds as the basis for lower bounds on the number of bends in infinite families of graphs. In Section 5 we present lower bounds for the number of bends in general position drawings. These results have important implications for the nature of any solution to the 2-bends problem, which are discussed in Section 6.

Some technical aspects of our proofs are presented in the appendices. In particular, in Appendix A we prove a number of results concerning the existence of cycles and other small subgraphs in graphs of a certain size; in Appendix B we establish the connectivity of the graphs used in our lower bounds; and in Appendix C we prove a result, which is not directly used in other parts of the paper, but may be of independent interest.

## 2    Drawings of Cycles

In this section we characterise the 0-bend drawings of the cycles $C_k$ ($k \leq 7$). We then show that if a drawing of a complete graph contains such a 0-bend drawing of a cycle then there are many edges with at least three bends in the drawing of the complete graph. These results are used in Section 3 in the proofs of our lower bounds on the total number of bends in drawings of complete graphs.

A straight-line path in a 0-bend drawing of a cycle is called a *side*. A side parallel to the $I$-axis for some $I \in \{X, Y, Z\}$ is called an $I$-side, and $I$ is called the *dimension* of the side. Clearly the dimension of adjacent sides is different. Thus in a 2-dimensional drawing the dimension of the sides alternate around the cycle. We therefore have the following observation.

**Observation 1.** *There is no 2-dimensional 0-bend drawing of a cycle with an odd number of sides.*

If there is an $I$-side in a drawing of a cycle for some $I \in \{X, Y, Z\}$ then clearly there is at least two $I$-sides. Therefore a drawing of a cycle with $X$-, $Y$- and $Z$-sides, which we call *truly 3-dimensional*, has at least six sides. Hence there is no truly 3-dimensional 3-, 4- or 5-sided 0-bend drawing of a cycle. By Observation 1 there is also no two-dimensional 3- or 5-sided 0-bend drawing of a cycle. We therefore have the following observations.

**Observation 2.** *There is no 3- or 5-sided 0-bend drawing of a cycle.*

**Observation 3.** *There is no 0-bend drawing of $C_3$.*

**Observation 4.** *All 0-bend drawings of $C_4$ and $C_5$ have four sides.*

**Lemma 1.** *If a drawing of a complete graph contains a 0-bend 4-cycle (respectively, 5-cycle) then at least two (four) chords of the cycle each have at least three bends.*

D. Wood, *Lower Bounds for 3-D Drawings*, JGAA, 7(1) 33–77 (2003)     40

*Proof.* By Observation 4 all 0-bend drawings of $C_4$ and of $C_5$ have four sides. As illustrated in Figure 3(a), the chord connecting diagonally opposite vertices in a 4-sided drawing of a cycle has at least three bends. Hence, if a drawing of a complete graph contains a 0-bend $C_4$, then the two chords each have at least three bends. Also, in the case of $C_5$, the edges from the vertex not at the intersection of two sides to the diagonally opposite vertices both have at least three bends, as illustrated in Figure 3(b). Hence, if a drawing of a complete graph contains a 0-bend $C_5$, then the four chords each have at least three bends. $\square$

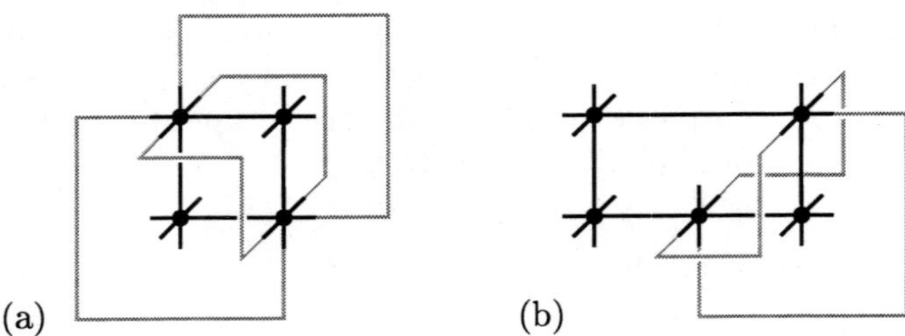

(a)        (b)

Figure 3: 3-bend edge 'across' the 4- and 5-cycle.

**Observation 5.** $K_{2,3}$ *does not have a 0-bend drawing.*

*Proof.* $K_{2,3}$ contains $C_4$. By Observation 4, all 0-bend drawing of $C_4$ have four sides. As in Lemma 1, an edge between the diagonally opposite vertices of a 4-sided cycle has at least three bends. Hence the 2-path in $K_{2,3}$ between the non-adjacent vertices of the 4-cycle has at least one bend, as illustrated in Figure 4(b). Hence $K_{2,3}$ does not have a 0-bend drawing. $\square$

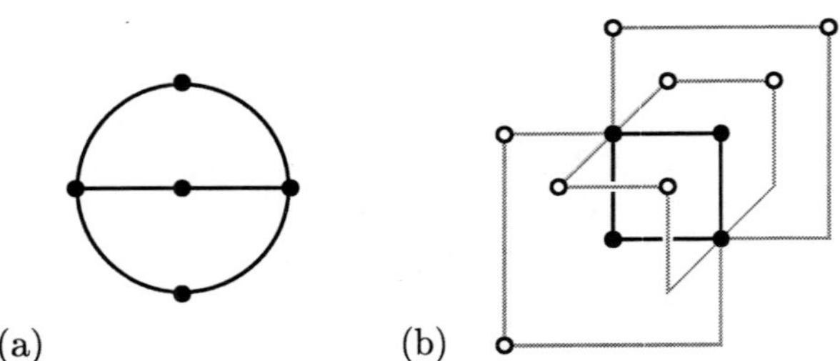

(a)        (b)

Figure 4: (a) The graph $K_{2,3}$. (b) $K_{2,3}$ does not have a 0-bend drawing.

The proof of the following lemma is almost identical to that of Observation 5 and is omitted.

**Observation 6.** *If a drawing of a graph contains a 0-bend 4-cycle* $(a, b, c, d)$ *with a chordal 2-path* $P \in \{(a, x, c), (b, x, d)\}$, *then* $P$ *has at least two bends.* $\square$

D. Wood, *Lower Bounds for 3-D Drawings*, JGAA, 7(1) 33–77 (2003)    41

We now classify the 0-bend drawings of $C_6$.

**Lemma 2.** *The only 6-sided 0-bend drawings of $C_6$ are those in Figure 5 (up to symmetry and the deletion of grid-planes not containing a vertex).*

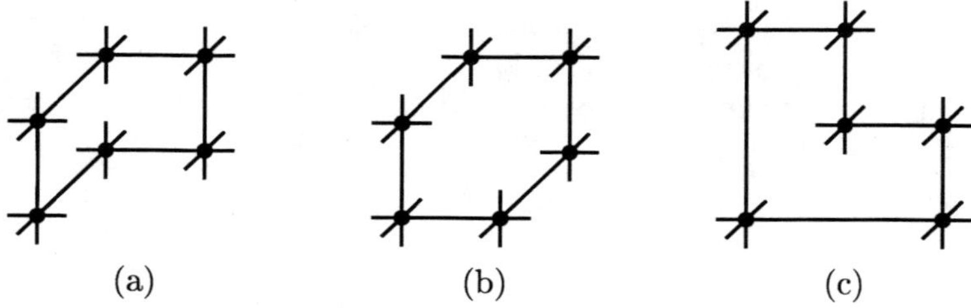

Figure 5: 6-sided 0-bend drawings of $C_6$.

*Proof.* Let $S$ be the cyclic sequence of dimensions of the sides around an arbitrary, but fixed, 6-sided 0-bend drawing of $C_6$.

First suppose the drawing is 2-dimensional. Since adjacent sides are perpendicular, without loss of generality three sides are $X$-sides and three sides are $Y$-sides. Therefore $S$ is $(X, Y, X, Y, X, Y)$. The length of one of the $X$-sides equals the sum of the lengths of the other two $X$-sides, and similarly for the $Y$-sides. Label these long sides $X^*$ and $Y^*$. If the long sides are adjacent then $S$ is $(X^*, Y^*, X, Y, X, Y)$, which corresponds to the drawing in Figure 5(c). If the long sides are not adjacent then $S$ is $(X^*, Y, X, Y^*, X, Y)$, which corresponds to the 'drawing' in Figure 6, which contains an edge crossing.

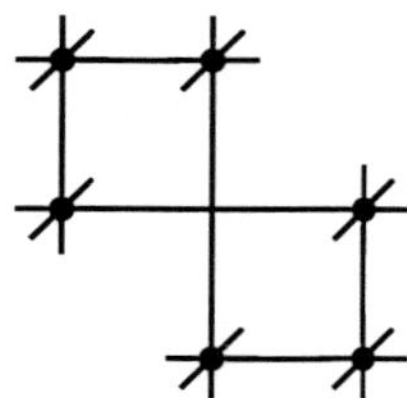

Figure 6: 6-sided 0-bend 'drawing' of $C_6$ with an edge crossing.

Now suppose the drawing is truly 3-dimensional. Clearly there are two $X$-sides, two $Y$-sides and two $Z$-sides. Let $x$ be the number of sides between the two $X$-sides in $S$. Clearly $x$ is one or two. Define $y$ and $z$ similarly for the $Y$- and $Z$-sides. We can assume without loss of generality that $x \leq y \leq z$.

If $x = 1$ and $y = 1$ then $S$ is $(X, Z, X, Y, Z, Y)$, and $z = 2$. This sequence corresponds to the drawing in Figure 5(a). If $x = 1$ and $y = 2$ then $S$ is $(X, Y, X, Z, Y, Z)$, and $z = 1$ which is a contradiction. Otherwise $x = y = z = 2$ and $S$ is $(X, Y, Z, X, Y, Z)$ without loss of generality, which corresponds to the drawing in Figure 5(b). $\square$

D. Wood, *Lower Bounds for 3-D Drawings*, JGAA, 7(1) 33–77 (2003)      42

**Lemma 3.** *If a drawing of a complete graph contains a 0-bend 6-cycle then there are at least six chords of the cycle each with at least three bends.*

*Proof.* We can assume without loss of generality that the complete graph in question is $K_6$. By Observation 2, all 0-bend drawings of $C_6$ are 4- or 6-sided. In a 4-sided 0-bend drawing of $C_6$ the two vertices not at the intersection of adjacent sides can be (a) on the same side, (b) on opposite sides, or (c) on adjacent sides, as illustrated in Figure 7. In each case there are at least six chords each with at least three bends if the 0-bend drawing of $C_6$ is contained in a drawing of $K_6$.

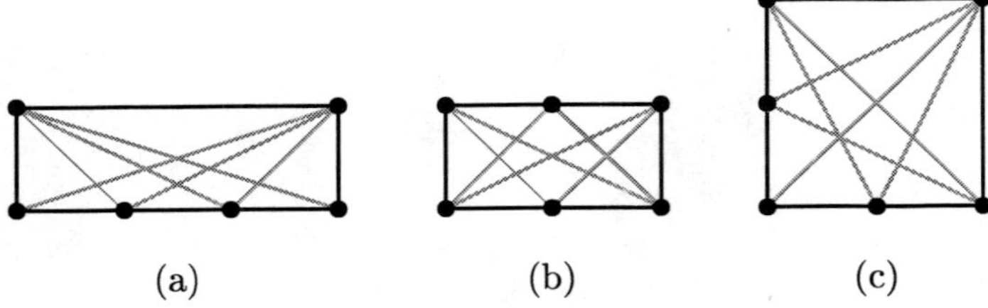

Figure 7: Edges with at least three bends in a drawing of $K_6$ containing a 4-sided 0-bend drawing of $C_6$.

By Lemma 2, the only 6-sided 0-bend drawings of $C_6$ (up to symmetry) are those in Figure 5. For each such drawing of $C_6$, if this is a sub-drawing of a drawing of $K_6$, then those chords of $C_6$ illustrated in Figure 8 each require at least three bends (compare with Figure 3). In the case of the drawing in Figure 8(c) there are at least six chords each requiring at least three bends.

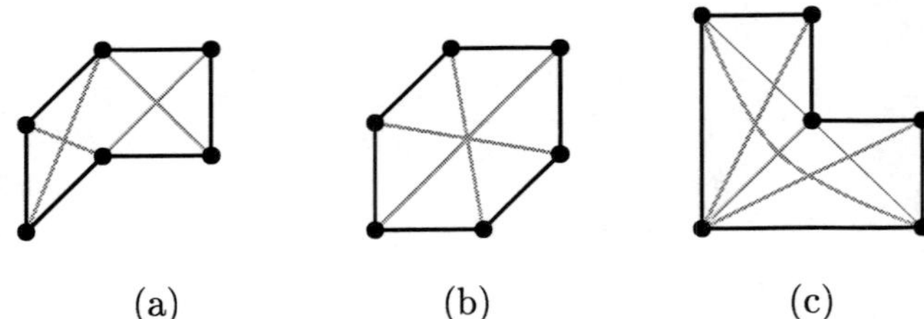

Figure 8: Edges with at least three bends in a drawing of $K_6$ containing a 6-sided 0-bend $C_6$.

Consider the drawing in Figure 8(a) which forces at least four chords to have at least three bends if a sub-drawing of a drawing of $K_6$. As illustrated in Figure 9(a), any drawing of the edges $vu$ and $vw$ with at most two bends per edge passes through the same point. Hence one of these edges has at least three bends. We can make the same argument for the edges $xw$ and $xu$. Hence if $K_6$ contains the sub-drawing of $C_6$ illustrated in Figure 8(a) then there are at least six chords each with at least three bends.

Now consider the drawing in Figure 8(b) which forces at least three chords to have at least three bends if a sub-drawing of a drawing of $K_6$. As illustrated in Figure 9(b), any drawing of the edges $vu$, $uw$ and $vw$ with at most two bends per edge passes through the same point. Hence two of these edges have at least three bends. We can make the same argument for three edges connecting the other three vertices. Hence if $K_6$ contains the sub-drawing of $C_6$ illustrated in Figure 8(b) then there are at least seven chords each with at least three bends. The result follows.      □

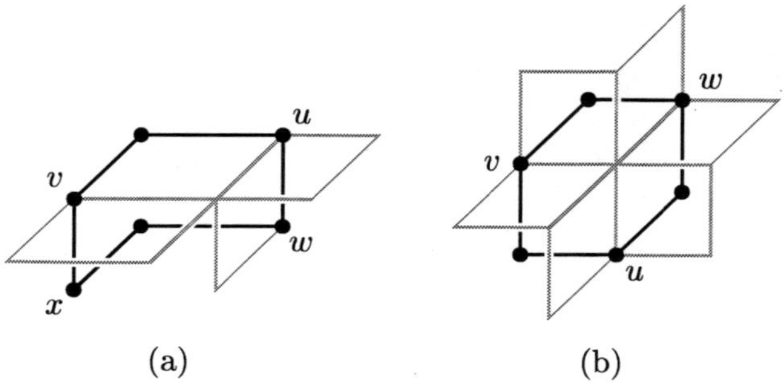


Figure 9: Intersecting 1- and 2-bend edges.

**Lemma 4.** *The only 7-sided 0-bend drawings of $C_7$ are those in Figure 10 (up to symmetry and the deletion of grid-planes not containing a vertex).*

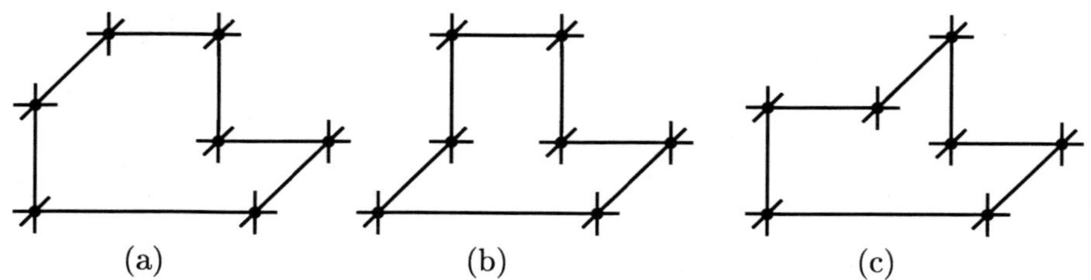


Figure 10: 7-sided 0-bend drawings of $C_7$.

*Proof.* Consider an arbitrary, but fixed, 7-sided 0-bend drawing of $C_7$. By Observation 2, there is no 2-dimensional 0-bend drawing of an odd cycle, and if there is an $I$-side in a drawing of a cycle for some $I \in \{X, Y, Z\}$, then there are at least two $I$-sides. Therefore in a 7-sided cycle, without loss of generality three of the sides are $X$-sides, two are $Y$-sides and two are $Z$-sides. Clearly the length of one of the $X$-sides equals the sum of the lengths of the other two $X$-sides. Label this long side $X^*$.

D. Wood, *Lower Bounds for 3-D Drawings*, JGAA, 7(1) 33–77 (2003)       44

Let $S$ be the cyclic sequence of the dimensions of the sides around $C_7$, which without loss of generality begins with the $X^*$-side. Therefore $S$ is (i) $(X^*, 1, X, 2, X, 3, 4)$, (ii) $(X^*, 1, X, 2, 3, X, 4)$, or (iii) $(X^*, 1, 2, X, 3, X, 4)$, where the numbered locations refer to a $Y$- or $Z$-side.

In case (i), the dimensions of the '3' and '4' sides are different, hence the dimensions of the '1' and '2' sides are also different. Without loss of generality '1' is a $Y$-side and '2' is a $Z$-side. Therefore $S$ is either $(X^*, Y, X, Z, X, Y, Z)$ or $(X^*, Y, X, Z, X, Z, Y)$, which correspond to the drawings in Figure 10(a) and Figure 10(b), respectively.

In case (ii), the dimensions of the '2' and '3' sides are different, hence the dimensions of the '1' and '4' sides are also different. Without loss of generality '1' is a $Y$-side and '4' is a $Z$-side. Therefore $S$ is either $(X^*, Y, X, Z, Y, X, Z)$, which corresponds to the drawing in Figure 10(c), or $(X^*, Y, X, Y, Z, X, Z)$ which corresponds to the 'drawing' in Figure 11 with an edge crossing.

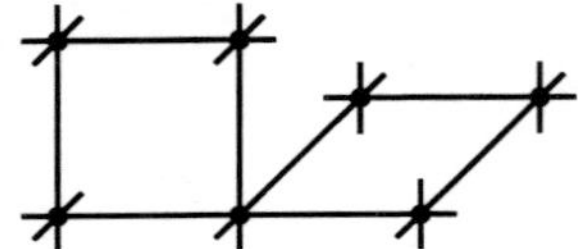

Figure 11: 7-sided 0-bend 'drawing' of $C_7$ with an edge intersection.

In case (iii), $S$ is simply the reverse sequence of $S$ in case (i). We therefore have classified all 7-sided 0-bend drawings of $C_7$ up to symmetry and after removing grid-planes not containing a vertex.  $\square$

**Lemma 5.** *If a drawing of $K_7$ contains a 0-bend 7-cycle then there are at least six chords of the cycle each with at least three bends.*

*Proof.* By Observation 2, a 0-bend drawing of $C_7$ has four, six or seven sides.

In a 4-sided 0-bend drawing of $C_7$, as illustrated in Figure 12, the three vertices not at the intersection of two adjacent sides can be (a) all on the same side, (b) two on one side and one on an adjacent side, (c) two on one side and one on the opposite side, or (d) all on different sides. For each drawing, if the 7-cycle is a sub-drawing of a drawing of $K_7$, then eight chords of the cycle have at least three bends (compare with Figure 3).

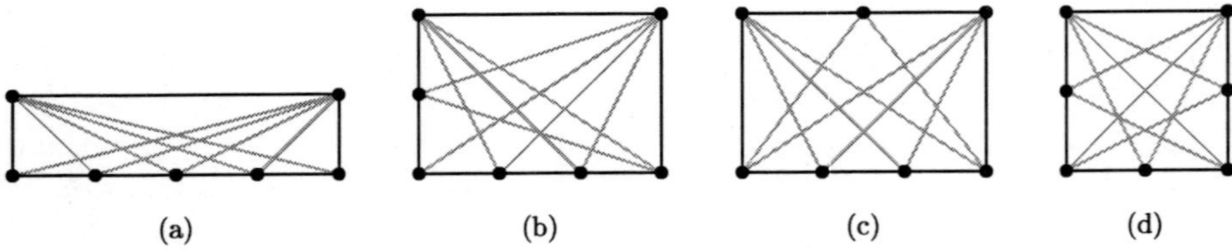

(a)　　　　　(b)　　　　　(c)　　　　　(d)

Figure 12: Edges with at least three bends in a 4-sided 0-bend drawing of $C_7$.

Any 6-sided 0-bend drawing of $C_7$ can be obtained by placing a new vertex on one side of a 6-sided 0-bend drawing of $C_6$. Thus, by Lemma 3 if a drawing of $K_7$ contains a 6-sided 0-bend drawing of $C_7$ then at least six of the chords have at least three bends.

By Lemma 4, the only 7-sided drawings of $C_7$ are those illustrated in Figure 10. For each such drawing, if the 7-cycle is a sub-drawing of a drawing of $K_7$, Figure 13 shows chords of the cycle which need at least three bends. The drawings in Figure 13(a), (b) and (c) have four, six, and four chords, respectively, which need at least three bends.

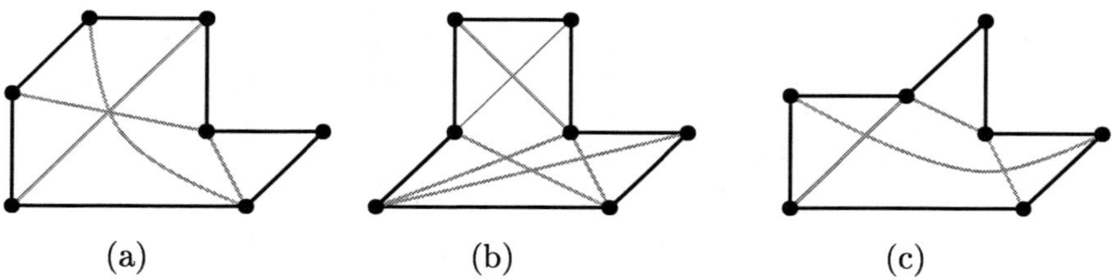

(a)        (b)        (c)

Figure 13: Edges with at least three bends in a 7-sided 0-bend drawing of $C_7$.

Consider the drawing in Figure 13(a). As illustrated in Figure 14(a), any drawing of the edges $vu$, $vw$ and $vx$ with at most two bends per edge passes through the same grid-point. Hence two of these edges have at least three bends. Therefore if $K_7$ contains the sub-drawing of $C_7$ illustrated in Figure 13(a) then there are at least six edges each with at least three bends.

Now consider the drawing in Figure 13(c). As illustrated in Figure 14(b), there is one route for the edge $vu$ with at most two bends, one route for the edge $vw$ with at most two bends, and three routes for the edge $vx$ with at most two bends. Any two of these edge routes for distinct edges pass through the same point. Hence two of these edges have at least three bends. Therefore if $K_7$ contains the sub-drawing of $C_7$ illustrated in Figure 13(c), then there are at least six edges each with at least three bends.

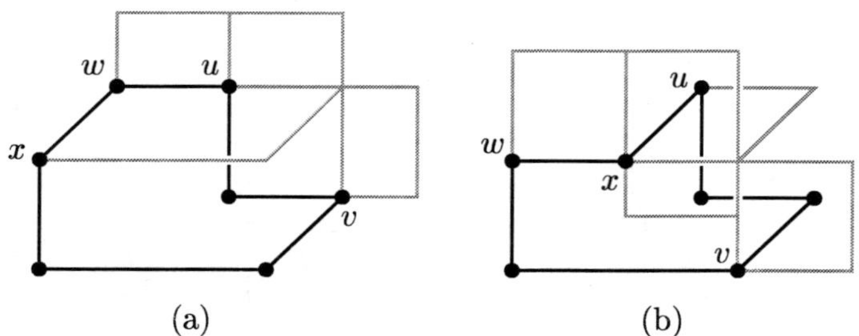

(a)        (b)

Figure 14: Intersecting 1- and 2-bend edges.

Therefore if a drawing of a complete graph contains a 0-bend 7-cycle, then

D. Wood, *Lower Bounds for 3-D Drawings*, JGAA, 7(1) 33–77 (2003)     46

there are at least six chords of the cycle each with at least three bends.     □

The results in this section are summarised by the following immediate corollary of Lemmata 1, 3 and 5.

**Theorem 1.** *If a drawing of $K_p \setminus M_j$ contains a 0-bend 4-cycle (respectively, 5-cycle, 6-cycle, or 7-cycle), then there are at least $2 - j$ $(4 - j,\ 6 - j,\ 6 - j)$ chords of the cycle each with at least three bends.*     □

## 3   Drawings of Complete Graphs

In this section we establish lower bounds for the total number of bends in drawings of the complete graphs $K_4$, $K_5$, $K_6$ and $K_7$, and the graphs obtained from these complete graphs by deleting a matching. We complete the section by establishing lower bounds for the number of bends in drawings of the multigraphs $j \cdot K_2$ for $2 \leq j \leq 6$.

The 0-*bend subgraph* of a given drawing consists of those edges drawn with no bends. The following proofs typically proceed by case analysis on the size of the 0-bend subgraph.

Figure 15 shows a drawing of $K_4$ with three bends. Deleting one of the 1-bend edges produces a drawing of $K_4 \setminus M_1$ with two bends. We now prove that both of these drawings are bend-minimum. This elementary result is indicative of the method of proof for the corresponding results for larger complete graphs which follow.

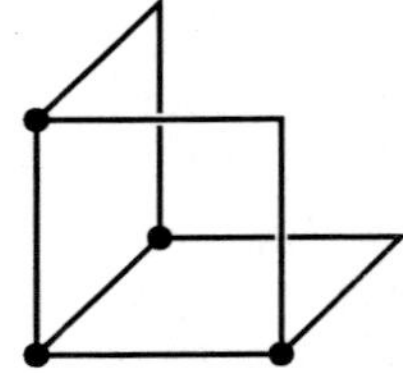

Figure 15: A drawing of $K_4$ with three bends.

**Theorem 2.** *Let $j \in \{0, 1\}$. Every drawing of $K_4 \setminus M_j$ has at least $3 - j$ bends.*

*Proof.* Let $k_0$ be the number of 0-bend edges in a drawing of $K_4 \setminus M_j$. If $k_0 \leq 3$ then there are at least $3 - j$ edges each with at least one bend, and we are done. Otherwise $k_0 \geq 4$. The 0-bend subgraph has no 3-cycle by Observation 3. The only 4-vertex graph with at least four edges and no 3-cycle is a 4-cycle. Thus the 0-bend subgraph is a 4-cycle. By Theorem 1, if $K_4 \setminus M_j$ (with $j \leq 1$) contains a 0-bend 4-cycle, then there is at least one chord with at least three bends.     □

We now establish tight lower bounds for the total number of bends in drawings of $K_5$ and the graphs obtained from $K_5$ by deleting a matching. To prove that the drawing of $K_5$ illustrated in Figure 16 is bend-minimum we use the following result, which may be of independent interest.

D. Wood, *Lower Bounds for 3-D Drawings*, JGAA, 7(1) 33–77 (2003)      47

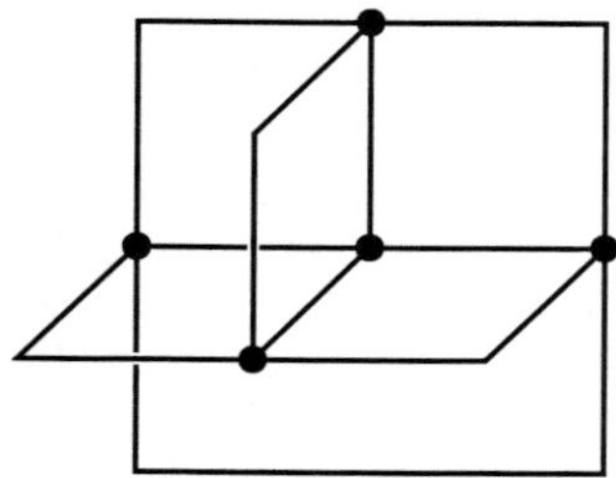

Figure 16: A 2-bend drawing of $K_5$ with seven bends.

**Lemma 6.** *For every set $S$ of grid-points in the 3-D orthogonal grid, either*
*(1) two points in $S$ are non-coplanar, or*
*(2) at least $\lceil 2|S|/3 \rceil$ grid-points are in a single grid-plane.*

*Proof.* Suppose (1) does not hold; that is, every pair of points in $S$ share at least one coordinate. Without loss of generality assume that there is a point $v \in S$ positioned at $(0,0,0)$. Thus every point in $S$ has at least one coordinate equal to 0. If some point $w \in S$ has exactly one coordinate equal to 0 then, supposing this is the $X$-coordinate, all points in $S$ have an $X$-coordinate of 0 (to be coplanar with $v$ and $w$); that is, all points are in a single grid-plane, and the result follows. Otherwise every point in $S$, except $v$, has exactly two coordinates equal to 0; that is, every point lies on an axis. Let $x$, $y$ and $z$ be the number of points in $S$, not counting $v$, on the $X$, $Y$ and $Z$ axes, respectively. Then $x+y+z = |S|-1$. Without loss of generality $x \geq y \geq z$. Clearly $x \geq \lceil (|S|-1)/3 \rceil$ and $y \geq \lfloor (|S|-1)/3 \rfloor$, which implies there are $x + y + 1 \geq \lceil 2|S|/3 \rceil$ points in $S$ on the $X$- or $Y$-axes, and thus in a single grid-plane. $\qquad\square$

Wood [32] shows that a 1-bend drawing of $K_n$ in a multi-dimensional orthogonal grid requires at least $n-1$ dimensions. We now provide a simple proof of an equivalent formulation of this result in the case of $n = 5$.

**Theorem 3.** *Every drawing of $K_5$ has an edge with at least two bends.*

*Proof.* In a layout of the vertices of $K_5$, if two vertices are non-coplanar, then the edge between them has at least two bends. By Lemma 6 with $S = V(K_5)$, if all the vertices are pairwise coplanar then four of the vertices are coplanar. Consider the $K_4$ subgraph $H$ induced by these four coplanar vertices. If any edge of $H$ leaves the plane containing the vertices then it has at least two bends. Otherwise we have a plane orthogonal drawing of $K_4$, which has an edge with at least two bends [14]. $\qquad\square$

We now prove that the drawing of $K_5$ in Figure 16 is bend-minimum.

**Theorem 4.** *Every drawing of $K_5$ has at least seven bends.*

*Proof.* Let $k_0$ be the number of 0-bend edges in a drawing of $K_5$. Since every subgraph of $K_5$ with at least eight edges contains a 3-cycle, and by Observation 3, the 0-bend subgraph has no 3-cycle, $k_0 \leq 7$. If $k_0 \leq 4$ then there are at least six edges each with at least one bend, and by Theorem 3, one of these edges has at least two bends, implying there are at least seven bends in total, and we are done. Now assume $5 \leq k_0 \leq 7$. Thus the 0-bend subgraph contains a cycle, which by Observation 3, is a 4-cycle or a 5-cycle $C$. By Lemma 1, $C$ has at least two chords each with at least three bends. Since $k_0 \leq 7$ there is at least one additional edge with at least one bend, implying a total of at least seven bends. □

By deleting the 2-bend edge in the drawing of $K_5$ illustrated in Figure 16, we obtain a drawing of $K_5 \setminus M_1$ with five bends in total. By deleting the appropriate 1-bend edge from this drawing, we obtain a drawing of $K_5 \setminus M_2$ with four bends in total. We now prove that both of these drawings are bend-minimum.

**Theorem 5.** *For each $j \in \{1, 2\}$, every drawing of $K_5 \setminus M_j$ has at least $6 - j$ bends.*

*Proof.* Let $k_0$ be the number of 0-bend edges in a drawing of $K_5 \setminus M_j$ for some $j \in \{1, 2\}$. If $k_0 \leq 4$ then there are at least $6 - j$ edges each with at least one bend, and we are done. Now assume $k_0 \geq 5$. By Lemma 13 in Appendix A, every subgraph of $K_5 \setminus M_j$ with at least six edges contains $C_3$ or $K_{2,3}$, and by Observations 3 and 5, the 0-bend subgraph does not contain $C_3$ or $K_{2,3}$. Thus $k_0 = 5$. Since every 5-edge subgraph of $K_5 \setminus M_j$ contains a cycle, and the 0-bend subgraph does not contain a 3-cycle (by Observation 3), there is a 0-bend 4-cycle or 5-cycle $C$.

Suppose $C$ has a chord, which is guaranteed by Theorem 1 in the case of $C$ being a 5-cycle. Then by Lemma 1 the chord has at least three bends. There are a further $4 - j$ edges each with at least one bend, giving a total of at least $7 - j$ bends. If $C$ is chordless then $j = 2$ and $C$ is a 4-cycle. Thus $C$ is spanned by two edge-disjoint chordal 2-paths, each of which has at least two bends by Observation 6. Thus the drawing has at least four bends, and we are done. □

We now establish tight lower bounds for the total number of bends in drawings of $K_6$, and the graphs obtained from $K_6$ by deleting a matching, except in the case of $K_6 \setminus M_3$ for which there is a difference of one bend between our lower bound and the best known drawing. Figure 17 shows the well-known drawing of $K_6$ with two 2-bend edges and a total of twelve bends.

**Theorem 6.** *Every drawing of $K_6$ has at least two edges each with at least two bends.*

*Proof.* By Theorem 3 every drawing of $K_5$, and thus $K_6$, has an edge with at least two bends. Suppose that there is a drawing of $K_6$ with exactly one edge $vw$ with at least two bends. By removing $v$ and all the edges incident to $v$ we obtain a 1-bend drawing of $K_5$, which contradicts Theorem 3. Thus every drawing of $K_6$ has at least two edges each with at least two bends. □

D. Wood, *Lower Bounds for 3-D Drawings*, JGAA, 7(1) 33–77 (2003)     49

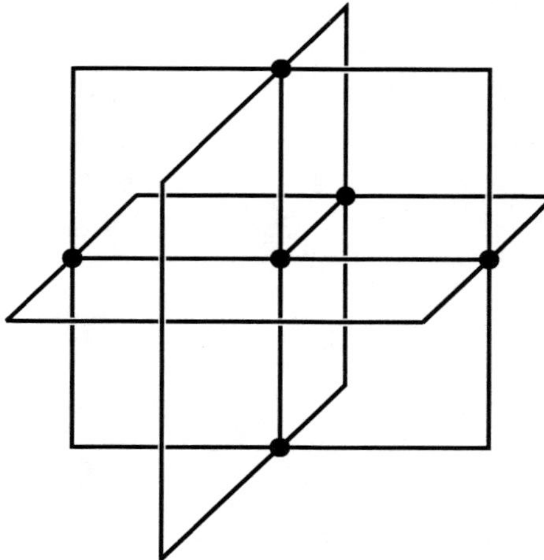

Figure 17: A 2-bend drawing of $K_6$ with 12 bends.

Note that the above result can be strengthened to say that in every drawing of $K_6$ there are two non-adjacent edges each with at least two bends. We now prove that the drawing of $K_6$ illustrated in Figure 17 is bend-minimum, as is the drawing obtained by deleting one or both of the 2-bend edges.

**Theorem 7.** *For each $j \in \{0, 1, 2\}$, every drawing of $K_6 \setminus M_j$ has at least $12 - 2j$ bends.*

*Proof.* Let $k_i$ $(i \geq 0)$ be the number of $i$-bend edges in a drawing of $K_6 \setminus M_j$ for some $j \in \{0, 1, 2\}$. Observe that $K_6 \setminus M_j$ has $15 - j$ edges. By Lemma 14 in Appendix A, every subgraph of $K_6 \setminus M_j$ with at least eight edges contains $C_3$ or $K_{2,3}$. By Observations 3 and 5, the 0-bend subgraph does not contain $C_3$ or $K_{2,3}$. Thus $k_0 \leq 7$, and hence there are at least $8 - j$ edges each with at least one bend.

If $k_0 \leq 5$ then at least $10 - j$ edges have at least one bend, and since there are at least $2 - j$ edges each with at least two bends (by Theorem 6), there are at least $12 - 2j$ bends, and we are done. Now assume $k_0 \geq 6$. Thus the 0-bend subgraph contains a cycle $C$, which by Observation 3, is not a 3-cycle. If $C$ has at least two chords then by Theorem 1, each chord has at least three bends, giving a total of at least $12 - j \geq 12 - 2j$ bends, and we are done. Also by Theorem 1, if $j = 0$ or $C$ is a 5- or 6-cycle, then $C$ has at least two chords. Thus, we now assume that $C = (a, b, c, d)$ is a 4-cycle and $j \in \{1, 2\}$.

If $j = 1$ then by Theorem 1, $C$ has exactly one chord, say $ac$. Let $x$ and $y$ be the other two vertices in $K_6 \setminus M_1$. The edge $ac$ has at least three bends, and each of the chordal 2-paths $axc$, $ayc$, $bxd$ and $byd$ have at least two bends, by Observation 6. Thus there is a total of at least $11 > 12 - 2j$ bends, and we are done. Now assume $j = 2$. If $C$ has one chord then this chord has at least three bends, giving a total of at least $8 = 12 - 2j$ bends, and we are done. Otherwise

D. Wood, *Lower Bounds for 3-D Drawings*, JGAA, 7(1) 33–77 (2003)　　　50

$C$ has no chords. Let $x$ and $y$ be the other two vertices in $K_6 \setminus M_2$. Each of the chordal 2-paths $axc$, $ayc$, $bxd$ and $byd$ have at least two bends by Observation 6, giving a total of at least $8 = 12 - 2j$ bends. This completes the proof. $\qquad\square$

We now prove a lower bound of seven on the number of bends in drawings of $K_6 \setminus M_3$, the octahedron graph. The drawing of $K_6 \setminus M_3$ obtained from the drawing of $K_6$ illustrated in Figure 17 by deleting the two 2-bend edges and the 0-bend edge has eight bends. We conjecture that every drawing of $K_6 \setminus M_3$ has at least eight bends.

**Theorem 8.** *Every drawing of $K_6 \setminus M_3$ has at least seven bends.*

*Proof.* Let $k_i$ $(i \geq 0)$ be the number of $i$-bend edges in a drawing of $K_6 \setminus M_3$. If $k_0 \leq 5$ then at least seven edges each have at least one bend, and we are done. Now assume $k_0 \geq 6$. Thus the 0-bend subgraph contains a cycle $C$ which is not a 3-cycle by Observation 3. Hence $C$ is a 4-, 5- or 6-cycle. Let $V(K_6 \setminus M_3) = \{a, b, c, 1, 2, 3\}$ with $a1, b2, c3 \notin E(K_6 \setminus M_3)$.

**Case 1.** $C$ **is a 4-cycle:** Since $K_4 \nsubseteq K_6 \setminus M_3$, $C$ has at most one chord. Initially suppose $C$ has no chords. Without loss of generality $C = (a, b, 1, 2)$. Then $ac1$, $bc2$, $a31$, and $b32$ are edge-disjoint chordal 2-paths between diagonally opposite vertices on $C$. By Observation 6, each of these chordal 2-paths have at least two bends, giving a total of eight bends, and we are done. Now suppose $C$ has one chord. Without loss of generality $C = (a, b, 3, 2)$, with the chord $a3$ having at least three bends (by Theorem 1). Thus $bc2$ and $b12$ are chordal 2-paths and $ac13$ is a chordal 3-path between diagonally opposite vertices on $C$. By Observation 6, these chordal 2-paths each have at least two bends and, similarly, the chordal 3-path has at least one bend. Therefore there is a total of at least eight bends.

**Case 2.** $C$ **is a 5-cycle:** Since $K_6 \setminus M_3$ is vertex-transitive, the vertices of $C$ induce the graph illustrated in Figure 18(a).

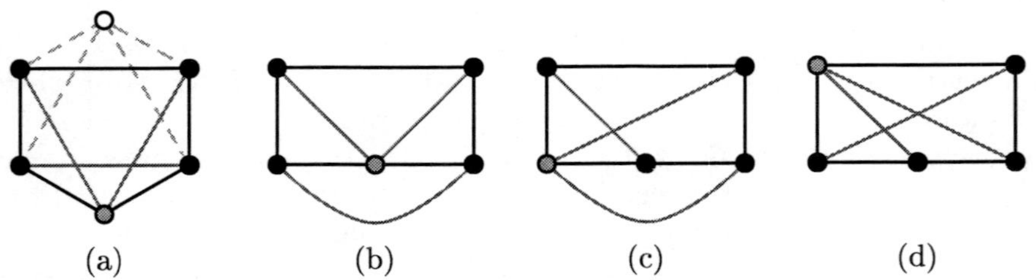

Figure 18: 0-bend 5-cycle in drawings of $K_6 \setminus M_3$.

Since there is only one drawing of a 0-bend 5-cycle, by symmetry there are three different ways to draw $C$, as illustrated in Figure 18(b), (c) and (d). In each case there are two chords of $C$ each with at least three bends, and a further chord with at least two bends, giving a total of at least eight bends.

D. Wood, *Lower Bounds for 3-D Drawings*, JGAA, 7(1) 33–77 (2003)    51

**Case 3. $C$ is a 6-cycle:** By Theorem 1, $C$ has at least three chords each with at least three bends. Thus the drawing has at least nine bends, and we are done.    □

We now establish lower bounds for the number of bends in drawings of $K_7 \setminus M_j$ for each $j \in \{0, 1, 2, 3\}$. Figure 19 shows a 4-bend drawing of $K_7$ with a total of 24 bends. (Compare this with the total of 42 bends in the 2-bend drawing of $K_7$ in Figure 1.) Deleting $j$ of the three 4-bend edges from this drawing produces a drawing of $K_7 \setminus M_j$ with $24 - 4j$ bends. The following lower bound is thus within $4 - j$ bends of being tight.

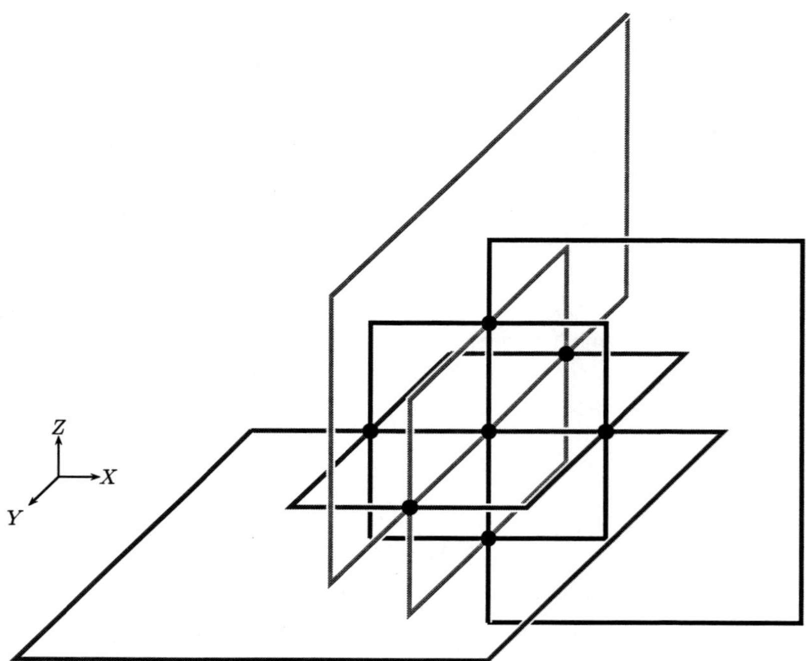


Figure 19: A 4-bend drawing of $K_7$ with 24 bends.

**Theorem 9.** *For each $j \in \{0, 1, 2, 3\}$, every drawing of $K_7 \setminus M_j$ has at least $20 - 3j$ bends.*

*Proof.* Suppose to the contrary, that for some $j \in \{0, 1, 2, 3\}$, there is a drawing of $K_7 \setminus M_j$ with at most $19 - 3j$ bends. Let $k_i$ $(i \geq 0)$ be the number of $i$-bend edges. Then

$$\sum_{i \geq 1} i k_i \leq 19 - 3j ,$$

D. Wood, *Lower Bounds for 3-D Drawings*, JGAA, 7(1) 33–77 (2003)    52

and

$$\sum_{i\geq 1} k_i \leq (19-3j) - \sum_{i\geq 2}(i-1)k_i \ .$$

Since $K_7 \setminus M_j$ has $21-j$ edges,

$$21-j = \sum_{i\geq 0} k_i \leq k_0 + (19-3j) - \sum_{i\geq 2}(i-1)k_i \ .$$

Hence,

$$\sum_{i\geq 2}(i-1)k_i \leq k_0 + (19-3j) - (21-j) = k_0 - 2j - 2 \ . \tag{1}$$

By Lemma 15 in Appendix A, every subgraph of $K_7 \setminus M_j$ with at least ten edges contains $C_3$ or $K_{2,3}$. By Observations 3 and 5, the 0-bend subgraph does not contain $C_3$ or $K_{2,3}$; thus $k_0 \leq 9$.

**Case 1. $k_0 = 8$ or $k_0 = 9$:** By Lemma 16 in Appendix A, every subgraph of $K_7 \setminus M_j$ with at least eight edges contains a cycle $C_k$ $(k \neq 4)$, two chord-disjoint cycles, or $K_{2,3}$. Therefore the 0-bend subgraph contains a cycle $C_k$ $(k \geq 5)$ or two chord-disjoint 4-cycles (since $C_3$ and $K_{2,3}$ do not have 0-bend drawings by Observations 3 and 5, respectively). In either case, by Theorem 1 there are at least $4-j$ chords of these cycles each with at least three bends. Thus $\sum_{i\geq 3} k_i \geq 4-j$, and hence,

$$k_2 + 2(4-j) \leq k_2 + \sum_{i\geq 3} 2k_i \leq \sum_{i\geq 2}(i-1)k_i \ .$$

By (1) with $k_0 \leq 9$, $k_2 + 8 - 2j \leq 9 - 2j - 2$ and thus $k_2 \leq -1$, which is a contradiction.

**Case 2. $k_0 \leq 7$:** Let $A$ be the set of edges of $K_7 \setminus M_j$ routed using an extremal port at exactly one end-vertex. Let $B$ be the set of edges routed using extremal ports in the same direction at its end-vertices. Let $C$ be the set of edges routed using extremal ports in differing directions at its end-vertices. Since, all but $2j$ ports in the drawing of $K_7 \setminus M_j$ are used, and there is at least one extremal port in each of the six directions, $|A| + |B| + 2|C| \geq 6 - 2j$. As illustrated in Figure 20(a) an edge in $A$ or $B$ has at least two bends, and an edge in $C$ has at least three bends, as illustrated in Figure 20(b). Hence,

$$k_2 + 2\sum_{i\geq 3} k_i \geq 6 - 2j \ ,$$

which implies,

$$\sum_{i\geq 2}(i-1)k_i \geq 6 - 2j \ . \tag{2}$$

D. Wood, *Lower Bounds for 3-D Drawings*, JGAA, 7(1) 33–77 (2003)      53

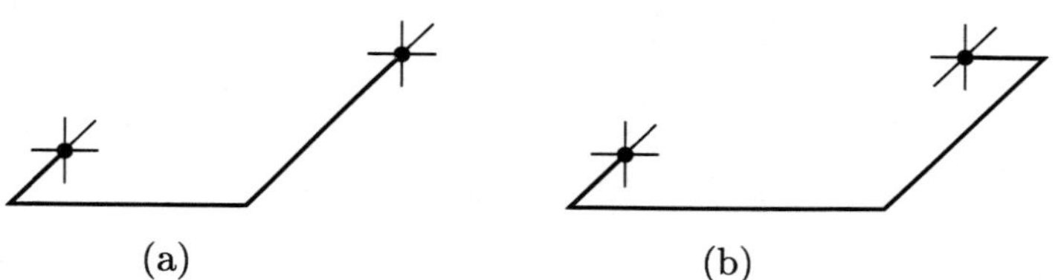

Figure 20: Edges using extremal ports have at least two bends.

However, by (1) with $k_0 \leq 7$,

$$\sum_{i \geq 2}(i - 1)k_i \leq 5 - 2j \; ,$$

which contradicts (2). The result follows. $\qquad\square$

## 3.1 Drawings of Multigraphs

We now prove tight bounds for the number of bends in drawings of the complete multigraphs $j \cdot K_2$ for $2 \leq j \leq 6$.

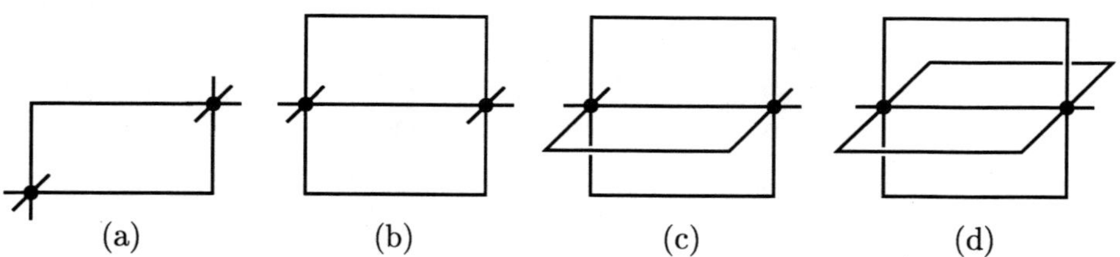

Figure 21: Bend-minimum drawings of (a) $2 \cdot K_2$, (b) $3 \cdot K_2$, (c) $4 \cdot K_2$ and (d) $5 \cdot K_2$.

We omit the proof of the following elementary result as the method is similar and simpler than the proofs of Theorems 10 and 11 for $6 \cdot K_2$ below.

**Lemma 7.** *For each of the graphs $j \cdot K_2$ ($2 \leq j \leq 5$), the drawings in Figure 21 have the minimum maximum number of bends per edge and the minimum total number of bends.* $\qquad\square$

In Figure 22 we show two drawings of $6 \cdot K_2$. We now provide a formal prove of the well-known result that the maximum number of bends per edge in the drawing in Figure 22(a) is optimal.

**Theorem 10.** *Every drawing of $6 \cdot K_2$ has an edge with at least three bends.*

*Proof.* Let the vertices of $6 \cdot K_2$ be $v$ and $w$. Since $6 \cdot K_2$ is 6-regular every port at $v$ and $w$ is used. The two vertices are either (a) collinear, (b) coplanar but not collinear, or (c) not coplanar, as illustrated in Figure 23.

D. Wood, *Lower Bounds for 3-D Drawings*, JGAA, 7(1) 33–77 (2003)      54

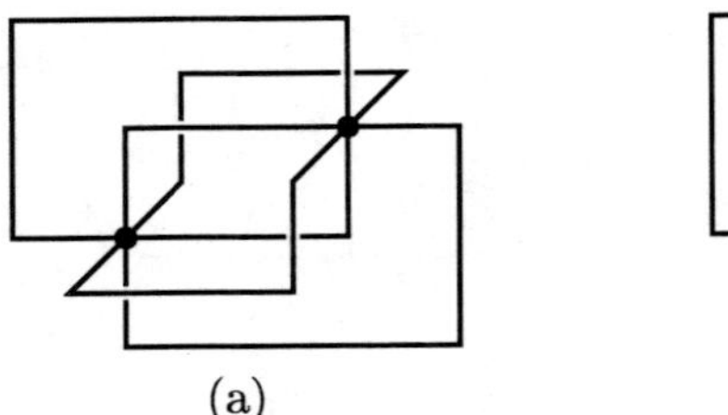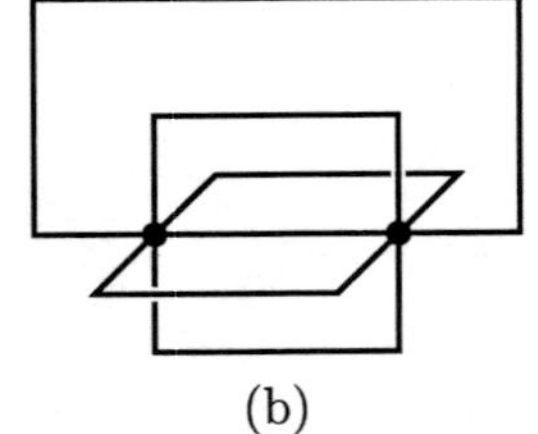

Figure 22: Drawings of $6 \cdot K_2$ with (a) a maximum of three bends per edge, and (b) a total of twelve bends.

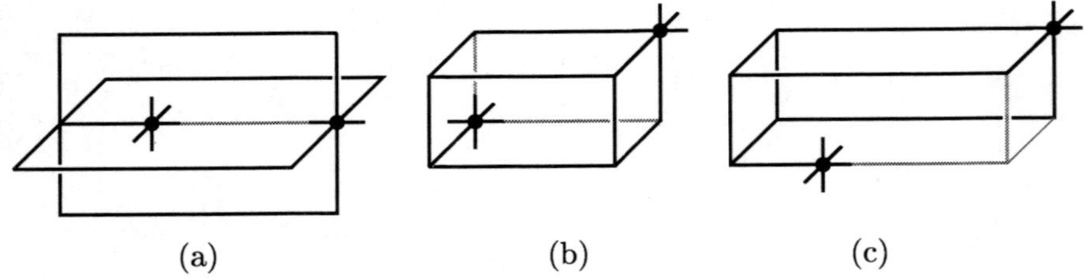

Figure 23: $6 \cdot K_2$ has a 3-bend edge.

In each case there is a port at $v$ pointing away from $w$ such that the edge using this port requires at least three bends to reach $w$.    □

We now prove that the drawing in Figure 22(b) is bend-minimum.

**Theorem 11.** *Every drawing of the multigraph $6 \cdot K_2$ has at least twelve bends.*

*Proof.* Let the vertices of $6 \cdot K_2$ be $v$ and $w$. Suppose that $v$ and $w$ are not coplanar. The edges using the three ports at $v$ pointing towards $w$ have at least two bends, and the other edges have at least three bends. Thus the drawing has at least 15 bends.

Suppose $v$ and $w$ are coplanar but not collinear. The edges using the two ports at $v$ pointing towards $w$ have at least one bend, the edges using the two opposite ports have at least two bends, and the remaining two edges have at least three bends. Thus the drawing has at least 12 bends.

Suppose $v$ and $w$ are collinear, and without loss of generality, that $v$ and $w$ lie in an $X$-axis parallel line, such that the $X$-coordinate of $v$ is less than the $X$-coordinate of $w$. The edge using the port $X_v^-$ has at least three bends, and the four edges using the other four ports at $v$ pointing away from $w$ have at least two bends. Thus the drawing has at least 11 bends. Suppose there is such a drawing with exactly 11 bends. Then there are four 2-bend edges, and one 3-bend edge. These four 2-bend edges use the $Y^\pm$ and $Z^\pm$ ports at each vertex. Therefore, the edge using the $X_v^-$ port and the $X_w^+$ port has four bends, and thus the drawing has 12 bends, which is a contradiction. The result follows.    □

D. Wood, *Lower Bounds for 3-D Drawings*, JGAA, 7(1) 33–77 (2003)      55

# 4   Constructing Large Graphs

In this section we use the lower bounds for the number of bends in drawings of the complete graphs established in Section 3 as building blocks to construct infinite families of $c$-connected graphs $(2 \leq c \leq 6)$ with maximum degree $\Delta$ $(2 \leq \Delta \leq 6)$, and with lower bounds on the number of bends in drawings of every graph in the family.

A graph is *c-connected* $(c \geq 1)$ if the removal of fewer than $c$ vertices results in neither a disconnected graph nor the trivial graph. To establish that our graphs are $c$-connected we use the following characterisation due to Whitney [31], which is part of the family of results known as 'Menger's Theorem'. A graph $G$ is $c$-connected if and only if for each pair $u, v$ of distinct vertices there are at least $c$ internally disjoint paths from $u$ to $v$ in $G$. Our proofs of connectivity are postponed until Appendix B.

We employ two methods for constructing new graphs from two given graphs. First, given graphs $G$ and $H$, we define $H\langle G \rangle$ to be the graph obtained by replacing each vertex of $H$ by a copy of $G$, and connecting the edges in $H$ incident to a particular vertex in $H$ to different vertices in the corresponding copy of $G$. In most cases, $H$ is $\Delta$-regular and $G$ is a complete graph $K_p$ for some $p \geq \Delta$; thus $H\langle G \rangle$ is well-defined. In other cases we shall specify which edges of $H$ are connected to which vertices in each copy of $G$.

Our second method for constructing large graphs is the *cartesian product* $G \times H$ of graphs $G$ and $H$. $G \times H$ has vertex set $V(H) \times V(G)$ with $(v_1, w_1)$ and $(v_2, w_2)$ adjacent in $G \times H$ if either $v_1 = v_2$ and $w_1 w_2 \in E(G)$, or $w_1 = w_2$ and $v_1 v_2 \in E(H)$. For example, $C_p \times C_q$ is the 4-regular $p \times q$ torus graph.

Our lower bounds for simple disconnected graphs are obtained by taking disjoint copies of $K_p$ for $4 \leq p \leq 7$. For consistency we denote these graphs by $I_r \langle K_p \rangle$, where $I_r$ is the $r$-vertex graph with no edges. Our lower bounds for disconnected multigraphs are obtained by $I_r \langle p \cdot K_2 \rangle$ for $3 \leq p \leq 6$, and we use $I_r \langle L_p \rangle$ with $1 \leq p \leq 3$ to obtain lower bounds for disconnected pseudographs.

As illustrated in Figure 24, to obtain lower bounds for simple 2-connected graphs, we use $C_r \langle K_p \rangle$ for $3 \leq p \leq 6$ and $r \geq 3$, and $C_r \langle K_p \setminus M_1 \rangle$ for $4 \leq p \leq 7$ and $r \geq 2$, where the non-adjacent vertices in each copy of $K_p \setminus M_1$ are incident to the edges of $C_r$. To obtain lower bounds for 2-connected multigraphs, we use $C_r \langle p \cdot K_2 \rangle$ with $2 \leq p \leq 5$, and we use $C_r \langle L_p \rangle$ with $1 \leq p \leq 2$ to obtain lower bounds for 2-connected pseudographs.

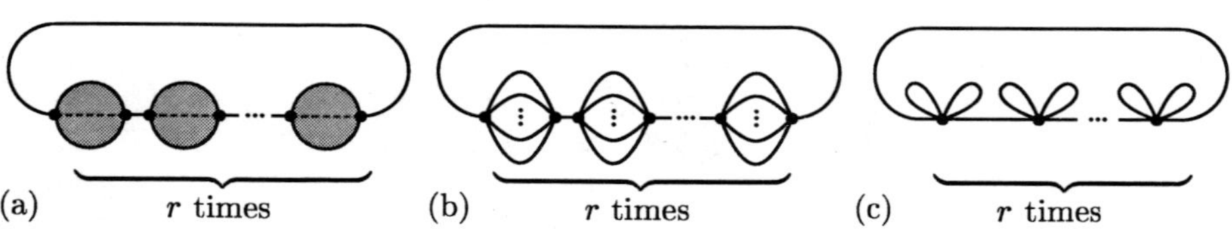


Figure 24: 2-connected graphs: (a) $C_r \langle K_p \setminus M_1 \rangle$, (b) $C_r \langle p \cdot K_2 \rangle$, and (c) $C_r \langle L_2 \rangle$.

As illustrated in Figure 25, to obtain lower bounds for simple 3-connected

D. Wood, *Lower Bounds for 3-D Drawings*, JGAA, 7(1) 33–77 (2003)      56

graphs, we use $(C_r \times K_2)\langle K_p \rangle$ for $3 \leq p \leq 6$ and $r \geq 3$, and $(C_r \times K_2)\langle K_p \setminus M_2 \rangle$ for $5 \leq p \leq 7$ and $r \geq 3$, where the non-adjacent pairs of vertices in each copy of $K_p \setminus M_2$ are incident to opposite edges of $C_r \times K_2$ where possible. To obtain lower bounds for 3-connected multigraphs, we use $C_r \times (p \cdot K_2)$ with $r \geq 3$ and $2 \leq p \leq 4$. We use $C_r \times ((p \cdot K_2)\langle L_1 \rangle)$ with $1 \leq p \leq 2$ to obtain lower bounds for 3-connected pseudographs.

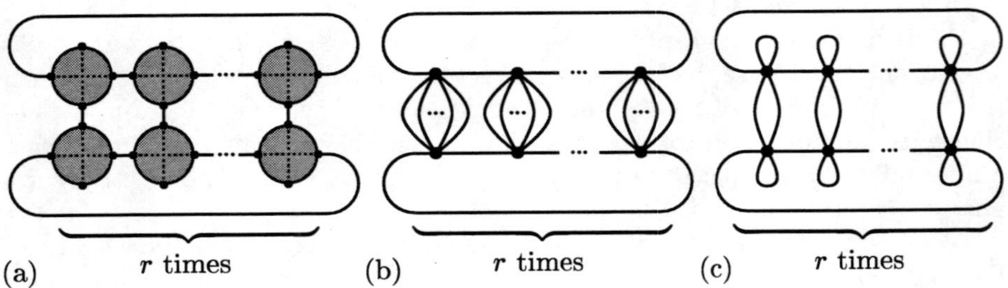

Figure 25: 3-connected graphs: (a) $(C_r \times K_2)\langle K_p \setminus M_2 \rangle$, (b) $C_r \times (p \cdot K_2)$, and (c) $C_r \times ((p \cdot K_2)\langle L_1 \rangle)$.

To obtain lower bounds for 4-connected simple graphs, we use $(C_r \times C_3)\langle K_p \rangle$ for $4 \leq p \leq 6$ and $r \geq 3$, and $(C_r \times C_3)\langle K_p \setminus M_2 \rangle$ for $5 \leq p \leq 7$ and $r \geq 3$, where the non-adjacent pairs of vertices in each copy of $K_p \setminus M_2$ are incident to opposite edges of $C_r \times C_3$, as illustrated in Figure 26(a). We use $(C_r \times C_3)\langle L_1 \rangle$ with $r \geq 3$ to obtain lower bounds for 4-connected pseudographs, as illustrated in Figure 26(b).

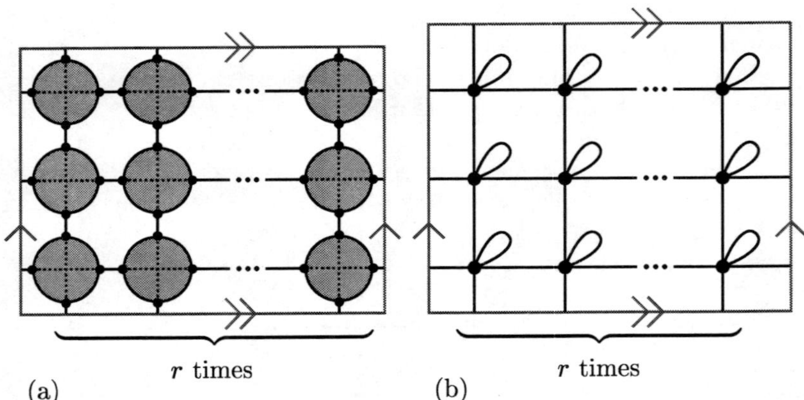

Figure 26: 4-connected graphs: (a) $(C_r \times C_3)\langle K_p \setminus M_2 \rangle$, and (b) $(C_r \times C_3)\langle L_1 \rangle$.

Let $2 \cdot C_m$ be the $m$-edge cycle with each edge having multiplicity 2, and let $\frac{3}{2} \cdot C_m$ for even $m$ be the $m$-edge cycle with alternating edges around the cycle having multiplicity 2. To obtain lower bounds for 4-connected multigraphs, we use the 6-regular multigraph $C_r \times (2 \cdot C_3)$ for some $r \geq 3$, as illustrated in Figure 27(b), and the 5-regular multigraph $C_r \times (\frac{3}{2} \cdot C_4)$ for $r \geq 3$, as illustrated in Figure 27(c).

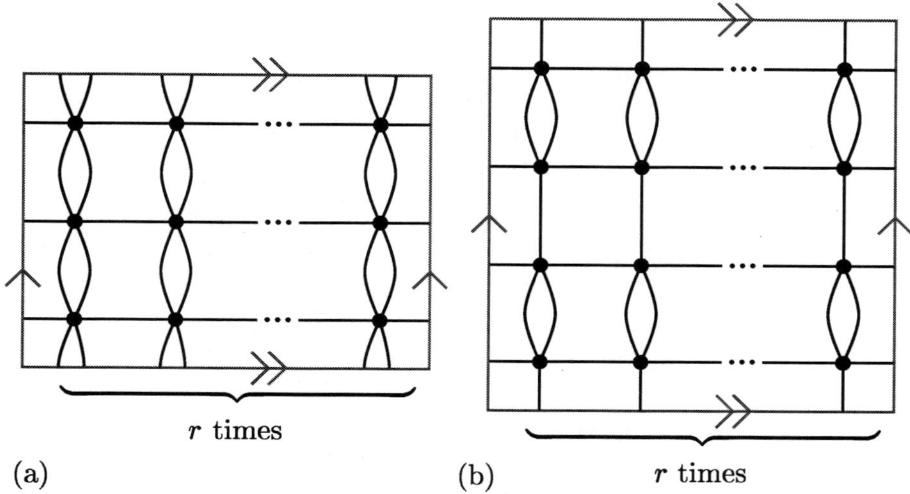

Figure 27: 4-connected graphs: (a) $C_r \times (2 \cdot C_3)$, (b) $C_r \times (\frac{3}{2} \cdot C_4)$.

To obtain lower bounds for 5-connected graphs, we use $(C_r \times C_3 \times K_2)\langle K_p \rangle$ for $5 \leq p \leq 6$ and $r \geq 3$, and $(C_r \times C_3 \times K_2)\langle K_7 \setminus M_3 \rangle$ for $r \geq 3$, where the non-adjacent pairs of vertices in each copy of $K_7 \setminus M_3$ are incident to opposite edges of $C_r \times C_3 \times K_2$ where possible, as illustrated in Figure 28(a). To obtain lower bounds for 5-connected multigraphs, we use $C_r \times C_3 \times (2 \cdot K_2)$, as illustrated in Figure 28(b).

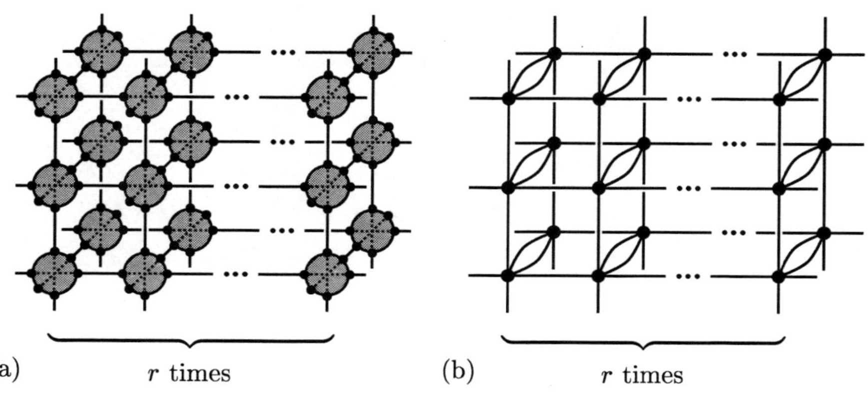

Figure 28: 5-connected graphs: (a) $(C_r \times C_3 \times K_2)\langle G \rangle$, (b) $C_r \times C_3 \times (2 \cdot K_2)$.

To obtain lower bounds for 6-connected graphs we use $(C_r \times C_3 \times C_3)\langle K_6 \rangle$ and $(C_r \times C_3 \times C_3)\langle K_7 \setminus M_3 \rangle$ for $r \geq 3$, where the non-adjacent pairs of vertices in each copy of $K_7 \setminus M_3$ are incident to opposite edges in $C_r \times C_3 \times C_3$, as illustrated in Figure 29.

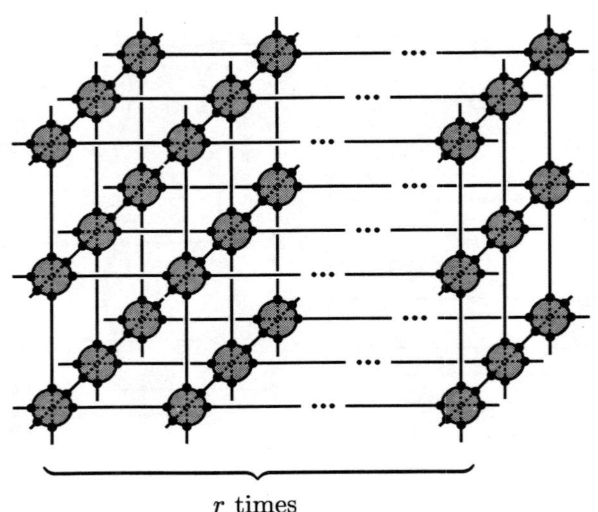



Figure 29:    6-connected   6-regular   graphs   $(C_r \times C_3 \times C_3)\langle K_6 \rangle$   and $(C_r \times C_3 \times C_3)\langle K_7 \setminus M_3 \rangle$.

In Table 3 we prove lower bounds on the number of bends in drawings of the above families of graphs. Each line of the table corresponds to one such family $H\langle G \rangle$ (or $H \times G$) parameterised by some value $r$, all of which have maximum degree $\Delta$ (shown in the first column). The third column shows the lower bounds on the number of bends in a drawing of $G$, as proved earlier in the paper. The fourth and fifth columns shows the number of edge-disjoint copies of $G$ and the number of edges in $H\langle G \rangle$ (or $H \times G$), respectively. The sixth column shows the lower bound on the average number of bends per edge in $H\langle G \rangle$ (or $H \times G$) obtained by

$$\text{average \# bends}(H\langle G \rangle \text{ or } H \times G) \geq \frac{\text{\# bends}(G) \times \text{\# copies}(G)}{\text{\# edges}(H\langle G \rangle \text{ or } H \times G)}.$$

A line marked with a $\star$ indicates the corresponding lower bound is the best out of those for graphs with a specific connectivity and maximum degree. These 'best known' lower bounds are those listed in Table 2 in Section 1.

D. Wood, *Lower Bounds for 3-D Drawings*, JGAA, 7(1) 33–77 (2003)    59

Table 3: Lower bounds for the average number of bends per edge.

| $\Delta$ | $H\langle G\rangle$ or $H \times G$ | #bends($G$) | | #copies($G$) | #edges | avg. #bends | |
|---|---|---|---|---|---|---|---|
| **Disconnected Simple Graphs** | | | | | | | |
| 6 | $I_r\langle K_7\rangle$ | 20 | (Thm. 9) | $r$ | $21r$ | $\frac{20}{21}$ | $\star$ |
| 5 | $I_r\langle K_6\rangle$ | 12 | (Thm. 7 ) | $r$ | $15r$ | $\frac{12}{15} = \frac{4}{5}$ | $\star$ |
| 4 | $I_r\langle K_5\rangle$ | 7 | (Thm. 4) | $r$ | $10r$ | $\frac{7}{10}$ | $\star$ |
| 3 | $I_r\langle K_4\rangle$ | 3 | (Thm. 2) | $r$ | $6r$ | $\frac{3}{6} = \frac{1}{2}$ | $\star$ |
| 2 | $I_r\langle K_3\rangle$ | 1 | (Obs. 3) | $r$ | $3r$ | $\frac{1}{3}$ | $\star$ |
| **Disconnected Multigraphs** | | | | | | | |
| 6 | $I_r\langle 6 \cdot K_2\rangle$ | 12 | (Thm. 11) | $r$ | $6r$ | $\frac{12}{6} = 2$ | $\star$ |
| 5 | $I_r\langle 5 \cdot K_2\rangle$ | 8 | (Lem. 7) | $r$ | $5r$ | $\frac{8}{5}$ | $\star$ |
| 4 | $I_r\langle 4 \cdot K_2\rangle$ | 6 | (Lem. 7) | $r$ | $4r$ | $\frac{6}{4} = \frac{3}{2}$ | $\star$ |
| 3 | $I_r\langle 3 \cdot K_2\rangle$ | 4 | (Lem. 7) | $r$ | $3r$ | $\frac{4}{3}$ | $\star$ |
| 2 | $I_r\langle 2 \cdot K_2\rangle$ | 2 | (Lem. 7) | $r$ | $2r$ | $\frac{2}{2} = 1$ | $\star$ |
| **Disconnected Pseudographs** | | | | | | | |
| 6 | $I_r\langle L_3\rangle$ | 9 | | $r$ | $3r$ | 3 | $\star$ |
| 4 | $I_r\langle L_2\rangle$ | 6 | | $r$ | $2r$ | 3 | $\star$ |
| 2 | $I_r\langle L_1\rangle$ | 3 | | $r$ | $r$ | 3 | $\star$ |
| **2-Connected Simple Graphs** | | | | | | | |
| 6 | $C_r\langle K_6\rangle$ | 12 | (Thm. 7) | $r$ | $16r$ | $\frac{12}{16} = \frac{3}{4}$ | |
| 6 | $C_r\langle K_7 \setminus M_1\rangle$ | 17 | (Thm. 9) | $r$ | $21r$ | $\frac{17}{21}$ | $\star$ |
| 5 | $C_r\langle K_5\rangle$ | 7 | (Thm. 4) | $r$ | $11r$ | $\frac{7}{11}$ | |
| 5 | $C_r\langle K_6 \setminus M_1\rangle$ | 10 | (Thm. 7) | $r$ | $15r$ | $\frac{10}{15} = \frac{2}{3}$ | $\star$ |
| 4 | $C_r\langle K_4\rangle$ | 3 | (Thm. 2) | $r$ | $7r$ | $\frac{3}{7}$ | |
| 4 | $C_r\langle K_5 \setminus M_1\rangle$ | 5 | (Thm. 5) | $r$ | $10r$ | $\frac{5}{10} = \frac{1}{2}$ | $\star$ |
| 3 | $C_r\langle K_3\rangle$ | 1 | (Obs. 3) | $r$ | $4r$ | $\frac{1}{4}$ | |
| 3 | $C_r\langle K_4 \setminus M_1\rangle$ | 2 | (Thm. 2) | $r$ | $6r$ | $\frac{2}{6} = \frac{1}{3}$ | $\star$ |
| **2-Connected Multigraphs** | | | | | | | |
| 6 | $C_r\langle 5 \cdot K_2\rangle$ | 8 | (Lem. 7) | $r$ | $6r$ | $\frac{8}{6} = \frac{4}{3}$ | $\star$ |
| 5 | $C_r\langle 4 \cdot K_2\rangle$ | 6 | (Lem. 7) | $r$ | $5r$ | $\frac{6}{5}$ | $\star$ |
| 4 | $C_r\langle 3 \cdot K_2\rangle$ | 4 | (Lem. 7) | $r$ | $4r$ | $\frac{4}{4} = 1$ | $\star$ |
| 3 | $C_r\langle 2 \cdot K_2\rangle$ | 2 | (Lem. 7) | $r$ | $3r$ | $\frac{2}{3}$ | $\star$ |

*continued on next page*

D. Wood, *Lower Bounds for 3-D Drawings*, JGAA, 7(1) 33–77 (2003)  60

Table 3: *continued*

| $\Delta$ $H\langle G\rangle$ or $H \times G$ | #bends($G$) | | #copies($G$) | #edges | avg. #bends | |
|---|---|---|---|---|---|---|
| **2-Connected Pseudographs** | | | | | | |
| 6 $C_r\langle L_2\rangle$ | 6 | | $r$ | $3r$ | 2 | $\star$ |
| 4 $C_r\langle L_1\rangle$ | 3 | | $r$ | $2r$ | $\frac{3}{2}$ | $\star$ |
| **3-Connected Simple Graphs** | | | | | | |
| 6 $(C_r \times K_2)\langle K_6\rangle$ | 12 | (Thm. 7) | $2r$ | $33r$ | $\frac{2\cdot 12}{33} = \frac{8}{11}$ | $\star$ |
| 6 $(C_r \times K_2)\langle K_7 \setminus M_2\rangle$ | 14 | (Thm. 9) | $2r$ | $41r$ | $\frac{2\cdot 14}{41} = \frac{28}{41}$ | |
| 5 $(C_r \times K_2)\langle K_5\rangle$ | 7 | (Thm. 4) | $2r$ | $23r$ | $\frac{2\cdot 7}{23} = \frac{14}{23}$ | $\star$ |
| 5 $(C_r \times K_2)\langle K_6 \setminus M_2\rangle$ | 8 | (Thm. 7) | $2r$ | $29r$ | $\frac{2\cdot 8}{29} = \frac{16}{29}$ | |
| 4 $(C_r \times K_2)\langle K_4\rangle$ | 3 | (Thm. 2) | $2r$ | $15r$ | $\frac{2\cdot 3}{15} = \frac{2}{5}$ | |
| 4 $(C_r \times K_2)\langle K_5 \setminus M_2\rangle$ | 4 | (Thm. 5) | $2r$ | $19r$ | $\frac{2\cdot 4}{19} = \frac{8}{19}$ | $\star$ |
| 3 $(C_r \times K_2)\langle K_3\rangle$ | 1 | (Obs. 3) | $2r$ | $9r$ | $\frac{2}{9}$ | $\star$ |
| **3-Connected Multigraphs** | | | | | | |
| 6 $C_r \times (4 \cdot K_2)$ | 6 | (Lem. 7) | $r$ | $6r$ | $\frac{6}{6} = 1$ | $\star$ |
| 5 $C_r \times (3 \cdot K_2)$ | 4 | (Lem. 7) | $r$ | $5r$ | $\frac{4}{5}$ | $\star$ |
| 4 $C_r \times (2 \cdot K_2)$ | 2 | (Lem. 7) | $r$ | $4r$ | $\frac{2}{4} = \frac{1}{2}$ | $\star$ |
| **3-Connected Pseudographs** | | | | | | |
| 6 $C_r \times ((2 \cdot K_2)\langle L_1\rangle)$ | 8 | | $r$ | $6r$ | $\frac{4}{3}$ | $\star$ |
| 5 $(C_r \times K_2)\langle L_1\rangle$ | 3 | | $2r$ | $5r$ | $\frac{6}{5}$ | $\star$ |
| **4-Connected Simple Graphs** | | | | | | |
| 6 $(C_r \times C_3)\langle K_6\rangle$ | 12 | (Thm. 7) | $3r$ | $(3\cdot 15 + 6)r$ | $\frac{3\cdot 12}{51} = \frac{12}{17}$ | $\star$ |
| 6 $(C_r \times C_3)\langle K_7 \setminus M_2\rangle$ | 14 | (Thm. 9) | $3r$ | $(3\cdot 19 + 6)r$ | $\frac{3\cdot 14}{63} = \frac{2}{3}$ | |
| 5 $(C_r \times C_3)\langle K_5\rangle$ | 7 | (Thm. 4) | $3r$ | $(3\cdot 10 + 6)r$ | $\frac{3\cdot 7}{36} = \frac{7}{12}$ | $\star$ |
| 5 $(C_r \times C_3)\langle K_6 \setminus M_2\rangle$ | 8 | (Thm. 7) | $3r$ | $(3\cdot 13 + 6)r$ | $\frac{3\cdot 8}{45} = \frac{8}{15}$ | |
| 4 $(C_r \times C_3)\langle K_4\rangle$ | 3 | (Thm. 2) | $3r$ | $(3\cdot 6 + 6)r$ | $\frac{3\cdot 3}{24} = \frac{3}{8}$ | |
| 4 $(C_r \times C_3)\langle K_5 \setminus M_2\rangle$ | 4 | (Thm. 5) | $3r$ | $(3\cdot 8 + 6)r$ | $\frac{3\cdot 4}{30} = \frac{2}{5}$ | $\star$ |
| **4-Connected Multigraphs** | | | | | | |
| 5 $C_r \times (\frac{3}{2} \cdot C_4)$ | 4 | (Lem. 7) | $r$ | $10r$ | $\frac{4}{10} = \frac{2}{5}$ | $\star$ |
| 6 $C_r \times (2 \cdot C_3)$ | 6 | (Lem. 7) | $r$ | $9r$ | $\frac{6}{9} = \frac{2}{3}$ | $\star$ |
| **4-Connected Pseudographs** | | | | | | |
| 6 $(C_r \times C_3)\langle L_1\rangle$ | 3 | | $3r$ | $(3\cdot 1 + 6)r$ | $\frac{9}{9} = 1$ | $\star$ |

*continued on next page*

Table 3: *continued*

| $\Delta$ | $H\langle G\rangle$ or $H\times G$ | #bends($G$) | #copies($G$) | #edges | avg. #bends | |
|---|---|---|---|---|---|---|
| **5-Connected Simple Graphs** | | | | | | |
| 6 | $(C_r\times C_3\times K_2)\langle K_6\rangle$ | 12 (Thm. 7) | $6r$ | $(6\cdot 15+15)r$ | $\frac{6\cdot 12}{105}=\frac{24}{35}$ | $\star$ |
| 6 | $(C_r\times C_3\times K_2)\langle K_7\setminus M_3\rangle$ | 11 (Thm. 9) | $6r$ | $(6\cdot 18+15)r$ | $\frac{6\cdot 11}{123}=\frac{22}{41}$ | |
| 5 | $(C_r\times C_3\times K_2)\langle K_5\rangle$ | 7 (Thm. 4) | $6r$ | $(6\cdot 10+15)r$ | $\frac{6\cdot 7}{75}=\frac{14}{25}$ | $\star$ |
| **5-Connected Multigraphs** | | | | | | |
| 6 | $C_r\times C_3\times(2\cdot K_2)$ | 2 (Lem. 7) | $3r$ | $18r$ | $\frac{2\cdot 3}{18}=\frac{1}{3}$ | $\star$ |
| **6-Connected Simple Graphs** | | | | | | |
| 6 | $(C_r\times C_3\times C_3)\langle K_6\rangle$ | 12 (Thm. 7) | $9r$ | $(9\cdot 15+27)r$ | $\frac{9\cdot 12}{162}=\frac{2}{3}$ | $\star$ |
| 6 | $(C_r\times C_3\times C_3)\langle K_7\setminus M_3\rangle$ | 11 (Thm. 9) | $9r$ | $(9\cdot 18+27)r$ | $\frac{9\cdot 11}{189}=\frac{11}{21}$ | |

# 5   Lower Bounds for General Position Drawings

Recall that a 3-D orthogonal graph drawing is said to be in *general position* if no two vertices lie in a common grid-plane. The general position model has been used for 3-D orthogonal graph drawing by Eades *et al.* [13] and Wood [33], Wood [36], Wood [37], and for 3-D orthogonal box-drawing of arbitrary degree graphs by Papakostas and Tollis [21], Biedl [8] and Wood [34]. In this section we establish lower bounds for the number of bends in general position drawings of 2-connected and 4-connected graphs. The next result will be crucial for the lower bounds to follow.

**Lemma 8.** *If the graph $G$ has at least $k$ bends in every general position drawing then for every edge $e$ of $G$, the graph $G\setminus e$ has at least $k-4$ bends in every general position drawing.*

*Proof.* Suppose $G\setminus e$ has a general position drawing with $b$ bends. Wood [37] proved that the edge $e$ can be inserted into the drawing of $G\setminus e$ with at most four bends (possibly introducing edge crossings), and that the edges can be rerouted to eliminate all edge crossings without increasing the total number of bends. Thus there is a (crossing-free) general position drawing of $G$ with $b+4$ bends. By assumption, every general position drawing of $G$ has at least $k$ bends. Thus $b+4\geq k$ and $b\geq k-4$. $\qquad\square$

Clearly every edge in a general position drawing has at least two bends. Observe that if an edge is routed using an extremal port, then this edge has at least three bends, as illustrated in Figure 30.

Since all ports are used in a drawing of a 6-regular $m$-edge graph, a general position drawing of such a graph requires at least $2m+6$ bends. Hence the graphs consisting of disjoint copies of $K_7$ provide the following lower bound.

D. Wood, *Lower Bounds for 3-D Drawings*, JGAA, 7(1) 33–77 (2003)     62

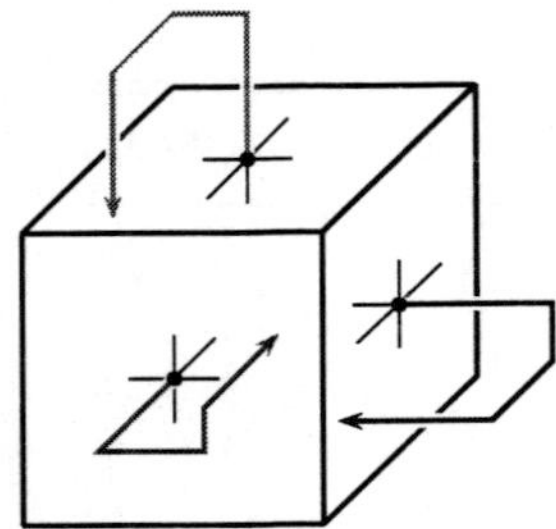

Figure 30: General position edges using extreme ports have at least three bends.

**Lemma 9.** *There exists an infinite family of n-vertex m-edge simple graphs, each with at least $2m + \frac{6}{7}n$ bends in every general position drawing.* $\square$

Note that for 6-regular graphs $m = 3n$; thus the above lower bound matches the upper bound of $\frac{16}{7}m$ for the total number of bends in general position drawings established by the DIAGONAL LAYOUT & MOVEMENT algorithm [37].

To obtain a lower bound for general position drawings of 2-connected graphs, we use the 6-regular graph $C_r\langle K_7 \setminus M_1\rangle$, where the non-adjacent vertices of each $K_7 \setminus M_1$ are incident to the edges of $C_r$, as illustrated in Figure 24(a).

**Lemma 10.** *There exists an infinite family of n-vertex m-edge simple 2-connected graphs, each with at least $2m + \frac{4}{7}n$ bends in every general position drawing.*

*Proof.* Clearly $C_r\langle K_7 \setminus M_1\rangle$ is 2-connected. $K_7$ has at least $2|E(K_7)|+6$ bends in any general position drawing. Thus by Lemma 8, a general position drawing of $K_7 \setminus M_1$ has at least $2|E(K_7)|+6-4 = 2|E(K_7 \setminus M_1)|+4$ bends. The edges of $C_r$ each have at least two bends. Thus $C_r\langle K_7 \setminus M_1\rangle$ has at least $2m + \frac{4}{7}n$ bends. $\square$

To obtain a lower bound for general position drawings of 4-connected graphs, we use the 6-regular graph $(C_r \times C_3)\langle K_7 \setminus M_2\rangle$ for $r \geq 3$, as illustrated in Figure 26(a).

**Lemma 11.** *There exists an infinite family of n-vertex m-edge simple 4-connected graphs, each with at least $2m + \frac{2}{7}n$ bends in every general position drawing.*

*Proof.* As proved in Appendix B, $(C_r \times C_3)\langle K_7 \setminus M_2\rangle$ is 4-connected. $K_7$ has at least $2|E(K_7)|+6$ bends in any general position drawing. Hence, by Lemma 8 a general position drawing of $K_7 \setminus M_2$ has at least $2|E(K_7)|+6-8 = 2|E(K_7 \setminus M_2)| + 2$ bends. Edges not in a $K_7 \setminus M_2$ have at least two bends. Thus $(C_r \times C_3)\langle K_7 \setminus M_2\rangle$ has at least $2m + \frac{2}{7}n$ bends. $\square$

# 6   On the 2-Bends Problem

We now look at the ramifications of the above general position lower bounds for the 2-bends problem. Edges with at most two bends can be classified as 0-bend,

D. Wood, *Lower Bounds for 3-D Drawings*, JGAA, 7(1) 33–77 (2003)     63

1-bend, 2-bend planar or 2-bend non-planar, as illustrated in Figure 31.

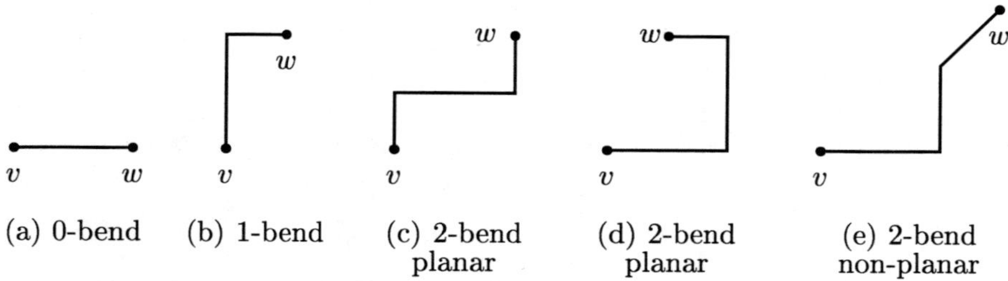

(a) 0-bend     (b) 1-bend     (c) 2-bend
planar     (d) 2-bend
planar     (e) 2-bend
non-planar

Figure 31: Edges $vw$ with at most two bends.

In a given 2-bend drawing of a graph $G$, we denote the number of 0-bend edges by $k_0$, and the number of 2-bend planar edges by $k_2'$. We now describe how to transform a given 2-bend drawing into a general position drawing.

**Lemma 12.** *If there is a 2-bend drawing of a graph $G$ then there exists a general position drawing of $G$ with $2m + k_0 + k_2'$ bends.*

*Proof.* We show that by inserting planes and adding bends to the edge routes a given 2-bend drawing can be transformed into a drawing with a general position vertex layout and the stated number of bends. Consider a grid plane $P$ containing $k$ vertices ($k > 1$). As illustrated in Figure 32, replace the plane by $k$ adjacent planes, and position each of the $k$ vertices in a unique plane.

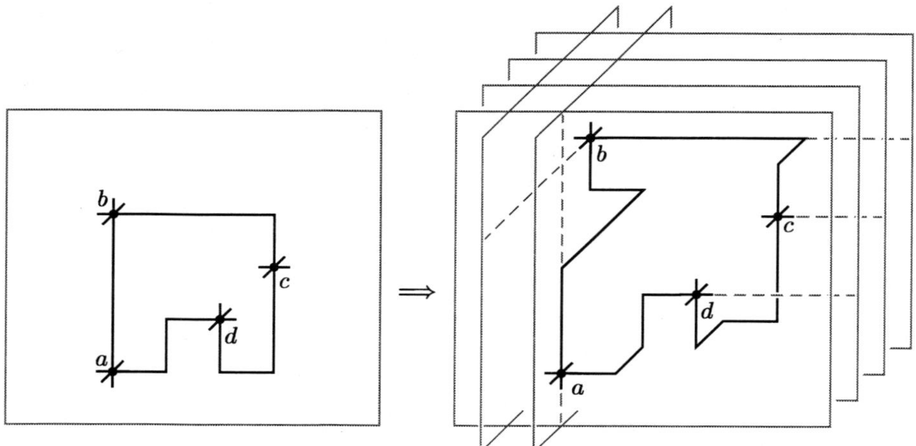

Figure 32: Removing a plane containing many vertices.

A 0-bend edge is split in the middle and replaced by the 2-bend planar edge illustrated in Figure 31(c). If the 0-bend edge has length one then an extra plane perpendicular to the 0-bend edge is also inserted.

D. Wood, *Lower Bounds for 3-D Drawings*, JGAA, 7(1) 33–77 (2003)      64

Edge segments from an edge with at least one bend and incident to a vertex $v$ are routed in the plane containing $v$. For a 1-bend edge $vw$ in the original plane, an extra segment is inserted perpendicular to $P$, running between the planes containing $v$ and $w$. Hence $vw$ is replaced by a 2-bend non-planar edge (for example, edge $bc$ in Figure 32).

For a 2-bend edge $vw$ in the original plane, the middle segment of $vw$ is routed arbitrarily in the plane containing $v$ or $w$, and a third segment is inserted perpendicular to $P$, running between the planes containing $v$ and $w$. Hence $vw$ is replaced by a 3-bend non-planar edge (for example, edges $ad$ and $cd$ in Figure 32).

For a 2-bend non-planar edge $vw$ incident to one of the $k$ vertices, the segment of $vw$ perpendicular to $P$ is extended in the obvious manner. Similarly, an edge passing through the original plane and not incident to any of the $k$ vertices, is extended so that it passes through all $k$ planes.

This process is continued until there are no grid planes containing more than one vertex. Note that a 0-bend edge will initially be replaced by a 2-bend planar edge, and in a second transformation will be replaced by a 3-bend edge route (for example, edge $ab$ in Figure 32). The resulting drawing has no crossings, has a general position vertex layout, and every edge has two bends except for the 0-bend and 2-bend planar edges in the original drawing, which now have three bends. Hence the new drawing has $2m + k_0 + k_2'$ bends.  $\square$

We now prove that for certain graphs any 2-bend drawing has many 0-bend or 2-bend planar edge routes.

**Corollary 1.** *There exists an infinite family of 6-regular $n$-vertex graphs, such that in any 2-bend drawing of any one of the graphs, $k_0 + k_2' \geq \frac{6}{7}n$.*

*Proof.* By Lemma 9, there exists an infinite family of graphs, each with at least $2m + \frac{6}{7}n$ bends in any general position drawing. If there is a 2-bend drawing of such a graph, then by Lemma 12 there exists a general position drawing with $2m + k_0 + k_2'$ bends. Hence $2m + k_0 + k_2' \geq 2m + \frac{6}{7}n$ and $k_0 + k_2' \geq \frac{6}{7}n$.  $\square$

The following two results are obtained using the same argument used in the proof of Corollary 1 applied with Lemma 10 and Lemma 11, respectively.

**Corollary 2.** *There exists an infinite family of 6-regular 2-connected $n$-vertex graphs, such that in any 2-bend drawing of any one of the graphs, $k_0 + k_2' \geq \frac{4}{7}n$.*  $\square$

**Corollary 3.** *There exists an infinite family of 6-regular 4-connected $n$-vertex graphs, such that in any 2-bend drawing of any one of the graphs, $k_0 + k_2' \geq \frac{2}{7}n$.*  $\square$

A natural variation of the general position model allows at most two vertices in any one grid-plane and with each vertex being coplanar with at most one other vertex. We now show that there exists graphs which do not have 2-bend drawings in this model.

**Theorem 12.** *There exists an infinite family of 2-connected graphs each of which does not have a 2-bend drawing with at most two vertices in any one grid-plane and with each vertex being coplanar with at most one other vertex.*

*Proof.* By Corollary 2 there exists an infinite family of 6-regular 2-connected $n$-vertex graphs, such that in any 2-bend drawing of any one of the graphs, $k_0 + k_2' \geq \frac{4}{7}n$. Assume, to the contrary, that for such a graph there is a 2-bend drawing with at most two vertices in any one grid-plane and with each vertex being coplanar with at most one other vertex. Then the number of pairs of vertices in a common grid-plane is at most $\frac{n}{2}$, and the number of planar edge routes is at most $\frac{n}{2}$; that is, $k_0 + k_1 + k_2' \leq \frac{n}{2}$. Hence $\frac{4}{7}n \leq k_0 + k_2' \leq \frac{n}{2} - k_1$, implying $k_1 < 0$, which is a contradiction, as required. $\square$

## 7 Conclusion and Open Problems

In this paper we have initiated the study of lower bounds for the number of bends in 3-D orthogonal drawings of maximum degree six graphs. As well as closing the gap between the established lower and upper bounds, the following are interesting open problems not already discussed in this paper.

- The sequence of lower bounds on the number of bends in general position drawings in Section 5 suggests the following open problem. Does every 6-connected 6-regular graph have a general position drawing with at most $2m + 6$ bends?

- Are there classes of graphs (besides maximum degree five simple graphs) which admit general position 2-bend drawings? For example, it is conceivable that planar graphs with maximum degree at most six admit general position 2-bend drawings.

- In the bend-minimum drawings of $K_4$, $K_5$ and $K_6$ the 0-bend subgraph is a tree. Is this the case for all graphs? It is easily seen that every tree has a 0-bend drawing.

- Does every graph with maximum degree at most three have a 1-bend drawing?

## Acknowledgements

The author gratefully acknowledges Therese Biedl, Antonios Symvonis, and Ben Lynn for helpful discussions. In particular, Lemmata 10, 12 and 19 were developed in conjunction with Therese Biedl, Antonios Symvonis, and Ben Lynn, respectively. Many thanks to Graham Farr for advice and encouragement. Thanks also to the anonymous referees for numerous helpful comments, in particular a simplified proof of Theorem 6.

D. Wood, *Lower Bounds for 3-D Drawings*, JGAA, 7(1) 33–77 (2003)    66

# References



[1] M. A. ABOELAZE AND B. W. WAH, Complexities of layouts in three-dimensional VLSI circuits. *Inform. Sci.*, **55(1-3)**:167–188, 1991.

[2] A. AGGARWAL, M. KLAWE, AND P. SHOR, Multilayer grid embeddings for VLSI. *Algorithmica*, **6(1)**:129–151, 1991.

[3] T. BIEDL AND T. CHAN, Cross-coloring: improving the technique by Kolmogorov and Barzdin. Tech. Rep. CS-2000-13, Department of Computer Science, University of Waterloo, Canada, 2000.

[4] T. BIEDL, J. R. JOHANSEN, T. SHERMER, AND D. R. WOOD, Orthogonal drawings with few layers. In P. MUTZEL, M. JÜNGER, AND S. LEIPERT, eds., *Proc. 9th International Symp. on Graph Drawing (GD '01)*, vol. 2265 of *Lecture Notes in Comput. Sci.*, pp. 297–311, Springer, 2002.

[5] T. BIEDL, T. THIELE, AND D. R. WOOD, Three-dimensional orthogonal graph drawing with optimal volume. In [20], pp. 284–295.

[6] T. C. BIEDL, Heuristics for 3D-orthogonal graph drawings. In *Proc. 4th Twente Workshop on Graphs and Combinatorial Optimization*, pp. 41–44, 1995.

[7] T. C. BIEDL, New lower bounds for orthogonal drawings. *J. Graph Algorithms Appl.*, **2(7)**:1–31, 1998.

[8] T. C. BIEDL, Three approaches to 3D-orthogonal box-drawings. In [30], pp. 30–43.

[9] T. C. BIEDL, T. SHERMER, S. WHITESIDES, AND S. WISMATH, Bounds for orthogonal 3-D graph drawing. *J. Graph Algorithms Appl.*, **3(4)**:63–79, 1999.

[10] M. CLOSSON, S. GARTSHORE, J. JOHANSEN, AND S. K. WISMATH, Fully dynamic 3-dimensional orthogonal graph drawing. *J. Graph Algorithms Appl.*, **5(2)**:1–34, 2001.

[11] G. DI BATTISTA, M. PATRIGNANI, AND F. VARGIU, A split&push approach to 3D orthogonal drawing. *J. Graph Algorithms Appl.*, **4(3)**:105–133, 2000.

[12] P. EADES, C. STIRK, AND S. WHITESIDES, The techniques of Kolmogorov and Barzdin for three dimensional orthogonal graph drawings. *Inform. Proc. Lett.*, **60(2)**:97–103, 1996.

[13] P. EADES, A. SYMVONIS, AND S. WHITESIDES, Three dimensional orthogonal graph drawing algorithms. *Discrete Applied Math.*, **103**:55–87, 2000.

D. Wood, *Lower Bounds for 3-D Drawings*, JGAA, 7(1) 33–77 (2003)    67


[14] S. EVEN AND G. GRANOT, Rectilinear planar drawings with few bends in each edge. Tech. Rep. 797, Computer Science Department, Technion, Israel Inst. of Tech., 1994.

[15] H. KOIKE, An application of three-dimensional visualization to object-oriented programming. In T. CATARCI, M. F. COSTABILE, AND S. LEVIALDI, eds., *Proc. Advanced Visual Interfaces (AVI '92)*, vol. 36 of *Series in Comput. Sci.*, pp. 180–192, World Scientific, 1992.

[16] H. KOIKE, The role of another spatial dimension in software visualization. *ACM Trans. Inf. Syst.*, **11(3)**:266–286, 1993.

[17] A. N. KOLMOGOROV AND Y. M. BARZDIN, On the realization of nets in 3-dimensional space. *Problems in Cybernetics*, **8**:261–268, 1967.

[18] F. T. LEIGHTON AND A. L. ROSENBERG, Three-dimensional circuit layouts. *SIAM J. Comput.*, **15(3)**:793–813, 1986.

[19] B. Y. S. LYNN, A. SYMVONIS, AND D. R. WOOD, Refinement of three-dimensional orthogonal graph drawings. In [20], pp. 308–320.

[20] J. MARKS, ed., *Proc. 8th International Symp. on Graph Drawing (GD '00)*, vol. 1984 of *Lecture Notes in Comput. Sci.*, Springer, 2001.

[21] A. PAPAKOSTAS AND I. G. TOLLIS, Algorithms for incremental orthogonal graph drawing in three dimensions. *J. Graph Algorithms Appl.*, **3(4)**:81–115, 1999.

[22] M. PATRIGNANI AND F. VARGIU, 3DCube: a tool for three dimensional graph drawing. In G. DI BATTISTA, ed., *Proc. 5th International Symp. on Graph Drawing (GD '97)*, vol. 1353 of *Lecture Notes in Comput. Sci.*, pp. 284–290, Springer, 1998.

[23] F. P. PREPARATA, Optimal three-dimensional VLSI layouts. *Math. Systems Theory*, **16**:1–8, 1983.

[24] S. REISS, 3-D visualization of program information. In R. TAMASSIA AND I. G. TOLLIS, eds., *Proc. DIMACS International Workshop on Graph Drawing (GD '94)*, vol. 894 of *Lecture Notes in Comput. Sci.*, pp. 12–24, Springer, 1995.

[25] G. G. ROBERTSON, J. D. MACKINLAY, AND S. K. CARD, Cone trees: Animated 3D visualizations of hierarchical information. In S. ROBERTSON, G. OLSEN, AND J. OLSEN, eds., *Proc. Human Factors in Computing Systems (CHI '91)*, pp. 189–194, ACM, 1991.

[26] A. L. ROSENBERG, Three-dimensional VLSI: A case study. *J. Assoc. Comput. Mach.*, **30(2)**:397–416, 1983.

[27] R. TAMASSIA, I. G. TOLLIS, AND J. S. VITTER, Lower bounds for planar orthogonal drawings of graphs. *Inform. Process. Lett.*, **39(1)**:35–40, 1991.

[28] C. WARE AND G. FRANCK, Viewing a graph in a virtual reality display is three times as good as a 2D diagram. In A. L. AMBLER AND T. D. KIMURA, eds., *Proc. IEEE Symp. Visual Languages (VL '94)*, pp. 182–183, IEEE, 1994.

[29] C. WARE AND G. FRANCK, Evaluating stereo and motion cues for visualizing information nets in three dimensions. *ACM Trans. Graphics*, **15(2)**:121–140, 1996.

[30] S. WHITESIDES, ed., *Proc. 6th International Symp. on Graph Drawing (GD '98)*, vol. 1547 of *Lecture Notes in Comput. Sci.*, Springer, 1998.

[31] H. WHITNEY, Congruent graphs and the connectivity of graphs. *Amer. J. Math.*, **54**:150–168, 1932.

[32] D. R. WOOD, On higher-dimensional orthogonal graph drawing. In J. HARLAND, ed., *Proc. Computing: the Australasian Theory Symp. (CATS '97)*, vol. 19(2) of *Austral. Comput. Sci. Comm.*, pp. 3–8, 1997.

[33] D. R. WOOD, An algorithm for three-dimensional orthogonal graph drawing. In [30], pp. 332–346.

[34] D. R. WOOD, Multi-dimensional orthogonal graph drawing with small boxes. In J. KRATOCHVIL, ed., *Proc. 7th International Symp. on Graph Drawing (GD '99)*, vol. 1731 of *Lecture Notes in Comput. Sci.*, pp. 311–222, Springer, 1999.

[35] D. R. WOOD, *Three-Dimensional Orthogonal Graph Drawing*. Ph.D. thesis, School of Computer Science and Software Engineering, Monash University, Melbourne, Australia, 2000.

[36] D. R. WOOD, Minimising the number of bends and volume in three-dimensional orthogonal graph drawings with a diagonal vertex layout, submitted. See Technical Report CS-AAG-2001-03, Basser Department of Computer Science, The University of Sydney, 2001.

[37] D. R. WOOD, Optimal three-dimensional orthogonal graph drawing in the general position model. *Theoret. Comput. Sci.*, to appear.

D. Wood, *Lower Bounds for 3-D Drawings*, JGAA, 7(1) 33–77 (2003)    69

# A    Existence of Small Subgraphs

In this appendix we prove a number of results concerning the existence of cycles and other small subgraphs in graphs of a certain size.

**Lemma 13.** *Every 5-vertex graph with at least six edges contains $C_3$ or $K_{2,3}$.*

*Proof.* Suppose to the contrary that there exists a 6-edge 5-vertex graph $G$ not containing $C_3$ or $K_{2,3}$. $G$ contains a cycle. Let $C$ be the cycle of maximum length in $G$. Then $|C| = 5$ or $|C| = 4$.

**Case 1.** $|C| = 5$: Since $G$ has six edges, $C$ has a chord, in which case $G$ contains a 3-cycle, as illustrated in Figure 33(a).

**Case 2.** $|C| = 4$: If $C$ has a chord then $G$ contains a 3-cycle, as illustrated in Figure 33(b). Thus $C$ does not have a chord. Hence the vertex $v$ not in $C$ is incident to two edges $vu$ and $vw$, where $u$ and $w$ are in $C$. If $u$ and $w$ are adjacent in $C$ then $G$ contains a 3-cycle, as illustrated in Figure 33(c). Thus $u$ and $w$ are not adjacent in $C$, which implies that $G$ contains $K_{2,3}$, as illustrated in Figure 33(d). □

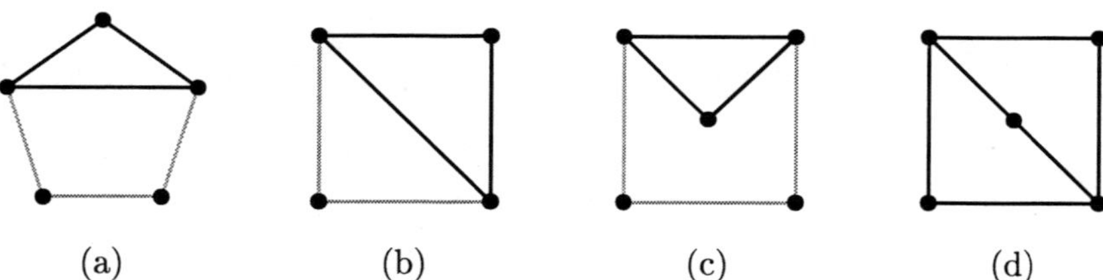

(a)          (b)          (c)          (d)

Figure 33: $C_3$ or $K_{2,3}$ in a 5-vertex 6-edge graph.

**Lemma 14.** *Every 6-vertex graph with at least eight edges contains $C_3$ or $K_{2,3}$.*

*Proof.* Suppose to the contrary that there exists an 8-edge 6-vertex graph $G$ not containing $C_3$ or $K_{2,3}$. Let $C$ be the longest cycle in $G$. Clearly $3 \leq |C| \leq 6$.

**Case 1.** $|C| = 6$: If $|C| = 6$ then $C$ has at least two chords. Any two chords of $C$ which do not induce a 3-cycle induce $K_{2,3}$, as illustrated in Figure 34(a).

**Case 2.** $|C| = 5$: If $|C| = 5$ then any chord of $C$ induces a 3-cycle, and we are done. Otherwise $C$ has no chords. Since $G$ has at least eight edges, the vertex $v$ not in $C$ is adjacent to three vertices $u$, $w$ and $x$ in $C$. Two of $u$, $w$ and $x$ are adjacent in $C$, which implies $G$ contains a 3-cycle, as illustrated in Figure 34(b).

**Case 3.** $|C| = 4$: If $|C| = 4$ then any chord of $C$ induces a 3-cycle, and we are done. Otherwise $C$ has no chords. Let $v$ and $w$ be the vertices not in $C$. Since $G$ has at least eight edges, there are at least four edges in $G$ incident with $v$ or $w$. Even if $G$ contains the edge $vw$, there are two edges from $v$ to vertices on $C$, or two edges from $w$ to vertices on $C$. In either case, $G$ contains $C_3$ or $K_{2,3}$, as illustrated in Figure 33(c) and Figure 33(d), respectively. □

D. Wood, *Lower Bounds for 3-D Drawings*, JGAA, 7(1) 33–77 (2003)      70

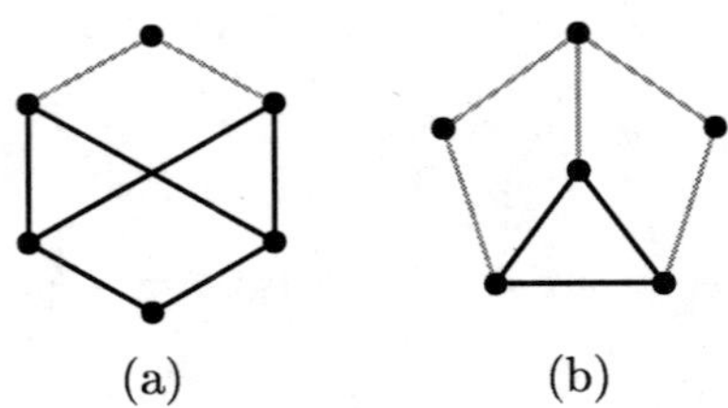

(a)          (b)

Figure 34: $C_3$ or $K_{2,3}$ in an 8-edge 6-vertex graph.

**Lemma 15.** *Every 7-vertex graph with at least ten edges contains $C_3$ or $K_{2,3}$.*

*Proof.* Suppose to the contrary that there exists a 10-edge 7-vertex graph $G$ not containing $C_3$ or $K_{2,3}$. Let $C$ be the longest cycle in $G$. Clearly $3 \leq |C| \leq 7$.

**Case 1.** $|C| = 7$: If $|C| = 7$ then there are three chords of $C$ in $G$. If two of these chords are incident to one vertex then, as illustrated in Figure 35(a), $G$ contains a 3-cycle. Thus each vertex is incident to at most one chord. Hence there exists a vertex not incident to any chords of $C$. It is easily seen that the only configuration of three chords of $C$ not inducing a 3-cycle is that illustrated in Figure 35(b); however in this case, $G$ contains $K_{2,3}$.

**Case 2.** $|C| = 6$: Suppose the vertex not in $C$ is $v$. A chord of $C$ which does not induce a 3-cycle is between vertices at distance two in $C$; that is, 'opposite' vertices. As illustrated in Figure 34(a) any two such chords induce an $K_{2,3}$ subgraph. Thus the number of chords of $C$ is at most one.

Suppose $C$ has one chord. Then there are three edges $vu$, $vw$ and $vx$ in $G$ incident to $v$. If two of $u$, $w$ and $x$ are adjacent in $C$ then there is a 3-cycle in $G$, as illustrated in Figure 35(c). Otherwise, one of $u$, $w$ or $x$ is incident to the chord of $C$, and hence $G$ contains $K_{2,3}$, as illustrated in Figure 35(d).

If $C$ has no chords then $v$ is adjacent to four vertices $u$, $w$, $x$ and $y$ in $C$. Two of $u$, $w$, $x$ and $y$ are adjacent in $C$; thus $G$ contains a 3-cycle, as illustrated in Figure 35(c).

**Case 3.** $|C| = 5$: Suppose the vertices not in $C$ are $v$ and $w$. Any chord of $C$ induces a 3-cycle, as illustrated in Figure 33(a). Thus $C$ has no chords, and there are five edges incident to $v$ and $w$. If $vw$ is an edge of $G$ and each of $v$ and $w$ are incident to two edges then $G$ contains a 3-cycle or $K_{2,3}$, as illustrated in Figure 35(e). Otherwise at least one of $v$ and $w$, say $v$, is adjacent to three vertices $u$, $x$ and $y$ in $C$. Two of $u$, $x$ and $y$ are adjacent in $C$. Thus $G$ contains a 3-cycle, as illustrated in Figure 34(b).

**Case 4.** $|C| = 4$: Suppose the vertices not on $C$ are $u$, $v$ and $w$. Any chord of $C$ induces a 3-cycle, as illustrated in Figure 33(b). Thus $C$ has no chords, and there are at least six edges incident to $u$, $v$ and $w$. Since $u$, $v$ and $w$ do not form a 3-cycle, there are at least four edges between $u$, $v$ or $w$ and vertices in $C$. Hence at least one of $u$, $v$ and $w$, say $v$, is adjacent to two vertices $x$ and $y$ in $C$. So that $v$, $x$ and $y$ do not form a 3-cycle, $x$ and $y$ are not adjacent. In this case, $G$ contains $K_{2,3}$, as illustrated in Figure 33(d). $\square$

D. Wood, *Lower Bounds for 3-D Drawings*, JGAA, 7(1) 33–77 (2003)    71

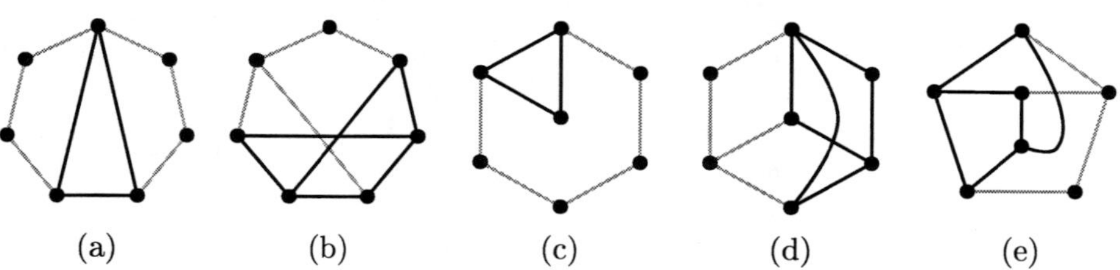

(a)    (b)    (c)    (d)    (e)

Figure 35: $C_3$ or $K_{2,3}$ in a 10-edge 7-vertex graph.

**Lemma 16.** *Every 7-vertex graph with at least eight edges contains a cycle $C_k$ $(k \neq 4)$, two chord-disjoint cycles, or $K_{2,3}$.*

*Proof.* Let $G$ be a 7-vertex graph with at least eight edges. Let $C$ be the longest cycle in $G$; thus $3 \leq |C| \leq 7$. If $|C| \neq 4$ then we are done, otherwise $|C| = 4$. If $C$ has a chordal path then $G$ either contains $K_{2,3}$ or a cycle $C_k$ $(k \neq 4)$, as illustrated in Figure 36(a) and Figure 36(b). Thus we now assume that $C$ has no chordal path.

There are at least four edges not in $C$. Let $X$ be the subgraph of $G$ induced by the vertices not in $C$. Then $X$ has at least three vertices, and the number of edges in $X$ is at most three. We proceed by considering the number of edges in $X$.

**Case 1.** $|E(X)| = 3$: Then $X$ is a 3-cycle, as illustrated in Figure 36(c), and we are done.

**Case 2.** $|E(X)| = 2$: If there are two edges in $X$ then $X$ is connected and there are at least two edges $e_1$ and $e_2$ between $X$ and $C$. Since $X$ is connected, $e_1$ and $e_2$ have the same end-vertex in $C$ for $C$ not to have a chordal path. In this case, $e_1$ and $e_2$ along with one or two of the edges in $X$ form a cycle which is chord-disjoint from $C$, as illustrated in Figure 36(d).

**Case 3.** $|E(X)| = 1$: If there is one edge in $X$ then there are at least three edges between $X$ and $C$. If one of the vertices in $X$ is incident to at least two edges between $X$ and $C$ then $C$ has a chordal-path. Thus every vertex in $X$ is incident to at most one edge between $X$ and $C$. Since $X$ has three vertices and there are at least three edges between $X$ and $C$, each vertex in $X$ is incident to exactly one edge between $X$ and $C$. Let $vw$ be the edge in $X$. For $C$ not to have a chordal path, $v$ and $w$ are incident to the same vertex in $C$, in which case $G$ contains a 3-cycle, as illustrated in Figure 36(d).

**Case 4.** $|E(X)| = 0$: If there are no edges in $X$ then there are at least four edges between $X$ and $C$. Thus one of the vertices in $X$ is incident to at least two edges between $X$ and $C$, in which case $C$ has a chordal-path. $\square$

D. Wood, *Lower Bounds for 3-D Drawings*, JGAA, 7(1) 33–77 (2003)          72

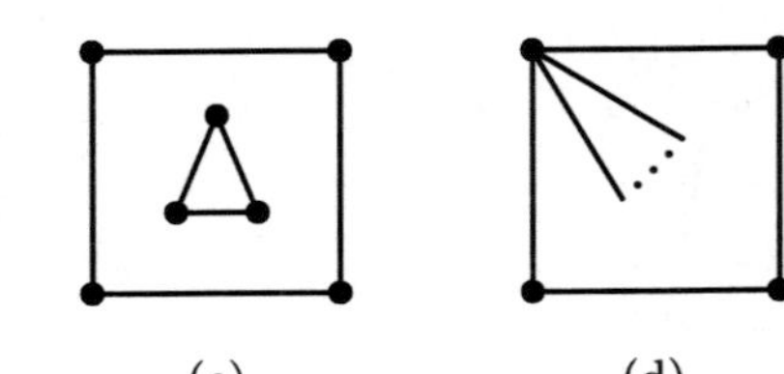

(a)          (b)          (c)          (d)

Figure 36: $C_k$ $(k \neq 4)$, $K_{2,3}$ or two chord-disjoint cycles in a 7-vertex 8-edge graph.

# B    Proofs of Connectivity

In this section we prove the connectivity of the graphs used to establish our main lower bounds. We first prove that the 'grid-graphs' have the desired connectivity. Of course $C_r$ with $r \geq 3$ is 2-connected.

**Observation 7.** $C_r \times K_2$ *with* $r \geq 3$ *is 3-connected.*

*Proof.* Let $v$ and $w$ be distinct vertices of $C_r \times K_2$. As illustrated in Figure 37, if $v$ and $w$ are (a) in the same 'row', (b) in the same 'column', or (c) 'non-collinear', there are three internally disjoint paths between $v$ and $w$ in $C_r \times K_2$. Since $r \geq 3$, in case (c) we can assume that $v$ and $w$ are at least two columns apart. By Menger's Theorem, $C_r \times K_2$ is 3-connected. $\qquad\Box$

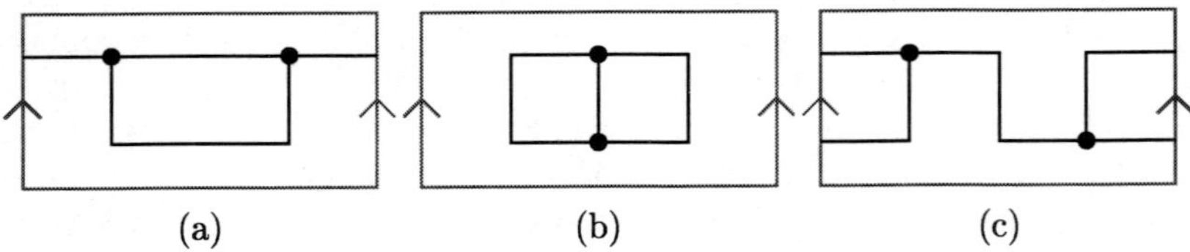

(a)                    (b)                    (c)

Figure 37: Three disjoint paths in $C_r \times K_2$.

**Observation 8.** $C_r \times C_3$ *with* $r \geq 3$ *is 4-connected.*

*Proof.* Let $v$ and $w$ be distinct vertices of $C_r \times C_3$. As illustrated in Figure 38, if $v$ and $w$ are (a) in the same 'row', (b) in the same 'column', or (c) 'non-collinear', there are four internally disjoint paths between $v$ and $w$ in $C_r \times C_3$. Since $r \geq 3$, in case (c) we can assume that $v$ and $w$ are one row apart and at least two columns apart. By Menger's Theorem, $C_r \times C_3$ is 4-connected. $\qquad\Box$

**Observation 9.** $C_r \times C_3 \times K_2$ *with* $r \geq 3$ *is 5-connected.*

*Proof.* Let $v$ and $w$ be distinct vertices of $C_r \times C_3 \times K_2$. As illustrated in Figure 39, if $v$ and $w$ (a) are in the same 'row', (b) are in the same 'column', or

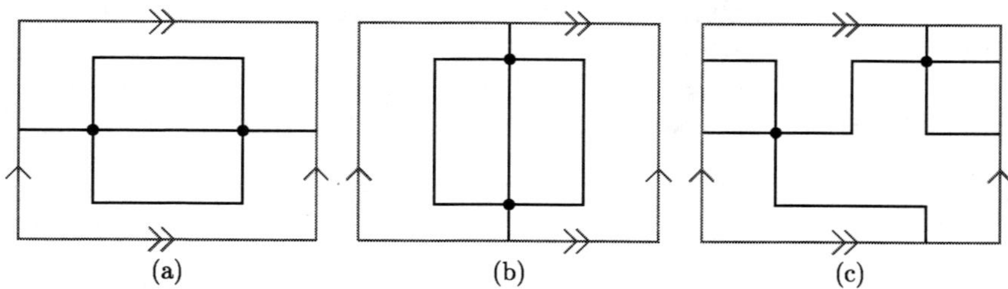

Figure 38: Four internally disjoint paths in $C_r \times C_3$.

(c) have the same 'depth', (d) have the same 'height', or (e) are 'non-coplanar', there are five internally disjoint paths between $v$ and $w$ in $C_r \times C_3 \times K_2$. Since $r \geq 3$, in cases (c), (d) and (e) we can assume that $v$ and $w$ are at least two columns apart. By Menger's Theorem, $C_r \times C_3 \times K_2$ is 5-connected. $\qquad\square$

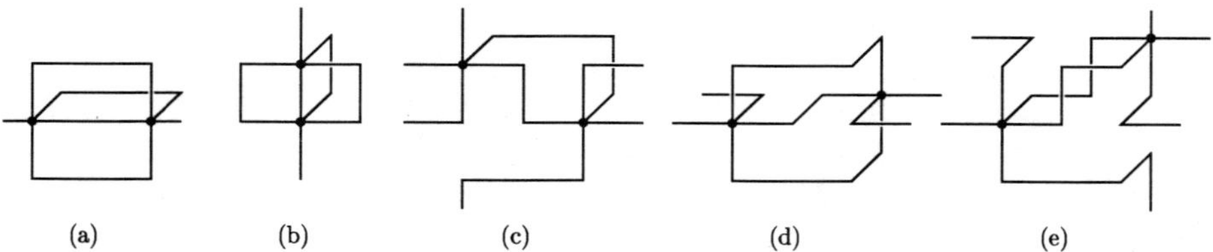

Figure 39: Five internally disjoint paths in $C_r \times C_3 \times K_2$.

**Observation 10.** $C_r \times C_3 \times C_3$ with $r \geq 3$ *is 6-connected.*

*Proof.* Let $v$ and $w$ be distinct vertices of $C_r \times C_3 \times C_3$. As illustrated in Figure 40, if $v$ and $w$ are (a) 'collinear', (b) 'coplanar' or (c) 'non-coplanar', there are six internally disjoint paths between $v$ and $w$ in $C_r \times C_3 \times C_3$. Since $r \geq 3$, in cases (b) and (c) we can assume that $v$ and $w$ are at least two columns apart. By symmetry these three cases suffice. By Menger's Theorem, $C_r \times C_3 \times C_3$ is 6-connected. $\qquad\square$

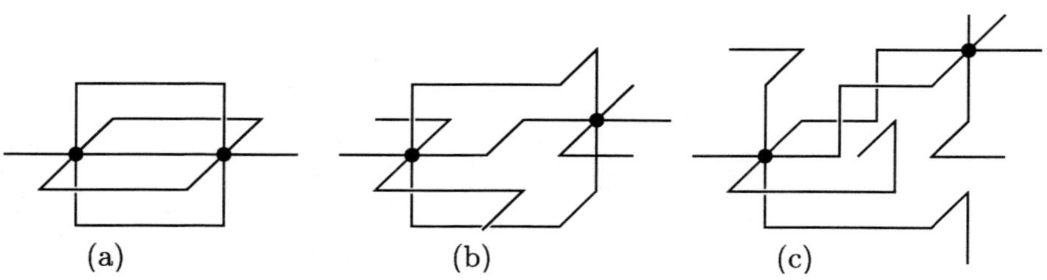

Figure 40: Six internally disjoint paths in $C_r \times C_3 \times C_3$.

D. Wood, *Lower Bounds for 3-D Drawings*, JGAA, 7(1) 33–77 (2003)      74

The next lemma combined with the above observations proves that $C_r\langle K_p\rangle$ for $p \geq 2$ is 2-connected, that $(C_r \times K_2)\langle K_p\rangle$ for $p \geq 3$ is 3-connected, that $(C_r \times C_3)\langle K_p\rangle$ for $p \geq 4$ is 4-connected, that $(C_r \times C_3 \times K_2)\langle K_p\rangle$ for $p \geq 5$ is 5-connected, and that $(C_r \times C_3 \times C_3)\langle K_p\rangle$ for $p \geq 6$ is 6-connected.

**Lemma 17.** *If a graph $G$ is $c$-connected for some $c \geq 2$, then $G\langle K_p\rangle$ is $c$-connected for all $p \geq c$.*

*Proof.* In each $K_p$ subgraph $H$, those vertices of $H$ adjacent to vertices not in $H$ are called *exit vertices*. Every vertex $v$ of $H$ is adjacent to every exit vertex of $H$ (except for itself). Hence there are $c$ internally disjoint paths between any two vertices of $G\langle K_p\rangle$ that are in distinct $K_p$ subgraphs, since $G$ itself is $c$-connected. Consider vertices $v$ and $w$ in the same $K_p$ subgraph $H$ whose original vertex in $G$ is $u$.

Suppose $v$ and $w$ are both exit vertices. Then there are $c - 2$ internally disjoint $vw$-paths via the other exit vertices of $H$. Let $x$ and $y$ be the original vertices of $G$ such that there are edges incident to $v$ and $w$ whose other endvertices are in the subgraphs corresponding to $x$ and $y$, respectively. Since $G$ is 2-connected, there is an $xy$-path in $G$ which avoids $u$. This path and the edge $vw$ gives a total of $c$ internally disjoint $vw$-paths.

Now suppose one of $v$ and $w$, say $v$, is not an exit vertex. Then there are at least $c - 1$ internally disjoint $vw$-paths via the other exit vertices. Along with the edge $vw$, there are at least $c$ disjoint $vw$-paths.

By Menger's Theorem, $G\langle K_p\rangle$ is $c$-connected. $\qquad\square$

**Lemma 18.** *[(a)]*

*$C_r\langle K_p \setminus M_1\rangle$ is 2-connected for all $p \geq 4$,*

*2. $(C_r \times K_2)\langle K_p \setminus M_2\rangle$ is 3-connected for all $p \geq 5$,*

*3. $(C_r \times C_3)\langle K_p \setminus M_2\rangle$ is 4-connected for all $p \geq 5$,*

*4. $(C_r \times C_3 \times K_2)\langle K_p \setminus M_3\rangle$ is 5-connected for all $p \geq 7$, and*

*5. $(C_r \times C_3 \times C_3)\langle K_p \setminus M_3\rangle$ is 6-connected for all $p \geq 7$.*

*Proof.* Each part of the lemma states that a graph of the form $G\langle K_p \setminus M_j\rangle$ is $c$-connected, where by the preceding observations, $G$ is a $c$-connected graph, $j = \lceil\frac{c}{2}\rceil$, and $p \geq 2j + 1$. Observe that $p \geq c + 1$. Moreover, in each $K_p \setminus M_j$ subgraph $H$, if $c$ is even then it is precisely the exit vertices in $H$ that are matched in $M_j$, and for odd $c$, all but one of the exit vertices are matched to each other in $M_j$, and the one remaining exit vertex is matched with one of the (at least two) non-exit vertices. By the same argument used in Lemma 17, for any two exit vertices $v$ and $w$ of a $K_p \setminus M_j$ subgraph $H$, there is a $vw$-path in $G\langle K_p \setminus M_j\rangle$ not using any edges in $H$.

Let $v$ be a vertex of $G\langle K_p \setminus M_j\rangle$ contained in a $K_p \setminus M_j$ subgraph $H$. We claim that there are $c$ internally disjoint (possibly empty) paths from $v$ to the exit vertices of $H$. If $v$ is an exit vertex then there are at least $c - 1$ other exit

D. Wood, *Lower Bounds for 3-D Drawings*, JGAA, 7(1) 33–77 (2003)      75

vertices in $H$, none of which are matched with $v$. Hence $v$ is adjacent to each such exit vertex, and counting the empty path from $v$ to $v$, the claim holds. Now suppose $v$ is not an exit vertex. If $v$ is adjacent to each exit vertex, which is guaranteed in the case of even $c$, then the claim holds. Otherwise $c$ is odd, and $v$ is matched with one of the exit vertices $x$. In this case $v$ is adjacent to the $c-1$ remaining exit vertices, and there is a 2-path from $v$ to $x$ via the other non-exit vertex in $H$, giving a total of $c$ internally disjoint paths from $v$ to the exit vertices of $H$. This proves our claim. It follows since $G$ is $c$-connected that between any two vertices in distinct $K_p \setminus M_j$ subgraphs, there are $c$ internally disjoint paths.

Now consider two vertices $v$ and $w$ contained in the same $K_p \setminus M_j$ subgraph $H$. First suppose $v$ and $w$ are both exit vertices. If $v$ and $w$ are matched then there are $c-2$ internally disjoint $vw$-paths via the other exit vertices, there is at least one path between $v$ and $w$ via the non-exit vertices of $H$, and there is a path between $v$ and $w$ not using the edges of $H$. Thus there are $c$ internally disjoint $vw$-paths in $G\langle K_p \setminus M_j \rangle$. Now suppose $v$ and $w$ are non-matched exit vertices of $H$. Thus $c \geq 4$. Suppose $c \in \{4, 6\}$. Let $vx$ and $wy$ be in $M_j$. By construction, $v$ is opposite to $x$, and $w$ is opposite to $y$ (with respect to the grid-structure of $G$). Thus there exists a $vx$-path $P$ disjoint from some $wy$-path $Q$ in $G\langle K_p \setminus M_j \rangle$, not using any edges in $H$. Hence $P \cup \{xw\}$ and $Q \cup \{vy\}$ are internally disjoint $vw$-paths. There are $c-4$ internally disjoint $vw$-paths via the other exit vertices of $H$. There is one $vw$-path via the non-exit vertex of $H$, and there is the edge $vw$, giving a total of $c$ internally disjoint $vw$-paths in $G\langle K_p \setminus M_j \rangle$. Now suppose $c = 5$. Either both of $v$ and $w$ are matched to other exit vertices, or one of $v$ and $w$ is matched with an exit vertex and the other is matched with a non-exit vertex. First suppose that $v$ is matched with an exit-vertex $x$ and $w$ is matched with an exit vertex $y$. There are $c-4$ internally disjoint $vw$-paths via the other exit vertices, there is the edge $vw$, there are two $vw$-paths via the two non-exit vertices, and there is the path $v$–$y$–$x$–$w$, giving a total of $c$ internally disjoint $vw$-paths. Now suppose $v$ is matched with an exit vertex $x$, but $w$ is matched with a non-exit vertex $y$. There is a $vx$-path $P$ not using any edges in $H$. There are $c-3$ $vw$-paths via the other exit vertices, there is the edge $vw$, there is one $vw$-path via the one remaining non-exit vertex, and $P \cup \{xw\}$ forms a $vw$-path, giving a total of $c$ internally disjoint $vw$-paths.

Now consider two vertices $v$ and $w$ contained in the same $K_p \setminus M_j$ subgraph $H$, where $v$ is an exit vertex and $w$ is not an exit vertex. First suppose $v$ and $w$ are matched, in which case $c$ is odd. There are $c-1$ $vw$-paths via the other exit vertices, and there is a $vw$-path via the other non-exit vertex of $H$, giving a total of $c$ internally disjoint $vw$-paths. Now suppose $v$ and $w$ are not matched. Let $x$ be the vertex matched with $v$. Suppose $x$ is an exit vertex. There is a $vx$-path $P$ not using any edge in $H$, and thus $P \cup \{wx\}$ forms a $vw$-path. There are $c-2$ $vw$-paths via the other exit vertices of $H$, and there is the edge $vw$, giving a total of $c$ internally disjoint $vw$-paths. If $x$ is not an exit vertex, then there are $c-1$ $vw$-paths via the other exit vertices, and the edge $vw$ gives a total of $c$ internally disjoint $vw$-paths.

The final case is when $v$ and $w$ are both not exit vertices contained in the same $K_p \setminus M_j$ subgraph $H$. At most one of $v$ and $w$ is matched with an exit vertex. Thus there are at least $c - 1$ $vw$-paths via the remaining exit vertices, which along with the edge $vw$, give $c$ internally disjoint $vw$-paths.

Thus for every pair of vertices $v$ and $w$ of $G\langle K_p \setminus M_j\rangle$, there are $c$ internally disjoint $vw$-paths. By Menger's Theorem, $G\langle K_p \setminus M_j\rangle$ is $c$-connected.    $\square$

The multigraphs and pseudo graphs constructed in Section 4 have the claimed connectivity, since each contains a simple subgraph that is proved in Lemmata 17 and 18 to have the same connectivity.

## C   Final Observation

Lemma 8 states that if a graph $G$ has at least $k$ bends in every general position drawing then for any edge $e$ of $G$ the graph $G \setminus e$ has at least $k - 4$ bends in every drawing. We now prove the analogue of this result for arbitrary (non general position) drawings.

**Lemma 19.** *If a graph $G$ has at least $k$ bends in every drawing then for any edge $e$ of $G$ the graph $G \setminus e$ has at least $k - 6$ bends in every drawing.*

*Proof.* Suppose there is a drawing of $G \setminus e$ with $b$ bends. Let $e = vw$. At each of $v$ and $w$ there is an unused port. Regardless of the relative directions of the unused ports at $v$ and $w$, by inserting at most two planes at each of $v$ and $w$, we can route $e$ with at most six bends and entirely within the inserted planes. Hence the edge route for $e$ does not intersect any existing edge routes in the drawing of $G \setminus e$. In Figure 41 we illustrate such an edge routing in the worst case scenario with $v$ and $w$ non-coplanar and the unused ports at $v$ and $w$ pointing away from each other. Note that in many other cases less than six bends are needed. Hence $G$ has a drawing with $b + 6$ bends. By assumption, every drawing of $G$ has at least $k$ bends. Thus $b + 6 \geq k$ and $b \geq k - 6$.    $\square$

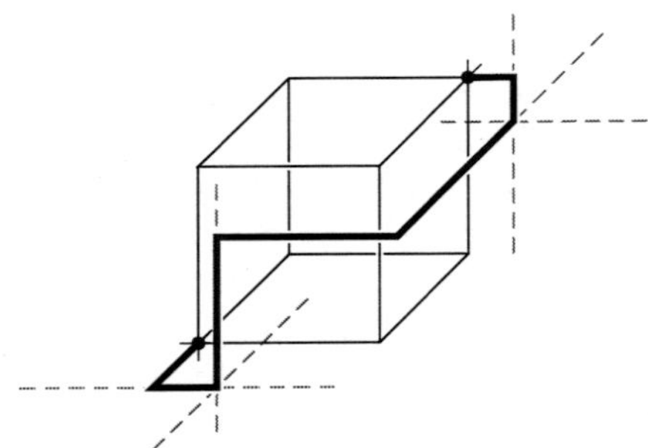

Figure 41: Inserting a 6-bend edge.

Note that this technique can also be used to provide an upper bound on the maximum number of bends per edge route in a given drawing. For example, the

D. Wood, *Lower Bounds for 3-D Drawings*, JGAA, 7(1) 33–77 (2003)     77

REDUCE FORKS algorithm of Di Battista *et al.* [11] does not provide a bound on the maximum number of bends per edge, and in many instances, edges are routed with more than six bends [11, 33]. By replacing each edge route with more than six bends by an edge route with at most six bends, as described in the proof of Lemma 19, the algorithm can be modified to produce drawings with an upper bound on the maximum number of bends per edge. Of course, doing so may increase the volume of the drawing.

# Journal of Graph Algorithms and Applications

http://jgaa.info/

vol. 7, no. 1, pp. 79-98 (2003)


# Hamilton Decompositions and $(n/2)$-Factorizations of Hypercubes


*Douglas W. Bass*

Graduate Programs in Software
University of St. Thomas
http://gps.stthomas.edu
dbass@stthomas.edu

*I. Hal Sudborough*

Department of Computer Science
University of Texas at Dallas
http://www.utdallas.edu/dept/cs
hal@utdallas.edu


## Abstract


Since $Q_n$, the hypercube of dimension $n$, is known to have $n$ link-disjoint paths between any two nodes, the links of $Q_n$ can be partitioned into multiple link-disjoint spanning subnetworks, or factors. We seek to identify factors which efficiently simulate $Q_n$, while using only a portion of the links of $Q_n$. We seek to identify $(n/2)$-factorizations, of $Q_n$ where 1) the factors have as small a diameter as possible, and 2) mappings (embeddings) of $Q_n$ to each of the factors exist, such that the maximum number of links in a factor corresponding to one link in $Q_n$ (dilation), is as small as possible. In this paper we consider two algorithms for generating Hamilton decompositions of $Q_n$, and three methods for constructing $(n/2)$-factorizations of $Q_n$ for specific values of $n$. The most notable $(n/2)$-factorization of $Q_n$ results in two mutually isomorphic factors, each with diameter $n + 2$, where an embedding exists which maps $Q_n$ to each of the factors with constant dilation.



Communicated by Balaji Raghavachari: submitted November 2001;
revised September 2002 and February 2003.

D. Bass, *Decompositions and Factorizations*, JGAA, 7(1) 79-98 (2003)    80

# 1   Introduction

## 1.1   Traditional Partitioning of Hypercubes

In traditional computing environments with a single processor, a large number of memory locations and multiple users, memory is a partitionable resource. Processes make requests for certain amounts of contiguous memory locations, and various allocation and collection strategies are used to minimize memory fragmentation.

In an environment with $2^n$ processors, connected in the form of the hypercube $Q_n$, and multiple users, the body of processors is also a partitionable resource. However, the hypercube is not partitioned in the same way as memory, as $Q_n$ has a recursive substructure. $Q_n$ consists of 2 copies of $Q_{n-1}$, with links connecting corresponding nodes in the two copies. When processes make requests for some of the processors, they traditionally request a complete hypercube of dimension smaller than $n$, known as a subcube. Significant research has been conducted into identifying, allocating, and recollecting subcubes in order to minimize subcube fragmentation [13, 19].

In this environment, if two or more processes require $Q_n$, only one of them can run at any given time. We seek to take advantage of the node-connectivity of $Q_n$, to increase the effective capacity of a hypercube-based computing environment, so that two processes requiring $Q_n$ can run concurrently.

## 1.2   Definitions

A *network G*, is a pair $(N, L)$, where $N$ is a set of distinct *nodes*, and $L$ is a set of *links*. $L$ is a set of two element subsets of $N$. In a network, the *degree* of a node $n$ is the number of elements of $L$ containing $n$ as an element. A network is *regular* if every node $n \in N$ has the same degree. The *degree* of a regular network is the degree of any node $n \in N$. A *path* is a sequence of nodes $n_1, n_2, \ldots, n_k$, such that $\forall\, i,\ 1 \leq i \leq k - 1,\ \{n_i, n_{i+1}\} \in L$. A network is *connected* if for all pairs of nodes $u$ and $v$, there exists a path from $u$ to $v$. The *node-connectivity* of a network $G$ is the minimal number of nodes which must be removed from $G$, in order to make $G$ no longer connected.

The *hypercube of dimension n*, or $Q_n$, is a network of $2^n$ nodes where each node is labeled by a bit string $b_0\, b_1 \ldots b_{n-1}$ of length $n$, and there is a link between two nodes in $Q_n$ if and only if their labels differ in exactly one bit. $Q_n$ is regular with degree $n$, and has $n2^{n-1}$ links. A *dimension k link* is a link in $Q_n$ which connects two nodes whose labels differ in the $k^{th}$ bit.

A *Hamilton cycle* of a network is a path $n_1, n_2, \ldots, n_k, n_1$, which visits each node in the network exactly once, and returns to its starting point $n_1$. A *Hamilton decomposition* of a network is a partitioning of the links of the network into link-disjoint Hamilton cycles [2]. A *matching* in a network is a set of node-disjoint links. A matching is *orthogonal* to a set of Hamilton cycles if it contains one and only one link from each Hamilton cycle. The *Cartesian product* $N_1 \times N_2$ of two networks $N_1$ and $N_2$ is the network where the nodes are ordered pairs

D. Bass, *Decompositions and Factorizations*, JGAA, 7(1) 79-98 (2003)          81

of the nodes of $N_1$ and $N_2$, and there are links between $\{u, v\}$ and $\{w, x\}$ in $N_1 \times N_2$ if $\{u, w\}$ is a link in $N_1$ or $\{v, x\}$ is a link in $N_2$. $C_n$ represents the simple cycle of n nodes.

A *spanning subnetwork S of a network N* is a connected network constructed from all the nodes of $N$ and a proper subset of the links of $N$, such that for every pair of nodes $u$ and $v$ in $N$, there is a path in $S$ between $u$ and $v$. A *perfect matching* in a network is a matching where every node in the network is incident to exactly one link. A *k-factorization* of a network is a partitioning of the links of the network into disjoint regular spanning subnetworks, or factors, of degree $k$ [3]. The *distance* between two vertices $u$ and $v$ in a network $G$, denoted by $distance_G(u, v)$, is the length of the shortest path between $u$ and $v$ in $G$. The *diameter* of a network $G$, denoted by $diameter(G)$, is the maximum value of $distance_G(u, v) \ \forall \ u, v \in N$.

An *embedding* of a network $G$ (commonly called the *guest*) into a network $H$ (commonly called the *host*) is a 1-1 function $f$ mapping the nodes of $G$ to the nodes of $H$. When $G$ and $H$ have the same set of nodes, the *identity embedding* $I$ is the embedding $I(u) = u \ \forall \ u$ in $G$. The *dilation* of an embedding $f$ is the largest value of $distance_H(f(u), f(v))$, $\forall$ edges $\{u, v\}$ in $G$.

## 1.3    An Alternate Partitioning of Hypercubes

We propose a different method for partitioning $Q_n$, with a number of potential benefits. Currently, when a process asks to use a subcube of $Q_n$, it gets full control of all the processors assigned to it. We propose that each node u of $Q_n$ be divided into $\frac{n}{k}$ virtual nodes, where $2 \leq k \leq \frac{n}{2}$ , and $n \bmod k = 0$. We propose that a $k$-factorization of $Q_n$ be constructed. In other words, let the links of $Q_n$ be divided into $\frac{n}{k}$ link-disjoint regular spanning subnetworks or factors $F_1, F_2, \ldots, F_{n/k}$ of $Q_n$ with degree $k$. The $j^{th}$ virtual node ($1 \leq j \leq \frac{n}{k}$) of some node $u$ in $Q_n$ is connected to the $j^{th}$ virtual node of another node $v$ in $Q_n$, by the links of $F_j$ only.

Figure 1 shows an example of this arrangement, where $n = 4$ and $k = 2$. The numbers on the nodes of $Q_4$ are numeric representations of their bit string labels. Each node of $Q_4$ has been divided into 2 virtual nodes, which are colored black and gray. The links of $Q_4$ have been divided into two factors $F_1$ and $F_2$ of degree $\frac{4}{2} = 2$. $F_1$ is the set of gray links, connecting the gray virtual nodes, and $F_2$ is the set of black links, connecting the black virtual nodes. $F_1$ and $F_2$ are a 2-factorization of $Q_4$.

Under this arrangement, up to $\frac{n}{k}$ different processes can access a virtual copy of $Q_n$ at the same time with no link interference between computations; since communication between virtual nodes is taking place over disjoint link sets. Another potential application for $k$-factorizations of $Q_n$ is in the area of fault-tolerant computing. If a factor could efficiently simulate $Q_n$, then $Q_n$ could tolerate the failure of all links not part of the factor. $k$-factorizations of $Q_n$ could also be used in the construction of adaptive routing algorithms for $Q_n$, which make routing decisions based on the traffic for a particular node [11, 12].

D. Bass, *Decompositions and Factorizations*, JGAA, 7(1) 79-98 (2003)     82

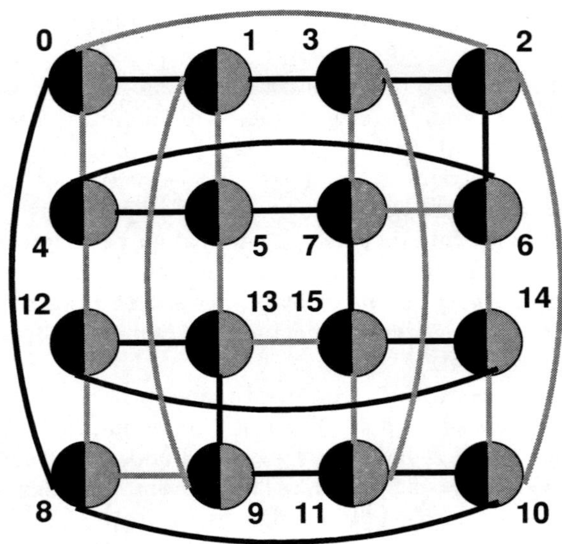

Figure 1: A 2-factorization of $Q_4$

| n | 2 | 3 | 4 | 5 | 6 | 7 | 8 |
|---|---|---|---|---|---|---|---|
| Links in $Q_n$ | 4 | 12 | 32 | 80 | 192 | 448 | 1024 |
| Removable Links | 0 | 3 | 14 | 45 | 123 | 311 | 752 |

Table 1: The number of links removeable from $Q_n$ without increasing diameter.

## 1.4   Progress in Finding Factors of Small Diameter

$Q_n$ is known to have node-connectivity of $n$ [17]. Menger's Theorem states if a network has node-connectivity $k$, then $k$ node-disjoint paths connect any two distinct nodes [15]. Therefore, perhaps some, or even most of the links can be removed from $Q_n$ without increasing its diameter. Discovering the number of removable links has been a subject of recent research [16, 10, 14]. It has been shown that $(n-2)2^{n-1} + 1 - \lceil \frac{2^n-1}{2n-1} \rceil$ links are removable from $Q_n$ without increasing its diameter [10]. Table 1 shows the number of links removable from $Q_n$ without increasing its diameter, for small values of n.

However, the resulting spanning subnetwork of $Q_n$ is not regular, and therefore cannot be part of any $k$-factorization. Regular spanning subnetworks of $Q_n$ are known to exist for specific values of $n$. These spanning subnetworks are described in Table 2. For example, the cube-connected cycles network of dimension n $CCC_n$ [23] is known to be a spanning subnetwork of $Q_{n+lgn}$, where

D. Bass, *Decompositions and Factorizations*, JGAA, 7(1) 79-98 (2003)      83

| Network | Spans | Degree | Diameter |
|---------|-------|--------|----------|
| $CCC_n$ | $Q_{n+lgn}$ | 3 | $\frac{5n}{2} - 2$ |
| $ACCC_n^1$ | $Q_{n+lgn}$ | 4 | $2n$ - 2 |
| $ACCC_n^2$ | $Q_{n+lgn}$ | 6 | $\frac{3n}{2}$ |
| $Subcube_n$ | $Q_{n-1}$ | $\lg n$ | $\frac{3n}{2}$ - 2 |
| $Q_{n,2,1}$ | $Q_n$ | $\frac{n}{2} + 1$ | $n$ |

Table 2: Known regular spanning subnetworks of $Q_n$.

$n = 2^k$ and $\lg n = \log_2 n$ [20]. The augmented cube-connected cycles networks $ACCC_n^1$ and $ACCC_n^2$ [8] are derived from adding links to $CCC_n$. The subcube network $Subcube_n$ [6] is both a spanning subnetwork of $Q_{n-1}$ and a subnetwork of the pancake network of dimension $n$. The spanning subnetwork $Q_{n,2,1}$ [7] contains all the links for dimensions 0 and 1, and uses the value of the first two bits of the label of each node to determine the dimensions of links incident to that node. $Q_{n,2,1}$ is the first regular spanning subnetwork of $Q_n$ with degree less than $n$ and diameter $n$.

However, none of the networks of Table 2 are part of any $k$-factorization of the hypercubes they span, as they use all of the links for a particular dimension of $Q_n$. We therefore seek to identify $k$-factorizations of $Q_n$, where the factors have certain properties. The factors should have minimal degree (preferably $\Theta(1)$), so as to maximize the number of factors. The factors should have minimal diameter (preferably n, the diameter of $Q_n$). The factors in this paper all have degree $\frac{n}{2}$. It is an open question as to whether factors exist with degree smaller than $\frac{n}{2}$ and diameter $n$. Finally, the factors should be constructed so that there exists an embedding $f$ of $Q_n$ into each of the factors, with minimal dilation (preferably $\Theta(1)$). An embedding can be considered as a high-level description of how one network simulates another [22]. The dilation of an embedding is a commonly used measure of the efficiency of the simulation. Since parallel algorithms on hypercubes involve communication between adjacent nodes, the path in each factor between $f(u)$ and $f(v)$, where $u$ and $v$ are adjacent nodes in $Q_n$, should have a length of at most a constant in order for each factor to efficiently simulate $Q_n$.

## 2   Creating Hamilton Decompositions of $Q_n$

### 2.1   Creating Hamilton Decompositions of $Q_{2n}$ from Hamilton Decompositions of $Q_n$

$Q_n$ has been known to have a Hamilton decomposition for some time [4]. That is, it is known that the links of $Q_n$ can be partitioned into disjoint Hamilton cycles. However, the proof did not readily result in an algorithm for producing the actual decomposition [1] [26].

Two algorithms are known for generating Hamilton decompositions of $Q_n$.

D. Bass, *Decompositions and Factorizations*, JGAA, 7(1) 79-98 (2003)          84

```
Algorithm HAMILTONCOMP1(n, input, output)
begin
    outCycle = 0
    for inCycle = 0 to n − 1 do
    begin
        for outCycleElement = 0 to 2^{2n} − 1 do
        begin
            Set the first n bits of output[outCycle][outCycleElement] to
                input[inCycle][outCycleElement div 2^n]
            Set the last n bits of output[outCycle][outCycleElement] to
                input[inCycle][(outCycleElement mod 2^n) −
                (outCycleElement div 2^n)]
        end
        outCycle = outCycle + 1
        for outCycleElement = 0 to 2^{2n} - 1 do
        begin
            Set the first n bits of output[outCycle][outCycleElement] to
                input[inCycle][(outCycleElement mod 2^n) −
                (outCycleElement div 2^n)]
            Set the last n bits of output[outCycle][outCycleElement] to
                input[inCycle][outCycleElement div 2^n]
        end
        outCycle = outCycle + 1
    end
end
```

Figure 2: Creating a Hamilton decomposition of $Q_{2n}$

The first, discovered by Ringel and given in [24], yields a Hamilton decomposition of $Q_{2n}$ from a Hamilton decomposition of $Q_n$. Each Hamilton cycle of the Hamilton decomposition of $Q_n$ is used to form two disjoint Hamilton cycles of the Hamilton decomposition of $Q_{2n}$. Let the Hamilton decomposition of $Q_n$ be stored in the two-dimensional array $input[n − 1][2^n − 1]$, where the first element for both dimensions (and for all dimensions of all arrays in this paper) is 0. The Hamilton decomposition of $Q_{2n}$ will be stored in the array $output[2n − 1][2^{2n} − 1]$. The algorithm is shown in Figure 2.

**Example**: The cycle 00, 01, 11, 10 is a Hamilton decomposition of $Q_2$. The algorithm yields the Hamilton decomposition of $Q_4$, {{0000, 0001, 0011, 0010, 0110, 0100, 0101, 0111, 1111, 1110, 1100, 1101, 1001, 1011, 1010, 1000}, {0000, 0100, 1100, 1000, 1001, 0001, 0101, 1101, 1111, 1011, 0011, 0111, 0110, 1110, 1010, 0010}}. This is the Hamilton decomposition of $Q_4$ shown in Figure 1.

## 2.2 Creating Hamilton Decompositions of $Q_{2n+2}$ from Hamilton Decompositions of $Q_{2n}$

The second Hamilton decomposition algorithm, based on [26] and given in [5], yields a Hamilton decomposition of $Q_{2n+2}$ from a Hamilton decomposition of $Q_{2n}$, and an orthogonal matching to the Hamilton decomposition of $Q_{2n}$. This algorithm is based on the following result:

**Theorem 1** *If a network $N$ can be decomposed into $n$ - 1 Hamilton cycles, and there exists a matching orthogonal to the set of Hamilton cycles, then $N \times C_{2k}$, $k \geq 2$, can be decomposed into $n$ Hamilton cycles [26].*

$C_4$, the cycle of 4 nodes, is but another way of describing $Q_2$. It is known that $Q_i \times Q_j = Q_{i+j}$ for all nonnegative integers $i$ and $j$. We take advantage of these facts to arrive at the following corollary:

**Corollary 1:** If $Q_{2n}$ can be decomposed into $n$ Hamilton cycles, and there exists a matching in $Q_{2n}$ orthogonal to the set of Hamilton cycles, then $Q_{2n} \times C_4 = Q_{2n} \times Q_2 = Q_{2n+2}$ can be decomposed into $n + 1$ Hamilton cycles.

The algorithm [25] generates two Hamilton cycles for $Q_{2n+2}$ from a selected Hamilton cycle for $Q_{2n}$, and one Hamilton cycle for $Q_{2n+2}$ from each of the remaining Hamilton cycles for $Q_{2n}$. Let $N = 2^{2n}$. Assume the nodes of $Q_{2n}$ are labeled by the integers 0, 1, ..., $N$ - 1, where two nodes are adjacent if they differ in exactly one bit in their binary representations. The $n$ Hamilton cycles of $Q_{2n}$ are stored in the array in$[n][N$ - 1$]$. The $n + 1$ Hamilton cycles of $Q_{2n+2}$ are stored in the array out$[n + 1][4N$ - 1$]$. We arrange the cycles so that the links of the orthogonal matching are $\{\{$in$[0][0]$, in$[0][N$ - 1$]\}$, $\{$in$[1][0]$, in$[1][N$ - 1$]\}$, ..., $\{$in$[n$ - 1$][0]$, in$[n$ - 1$][N$ - 1$]\}\}$. Furthermore, flipping cycles if necessary, we arrange that for $1 \leq i \leq n$ - 1, in$[i][0]$ occurs before in$[i][N$ - 1$]$ in the list in$[0][0]$, in$[0][1]$, ..., in$[0][N$ - 1$]$. The purpose of arranging the edges in the orthogonal matching in this manner is to simplify the algorithm. We also use an array flag$[n]$ to keep track of the paths taken through nodes of $Q_{2n}$, which are related to endpoints of links in the matching. The algorithm is shown in Figures 3, 4 and 5 and 6.

**Example**   Let $C_1$ and $C_2$ be the Hamilton cycles of $Q_4$ shown in Figure 1. Let $C_1$ be the black links, and $C_2$ be the gray links. In this example, $n = 2$ and $N = 16$. $C_1$ can be expressed as $\{0, 1, 3, 2, 6, 4, 5, 7, 15, 14, 12, 13, 9, 11, 10, 8\}$, and $C_2$ can be expressed as $\{4, 0, 2, 10, 14, 6, 7, 3, 11, 15, 13, 5, 1, 9, 8, 12\}$. The orthogonal matching we will use will be the two links $\{0, 8\}$ and $\{4, 12\}$. The cycles have been arranged so that the links of the orthogonal matching are in the proper position.

The algorithm FIRST-OUTPUT-CYCLE produces the Hamilton cycle for $Q_6$ $\{0, 1, 3, 2, 6, 4, 5, 7, 15, 14, 12, 13, 9, 11, 10, 8, 40, 56, 48, 16, 24, 26, 58, 42, 43, 59, 27, 25, 57, 41, 45, 61, 29, 28, 30, 62, 46, 47, 63, 31, 23, 55, 39, 37, 53, 21, 20, 52, 60, 44, 36, 38, 54, 22, 18, 50, 34, 35, 51, 19, 17, 49, 33, 32\}$.

D. Bass, *Decompositions and Factorizations*, JGAA, 7(1) 79-98 (2003)     86

**Algorithm** FIRST-OUTPUT-CYCLE($n$, $N$, in, out)
**begin**
    Set out[0][0] through out[0][$N$ - 1] to
        in[0][0] through in[0][$N$ - 1], respectively
    **if** $n$ is even **then begin**
        Set out[0][$N$] through out[0][$N$ + 4] to
            in[0][$N$ - 1] + 2$N$, in[0][N - 1] + 3$N$,
                in[0][0] + 3$N$, in[0][0] + $N$, in[0][$N$ - 1] + $N$, respectively
        count = $N$ + 5
    **end**
    **else begin**
        out[0][$N$] = in[0][$N$ - 1] + $N$
        count = $N$ + 1
    **end**
    **for** $j = N$ - 2 **downto** 1 **do**
    **begin**
        **if** out[0][count - 1] is of the form in[0][$j$ + 1] + $N$ and
            in[0][j] is not an endpoint of a link in the matching **then**
        **begin**
            Set out[0][count] through out[0][count + 2] to in[0][$j$] + $N$,
                in[0][$j$] + 3$N$, in[0][$j$] + 2$N$, respectively
            count = count + 3
        **end**
        **else if** out[0][count - 1] is of the form in[0][$j$ + 1] + 2$N$ and
            in[0][j] is not an endpoint of an edge in the matching **then**
        **begin**
            Set out[0][count] through out[0][count + 2] to in[0][$j$] + 2$N$,
                in[0][$j$] + 3$N$, in[0][$j$] + $N$, respectively
            count = count + 3
         **end**
        **else if** out[0][count - 1] is of the form in[0][$j$ + 1] + $N$ and
            in[0][j] = in[k][$N$ - 1] for some k **then**
        **begin**
            out[0][count] = in[0][j] + N
            count = count + 1
            flag[k] = 0
        **end**
        **else if** out[0][count - 1] is in[0][$j$ + 1] + 2$N$ and
            in[0][$j$] = in[$k$][$N$-1] for some $k$ **then**
        **begin**
            out[0][count] = in[0][j] + 2N
            count = count + 1
            flag[k] = 1
        **end**

Figure 3: Creating the first cycle of the Hamilton decomposition.

```
    else if out[0][count - 1] is of the form in[0][j + 1] + N and
        in[0][j] = in[k][0] for some k then
        if flag[k] = 0 then
        begin
            Set out[0][count] to out[0][count + 4] to
                in[0][j]+ N, in[0][j] + 3N,
                in[k][N - 1] + 3N, in[k][N - 1] + 2N, in[0][j] + 2N,
                respectively
            count = count + 5
        end
        else
        begin
            Set out[0][count] to
                out[0][count + 4] to in[0][j] + N, in[k][N - 1] + N,
                in[k][N - 1] + 3N, in[0][j] + 3N, in[0][j] + 2N, respectively
            count = count + 5
        end
    else if out[0][count - 1] is of the form in[0][j + 1] + 2N and
        in[0][j] = in[k][0] for some k then
        if flag[k] = 1 then
        begin
            Set out[0][count] to
                out[0][count + 4] to in[0][j] + 2N, in[0][j]+ 3N,
                in[k][N - 1] + 3N, in[k][N - 1] + N, in[0][j] + N, respectively
            count = count + 5
        end
        else
        begin
            Set out[0][count] to
                out[0][count + 4] to in[0][j] + 2N, in[k][N - 1] + 2N,
                in[k][N - 1] + 3N, in[0][j] + 3N, in[0][j] + N, respectively
            count = count + 5
        end
    end
    if n is even then
        out[0][count] = in[0][0] + 2N
    else
        Set out[0][4N - 5] through out[0][4N - 1] to in[0][0] + N, in[0][0] + 3N,
            in[0][N - 1] + 3N, in[0][N - 1] + 2N, in[0][0] + 2N, respectively
end
```

Figure 4: Creating the first cycle (Continued).

D. Bass, *Decompositions and Factorizations*, JGAA, 7(1) 79-98 (2003)      88

**Algorithm** SECOND-OUTPUT-CYCLE($n$, $N$, in, out)
**begin**
    Set out[1][0] through out[1][$N$ - 1] to in[0][0] + 3$N$
        through in[0][$N$ - 1] + 3$N$, respectively
    if n is even **then**
    **begin**
        Set out[1][$N$] through out [1][$N$ + 3] to in[0][$N$ - 1] + $N$,
            in[0][$N$ - 1], in[0][0], in[0][0] + $N$, respectively
        **for** count = $N$ + 4 **to** 4N - 2 **do**
            output[1][count] = output[0][5$N$ + 2 - count] XOR 3$N$
        output[1][4$N$ - 1] = input[0][0] + 2$N$
    **end**
    **else**
    **begin**
        Set out[1][$N$] through out[1] to in[0][$N$ - 1] + $N$, in[0][0] + $N$, in[0][0],
            in[0][$N$ - 1], in[0][$N$ - 1] + 2$N$
        **for** count = $N$ + 5 **to** 4$N$ - 1 **do**
            out[1][count] = out[0][count - 4] XOR 3$N$
    **end**
**end**

Figure 5: Creating the second cycle of the Hamilton decomposition

**Algorithm** REMAINING-CYCLES($n$, in, out, $N$)
**begin**
    **for** cycle = 1 **to** n - 1 **do**
        **for** node = 0 **to** N - 1 **do**
        **begin**
            out[cycle + 1][node] = in[cycle][node]
            out[cycle + 1][node + 2$N$] = in[cycle][node] + 3$N$
            out[cycle + 1][node + $N$] = in[cycle][$N$ - 1 - node] +
                ((2 - flag[cycle]) * $N$)
            out[cycle + 1][node + 3$N$] = in[cycle][N - 1 - node] +
                ((1 + flag[cycle]) * $N$)
        **end**
**end**

Figure 6: Creating the remaining cycles of the Hamilton decomposition.

D. Bass, *Decompositions and Factorizations*, JGAA, 7(1) 79-98 (2003)      89

The algorithm SECOND-OUTPUT-CYCLE produces the Hamilton cycle for $Q_6$ {48, 49, 51, 50, 54, 52, 53, 55, 63, 62, 60, 61, 57, 59, 58, 56, 24, 8, 0, 16, 17, 1, 33, 35, 3, 19, 18, 2, 34, 38, 6, 22, 20, 28, 12, 4, 36, 37, 5, 21, 23, 7, 39, 47, 15, 31, 30, 14, 46, 44, 45, 13, 29, 25, 9, 41, 43, 11, 27, 26, 10, 42, 40, 32}. Taking elements of the first output cycle, and performing an exclusive or with $3N$ has the effect of toggling (changing 0 to 1 and 1 to 0) the first two bits of those elements.

Finally, the algorithm REMAINING-CYCLES produces the Hamilton cycle for $Q_6$ {4, 0, 2, 10, 14, 6, 7, 3, 11, 15, 13, 5, 1, 9, 8, 12, 44, 40, 41, 33, 37, 45, 47, 43, 35, 39, 38, 46, 42, 34, 32, 36, 52, 48, 50, 58, 62, 54, 55, 51, 59, 63, 61, 53, 49, 57, 56, 60, 28, 24, 25, 17, 21, 29, 31, 27, 19, 23, 22, 30, 26, 18, 16, 20}.

Variable flag[1] was set to 0 in the course of executing algorithm FIRST-OUTPUT-CYCLE. REMAINING-CYCLES creates four copies of $C_2$ with one edge removed, within $Q_6$, by adding either 0, $N$, $2N$ or $3N$. The Hamilton paths where $N$ and $3N$ are added are traversed in the opposite direction of the Hamilton paths where 0 and $2N$ are added. Since REMAINING-CYCLES uses $n-1$ disjoint Hamilton cycles of $Q_{2n}$ as input, it creates $n-1$ disjoint Hamilton cycles for $Q_{2n+2}$.

The Hamilton decomposition of $Q_{2n+2}$ generated by this algorithm is partially determined by the Hamilton cycle of $Q_{2n}$, which is selected as input[0], the input to FIRST-OUTPUT-CYCLE and SECOND-OUTPUT-CYCLE. It is also partially determined by the edges selected for the orthogonal matching required by the algorithm. It is therefore possible that a large number of distinct Hamilton decompositions of $Q_{2n+2}$ can be generated using this algorithm. For example, four distinct Hamilton decompositions of $Q_4$ were generated using this algorithm.

# 3   Constructing (n/2)-Factorizations of $Q_n$

## 3.1   Constructing Factorizations from Perfect Matchings Derived From Hamilton Decompositions

In Section 1.3, we proposed creating $k$-factorizations of $Q_n$. In this section and the next, we use Hamilton decompositions of $Q_n$ to create $(n/2)$-factorizations of $Q_n$ for certain values of $n$. One method of constructing $(n/2)$-factorizations from Hamilton decompositions is uniting perfect matchings derived from the cycles of the Hamilton decomposition.

Suppose $C_1, C_2, \ldots, C_{n/2}$ is a Hamilton decomposition of $Q_n$. If the links of any cycle were numbered, the even-numbered links would form a perfect matching of $Q_n$, as would the odd-numbered links. $n$ link-disjoint perfect matchings of $Q_n$ can be constructed in this manner. When n mod k = 0, and $\frac{n}{k}$ unions of $k$ perfect matchings are selected, the result is a $k$-factorization of $Q_n$. However, not all unions of $k$ perfect matchings are connected. Table 3 shows the results of a computer search for the $(n/2)$-factorizations of $Q_n$, whose factors had the smallest diameter.

D. Bass, *Decompositions and Factorizations*, JGAA, 7(1) 79-98 (2003)     90

| Network | Degree of Factors | Diameter |
|---------|-------------------|----------|
| $Q_4$ | 2 | 8, 8 |
| $Q_6$ | 3 | 8, 10 |
| $Q_8$ | 4 | 8, 8 |
| $Q_{12}$ | 6 | 12, 12 |

Table 3: Degrees and diameters of factors in factorization of $Q_n$

The 2-factorization of $Q_4$ is the same as a Hamilton decomposition of $Q_4$, where the cycles of 16 nodes have diameter 8. The difference in the diameters of the factors in the 3-factorization of $Q_6$ reflects the fact that the algorithm of Section 2.2 generates two Hamilton cycles for $Q_6$ from one of the Hamilton cycles for $Q_4$, and one Hamilton cycle for $Q_6$ from the other Hamilton cycle for $Q_4$. Table 3 shows that half the links incident to each node of both $Q_8$ and $Q_{12}$ can be removed without increasing their diameters. Furthermore, the removed links themselves form a spanning subnetwork with the same diameter as the original hypercube.

## 3.2 Constructing Factorizations from Hamilton Cycles of Hamilton Decompositions

Observation of Algorithm HAMILTONCOMP1 reveals that the algorithm creates two disjoint Hamilton cycles of $Q_{2n}$ for each Hamilton cycle of $Q_n$. If $C$ is a Hamilton cycle of $Q_n$, then let these Hamilton cycles of $Q_{2n}$ be called the *children* of $C$. Let the *descendants of $C$ at level $k$* be the $2^k$ disjoint Hamilton cycles of $Q_{n2^k}$, obtained by repeatedly applying the algorithm. Let $D(C, k)$ represent the union of the descendants of C at level k.

We observe the following regarding the children of C. One of the children has a pattern of changing the first $n$ bits $2^n$ - 1 times, then changing the last $n$ bits once. This pattern is repeated $2^n$ times. The other child has a pattern of changing the last $n$ bits $2^n$ - 1 times, then changing the first $n$ bits once, a pattern which is repeated $2^n$ times. In general, each descendant of $C$ at level $k$ has a pattern of changing a unique block of $n$ bits $2^n$ - 1 times, then changing some other bit once.

**Lemma 1** *If $C$ is a Hamilton cycle in $Q_n$, then $D(C, k)$ is a spanning subnetwork of $Q_{n2^k}$, with degree $2^{k+1}$ and diameter $2^{n-1+k}$.*

**Proof:** The diameter of C is $2^{n-1}$. Since $D(C, k)$ is the union of $2k$ Hamilton cycles of $Q_{n2^k}$, $D(C, k)$ is a spanning subnetwork of $Q_{n2^k}$. Since $D(C, k)$ is the union of $2^k$ disjoint Hamilton cycles of $Q_{n2^k}$, $D(C, k)$ is of degree $2^{k+1}$. Let $w$ and $w'$ be the labels of two nodes in $D(C, k)$. Algorithm ROUTE provides a route in $D(C, k)$ between $w$ and $w'$, and is shown in Figure 7. It takes at most $2^{n-1}$ nodes to arrange the bits of each of the $2^k$ blocks of $n$ bits, therefore the diameter is $2^{n-1+k}$. $\qquad\square$

D. Bass, *Decompositions and Factorizations*, JGAA, 7(1) 79-98 (2003)        91

**Algorithm** ROUTE($w$, $w$')
**begin**
    **for** $i = 0$ **to** $2k$ - 1 **do**
    **begin**
        Select the descendant of $C$ at level $k$ which changes the $i^{th}$ block of $n$
            bits $2n$ - 1 times
        Set the $i^{th}$ block of $n$ bits of $w$ to the $i^{th}$ block of $n$ bits of $w$' by
            traversing that descendant, using as few nodes as possible
    **end**
**end**

Figure 7: A routing algorithm for $D(C_0, k)$

**Theorem 2** *Let $C$ be the Hamilton cycle for $Q_2$. Then $D(L(C), k)$ and $D(R(C), k)$ are mutually isomorphic.*

**Proof:** Consider any of the cycles in $D(L(C), k)$. If the labels of the nodes of this cycle are reversed, then the labels for one of the cycles in $D(R(C), k)$ are obtained. This is because in the cycles of $D(L(C), k)$, some portion of the first half of the $2^{k+2}$ bits of the labels are changed most rapidly while traversing the cycle, while in the cycles of $D(R(C), k)$, some portion of the second half of the bits of the labels are changed most rapidly. $\square$

**Theorem 3** *For $j \geq 2$, there exists an $2^{j-1}$-factorization of $Q_{2^j}$ where the two factors have diameter $2^{j+1}$.*

**Proof:** Let $A$ and $B$ be any Hamilton decomposition of $Q_4$. $A$ and $B$ are a 2-factorization of $Q_4$, where each of the factors have diameter 8. $D(A, k)$ and $D(B, k)$, k $\geq 1$, form a $2^{k+1}$-factorization of $Q_{4*2^k} = Q_{2^{k+2}}$ , because they are comprised of all the Hamilton cycles of the Hamilton decomposition of $Q_{2^{k+2}}$. $D(A, k)$ and $D(B, k)$ have degree $2^{k+1}$ and diameter $2^{4-1+k} = 2^{k+3}$ by Lemma 1, which is twice the diameter of $Q_{2^{k+2}}$. $\square$

In Section 1.4, we mentioned that in order for a factor to effectively simulate $Q_n$, an embedding of $Q_n$ into the factor must exist with no more than constant dilation.

**Theorem 4** *Let $A$ and $B$ represent the Hamilton cycles of any Hamilton decomposition of $Q_4$. For $j > 2$, The identity embedding embeds $Q_{2^j}$ into $D(A, j - 2)$ and $D(B, j - 2)$ with $\Theta(1)$ dilation.*

**Proof:** Suppose we wish to route in either $D(A, j$ - 2$)$ or $D(B, j$ - 2$)$ between adjacent nodes in $u$ and $v$ differ in some bit in some block of 4 bits. Without loss of generality, we select $D(A, j$ - 2$)$. There exists a descendant of $A$ at level $k$ - 2, which changes the block of 4 bits, containing the bit in which $u$ and $v$

D. Bass, *Decompositions and Factorizations*, JGAA, 7(1) 79-98 (2003)      92

differ, $2^4$ - 1 times before changing another bit. By using that descendant, we can route from $u$ to $v$ in at most $2^4$ - 1 = 15 = $\Theta(1)$ links.      $\square$

In summary, it is possible to create $(n/2)$-factorizations from Hamilton decompositions of $Q_n$ in two ways; by uniting perfect matchings derived from Hamilton cycles, and by uniting the Hamilton cycles themselves. It is currently unknown whether the factors in the factorizations of Section 3.1 have a Hamilton cycle. Since the factors mentioned in this section are composed of the union of Hamilton cycles, they have Hamilton cycles of their own. Furthermore, Hamilton decompositions exist for the factors as well.

## 3.3 Constructing Factorizations of $Q_n$ from Variations on Reduced and Thin Hypercubes

Many reduced-degree variations on hypercubes have been proposed. Some of these variants use the values of portions of the labels of nodes, to determine the dimensions of the links incident to those nodes. Examples include the reduced hypercube [27], and the thin hypercube [7, 18]. The motivation for these networks was to construct a subnetwork of a hypercube with a smaller degree than the original hypercube, and a diameter which is either the same (thin hypercube) or only slightly larger (reduced hypercube) than the original hypercube. However, neither the reduced hypercube nor the thin hypercube are part of a $k$-factorization. We use the idea behind reduced and thin hypercubes, to construct an $(n/2)$-factorization of $Q_n$, where n is even, where the factors have diameter $n + \Theta(1)$. This factorization was first given in [9].

Consider $Q_n$, where $n$ is even. Recall that the label of each node is a bit string $b_0 b_1 \ldots b_{n-1}$. Let the substring $b_0 b_1$ represent the first two bits of the label of a node. Let the parity of a bit string signify the number of 1's in the bit string. Let $F_1$ be a degree $\frac{n}{2}$ spanning subnetwork, defined as follows:

- If a label of a node has the value 00 in $b_0 b_1$, then that node is incident to links in dimensions $n$ - 4 and $n$ - 3

- If a label of a node has the value 01 in $b_0 b_1$, then that node is incident to links in dimensions $n$ - 3 and $n$ - 2

- If a label of a node has the value 11 in $b_0 b_1$, then that node is incident to links in dimensions $n$ - 2 and $n$ - 1

- If a label of a node has the value 10 in $b_0 b_1$, then that node is incident to links in dimensions $n$ - 1 and $n$ - 4

- If a label of a node has even parity in $b_{n-4} b_{n-3} \, b_{n-2} b_{n-1}$, then that node is incident to a link in dimensions 0, 2, $\ldots$, $n$ - 6. Otherwise, that node is incident to a link in dimension 1, 3, $\ldots$, $n$ - 5.

For example, the node with the label 010010, is incident to links of dimensions 3 and 4, because 01 is the value of $b_0 b_1$, and is connected to nodes with

D. Bass, *Decompositions and Factorizations*, JGAA, 7(1) 79-98 (2003)      93

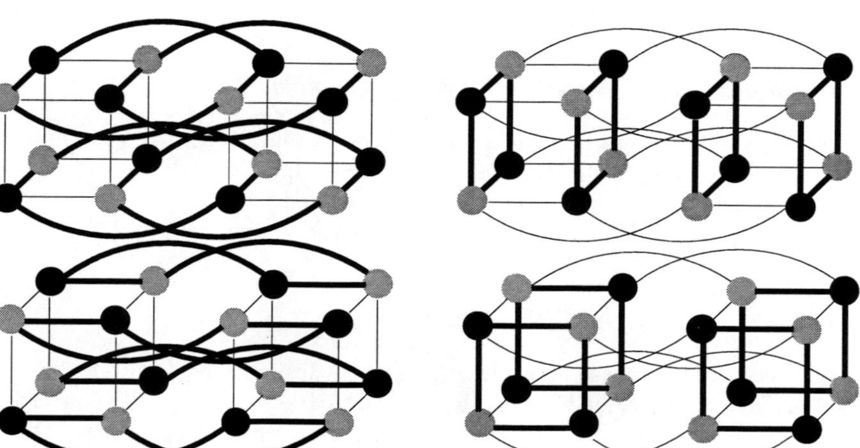

Figure 8: $F_1$, a spanning subnetwork of $Q_6$, for $n = 6$.

labels 010110 and 010000. This node is incident to a link of dimension 1, since 0010, the value of $b_2 b_3 b_4 b_5$, has odd parity. Therefore, this node is connected to the node with label 000010. $F_1$ is shown in Figure 8, with the dimension 0 and 1 links not shown. The bold links in Figure 8 are the links of $F_1$. The black nodes are incident to dimension 0 links, and the gray nodes are incident to dimension 1 links.

Let $F_2$ be a degree $\frac{n}{2}$ spanning subnetwork, defined as follows:

- If a label of a node has the value 00 in $b_0 b_1$, then that node is incident to links in dimensions $n$ - 2 and $n$ - 1

- If a label of a node has the value 01 in $b_0 b_1$, then that node is incident to links in dimensions $n$ - 1 and $n$ - 4

- If a label of a node has the value 11 in $b_0 b_1$, then that node is incident to links in dimensions $n$ - 4 and $n$ - 4

- If a label of a node has the value 10 in $b_0 b_1$, then that node is incident to links in dimensions $n$ - 3 and $n$ - 2

- If a label of a node has odd parity in $b_{n-4} b_{n-3}\, b_{n-2} b_{n-1}$, then that node is incident to a link in dimensions 0, 2, ..., $n$ - 6. Otherwise, that node is incident to a link in dimension 1, 3, ..., $n$ - 5.

$F_1$ and $F_2$ form an $\frac{n}{2}$-factorization of $Q_n$. $F_1$ and $F_2$ contain exactly half of the links in each dimension. If a node $u$ in $Q_n$ is incident to links in dimensions

D. Bass, *Decompositions and Factorizations*, JGAA, 7(1) 79-98 (2003)    94

0, 2, ..., $n$ - 6, then any node $v$ adjacent to $u$ by dimensions $n$ - 4 through $n$ - 1, is incident to links in dimensions 1, 3, ..., $n$ - 5.

**Theorem 5** $F_1$ *and* $F_2$ *are isomorphic to each other.*

**Proof:** $F_1$ and $F_2$ are isomorphic if there exists a mapping g from the nodes of $F_1$ to the nodes of $F_2$, such that for all pairs of adjacent nodes $u$ and $v$ in $F_1$, $g(u)$ is adjacent to $g(v)$ in $F_2$. Let $g(u)$ be obtained from $u$ by toggling $b_0$ and $b_1$. If $u$ and $v$ are adjacent in $F_1$ by some dimension, then $g(u)$ will be adjacent to $g(v)$ in $F_2$ by the same dimension. $\square$

**Theorem 6** *Both* $F_1$ *and* $F_2$ *have diameter* $n + 2$.

**Proof:** Let $u$ and $v$ be two nodes in $F_1$. Without loss of generality, let the value of $b_0b_1$ in $u$ be 00, and let $b_{n-4}b_{n-3}b_{n-2}b_{n-1}$ have even parity. Assume bits $b_2$ through $b_{n-1}$ of $u$ are to be toggled to form bits $b_2$ through $b_{n-1}$ of $v$.

Case I: $b_0b_1$ in $v$ is 00. Toggle $b_0, b_2, \ldots, b_{n-6}$, causing $b_0b_1$ to be 10. Now $b_{n-1}$ can be toggled. Toggle $b_1, b_3, \ldots, b_{n-5}$, causing $b_0b_1$ to be 11. Now $b_{n-2}$ can be toggled. Toggle $b_0$, causing $b_0b_1$ to be 01. Now $b_{n-3}$ can be toggled. Toggle $b_1$, causing $b_0b_1$ to be 00. Now $b_{n-4}$ can be toggled. $b_0$ and $b_1$ were toggled twice, while the remaining bits were toggled once, for a total of $n + 2$ links.

Case II: $b_0b_1$ in $v$ is 01. Toggle $b_0, b_2, \ldots, b_{n-6}$, causing $b_0b_1$ to be 10. Now $b_{n-1}$ and $b_{n-4}$ can be toggled. Toggle $b_0$, causing $b_0b_1$ to be 00. Now $b_{n-2}$ can be toggled. Toggle $b_1, b_3, \ldots, b_{n-5}$, causing $b_0b_1$ to be 11. Now $b_{n-3}$ can be toggled. $b_0$ was toggled twice, while the remaining bits were toggled once, for a total of $n + 1$ links.

Case III: $b_0b_1$ in $v$ is 11. Toggle $b_{n-4}$ and $b_{n-3}$. Toggle $b_0, b_2, \ldots, b_{n-6}$, causing $b_0b_1$ to be 10. Now $b_{n-1}$ can be toggled. Toggle $b_1, b_3, \ldots, b_{n-5}$, causing $b_0b_1$ to be 11. Now $b_{n-2}$ can be toggled. No bit was toggled more than once for a total of $n$ links.

Case IV: $b_0b_1$ in $v$ is 10. Toggle $b_{n-3}$. Toggle $b_1, b_3, \ldots, b_{n-5}$, causing $b_0b_1$ to be 01. Now $b_{n-2}$ can be toggled. Toggle $b_0, b_2, \ldots, b_{n-6}$, causing $b_0b_1$ to be 11. Now $b_{n-1}$ can be toggled. Toggle $b_1$, causing $b_0b_1$ to be 10. Now $b_{n-4}$ can be toggled. $b_1$ was toggled twice, while the remaining bits were toggled once, for a total of $n + 1$ links.

Let $u$ and $v$ be two nodes in $F_2$. Without loss of generality, let the value of $b_0b_1$ in $u$ be 00, and let $b_{n-4}b_{n-3}b_{n-2}b_{n-1}$ have odd parity. Assume bits $b_2$ through $b_{n-1}$ of $u$ are to be toggled to form bits $b_2$ through $b_{n-1}$ of $v$.

Case I: $b_0b_1$ in $v$ is 00. Toggle $b_1, b_3, \ldots, b_{n-5}$, causing $b_0b_1$ to be 01. Now $b_{n-1}$ can be toggled. Toggle $b_0, b_2, \ldots b_{n-6}$, causing $b_0b_1$ to be 11. Now $b_{n-4}$ can be toggled. Toggle $b_1$, causing $b_0b_1$ to be 10. Now $b_{n-3}$ can be toggled. Toggle $b_0$, causing $b_0b_1$ to be 00. Now $b_{n-2}$ can be toggled. $b_0$ and $b_1$ were toggled twice, while the remaining bits were toggled once, for a total of $n + 2$ links.

Case II: $b_0b_1$ in $v$ is 01. Toggle $b_{n-2}$. Toggle $b_0, b_2, \ldots, b_{n-6}$, causing $b_0b_1$ to be 10. Now $b_{n-3}$ can be toggled. Toggle $b_1, b_3, \ldots, b_{n-5}$, causing $b_0b_1$ to be

11. Now $b_{n-4}$ can be toggled. Toggle $b_0$, causing $b_0b_1$ to be 01. Now $b_{n-1}$ can be toggled. $b_0$ was toggled twice, while the remaining bits were toggled once, for a total of $n + 1$ links.

Case III: $b_0b_1$ in $v$ is 11. Toggle $b_{n-2}$ and $b_{n-1}$. Toggle $b_1$, $b_3$, ..., $b_{n-5}$, causing $b_0b_1$ to be 01. Now $b_{n-4}$ can be toggled. Toggle $b_0$, $b_2$, ..., $b_{n-6}$, causing $b_0b_1$ to be 11. Now $b_{n-3}$ can be toggled. No bit was toggled more than once for a total of $n$ links.

Case IV: $b_0b_1$ in $v$ is 10. Toggle $b_1$, $b_3$, ..., $b_{n-5}$, causing $b_0b_1$ to be 01. Now $b_{n-1}$ and $b_{n-4}$ can be toggled. Toggle $b_1$, causing $b_0b_1$ to be 00. Now $b_{n-2}$ can be toggled. Toggle $b_0$, $b_2$, ..., $b_{n-6}$, causing $b_0b_1$ to be 10. Now $b_{n-3}$ can be toggled. $b_0$ was toggled twice, while the remaining bits were toggled once, for a total of $n + 1$ links. $\square$

**Theorem 7** *The identity embedding embeds $Q_n$ into both $F_1$ and $F_2$ with dilation 5.*

**Proof:** Let $u$ and $v$ be two nodes in $Q_n$, which differ in exactly 1 bit. Without loss of generality, let the value of $b_0b_1$ in $u$ be 00, and let $b_{n-4}b_{n-3}b_{n-2}b_{n-1}$ have even parity. We show that the maximum distance in $F_1$ between $u$ and $v$ is 5.

Case I: $u$ and $v$ differ in either $b_0$, $b_2$, ..., $b_{n-6}$, $b_{n-4}$ or $b_{n-3}$. In this case, $u$ and $v$ are adjacent in $F_1$.

Case II: $u$ and $v$ differ in $b_1$, $b_3$, ..., or $b_{n-5}$. In this case, toggle $b_{n-3}$, toggle the bit in which $u$ and $v$ differ, then toggle $b_{n-3}$ again, for a total of three links.

Case III: $u$ and $v$ differ in $b_{n-2}$. In this case, toggle $b_{n-3}$, toggle $b_1$, toggle $b_{n-2}$, toggle $b_{n-3}$, and toggle $b_1$, for a total of five links.

Case IV: $u$ and $v$ differ in $b_{n-1}$. In this case, toggle $b_0$, toggle $b_{n-1}$, toggle $b_{n-4}$, toggle $b_0$, and toggle $b_{n-4}$, for a total of five links.

We now show that the maximum distance in $F_2$ between $u$ and $v$ is also 5.

Case I: $u$ and $v$ differ in either $b_1$, $b_3$, ..., $b_{n-5}$, $b_{n-2}$ or $b_{n-1}$. In this case, $u$ and $v$ are adjacent in $F_1$.

Case II: $u$ and $v$ differ in $b_0$, $b_2$, ..., or $b_{n-6}$. In this case, toggle $b_{n-2}$, toggle the bit in which $u$ and $v$ differ, then toggle $b_{n-2}$ again, for a total of three links.

Case III: $u$ and $v$ differ in $b_{n-4}$. In this case, toggle $b_1$, toggle $b_{n-4}$, toggle $b_{n-1}$, toggle $b_1$, and toggle $b_{n-1}$, for a total of five links.

Case IV: $u$ and $v$ differ in $b_{n-3}$. In this case, toggle $b_{n-2}$, toggle $b_0$, toggle $b_{n-3}$, toggle $b_{n-2}$, and toggle $b_0$, for a total of five links. $\square$

# 4  Conclusions and Future Research

Table 4 summarizes our findings. The consequence of our findings is that the links of $Q_n$ can be partitioned into two factors, each having a diameter close to that of $Q_n$. The factorizations can be produced either from Hamilton decompositions or directly. Furthermore, since there is an embedding of $Q_n$ into these factors with constant dilation, the factors can efficiently simulate the operation of $Q_n$.

| | Best Possible | Section 3.1 | Section 3.2 | Section 3.3 |
|---|---|---|---|---|
| Original Hypercube | $Q_n$ | $Q_4$, $Q_6$, $Q_8$, $Q_{12}$ | $Q_n$, $n = 2^k$ | $Q_n$, $n$ is even |
| Degree of Factors | $\Theta(1)$ | 2, 3, 4, 6 | $\frac{n}{2}$ | $\frac{n}{2}$ |
| Mutually Isomorphic? | Yes | Unknown | Yes | Yes |
| Diameter of factors | n | $\{8, 8\}$, $\{8, 10\}$, $\{8, 8\}$, $\{12, 12\}$ | $2n$ | $n + 2$ |
| Best Dilation | $\Theta(1)$ | Unknown | $\Theta(1)$ | $\Theta(1)$ |
| Hamilton Cycle? | Yes | Unknown | Yes | Unknown |
| Hamilton Decomposition? | Yes | Unknown | Yes | Unknown |

Table 4: Properties of factorizations of hypercubes.

Possible directions for future research into Hamilton decompositions include identifying Hamilton decompositions for other well-known networks, determining if a given Hamilton cycle is part of a Hamilton decomposition and using Hamilton decompositions for solutions to various graph problems [21].

Possible directions for future research into $k$-factorizations include 1) determining the existence of a $k$-factorization of $Q_n$, constructed from perfect matchings, where the factors have diameter $n$, 2) determining if $k$-factorizations of $Q_n$ exist where $k < \frac{n}{2}$, and the diameters of the factors is $n$, 3) finding embeddings of minimal dilation of $Q_n$ into its factors.

## Acknowledgments

The authors thank Brian Alspach of the University of Regina, for his assistance regarding Hamilton decompositions and k-factorizations, Richard Stong of Rice University for providing the Hamilton decomposition algorithm of Section 2.2, and the referees for several useful suggestions.

D. Bass, *Decompositions and Factorizations*, JGAA, 7(1) 79-98 (2003)     97

# References


[1] B. Alspach. private communication.

[2] B. Alspach, J.-C. Bermond, and D. Sotteau. Decomposition into cycles 1: Hamilton decompositions. In G. Hahn, G. Sabidussi, and R. Woodrow, editors, *Cycles and Rays*, pages 9–18. Kluwer Academic Publishers, 1990.

[3] B. Alspach, K. Heinrich, and G. Liu. Orthogonal factorizations of graphs. In J. Dinitz and D. Stinson, editors, *Contemporary Design Theory: A Collection of Surveys*, pages 13–40. John Wiley and Sons, 1992.

[4] J. Aubert and B. Schneider. Dcompositions de la somme cartsienne d'un cycle et l'union de deux cycles hamiltoniens en cycles hamiltoniens. *Discrete Mathematics*, 38:7–16, 1982.

[5] D. Bass and I. H. Sudborough. Link-disjoint regular spanning subnetworks of hypercubes. In *Proceedings of the 2nd IASTED International Conference on Parallel and Distributed Computers and Networks*, pages 182–185, 1998.

[6] D. Bass and I. H. Sudborough. Pancake problems with restricted prefix reversals and some corresponding cayley networks. In *Proceedings of the 27th International Conference on Parallel Processing*, pages 11–17, 1998.

[7] D. Bass and I. H. Sudborough. Vertex-symmetric spanning subnetworks of hypercubes with small diameter. In *Proceedings of the 11th IASTED International Conference on Parallel and Distributed Computers and Systems*, pages 7–12, 1999.

[8] D. Bass and I. H. Sudborough. Removing edges from hypercubes to obtain vertex-symmetric networks with small diameter. *Telecommunications Systems*, 13(1):135–146, 2000.

[9] D. Bass and I. H. Sudborough. Symmetric k-factorizations of hypercubes with factors of small diameter. In *Proceedings of the 6th International Symposium on Parallel Architectures, Algorithms and Networks*, pages 219–224, 2002.

[10] A. Bouabdallah, C. Delorme, and S. Djelloul. Edge deletion preserving the diameter of the hypercube. *Discrete Applied Mathematics*, 63:91–95, 1995.

[11] B. D'Auriol. private communication.

[12] J. Duato. A new theory of deadlock-free adaptive routing in wormhole networks. *IEEE Transactions on Parallel and Distributed Systems*, 4(12):1320–1331, 1993.

[13] S. Dutt and J. P. Hayes. Subcube allocation in hypercube computers. *IEEE Transactions on Computers*, C-40:341–352, 1991.

D. Bass, *Decompositions and Factorizations*, JGAA, 7(1) 79-98 (2003)    98


[14] P. Erdos, P. Hamburger, R. Pippert, and W. Weakley. Hypercube subgraphs with minimal detours. *Journal of Graph Theory*, 23(2):119–128, 1996.

[15] A. Gibbons. *Algorithmic Graph Theory*. Cambridge University Press, 1985.

[16] N. Graham and F. Harary. Changing and unchanging the diameter of a hypercube. *Discrete Applied Mathematics*, 37/38:265–274, 1992.

[17] M.-C. Heydemann. Cayley graphs and interconnection networks. In G. Hahn and G. Sabidussi, editors, *Graph Symmetry*, pages 167–224. Kluwer Academic Publishers, 1997.

[18] O. Karam. Thin hypercubes for parallel computer architecture. In *Proceedings of the 11th IASTED International Conference on Parallel and Distributed Computers and Systems*, pages 66–71, 1999.

[19] S. Latifi. Distributed subcube identification algorithms for reliable hypercubes. *Information Processing Letters*, 38(6):315–321, 1991.

[20] F. T. Leighton. *Introduction to Parallel Algorithms and Architectures: Arrays, Trees, Hypercubes*. Morgan Kaufmann Publishers, 1992.

[21] U. Meyer and J. F. Sibeyn. Time-independent gossiping on full-port tori, research report mpi-i-98-1014. Technical report, Max-Planck Institut fr Informatik, 1998.

[22] B. Monien and I. H. Sudborough. Embedding one interconnection network in another. *Computing Supplement*, 7:257–282, 1990.

[23] F. P. Preparata and J. Vuillemin. The cube connected cycles: A versatile network for parallel computation. *Communications of the ACM*, 24(5):300–309, 1981.

[24] G. Ringel. Über drei kombinatorische probleme am n-dimensionalen würfel und würfelgitter. *Abh. Math. Sem. Univ. Hamburg*, 20:10–19, 1956.

[25] R. Stong. private communication.

[26] R. Stong. Hamilton decompositions of cartesian products of graphs. *Discrete Mathematics*, 90:169–190, 1991.

[27] S. G. Ziavras. Rh: a versatile family of reduced hypercube interconnection networks. *IEEE Transactions on Parallel and Distributed Systems*, 5(11):1210–1220, 1994.

Volume 7:2 (2003)

# Journal of Graph Algorithms and Applications

http://jgaa.info/

vol. 7, no. 2, pp. 101–103 (2003)

## Advances in Graph Algorithms

## Special Issue on Selected Papers from the Seventh International Workshop on Algorithms and Data Structures, WADS 2001

## Guest Editors' Foreword

*Giuseppe Liotta*

Dipartimento di Ingegneria Elettronica e dell'Informazione
Università di Perugia
via G. Duranti 93, 06125 Perugia, Italy
http://www.diei.unipg.it/~liotta/
liotta@diei.unipg.it

*Ioannis G. Tollis*

Department of Computer Science
The University of Texas at Dallas
Richardson, TX 75083-0688, USA
http://www.utdallas.edu//~tollis/
tollis@utdallas.edu

G.Liotta and I.G. Tollis, *Guest Editors' Foreword*, JGAA, 7(2) 101–103 (2003)102

# Introduction

This Special Issue brings together papers based on work presented at the Seventh International Workshop on Algorithms and Data Structures (WADS 2001) which was held August 8-10, 2001, at Brown University Providence, USA. Preliminary versions of the results presented at WADS 2001 have appeared in the conference proceedings published by Springer-Verlag, Lecture Notes in Computer Science, volume 2125.

As editors of this JGAA special issue, we chose to invite papers from WADS 2001 that reflect the broad nature of the workshop, which showcase theoretical contributions as well as experimental work in the field of algorithms and data structures. The issue collects six papers. The first three papers present new algorithms that either revisit basic graph algorithms within new realistic models of computation or that deal with fundamental questions about the combinatorial nature of graphs. The remaining three papers are devoted to the growing area of graph drawing and collect results that have a variety of applications from Web searching to Software Engineering.

All contributions in this Special Issue have gone through a rigorous review process. We thank the authors, the referees, and the editorial board of the journal for their careful work and for their patience, generosity, and support. We hope that we have captured a bit of the dynamic quality that the range of research interests presented at the workshop imparts.

# Scanning the Issue

External memory graph algorithms have received considerable attention lately because massive graphs arise naturally in many applications. Breadth-first search (BFS) and depth-first search (DFS) are the two most fundamental graph searching strategies. They are extensively used in many graph algorithms. Unfortunately no I/O-efficient BFS or DFS-algorithms are known for arbitrary sparse graphs, while known algorithms perform reasonably well on dense graphs. The paper "On External-Memory Planar Depth-First Search" by L. Arge, U. Meyer, L. Toma, and N. Zeh presents two new results on I/O-efficient depth-first search in an important class of sparse graphs, namely undirected embedded planar graphs.

One of the most studied problems in the area of worst-case analysis of NP-hard problems is graph coloring. An early paper by Lawler, dated 1976, contains two results: an algorithm for finding a 3-coloring of a graph (if the graph is 3-chromatic) and an algorithm for finding the chromatic number of an arbitrary graph. Since then, the area has grown and there has been a sequence of papers improving Lawler's 3-coloring algorithm. However, there has been no improvement to Lawler's chromatic number algorithm. The paper by David Eppstein titled "Faster Exact Graph Coloring" provides the first improvement to Lawler's

G.Liotta and I.G. Tollis, *Guest Editors' Foreword*, JGAA, 7(2) 101–103 (2003)103

chromatic number algorithm by showing how to compute the exact chromatic number of a graph in time $O((4/3 + 3^{4/3}/4)^n) \approx 2.4150^n$.

The clique-width of a graph is defined by a composition mechanism for vertex-labeled graphs. Graphs of bounded clique-width are interesting from an algorithmic point of view. A lot of NP-complete graph problems can be solved in polynomial time for graphs of bounded clique-width if the composition of the graph is explicitly given. The paper by W. Espelage, F. Gurski, and E. Wanke titled "Deciding clique-width for graphs of bounded tree-width" shows a linear time algorithm for deciding "clique-width at most $k$" for graphs of bounded tree-width and for some fixed integer $k$.

Methods for ranking World Wide Web resources according to their position in the link structure of the Web are receiving considerable attention, because they provide the first effective means for search engines to cope with the explosive growth and diversification of the Web. The paper titled "Visual Ranking of Link Structures" and authored by U. Brandes and S. Cornelsen proposes a visualization method that supports the simultaneous exploration of a link structure and a ranking of its nodes by showing the result of the ranking algorithm in one dimension and using graph drawing techniques in the remaining one or two dimensions to show the underlying structure. These techniques are useful for the analysis of query results, maintenance of search engines, and evaluation of Web graph models.

The paper "An Approach for Mixed Upward Planarization" by M. Eiglsperger, F. Eppinger, and M. Kaufmann considers the problem of finding a mixed upward planarization of a mixed graph, i.e., a graph with directed and undirected edges. Mixed drawings arise in applications where the edges of the graph can be partitioned into a set which denotes structural information and a another set which does not carry structural information. An example is UML class diagrams arising in software engineering. In these diagrams, the vertices of the graph represent classes in an object-oriented software system, and edges represent relations between these classes. In these diagrams hierarchies of subclasses are drawn upward, whereas relations can have arbitrary directions. The authors present a heuristic approach for this problem which provides good quality and reasonable running time in practice, even for large graphs.

Upward planar drawings are the topic of "Upward Embeddings and Orientations of Undirected Planar Graphs" by W. Didimo and M. Pizzonia. The paper characterizes the set of all upward embeddings and orientations of an embedded planar graph by using a simple flow model, which is related to that described by Bousset to characterize bipolar orientations. The authors take advantage of such a flow model to compute upward orientations with the minimum number of sources and sinks of 1-connected embedded planar graphs. A new algorithm that computes visibility representations of 1-connected planar graphs is also presented.

# Journal of Graph Algorithms and Applications

http://jgaa.info/

vol. 7, no. 2, pp. 105–129 (2003)

# On External-Memory Planar Depth-First Search

*Lars Arge*[1], *Ulrich Meyer*[2], *Laura Toma*[1], *Norbert Zeh*[3]

[1]Department of Computer Science
Duke University, Durham, NC 27708, USA
{large,laura }@cs.duke.edu

[2]Max-Planck-Institut für Informatik
Saarbrücken, Germany
umeyer@mpi-sb.mpg.de

[3]School of Computer Science
Carleton University, Ottawa, Canada
nzeh@scs.carleton.ca

### Abstract

Even though a large number of I/O-efficient graph algorithms have been developed, a number of fundamental problems still remain open. For example, no space- and I/O-efficient algorithms are known for depth-first search or breath-first search in sparse graphs. In this paper, we present two new results on I/O-efficient depth-first search in an important class of sparse graphs, namely undirected embedded planar graphs. We develop a new depth-first search algorithm that uses $O(\text{sort}(N)\log(N/M))$ I/Os, and show how planar depth-first search can be reduced to planar breadth-first search in $O(\text{sort}(N))$ I/Os. As part of the first result, we develop the first I/O-efficient algorithm for finding a simple cycle separator of an embedded biconnected planar graph. This algorithm uses $O(\text{sort}(N))$ I/Os.

Communicated by Giuseppe Liotta and Ioannis G. Tollis: submitted November 2002; revised January 2003.

---

Lars Arge and Laura Toma supported in part by the National Science Foundation through ESS grant EIA–9870734, RI grant EIA–9972879 and CAREER grant CCR–9984099. Ulrich Meyer supported in part by the IST Programme of the EU under contract number IST–1999–14186 (ALCOM–FT). Norbert Zeh supported in part by NSERC and NCE GEOIDE research grants.

L. Arge et al., *On External-Memory Planar DFS*, JGAA, 7(2) 105–129 (2003)106

# 1   Introduction

External memory graph algorithms have received considerable attention lately because massive graphs arise naturally in many applications. Recent web crawls, for example, produce graphs with on the order of 200 million vertices and 2 billion edges [11]. Recent work in web modeling uses depth-first search, breadth-first search, shortest path and connected component computations as primitive routines for investigating the structure of the web [9]. Massive graphs are also often manipulated in Geographic Information Systems (GIS), where many common problems can be formulated as basic graph problems. Yet another example of a massive graph is AT&T's 20 TB phone-call data graph [11]. When working with such massive data sets, the I/O-communication, and not the internal memory computation, is often the bottleneck. I/O-efficient algorithms can thus lead to considerable run-time improvements.

Breadth-first search (BFS) and depth-first search (DFS) are the two most fundamental graph searching strategies. They are extensively used in many graph algorithms. The reason is that in internal memory both strategies are easy to implement in linear time; yet they reveal important information about the structure of the given graph. Unfortunately no I/O-efficient BFS or DFS-algorithms are known for arbitrary sparse graphs, while known algorithms perform reasonably well on dense graphs. The problem with the standard implementations of DFS and BFS is that they decide which vertex to visit next one vertex at a time, instead of predicting the sequence of vertices to be visited. As a result, vertices are visited in a random fashion, which may cause the algorithm to spend one I/O per vertex. Unfortunately it seems that in order to predict the order in which vertices are visited, one essentially has to solve the searching problem at hand. For dense graphs, the I/Os spent on accessing vertices in a random fashion can be charged to the large number of edges in the graph; for sparse graphs, such an amortization argument cannot be applied.

In this paper, we consider an important class of sparse graphs, namely *undirected embedded planar graphs*: A graph $G$ is *planar* if it can be drawn in the plane so that its edges intersect only at their endpoints. Such a drawing is called a *planar embedding* of $G$. If graph $G$ is given together with an embedding, we call it *embedded*. The class of planar graphs is restricted enough, and the structural information provided by a planar embedding is rich enough, to hope for more efficient algorithms than for arbitrary sparse graphs. Several such algorithms have indeed been obtained recently [6, 16, 22, 24]. We develop an improved DFS-algorithm for embedded planar graphs and show that planar DFS can be reduced to planar BFS in an I/O-efficient manner.

L. Arge et al., *On External-Memory Planar DFS*, JGAA, 7(2) 105–129 (2003)107

## 1.1 I/O-Model and Previous Results

We work in the standard disk model proposed in [3]. The model defines the following parameters:

$$
\begin{aligned}
N &= \text{number of vertices and edges } (N = |V| + |E|), \\
M &= \text{number of vertices/edges that can fit into internal memory, and} \\
B &= \text{number of vertices/edges per disk block,}
\end{aligned}
$$

where $2B < M < N$. In an *Input/Output* operation (or simply *I/O*) one block of data is transferred between disk and internal memory. The measure of performance of an algorithm is the number of I/Os it performs. The number of I/Os needed to read $N$ contiguous items from disk is $\text{scan}(N) = \Theta\left(\frac{N}{B}\right)$ (the *linear* or *scanning* bound). The number of I/Os required to sort $N$ items is $\text{sort}(N) = \Theta\left(\frac{N}{B} \log_{M/B} \frac{N}{B}\right)$ (the *sorting* bound) [3]. For all realistic values of $N$, $B$, and $M$, $\text{scan}(N) < \text{sort}(N) \ll N$. Therefore the difference between the running times of an algorithm performing $N$ I/Os and one performing $\text{scan}(N)$ or $\text{sort}(N)$ I/Os can be considerable [8].

I/O-efficient graph algorithms have been considered by a number of authors [1, 2, 4, 5, 6, 10, 12, 14, 16, 19, 20, 21, 22, 23, 24, 26, 30]. We review the previous results most relevant to our work (see Table 1). The best known general DFS-algorithms on undirected graphs use $O\left(|V| + \text{scan}(|E|)\right) \cdot \log_2 |V|)$ [19] or $O\left(|V| + \frac{|V|}{M} \cdot \text{scan}(E)\right)$ I/Os [12]. Since the best known BFS-algorithm for general graphs uses only $O\left(\sqrt{\frac{|V|(|V|+|E|)}{B}} + \text{sort}(|V| + |E|)\right)$ I/Os [23], this suggests that on undirected graphs, DFS may be harder than BFS. For directed graphs, the best known algorithms for both problems use $O\left(\left(|V| + \frac{|E|}{B}\right) \cdot \log_2 \frac{|V|}{B} + \text{sort}(|E|)\right)$ I/Os [10]. For most graph problems $\Omega(\min\{|V|, \text{sort}(|V|)\})$ is a lower bound [5, 12], and, as discussed above, this is $\Omega(\text{sort}(|V|))$ in all practical cases. Still, all of the above algorithms, except the recent BFS-algorithm of [23], use $\Omega(|V|)$ I/Os. For sparse graphs, the same I/O-complexity can

| Problem | General graphs | | Planar graphs | |
|---------|----------------|---|---------------|---|
| DFS | $O\left(|V| + \frac{|V|}{M} \cdot \text{scan}(E)\right)$ | [12] | $O(N)$ | |
| | $O\left((|V| + \text{scan}(|E|)) \cdot \log_2 |V|\right)$ | [19] | | |
| BFS | $O\left(\sqrt{\frac{|V|(|V|+|E|)}{B}} + \text{sort}(|V| + |E|)\right)$ | [23] | $O(N/\sqrt{B})$ | [23] |

Table 1: Best known upper bounds for BFS and DFS on undirected graphs (and linear space).

L. Arge et al., *On External-Memory Planar DFS*, JGAA, 7(2) 105–129 (2003)108

be achieved much easier using the standard internal memory algorithm. Improved algorithms have been developed for special classes of planar graphs. For undirected planar graphs the first $o(N)$ DFS and BFS algorithms were developed by [24]. These algorithms use $O(\frac{N}{\gamma \log B} + \text{sort}(NB^\gamma))$ I/Os and $O(NB^\gamma)$ space, for any $0 < \gamma \le 1/2$. BFS and DFS can be solved in $O(\text{sort}(N))$ I/Os on trees [10, 12] and outerplanar graphs [20]. BFS can also be solved in $O(\text{sort}(N))$ I/Os on $k$-outerplanar graphs [21].

## 1.2  Our Results

The contribution of this paper is two-fold. In Section 3, we present a new DFS-algorithm for undirected embedded planar graphs that uses $O(\text{sort}(N) \log(N/M))$ I/Os and linear space. For most practical values of $B$, $M$ and $N$ this algorithm uses $o(N)$ I/Os and is the first algorithm to do so using linear space. The algorithm is based on a divide-and-conquer approach first proposed in [27]. It utilizes a new $O(\text{sort}(N))$ I/O algorithm for finding a simple cycle in a biconnected planar graph such that neither the subgraph inside nor the one outside the cycle contains more than a constant fraction of the vertices of the graph. Previously, no such algorithm was known.

In Section 4 we obtain an $O(\text{sort}(N))$ I/O reduction from DFS to BFS on undirected embedded planar graphs using ideas similar to the ones in [15]. Contrary to what has been conjectured for general graphs, this shows that for planar graphs, BFS is as hard as DFS. Together with two recent results [6, 22], this implies that planar DFS can be solved in $O(\text{sort}(N))$ I/Os. In particular, Arge *et al.* [6] show that BFS and the single source shortest path problem can be solved in $O(\text{sort}(N))$ I/Os, given a multi-way separator of a planar graph. Maheshwari and Zeh [22] show that such a separator can be computed in $O(\text{sort}(N))$ I/Os.

A preliminary version of this paper appeared in [7].

## 2  Basic Graph Operations

In the algorithms described in Sections 3 and 4 we make use of previously developed $O(\text{sort}(N))$ I/O solutions for a number of basic graph problems. We review these problems below. Most of the basic computations we use require a total order on the vertex set $V$ and on the edge set $E$ of the graph $G = (V, E)$. For the vertex set $V$, such a total order is trivially provided by a unique numbering of the vertices in $G$. For the edge set $E$, we assume that an edge $\{v, w\}$ is stored as the pair $(v, w)$, $v < w$, and we define $(v, w) < (x, y)$ for edges $(v, w)$ and $(x, y)$ in $E$ if either $v < x$ or, $v = x$ and $w < y$. We call this ordering the *lexicographical order* of $E$. Another ordering, which we call the *inverted lexicographical order* of $E$, defines $(v, w) < (x, y)$ if either $w < y$, or $w = y$ and $v < x$.

L. Arge et al., *On External-Memory Planar DFS*, JGAA, 7(2) 105–129 (2003)109

**Set difference:**   Even though strictly speaking set difference is not a graph operation, we often apply it to the vertex and edge sets of a graph. To compute the difference $X \setminus Y$ of two sets $X$ and $Y$ drawn from a total order, we first sort $X$ and $Y$. Then we scan the two resulting sorted lists simultaneously, in a way similar to merging them into one sorted list. However, elements from $Y$ are not copied to the output list, and an element from $X$ is copied only if it does not match the current element in $Y$. This clearly takes $O(\text{sort}(N))$ I/Os, where $N = |X| + |Y|$. We use SETDIFFERENCE as a shorthand for this operation.

**Duplicate removal:**   Given a list $X = \langle x_1, \ldots, x_N \rangle$ with some entries potentially occurring more than once, the DUPLICATEREMOVAL operation computes a list $Y = \langle y_1, \ldots, y_q \rangle$ such that $\{x_1, \ldots, x_N\} = \{y_1, \ldots, y_q\}$, $y_j = x_{i_j}$, for indices $i_1 < \cdots < i_q$, and $x_l \neq y_j$, for $1 \leq l < i_j$. That is, list $Y$ contains the first occurrences of all elements in $X$ in sorted order. (Alternatively we may require list $Y$ to store the last occurrences of all elements in $X$.) To compute $Y$ in $O(\text{sort}(N))$ I/Os, we scan $X$ and replace every element $x_i$ with the pair $(x_i, i)$. We sort the resulting list $X'$ lexicographically. Now we scan list $X'$ and discard for every $x$, all pairs that have $x$ as their first component, except the first such pair. List $Y$ can now be obtained by sorting the remaining pairs $(x, y)$ by their indices $y$ and scanning the resulting list to replace every pair $(x, y)$ with the single element $x$.

**Computing incident edges:**   Given a set $V$ of vertices and a set $E$ of edges, the INCIDENTEDGES operation computes the set $E'$ of edges $\{v, w\} \in E$ such that $v \in V$ and $w \notin V$. To compute $E'$ in $O(\text{sort}(N))$ I/Os where $N = |V| + |E|$, we sort $V$ in increasing order and $E$ in lexicographical order. We scan $V$ and $E$ and mark every edge in $E$ that has its first endpoint in $V$. We sort $E$ in inverted lexicographical order and scan $V$ and $E$ again to mark every edge in $E$ that has its second endpoint in $V$. Finally we scan $E$ and remove all edges that have not been marked or have been marked twice.

**Copying labels from edges to vertices:**   Given a graph $G = (V, E)$ and a labeling $\lambda : E \to X$ of the edges in $E$, the SUMEDGELABELS operation computes a labeling $\lambda' : V \to X$ of the vertices in $V$, where $\lambda'(v) = \bigoplus_{e \in E_v} \lambda(e)$, $E_v$ is the set of edges incident to $v$, and $\oplus$ is any given associative and commutative operator on $X$. To compute labeling $\lambda'$ in $O(\text{sort}(N))$ I/Os, we sort $V$ in increasing order and $E$ lexicographically. We scan $V$ and $E$ and compute a label $\lambda''(v) = \bigoplus_{e \in E'_v} \lambda(e)$, for each $v$, where $E'_v$ is the set of edges that have $v$ as their first endpoint. Then we sort $E$ in inverted lexicographical order and scan $V$ and $E$ to compute the label $\lambda'(v) = \lambda''(v) + \bigoplus_{e \in E''_v} \lambda(e)$, for each $v$, where $E''_v$ is the set of edges that have $v$ as their second endpoint.

**Copying labels from vertices to edges:**   Given a graph $G = (V, E)$ and a labeling $\lambda : V \to X$ of the vertices in $V$, the COPYVERTEXLABELS operation computes a labeling $\lambda' : E \to X \times X$, where $\lambda'(\{v, w\}) = (\lambda(v), \lambda(w))$. We can

L. Arge et al., *On External-Memory Planar DFS*, JGAA, 7(2) 105–129 (2003)110

compute this labeling in $O(\text{sort}(N))$ I/Os using a procedure similar to the one implementing operation SUMEDGELABELS.

**Algorithms for lists and trees:**  Given a list stored as an unordered sequence of edges $\{(u, next(u))\}$, *list ranking* is the problem of determining for every vertex $u$ in the list, the number of edges from $u$ to the end of the list. List ranking can be solved in $O(\text{sort}(N))$ I/Os [4, 12] using techniques similar to the ones used in efficient parallel list ranking algorithms [18]. Using list ranking and PRAM techniques, $O(\text{sort}(N))$ I/O algorithms can also be developed for most problems on trees, including Euler tour computation, BFS and DFS-numbering, and lowest common ancestor queries ($Q$ queries can be answered in $O(\text{sort}(Q + N))$ I/Os) [12]. Any computation that can be expressed as a "level-by-level" traversal of a tree, where the value of every vertex is computed either from the values of its children or from the value of its parent, can also be carried out in $O(\text{sort}(N))$ I/Os [12].

**Algorithms for planar graphs:**  Even though no $O(\text{sort}(N))$ I/O algorithms for BFS or DFS in planar graphs have been developed, there exist $O(\text{sort}(N))$ I/O solutions for a few other problems on planar graphs, namely computing the connected and biconnected components, spanning trees and minimum spanning trees [12]. All these algorithms are based on edge-contraction, similar to the PRAM algorithms for these problems [13, 29]. We make extensive use of these algorithms in our DFS-algorithms.

# 3 Depth-First Search using Simple Cycle Separators

## 3.1 Outline of the Algorithm

Our new algorithm for computing a DFS-tree of an embedded planar graph in $O(\text{sort}(N)\log(N/M))$ I/Os and linear space is based on a divide-and-conquer approach first proposed in [27]. First we introduce some terminology used in this section.

A *cutpoint* of a graph $G$ is a vertex whose removal disconnects $G$. A connected graph $G$ is *biconnected* if it does not have any cutpoints. The *biconnected components* or *bicomps* of a graph are its maximal biconnected subgraphs. A *simple cycle $\alpha$-separator* $C$ of an embedded planar graph $G$ is a simple cycle such that neither the subgraph inside nor the one outside the cycle contains more than $\alpha|V|$ vertices. Such a cycle is guaranteed to exist only if $G$ is biconnected.

The main idea of our algorithm is to partition $G$ using a simple cycle $\alpha$-separator $C$, recursively compute DFS-trees for the connected components of $G \setminus C$, and combine them to obtain a DFS-tree for $G$. If each recursive step can be carried out in $O(\text{sort}(N))$ I/Os, it follows that the whole algorithm takes $O(\text{sort}(N)\log(N/M))$ I/Os because the sizes of the subgraphs of $G$ we recurse

L. Arge et al., *On External-Memory Planar DFS*, JGAA, 7(2) 105–129 (2003)111

on are geometrically decreasing, and we can stop the recursion as soon as the current graph fits into main memory. Below we discuss our algorithm in more detail, first assuming that the graph is biconnected.

Given a biconnected embedded planar graph $G$ and some vertex $s \in G$, we construct a DFS-tree $T$ of $G$ rooted at $s$ as follows (see Figure 1):

1. *Compute a simple cycle $\frac{2}{3}$-separator $C$ of $G$.*

   In Section 3.2, we show how to do this in $O(\text{sort}(N))$ I/Os.

2. *Find a path $P$ from $s$ to some vertex $v$ in $C$.*

   To do this, we compute an arbitrary spanning tree $T'$ of $G$, rooted at $s$, and find a vertex $v \in C$ whose distance to $s$ in $T'$ is minimal. Path $P$ is the path from $s$ to $v$ in $T'$. The spanning tree $T'$ can be computed in $O(\text{sort}(N))$ I/Os [12]. Given tree $T'$, vertex $v$ can easily be found in $O(\text{sort}(N))$ I/Os using a BFS-traversal [12] of $T'$. Path $P$ can then be identified by extracting all ancestors of $v$ in $T'$. This takes $O(\text{sort}(N))$ I/Os using standard tree computations [12].

3. *Extend $P$ to a path $P'$ containing all vertices in $P$ and $C$.*

   To do this, we identify one of the two neighbors of $v$ in $C$. Let $w$ be this neighbor, and let $C'$ be the path obtained by removing edge $\{v, w\}$ from $C$. Then path $P'$ is the concatenation of paths $P$ and $C'$. This computation can easily be carried out in $O(\text{scan}(N))$ I/Os: First we scan the edge list of $C$ and remove the first edge we find that has $v$ as an endpoint. Then we concatenate the resulting edge list of $C'$ and the edge list of $P$.

4. *Compute the connected components $H_1, \ldots, H_k$ of $G \setminus P'$. For each component $H_i$, find the vertex $v_i \in P'$ furthest away from $s$ along $P'$ such that there is an edge $\{u_i, v_i\}$, $u_i \in H_i$.*

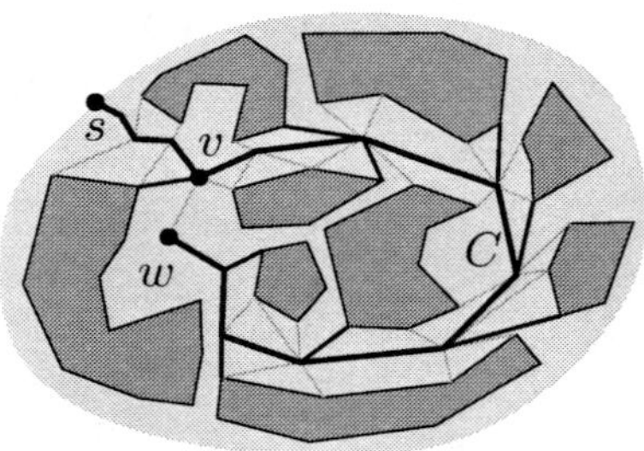


Figure 1: The path $P'$ is shown in bold. The connected components of $G \setminus P'$ are shaded dark gray. Medium edges are edges $\{u_i, v_i\}$. Light edges are non-tree edges.

L. Arge et al., *On External-Memory Planar DFS*, JGAA, 7(2) 105–129 (2003)112

The connected components $H_1, \ldots, H_k$ can be computed in $O(\text{sort}(N))$ I/Os [12]. We find vertices $v_1, \ldots, v_k$ in $O(\text{sort}(N))$ I/Os as follows: First we mark every vertex in $P'$ with its distance from $s$ along $P'$. These distances can be computed in $O(\text{sort}(N))$ I/Os using the Euler tour technique and list ranking [12]. Then we apply operation INCIDENTEDGES to $V(P')$ and $E(G)$, to find all edges in $E(G) \setminus E(P')$ incident to $P'$. We sort the resulting edge set so that edge $\{v, w\}$, $v \in H_i$, $w \in P'$, precedes edge $\{x, y\}$, $x \in H_j$, $y \in P'$, if either $i < j$ or $i = j$ and the distance from $s$ to $w$ is no larger than the distance from $s$ to $y$. Ties are broken arbitrarily. We scan the resulting list and extract for every $H_i$ the last edge $\{u_i, v_i\}$, $u_i \in H_i$, in this list.

5. *Recursively compute DFS-trees $T_1, \ldots, T_k$ for components $H_1, \ldots, H_k$, rooted at vertices $u_1, \ldots, u_k$, and construct a DFS-tree $T$ for $G$ as the union of trees $T_1, \ldots, T_k$, path $P'$, and edges $\{u_i, v_i\}$, $1 \le i \le k$. Note that components $H_1, \ldots, H_k$ are not necessarily biconnected. Below we show how to deal with this case.*

To prove the correctness of our algorithm, we have to show that $T$ is indeed a DFS-tree for $G$. To do this, the following classification of the edges in $E(G) \setminus E(T)$ is useful: An edge $e = (u, v)$ in $E(G) \setminus E(T)$ is called a *back-edge* if $u$ is an ancestor of $v$ in $T$, or vice versa; otherwise $e$ is called a *cross-edge*. In [28] it is shown that a spanning tree $T$ of a graph $G$ is a DFS-tree of $G$ if and only if all edges in $E(G) \setminus E(T)$ are back-edges.

**Lemma 1** *The tree $T$ computed by the above algorithm is a DFS-tree of $G$.*

**Proof:** It is easy to see that $T$ is a spanning tree of $G$. To prove that $T$ is a DFS-tree, we have to show that all non-tree edges in $G$ are back-edges. First note that there are no edges between components $H_1, \ldots, H_k$. All non-tree edges with both endpoints in a component $H_i$ are back-edges because tree $T_i$ is a DFS-tree of $H_i$. All non-tree edges with both endpoints in $P'$ are back-edges because $P'$ is a path. For every non-tree edge $\{v, w\}$ with $v \in P'$ and $w \in H_i$, $w$ is a descendant of the root $u_i$ of the DFS-tree $T_i$. Tree $T_i$ is connected to $P'$ through edge $\{v_i, u_i\}$. By the choice of vertex $v_i$, $v$ is an ancestor of $v_i$ and thus an ancestor of $u_i$ and $w$. Hence, edge $\{v, w\}$ is a back-edge. $\qquad\square$

In the above description of our algorithm we assume that $G$ is biconnected. If this is not the case, we find the bicomps of $G$, compute DFS-trees for all bicomps, and join these trees at the cutpoints of $G$. More precisely, we compute the *bicomp-cutpoint-tree* $T_G$ of $G$ containing all cutpoints of $G$ and one vertex $v(C)$ per bicomp $C$ (see Figure 2). There is an edge between a cutpoint $v$ and a bicomp vertex $v(C)$ if $v$ is contained in $C$. We choose the bicomp vertex $v(C_r)$ corresponding to a bicomp $C_r$ that contains vertex $s$ as the root of $T_G$. The *parent cutpoint* of a bicomp $C \neq C_r$ is the parent $p(v(C))$ of $v(C)$ in $T_G$. $T_G$ can be constructed in $O(\text{sort}(N))$ I/Os, using the algorithms discussed in Section 2. We compute a DFS-tree of $C_r$ rooted at vertex $s$. For every bicomp $C \neq C_r$,

L. Arge et al., *On External-Memory Planar DFS*, JGAA, 7(2) 105–129 (2003)113

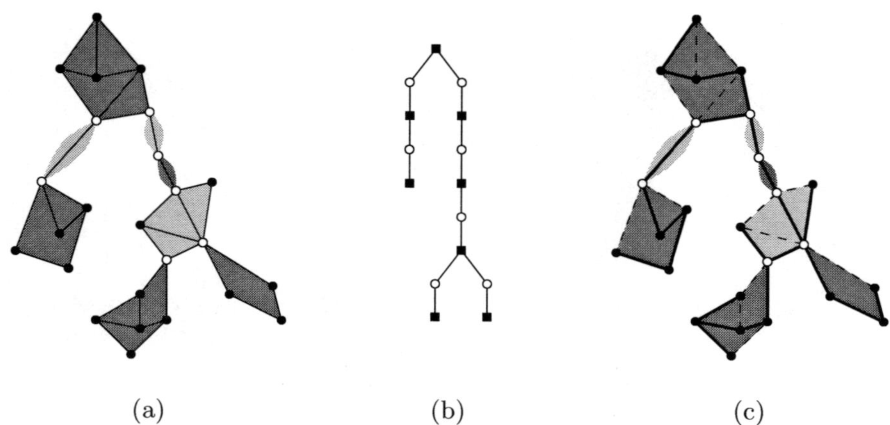

(a)          (b)          (c)

Figure 2: (a) A connected graph $G$ with its bicomps shaded. Cutpoints are hollow. Other vertices are solid. (b) The bicomp-cutpoint-tree of $G$. Bicomp vertices are squares. (c) The DFS tree of $G$ obtained by "gluing together" DFS-trees of its bicomps.

we compute a DFS-tree rooted at the parent cutpoint of $C$. The union of the resulting DFS-trees (see Figure 2c) is a DFS-tree for $G$ rooted at $s$, since there are no edges between different bicomps. Thus, we obtain our first main result.

**Theorem 1** *A DFS-tree of an embedded planar graph can be computed in $O(sort(N) \log(N/M))$ I/O operations and linear space.*

## 3.2   Finding a Simple Cycle Separator

In this section, we show how to compute a simple cycle $\frac{2}{3}$-separator of an embedded biconnected planar graph, utilizing ideas similar to the ones used in [17, 25]. As in the previous section, we start by introducing the necessary terminology.

Given an embedded planar graph $G$, the *faces* of $G$ are the connected regions of $\mathbb{R}^2 \setminus G$. We use $F$ to denote the set of faces of $G$. The *boundary* of a face $f$ is the set of edges contained in the closure of $f$. For a set $F'$ of faces of $G$, let $G_{F'}$ be the subgraph of $G$ defined as the union of the boundaries of the faces in $F'$ (see Figure 3a). The *complement* $\overline{G}_{F'}$ of $G_{F'}$ is the graph obtained as the union of the boundaries of all faces in $F \setminus F'$ (see Figure 3b). The *boundary* of $G_{F'}$ is the intersection between $G_{F'}$ and its complement $\overline{G}_{F'}$ (see Figure 3c). The *dual* $G^*$ of $G$ is the graph containing one vertex $f^*$ per face $f \in F$, and an edge between two vertices $f_1^*$ and $f_2^*$ if faces $f_1$ and $f_2$ share an edge (see Figure 3d). We use $v^*$, $e^*$, and $f^*$ to refer to the face, edge, and vertex that is dual to vertex $v$, edge $e$, and face $f$, respectively. The dual $G^*$ of a planar graph $G$ is planar and can be computed in $O(sort(N))$ I/Os [16].

The idea in our algorithm is to find a set of faces $F' \subset F$ such that the

L. Arge et al., *On External-Memory Planar DFS*, JGAA, 7(2) 105–129 (2003)114

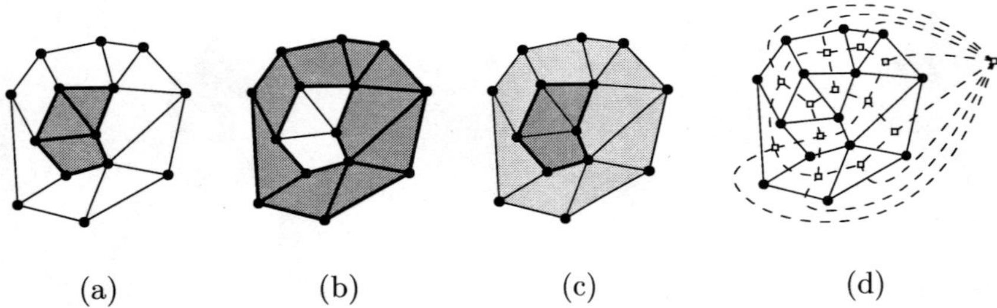

(a)            (b)            (c)            (d)

Figure 3: (a) A graph $G$ and a set $F'$ of faces shaded gray. The edges in $G_{F'}$ are shown in bold. (b) The shaded faces are the faces in $F \setminus F'$. The bold edges are in $\bar{G}_{F'}$. (c) The boundary of $G_{F'}$ shown in bold. (d) The dual of $G$ represented by hollow squares and dashed edges.

boundary of $G_{F'}$ is a simple cycle $\frac{2}{3}$-separator. The main difficulty is to ensure that the boundary of $G_{F'}$ is a simple cycle. We compute $F'$ as follows:

1. **Checking for heavy faces:** We check whether there is a single face whose boundary has size at least $\frac{|V|}{3}$ (Figure 4a). If we find such a face, we report its boundary as the separator $C$, as there are no vertices inside $C$ and at most $\frac{2}{3}|V|$ vertices outside $C$.

2. **Checking for heavy subtrees:** If there is no heavy face, we compute a spanning tree $T^*$ of the dual $G^*$ of $G$, and choose an arbitrary vertex $r$ as its root. Every vertex $v \in T^*$ defines a subtree $T^*(v)$ of $T^*$ that contains $v$ and all its descendants. The vertices in this subtree correspond to a set of faces in $G$ whose boundaries define a graph $G(v)$. Below we show that the boundary of $G(v)$ is a simple cycle. We try to find a vertex $v$ such that $\frac{1}{3}|V| \leq |G(v)| \leq \frac{2}{3}|V|$, where $|G(v)|$ is the number of vertices in $G(v)$ (Figure 4b). If we succeed, we report the boundary of $G(v)$ as the separator $C$.

3. **Splitting a heavy subtree:** If Steps 1 and 2 fail to produce a simple cycle $\frac{2}{3}$-separator of $G$, we are left in a situation where for every leaf $l \in T^*$ (face in $G$), we have $|G(l)| < \frac{1}{3}|V|$; for the root $r$ of $T^*$, we have $|G(r)| = |V|$; and for every other vertex $v \in T^*$, either $|G(v)| < \frac{1}{3}|V|$ or $|G(v)| > \frac{2}{3}|V|$. Thus, there has to be a vertex $v$ with $|G(v)| > \frac{2}{3}|V|$ and $|G(w_i)| < \frac{1}{3}|V|$, for all children $w_1, \ldots, w_k$ of $v$. We show how to compute a subgraph $G'$ of $G(v)$ consisting of the boundary of the face $v^*$ and a subset of the graphs $G(w_1), \ldots, G(w_k)$ such that $\frac{1}{3}|V| \leq |G'| \leq \frac{2}{3}|V|$, and the boundary of $G'$ is a simple cycle (Figure 4c).

Below we describe our algorithm in detail and show that all of the above steps can be carried out in $O(\text{sort}(N))$ I/Os. This proves the following theorem.

**Theorem 2** *A simple cycle $\frac{2}{3}$-separator of an embedded biconnected planar graph can be computed in $O(\text{sort}(N))$ I/O operations and linear space.*

L. Arge et al., *On External-Memory Planar DFS*, JGAA, 7(2) 105–129 (2003)115

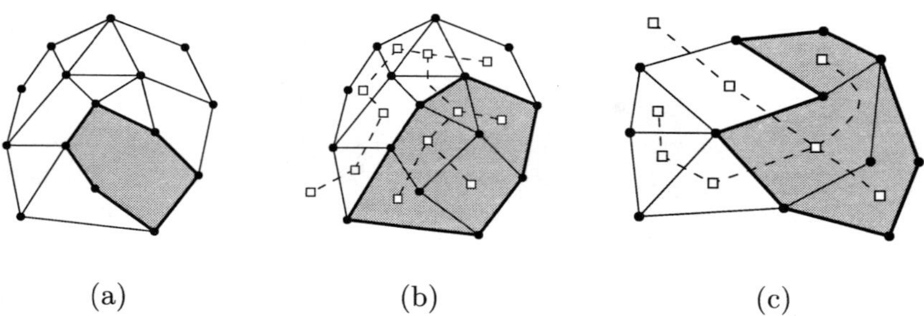

(a)          (b)          (c)

Figure 4: (a) A heavy face. (b) A heavy subtree. (c) Splitting a heavy subtree.

### 3.2.1 Checking for Heavy Faces

In order to check whether there exists a face $f$ in $G$ with a boundary of size at least $\frac{1}{3}|V|$, we represent each face of $G$ as a list of vertices along its boundary. Computing such a representation takes $O(\text{sort}(N))$ I/Os [16]. Then we scan these lists to see whether any of them has length at least $\frac{1}{3}|V|$. In total, this step uses $O(\text{sort}(N))$ I/Os.

### 3.2.2 Checking for Heavy Subtrees

First we prove that the boundary of $G(v)$ defined by the vertices in $T^*(v)$ is a simple cycle. Consider a subset $F'$ of the faces of an embedded planar graph $G$, and let $H$ be the subgraph of $G$ that is the union of the boundaries of the faces in $F'$. Let $H^*$ be the subgraph of the dual $G^*$ of $G$ induced by the vertices that are dual to the faces in $F'$. We call $H^*$ the *dual* of $H$. We call graph $H$ *uniform* if $H^*$ is connected. Since for every vertex $v \in T^*$, $T^*(v)$ and $T^* \setminus T^*(v)$ are both connected, $G(v)$ and its complement $\overline{G(v)}$ are both uniform. Using the following lemma, this implies that the boundary of $G(v)$ is a simple cycle.

**Lemma 2 (Smith [27])** *Let $G'$ be a subgraph of a biconnected planar graph $G$. The boundary of $G'$ is a simple cycle if and only if $G'$ and its complement are both uniform.*

The main difficulty in finding a vertex $v \in T^*$ such that $\frac{1}{3}|V| \leq |G(v)| \leq \frac{2}{3}|V|$ is the computation of the sizes $|G(v)|$ of graphs $G(v)$ for all vertices $v \in T^*$. Once this information has been computed, a single scan of the vertex set of $T^*$ is sufficient to decide whether there is a vertex $v \in T^*$ with $\frac{1}{3}|V| \leq |G(v)| \leq \frac{2}{3}|V|$. As $|T^*| = O(N)$, this takes $O(\text{scan}(N))$ I/Os. Given vertex $v$, the vertices in $T^*(v)$ can be reported in $O(\text{sort}(N))$ I/Os using standard tree computations [12]. Given these vertices, we can apply operation INCIDENTEDGES to find the set $E'$ of edges in $G^*$ with exactly one endpoint in $T^*(v)$. The set $\{e^* : e \in E'\}$ is the boundary of $G(v)$. All that remains is to describe how to compute the sizes of graphs $G(v)$ I/O-efficiently.

L. Arge et al., *On External-Memory Planar DFS*, JGAA, 7(2) 105–129 (2003)116

Assume that every vertex $v \in T^*$ stores the number $|v^*|$ of vertices on the boundary of face $v^*$. The basic idea in our algorithm for computing $|G(v)|$ is to sum $|w^*|$ for all descendants $w$ of $v$ in $T^*$. This can be done by processing $T^*$ level-by-level, bottom-up, and computing for every vertex $v$, the value $|G(v)| = |v^*| + \sum_{i=1}^{k} |G(w_i)|$, where $w_1, \ldots w_k$ are the children of $v$. By doing this, however, we count certain vertices several times. Below we discuss how to modify the above idea in order to make sure that every vertex is counted only once.

We define the lowest common ancestor $LCA(e)$ of an edge $e \in G$ to be the lowest common ancestor of the endpoints of its dual edge $e^*$ in $T^*$. For a vertex $v \in T^*$, we define $E(v)$ to be the set of edges in $G$ whose duals have $v$ as their lowest common ancestor. For a vertex $v$ with children $w_1, w_2, \ldots w_k$, $E(v)$ consists of all edges on the boundary between $v^*$ and graphs $G(w_1), G(w_2), \ldots, G(w_k)$, as well as the edges on the boundary between graphs $G(w_1), G(w_2), \ldots, G(w_k)$. Every endpoint of such an edge is contained in more than one subgraph of $G(v)$, and thus counted more than once by the above procedure. The idea in our modification is to define an *overcount* $c_{v,u}$, for every endpoint $u$ of an edge in $E(v)$, which is one less than the number of times vertex $u$ is counted in the sum $S = |v^*| + \sum_{i=1}^{k} |G(w_i)|$. The sum of these overcounts is then subtracted from $S$ to obtain the correct value of $|G(v)|$.

Let $V(v)$ denote the set of endpoints of edges in $E(v)$. A vertex $u \in V(v)$ is counted once for each subgraph in $\{v^*, G(w_1), G(w_2), \ldots, G(w_k)\}$ having $u$ on its boundary. Let $l$ be the number of edges in $E(v)$ incident to $u$. Each such edge is part of the boundary between two of the subgraphs $v^*, G(w_1), \ldots, G(w_k)$. Thus, if $u$ is an internal vertex of $G(v)$ (i.e., not on its boundary), there are $l$ such subgraphs, and $u$ is counted $l$ times (see vertex $u_1$ in Figure 5). Otherwise, if $u$ is on the boundary of $G(v)$, it follows from the uniformity of $G(v)$ and $\bar{G}(v)$ that two of the edges in $G(v)$ incident to $v$ are on the boundary of $G(v)$ (see vertex $u_2$ in Figure 5). Hence, $l+1$ of the subgraphs $v^*, G(w_1), \ldots, G(w_k)$ contain $u$, and $u$ is counted $l+1$ times. Therefore the overcount $c_{v,u}$ for vertex $u \in V(v)$ is defined as follows:

$$c_{v,u} = \begin{cases} l - 1 & \text{if all edges incident to } u \text{ have their LCA in } T^*(v) \\ l & \text{otherwise} \end{cases}$$

We can now compute $|G(v)|$ using the following lemma.

**Lemma 3** *For every vertex $v \in T^*$,*

$$|G(v)| = \begin{cases} |v^*| + \sum_{i=1}^{k} |G(w_i)| - \sum_{u \in V(v)} c_{v,u} & \text{if } v \text{ is an internal vertex} \\ & \text{with children } w_1, \ldots, w_k. \\ |v^*| & \text{if } v \text{ is a leaf} \end{cases}$$

**Proof:** The lemma obviously holds for the leaves of $T^*$. In order to prove the lemma for an internal vertex $v$ of $T^*$, we have to show that we count every

L. Arge et al., *On External-Memory Planar DFS*, JGAA, 7(2) 105–129 (2003)117

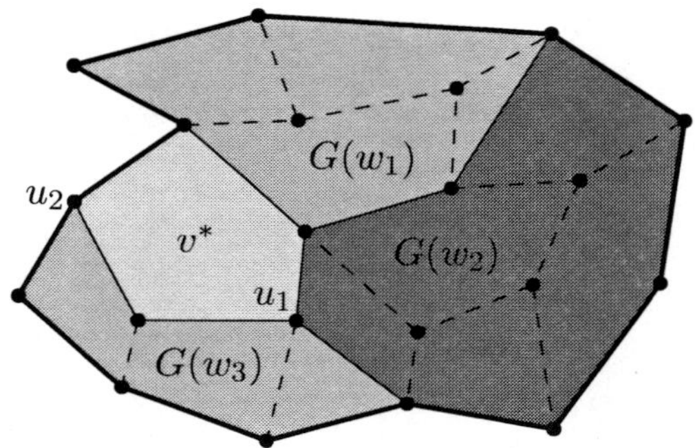


Figure 5: The boundary of graph $G(v)$ is shown in bold. The LCAs of these edges are ancestors of $v$ in $T^*$. Thin solid edges are those on the boundary between graphs $v^*$, $G(w_1)$, $G(w_2)$, and $G(w_3)$. The LCA of these edges in $T^*$ is $v$. The LCAs of dashed edges are descendants of $v^*$. Vertex $u_1$ is counted three times in the sum $S = |v^*| + |G(w_1)| + |G(w_2)| + |G(w_3)|$ because it is in $v^*$, $G(w_2)$, and $G(w_3)$. It has three incident edges with LCA $v$, and all edges incident to $u_1$ have their LCA in $T^*(v)$. Hence, its overcount $c_{v,u_1}$ is 2, so that by subtracting $c_{v,u_1}$ from $S$, vertex $u_1$ is counted only once. Vertex $u_2$ is counted twice in $S$, because it is in $v^*$ and $G(w_3)$. It has one incident edge with LCA $v$, but not all of its incident edges have their LCA in $T^*(v)$ (it is on the boundary of $G(v)$). Hence, its overcount $c_{v,u_2}$ is one, so that by subtracting $c_{v,u_2}$ from $S$, vertex $u_2$ is counted only once.

vertex in $G(v)$ exactly once in the sum $|v^*| + \sum_{i=1}^{k} |G(w_i)| - \sum_{u \in V(v)} c_{v,u}$. A vertex in $G(v) \setminus V(v)$ is counted once, since it is contained in only one of the graphs $v^*, G(w_1), \ldots, G(w_k)$. A vertex $u \in V(v)$ is included in the sum $|v^*| + \sum_{i=1}^{k} |G(w_i)|$ once for every graph $v^*$ or $G(w_i)$ containing it. If all edges incident to $u$ have their LCA in $T^*(v)$, then all faces around $u$ are in $G(v)$. That is, $G(v)$ is an internal vertex of $G(v)$. As argued above, $u$ is counted $l$ times in this case, where $l$ is the number of edges in $E(v)$ incident to $u$. Thus, it is overcounted $l - 1$ times, and we obtain the exact count by subtracting $c_{v,u} = l - 1$. Otherwise, $u$ is on the boundary of $G(v)$ and, as argued above, it is counted $l + 1$ times. Thus, we obtain the correct count by subtracting $c_{v,u} = l$.

$\square$

We are now ready to show how to compute $|G(v)|$, for all $v \in T^*$, I/O-efficiently. Assuming that every vertex $v \in T^*$ stores $|v^*|$ and $c_v = \sum_{u \in V(v)} c_{v,u}$, the graph sizes $|G(v)|$, $v \in T^*$, can be computed in $O(\text{sort}(N))$ I/Os basically as described earlier: For the leaves of $T^*$, we initialize $|G(v)| = |v^*|$. Then we process $T^*$ level by level, from the leaves towards the root, and compute for every internal vertex $v$ with children $w_1, \ldots, w_k$, $|G(v)| = |v^*| + \sum_{i=1}^{k} |G(w_i)| - c_v$. It remains to show how to compute $c_v$, $v \in T^*$.

L. Arge et al., *On External-Memory Planar DFS*, JGAA, 7(2) 105–129 (2003)118

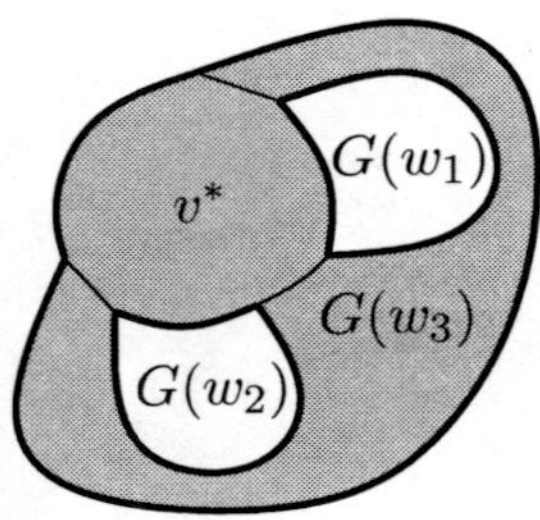

Figure 6: The boundary of $v^* \cup G(w_3)$ is not a simple cycle.

By the definition of overcounts $c_{v,u}$, $c_v = 2|E(v)| - |V'(v)|$, where $V'(v)$ is the set of vertices $u \in V(v)$ so that all edges in $G$ incident to $u$ have their LCAs in $T^*(v)$. To compute sets $E(v)$, for all $v \in T^*$, we compute the LCAs of all edges in $G$. As discussed in Section 2, we can do this in $O(\text{sort}(N))$ I/Os [12] because there are $O(N)$ edges in $G$ and $O(N)$ vertices in $T^*$. By sorting the edges of $G$ by their LCAs, we obtain the concatenation of lists $E(v)$, $v \in T^*$, which we scan to determine $|E(v)|$, for all $v \in T^*$. To compute sets $V'(v)$, for all $v \in T^*$, we apply operation SUMEDGELABELS to find for every vertex $u \in G$, the edge incident to $u$ whose LCA is closest to the root. We call the LCA of this edge the MAX-LCA of $u$. By sorting the vertices in $G$ by their MAX-LCAs, we obtain the concatenation of lists $V'(v)$, $v \in T^*$, which we scan to determine $|V'(v)|$, for all $v \in T^*$.

### 3.2.3 Splitting a Heavy Subtree

If the previous two steps did not produce a simple cycle $\frac{2}{3}$-separator of $G$, we have to deal with the case where no vertex $v \in T^*$ satisfies $\frac{1}{3}|V| \leq |G(v)| \leq \frac{2}{3}|V|$. In this case, there must be a vertex $v \in T^*$ with children $w_1, \ldots, w_k$ such that $|G(v)| > \frac{2}{3}|V|$ and $|G(w_i)| < \frac{1}{3}|V|$, for $1 \leq i \leq k$. Our goal is to compute a subgraph of $G(v)$, consisting of the boundary of $v^*$ and a subset of the graphs $G(w_i)$, whose size is between $\frac{1}{3}|V|$ and $\frac{2}{3}|V|$ and whose boundary is a simple cycle $C$.

In [17] it is claimed that the boundary of the graph defined by $v^*$ and any subset of graphs $G(w_i)$ is a simple cycle. Unfortunately, as illustrated in Figure 6, this is not true in general. However, as we show below, we can compute a permutation $\sigma : [1, k] \to [1, k]$ such that the boundary of each of the graphs obtained by incrementally "gluing" graphs $G(w_{\sigma(1)}), \ldots, G(w_{\sigma(k)})$ onto face $v^*$ is a simple cycle. More formally, we define graphs $H_\sigma(1), \ldots, H_\sigma(k)$ as $H_\sigma(i) = v^* \cup \bigcup_{j=1}^{i} G(w_{\sigma(j)})$. Then we show that $H_\sigma(i)$ and $\overline{H_\sigma(i)}$ are both uniform, for all $1 \leq i \leq k$. This implies that the boundary of $H_\sigma(i)$ is a simple cycle, by Lemma 2. Given the size $|v^*|$ of face $v^*$ and the sizes $|G(w_1)|, \ldots, |G(w_k)|$ of graphs $G(w_1), \ldots, G(w_k)$, the sizes $|H_\sigma(1)|, \ldots, |H_\sigma(k)|$

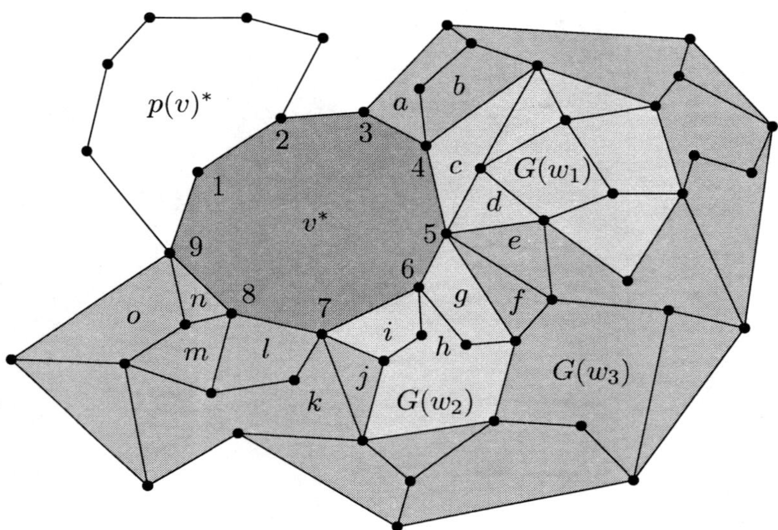


Figure 7: The graph $G(v)$ is shaded. Let $\sigma(i) = i$. Different shades represent the different subgraphs $G(w_1)$, $G(w_2)$ and $G(w_3)$ of $G(v)$. The vertices on the boundary of $v^*$ are numbered clockwise around $v^*$, starting at the endpoint of an edge shared by $v^*$ and $p(v)^*$. The faces in $G(v)$ incident to the boundary of $v^*$ are labeled with small letters.

of graphs $H_\sigma(1), \ldots, H_\sigma(k)$ can be computed in $O(\mathrm{sort}(N))$ I/Os using a procedure similar to the one applied in the previous section for computing the sizes $|G(v)|$ of graphs $G(v)$, $v \in T^*$. Since $|G(v)| > \frac{2}{3}|V|$ and $|G(w_i)| < \frac{1}{3}|V|$ for all $1 \leq i \leq k$, there must exist a graph $H_\sigma(i)$ such that $\frac{1}{3}|V| \leq |H_\sigma(i)| \leq \frac{2}{3}|V|$. It remains to show how to compute the permutation $\sigma$ I/O-efficiently.

To construct $\sigma$, we extract $G(v)$ from $G$, label every face in $G(w_i)$ with $i$, and all other faces of $G(v)$ with 0. This labeling can be computed in $O(\mathrm{sort}(N))$ I/Os by processing $T^*(v)$ from the root towards the leaves. Next we label every edge in $G(v)$ with the labels of the two faces on each side of it. Given the above labeling of the faces in $G(v)$ (or vertices in $T^*(v)$), this labeling of the edges in $G(v)$ can be computed in $O(\mathrm{sort}(N))$ I/Os by applying operation Copy-VertexLabels to the dual graph $G^*(v)$ of $G(v)$. Now consider the vertices $v_1, \ldots, v_t$ on the boundary of $v^*$ in their order of appearance clockwise around $v^*$, starting at an endpoint of an edge shared by $v^*$ and the face corresponding to $v$'s parent $p(v)$ in $T^*$ (see Figure 7). As in Section 3.1, we can compute this order in $O(\mathrm{sort}(N))$ I/Os using the Euler tour technique and list ranking [12]. For every vertex $v_i$, we construct a list $L_i$ of edges around $v_i$ in clockwise order, starting with edge $\{v_{i-1}, v_i\}$ and ending with edge $\{v_i, v_{i+1}\}$. These lists can be extracted from the embedding of $G$ in $O(\mathrm{sort}(N))$ I/Os. Let $L$ be the concate-

L. Arge et al., *On External-Memory Planar DFS*, JGAA, 7(2) 105–129 (2003)120

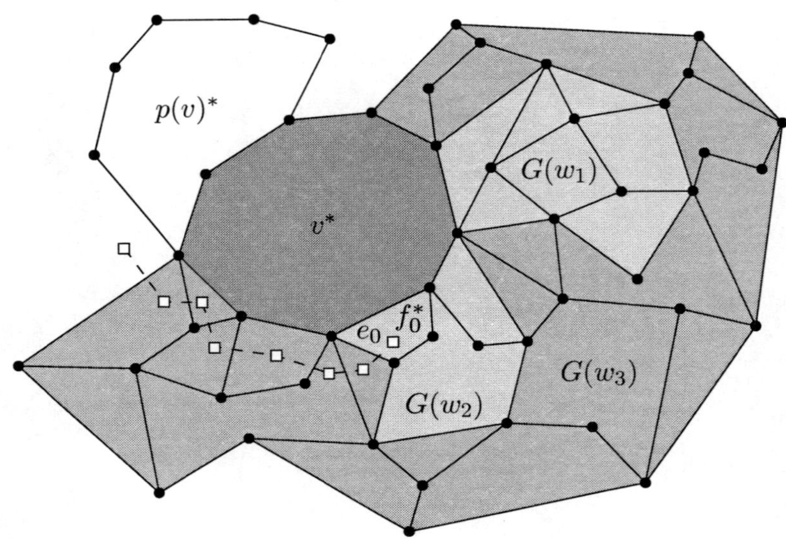


Figure 8: The dashed path is the path from vertex $f_0^*$ to a vertex in the dual of $\overline{G(v)}$ constructed in the proof of Lemma 4, assuming that $j = 2$.

nation of lists $L_1$, $L_2$, ..., $L_t$. For an edge $e$ in $L$ incident to a vertex $v_i$, let $f_1$ and $f_2$ be the two faces on each side of $e$, where $f_1$ precedes $f_2$ in clockwise order around $v_i$. We construct a list $F$ of face labels from $L$ by considering the edges in $L$ in their order of appearance and appending the non-zero labels of faces $f_1$ and $f_2$ in this order to $F$. (Recall that faces in $G(w_i)$ are labeled with number $i$.) This takes $O(\text{scan}(N))$ I/Os. List $F$ consists of integers between 1 and $k$. Some integers may appear more than once, and the occurrences of integer $i$ are not necessarily consecutive. (This happens if the union of $v^*$ with a subgraph $G(w_i)$ encloses another subgraph $G(w_j)$.) For the graph $G(v)$ in Figure 7,

$$F \;=\; \langle \underbrace{3,3}_{\text{v. 3}}, \underbrace{3,3,3,3,1,1}_{\text{vertex 4}}, \underbrace{1,1,1,1,3,3,3,3,2,2}_{\text{vertex 5}}, \underbrace{2,2,2,2,2,2}_{\text{vertex 6}}, \underbrace{2,2,3,3,3,3,3,3}_{\text{vertex 7}},$$

$$\underbrace{3,3,3,3,3,3}_{\text{vertex 8}}, \underbrace{3,3,3,3}_{\text{vertex 9}} \rangle.$$

We construct a final list $S$ by removing all but the last occurrence of each integer from $F$. (Intuitively, this ensures that if the union of $v^*$ and $G(w_i)$ encloses another subgraph $G(w_j)$, then $j$ appears before $i$ in $S$; for the graph in Figure 7, $S = \langle 1, 2, 3 \rangle$.) List $S$ can be computed from list $F$ in $O(\text{sort}(N))$ I/Os using operation DUPLICATEREMOVAL. List $S$ contains each of the integers 1 through $k$ exactly once and thus defines a permutation $\sigma : [1, k] \rightarrow [1, k]$, where $\sigma(i)$ equals the $i$-th element in $S$. It remains to show the following lemma.

L. Arge et al., *On External-Memory Planar DFS*, JGAA, 7(2) 105–129 (2003)121

**Lemma 4** *For all $1 \leq i \leq k$, graphs $H_\sigma(i)$ and $\overline{H_\sigma(i)}$ are both uniform.*

**Proof:** First note that every graph $H_\sigma(i)$ is uniform because every subgraph $G(w_j)$ is uniform and there is an edge between $v$ and $w_j$ in $G^*$, for $1 \leq j \leq k$. To show that every graph $\overline{H_\sigma(i)}$ is uniform, i.e., that its dual is connected., we first observe that $\overline{G(v)}$ is uniform and every subgraph $G(w_j)$, $1 \leq j \leq k$, is uniform. Graph $\overline{H_\sigma(i)}$ is the union of a subset of these graphs. Hence, if its dual is disconnected, there has to be a subgraph $G(w_j) \subseteq \overline{H_\sigma(i)}$ so that its dual and the dual of $\overline{G(v)}$ are in different connected components of the dual of $\overline{H_\sigma(i)}$. Since $G(w_j) \subseteq \overline{H_\sigma(i)}$, $j = \sigma(h)$, for some $h > i$.

Now recall the computation of permutation $\sigma$ (see Figure 8). Let $L$ be the list of edges clockwise around face $v^*$, as in our construction, let $e_0$ be the last edge of $G(w_j)$ in $L$, and let $f_0$ be the face of $G(w_j)$ that precedes edge $e_0$ in the clockwise order around $v^*$. Then for every subgraph $G(w_{j'})$ that contains an edge that succeeds $e_0$ in $L$, $j' = \sigma(h')$, for some $h' > h$. Hence, the following path in $G^*$ from $f_0^*$ to a vertex in the dual of $\overline{G(v)}$ is completely contained in the dual of $\overline{H_\sigma(i)}$: We start at vertex $f_0^*$ and follow the edge $e_0^*$ dual to $e_0$. For every subsequent vertex $f^*$ that has been reached through an edge $e^*$, where $e \in L$, either $f$ is a face of $\overline{G(v)}$, and we are done, or we follow the dual of the edge $e'$ that succeeds $e$ in $L$. This traversal of $G^*$ finds a vertex in the dual of $\overline{G(v)}$ because if it does not encounter a vertex in the dual of $\overline{G(v)}$ before, it will ultimately reach vertex $p(v)$, which is in the dual of $\overline{G(v)}$.

This shows that the duals of $\overline{G(v)}$ and $G(w_j)$ are in the same connected component of the dual of $\overline{H_\sigma(i)}$, for every graph $G(w_j) \subseteq \overline{H_\sigma(i)}$, so that the dual of $\overline{H_\sigma(i)}$ is connected and $\overline{H_\sigma(i)}$ is uniform. $\qquad\square$

# 4 Reducing Depth-First Search to Breadth-First Search

In this section, we give an I/O-efficient reduction from DFS in an embedded planar graph $G$ to BFS in its "vertex-on-face graph", using ideas from [15]. The idea is to use BFS to partition the faces of $G$ into levels around a source face that has the source $s$ of the DFS on its boundary, and then "grow" the DFS-tree level by level around that face.

In order to obtain a partition of the faces of $G$ into levels around the source face, we define a graph which we call the *vertex-on-face graph* $G^\dagger$ of $G$. As before, let $G^* = (V^*, E^*)$ denote the dual of graph $G$; recall that each vertex $f^*$ in $V^*$ corresponds to a face $f$ in $G$. The vertex set of the vertex-on-face graph $G^\dagger$ is $V \cup V^*$; the edge set contains an edge $(v, f^*)$ if vertex $v$ is on the boundary of face $f$ (see Figure 10a). We will show how a BFS-tree of $G^\dagger$ can be used to obtain a partition of the faces in $G$ such that the source face is at level 0, all faces sharing a vertex with the source face are at level 1, all faces sharing a vertex with a level-1 face—but not with the source face—are at level 2, and so on (Figure 9a). Let $G_i$ be the subgraph of $G$ defined as the union of the

L. Arge et al., *On External-Memory Planar DFS*, JGAA, 7(2) 105–129 (2003)122

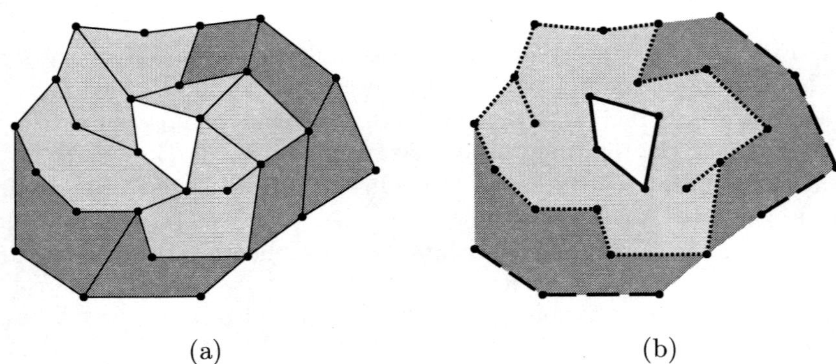

(a)            (b)

Figure 9: (a) A graph $G$ with its faces colored according to their levels; The level-0 face is white, level-1 faces are light gray, level-2 faces are dark gray. (b) Graphs $H_0$ (solid), $H_1$ (dotted), and $H_2$ (dashed).

boundaries of faces at level at most $i$, and let $H_i = G_i \setminus G_{i-1}$, for $i > 0$. (The difference $G_i \setminus G_{i-1}$ of graphs $G_i$ and $G_{i-1}$ is the subgraph of $G_i$ with vertex set $V(G_i) \setminus V(G_{i-1})$ and whose edge set contains all edges of $G_i$ that have both endpoints in $V(G_i) \setminus V(G_{i-1})$; see Figure 9b.) For $i = 0$, we define $H_0 = G_0$. We call the vertices and edges of $H_i$ *level-i vertices* and *edges*. An edge $\{v, w\}$ connecting two vertices $v \in H_i$ and $w \in G_{i-1}$ is called an *attachment edge* of $H_i$. The edges of $G_{i-1}$ and $H_i$ together with the attachment edges of $H_i$ form a partition of the edges of $G_i$. The basic idea in our algorithm is to grow the DFS-tree by walking clockwise[1] from $s$ around the level-0 face $G_0$ until we reach the counterclockwise neighbor of $s$. The resulting path is a DFS tree $T_0$ for $G_0$. Next we build a DFS-tree for $H_1$ and attach it to $T_0$ through an attachment edge of $H_1$ in a way that does not introduce cross-edges. Hence, the result is a DFS-tree $T_1$ for $G_1$. We repeat this process until we have processed all levels $H_0, \ldots, H_r$ obtaining a DFS-tree $T$ for $G$ (see Figure 11). The key to the efficiency of the algorithm lies in the simple structure of graphs $H_0, \ldots, H_r$. Below we give the details of our algorithm and prove the following theorem.

**Theorem 3** *Let $G$ be an undirected embedded planar graph, $G^\dagger$ its vertex-on-face graph, and $f_s$ a face of $G$ containing the source vertex $s$. Given a BFS-tree of $G^\dagger$ rooted at $f_s^*$, a DFS tree of $G$ rooted at $s$ can be computed in $O(sort(N))$ I/Os and linear space.*

First consider the computation of graphs $G_1, \ldots, G_r$ and $H_1, \ldots, H_r$. We start by computing graph $G^\dagger$ in $O(sort(N))$ I/Os as follows: First we compute a representation of $G$ consisting of a list of vertices clockwise around each face of $G$. Such a representation can be computed in $O(sort(N))$ I/Os [16]. Then we add a face vertex $f^*$, for every face $f$ of $G$, and connect $f^*$ to all vertices

---

[1] A *clockwise* walk on the boundary of a face means walking so that the face is to our right.

L. Arge et al., *On External-Memory Planar DFS*, JGAA, 7(2) 105–129 (2003)123

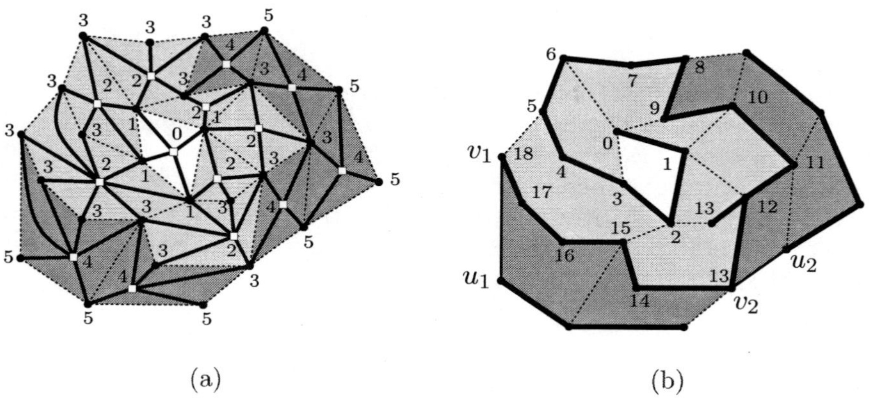

Figure 10: (a) $G^\dagger$ shown in bold; numbers represent BFS-depths in $G^\dagger$. (b) $T_1$, $H_2$ and attachment edges $\{u_i, v_i\}$.Vertices in $T_1$ are labeled with their DFS-depths.

in the vertex list representing $f$. This requires a single scan of the vertex lists representing the faces of $G$. The levels of the faces of $G$ can now be obtained from a BFS-tree of the vertex-on-face graph $G^\dagger$, rooted at the dual vertex $f_s^*$ of a face $f_s$ that contains $s$: Every vertex of $G$ is at an odd level in the BFS-tree; every dual vertex corresponding to a face of $G$ is at an even level (Figure 10a). The level of a face is the level of the corresponding vertex in the BFS-tree divided by two. The vertex set $V(H_i)$ of graph $H_i$ contains all vertices of $G$ at distance $2i + 1$ from $f_s^*$ in $G^\dagger$. Hence, we can obtain a partition of $V(G)$ into vertex sets $V(H_1), \ldots, V(H_r)$ by sorting the vertices in $V(G)$ by their distances from $f_s^*$ in $G^\dagger$. The vertex set $V(G_i)$ of graph $G_i$ is $V(G_i) = \bigcup_{j=0}^{i} V(H_i)$. An edge $e \in G$ is in $H_i$ if both its endpoints are at distance $2i + 1$ from $f_s^*$ in $G^\dagger$. For an attachment edge $\{v, w\}$ of $H_i$, $v$ is at distance $2i + 1$, and $w$ is at distance $2i - 1$ from $f_s^*$ in $G^\dagger$. Thus, we can obtain a partition of $E(G)$ into sets $E(H_0), \ldots, H(H_r)$ and the sets of attachment edges of graphs $H_1, \ldots, H_r$ by sorting the edges in $E(G)$ in inverted lexicographical order defined by the distances of their endpoints from $f_s^*$ in $G^\dagger$.

Next we discuss a few simple properties of graphs $G_i$ and $H_i$, which we use to prove the correctness of our algorithm. For every edge in $G_{i-2}$, as well as for every attachment edge of $H_{i-1}$, the two faces on both sides of the edge are at level at most $i - 1$. Thus, they cannot be boundary edges for $G_{i-1}$. It follows that the boundary edges of $G_{i-1}$ are in $E(H_{i-1})$. Consequently, all boundary vertices of $G_{i-1}$ are in $V(H_{i-1})$. As $G_{i-1}$ is a union of faces, its boundary consists of a set of cycles, called the *boundary cycles* of $G_{i-1}$. Graph $H_i$ lies entirely "outside" the boundary of $G_{i-1}$, i.e., in $\overline{G_{i-1}}$. Hence, all attachment edges of $H_i$ are connected only to boundary vertices of $G_{i-1}$, i.e., vertices of $H_{i-1}$. Finally, note that graph $G_i$ is uniform. This can be shown as follows:

L. Arge et al., *On External-Memory Planar DFS*, JGAA, 7(2) 105–129 (2003)124

Graph $G_i$ corresponds to the first $2i$ levels of the BFS-tree of $G^\dagger$. For a level-$(i-1)$ face $f_1$ and a level-$i$ face $f_2$ that share a vertex $v$, graph $G_i$ contains all faces incident to $v$. Hence, there is a path from $f_1^*$ to $f_2^*$ in $G_i^*$. Applying this argument inductively, we obtain that there is a path in $G_i^*$ from $f_s^*$ to every vertex of $G_i^*$, which shows that $G_i$ is uniform. On the other hand, graph $\overline{G_{i-1}}$, and thus $H_i$, is not necessarily uniform.

We are now ready to describe the details of our algorithm for constructing a DFS-tree for $G$ by repeatedly growing a DFS-tree $T_i$ for $G_i$ from a DFS-tree $T_{i-1}$ for $G_{i-1}$, starting with the DFS-tree $T_0$ for $G_0$. During the algorithm we maintain the following two invariants (see Figure 10b):

(i) Every boundary cycle $C$ of $G_{i-1}$ contains exactly one edge $e$ not in $T_{i-1}$. One of the two endpoints of that edge is an ancestor in $T_{i-1}$ of all other vertices in $C$.

(ii) The depth of each vertex in $G_{i-1}$, defined as the distance from $s$ in $T_{i-1}$, is known.

Assume we have computed a DFS-tree $T_{i-1}$ for $G_{i-1}$. Our goal is to compute a DFS-forest for $H_i$ and link it to $T_{i-1}$ through attachment edges of $H_i$ without introducing cross-edges, in order to obtain a DFS-tree $T_i$ for $G_i$. If we can compute a DFS-forest of $H_i$ in $O(\text{sort}(|H_i|))$ I/Os and link it to $T_{i-1}$ in $O(\text{sort}(|H_{i-1}| + |H_i|))$ I/Os, the overall computation of a DFS-tree $T$ for $G$ uses $O\left(\text{sort}(|H_0|) + \sum_{i=1}^{r} \text{sort}(|H_{i-1}| + |H_i|)\right) = O\left(\sum_{i=0}^{r} \frac{2|H_i|}{B} \log_{M/B} \frac{N}{B}\right) = O(\text{sort}(N))$ I/Os. Next we show how to perform both computations in the desired number of I/Os.

Let $H_1', \ldots, H_k'$ be the connected components of $H_i$. They can be computed in $O(\text{sort}(|H_i|))$ I/Os [12]. For every component $H_j'$, we find the deepest vertex $v_j$ on the boundary of $G_{i-1}$ such that there is an attachment edge $\{u_j, v_j\}$ of $H_i$ with $u_j \in H_j'$. Then we compute a DFS-tree $T_j'$ of $H_j'$ rooted at $u_j$ and attach $T_j'$ to $T_{i-1}$ using edge $\{u_j, v_j\}$. Let $T_i$ be the resulting tree.

**Lemma 5** *Tree $T_i$ is a DFS-tree of $G_i$.*

**Proof:** Tree $T_i$ is a spanning tree of $G_i$, since $T_{i-1}$ is a DFS-tree for $G_{i-1}$, trees $T_1', \ldots, T_k'$ are DFS-trees of the connected components of $H_i$, and each tree $T_j'$ is connected to $T_{i-1}$ by a single edge. Now let $\{v, w\}$ be a non-tree edge of $G_i$. As there are no edges between different connected components of $H_i$ in $G_i$, either $v, w \in H_j'$, for some $1 \le j \le k$, $v, w \in G_{i-1}$, or w.l.o.g. $v \in H_j'$, for some $1 \le j \le k$, and $w \in G_{i-1}$. In the first two cases, edge $\{v, w\}$ is a back edge, since trees $T_{i-1}$ and $T_j'$ are DFS-trees for $G_{i-1}$ and $H_j'$, respectively. In the latter case, $\{v, w\}$ is a back-edge because $v$ is a descendant of $u_j$, and, by Invariant (i), $w$ must be an ancestor of $v_j$ on the boundary cycle of $G_{i-1}$ enclosing $H_j'$. $\quad\square$

We can compute tree $T_i$ from tree $T_{i-1}$ in $O(\text{sort}(|H_{i-1}| + |H_i|))$ I/Os: First we find the attachment edges $\{u_1, v_1\}, \ldots, \{u_k, v_k\}$ connecting graphs $H_1', \ldots, H_k'$ to $G_{i-1}$. This can be done using a procedure similar to the one used

L. Arge et al., *On External-Memory Planar DFS*, JGAA, 7(2) 105–129 (2003)125

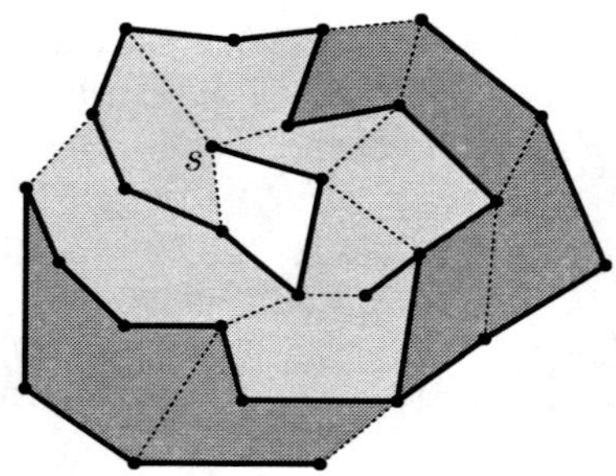

Figure 11: The DFS-tree of $G$ (Figure 9 and Figure 10)

in Section 3.1. As the attachment edges of $H_i$ are the edges of a planar graph with vertex set $V(H_{i-1}) \cup V(H_i)$, this procedure takes $O(\text{sort}(|H_{i-1}| + |H_i|))$ I/Os. All that remains is to show how to compute a DFS-tree $T'_j$ rooted at $u_j$, for each connected component $H'_j$ of $H_i$. The key to doing this I/O-efficiently is the following lemma, which shows that $H_i$ has a simple structure.

**Lemma 6** *The non-trivial bicomps of $H_i$ are the boundary cycles of $G_i$.*

**Proof:** Consider a cycle $C$ in $H_i$. All faces incident to $C$ are at level $i$ or greater. Thus, since $G_{i-1}$ is uniform, all its faces are either inside or outside $C$. Assume w.l.o.g. that $G_{i-1}$ is inside $C$. Then none of the faces outside $C$ shares a vertex with a level-$(i-1)$ face. That is, all faces outside $C$ must be at level at least $i+1$, which means that $C$ is a boundary cycle of $G_i$.

Every bicomp that is not a cycle contains at least three internally vertex-disjoint paths $P_1$, $P_2$, and $P_3$ with the same endpoints $v$ and $w$. As we have just shown, the graph $C_1 = P_1 \cup P_3$ is a boundary cycle of $G_i$, as is the graph $C_2 = P_1 \cup P_2$. Let $\{v, x\}$ be the first edge of $P_2$, and $\{y, w\}$ be the last edge of $P_2$. Since $C_1$ is a boundary cycle of $G_i$, $G_i$ is either completely inside or completely outside $C_1$. Since $C_1$ is a subgraph of $H_i$, all faces incident to $C_1$ that are on the same side of $C_1$ as $G_i$ are at level $i$ because all faces on the other side of $C_1$ are at level at least $i+1$. Hence, if $P_2$ is on the same side of $C_1$ as $G_i$, the four faces incident to edges $\{v, x\}$ and $\{y, w\}$ are at level $i$, which contradicts the fact that $C_2$ is a boundary cycle of $G_i$. If $P_2$ is on the other side of $C_1$, the four faces incident to edges $\{v, x\}$ and $\{y, w\}$ are at level at least $i+1$, which contradicts the fact that edges $\{v, x\}$ and $\{y, w\}$ are at level $i$. Thus, every bicomp of $H_i$ consists of a single boundary cycle. $\qquad \square$

In order to compute a DFS-tree of $H'_j$ rooted at $u_j$, we first partition $H'_j$ into its bicomps. This takes $O(\text{sort}(|H'_j|))$ I/Os [12]. Then, as in Section 3, we construct the bicomp-cutpoint-tree of $H'_j$, rooted at the bicomp containing $u_j$. For each bicomp $K$, we determine the parent cutpoint $x$. If $K$ is a trivial bicomp (i.e., consists of a single edge), the DFS-tree $T_K$ of $K$ consists of the

L. Arge et al., *On External-Memory Planar DFS*, JGAA, 7(2) 105–129 (2003)126

single edge in $K$. Otherwise, by Lemma 6, $K$ is a cycle. Let $y$ be a neighbor of $x$ in $K$. This neighbor can be computed in a single scan of the edge set of $K$. To obtain a DFS-tree $T_K$ of $K$ rooted at $x$, we remove edge $\{x, y\}$ from $K$. The DFS-tree $T'_j$ of $H'_j$ is the union of DFS-trees $T_K$ of all bicomps $K$ of $H'_j$. Note that $T_K$ is a path from $x$ to $y$, and all vertices along this path are descendants of $x$. Since the non-trivial bicomps of $H_i$ are the boundary cycles of $G_i$, Invariant (i) is hence maintained after attaching the resulting DFS-trees $T'_1, \ldots, T'_k$ to $T_{i-1}$.

Finally, to maintain Invariant (ii), we have to determine the depth of each vertex in $T_i$. The depth of vertices in $T_{i-1} \subseteq T_i$ do not change by adding trees $T'_j, \ldots, T'_k$ to $T_{i-1}$. The depths of the vertices in $H_i$ can be computed as follows: Every vertex $u_j$ has depth one more than the depth of $v_j \in T_{i-1}$. The depths of all other vertices in $T'_j$ can be computed from the depth of $u_j$ in $O(\text{sort}(|T'_j|))$ I/Os by performing a DFS-traversal of $T'_j$. Hence, this computation takes $O(\text{sort}(|H_i|))$ I/Os, for all trees in the DFS-forest of $H_i$.

This concludes the description of our reduction from planar DFS to planar BFS, and thus the proof of Theorem 3. The following corollary is an immediate consequence of Theorem 3 and recent results of [6, 22].

**Corollary 1** *A DFS-tree of an embedded planar graph can be computed in $O(sort(N))$ I/O operations and linear space.*

## 5  Conclusions

In this paper, we have developed the first $o(N)$ I/O and linear space algorithm for DFS in embedded planar graphs. We have also designed an $O(\text{sort}(N))$ I/O reduction from planar DFS to planar BFS, proving that external memory planar DFS is not harder than planar BFS. Together with recent results of [6, 22], this leads to an algorithm that computes a DFS-tree of an embedded planar graph in $O(\text{sort}(N))$ I/Os.

## References


[1] J. Abello, A. L. Buchsbaum, and J. R. Westbrook. A functional approach to external graph algorithms. *Algorithmica*, 32(3):437–458, 2002.

[2] P. Agarwal, L. Arge, M. Murali, K. Varadarajan, and J. Vitter. I/O-efficient algorithms for contour-line extraction and planar graph blocking. In *Proceedings of the 9th Annual ACM-SIAM Symposium on Discrete Algorithms*, pages 117–126, 1998.

[3] A. Aggarwal and J. S. Vitter. The input/output complexity of sorting and related problems. *Communications of the ACM*, pages 1116–1127, September 1988.

L. Arge et al., *On External-Memory Planar DFS*, JGAA, 7(2) 105–129 (2003)127


[4] L. Arge. The buffer tree: A new technique for optimal I/O-algorithms. In *Proceedings of the Workshop on Algorithms and Data Structures*, volume 955 of *Lecture Notes in Computer Science*, pages 334–345, 1995.

[5] L. Arge. The I/O-complexity of ordered binary-decision diagram manipulation. In *Proceedings of the International Symposium on Algorithms and Computation*, volume 1004 of *Lecture Notes in Computer Science*, pages 82–91, 1995.

[6] L. Arge, G. S. Brodal, and L. Toma. On external memory MST, SSSP, and multi-way planar separators. In *Proc. Scandinavian Workshop on Algorithms Theory 2000*, volume 1851 of *Lecture Notes in Computer Science*, pages 433–447, 2000.

[7] L. Arge and U. Meyer and L. Toma and N. Zeh. On External-Memory Planar Depth First Search, In *Proceedings of the Workshop on Algorithms and Data Structures*, volume 2125 of *Lecture Notes in Computer Science*, pages 471–482, 2001.

[8] L. Arge, L. Toma, and J. Vitter. I/O-efficient algorithms for problems on grid-based terrains. In *Proceedings of the Workshop on Algorithm Engineering and Experimentation*, 2000.

[9] A. Broder, R. Kumar, F. Manghoul, P. Raghavan, S. Rajagopalan, R. Stata, A. Tomkins, and J. Wiener. Graph structure in the web: Experiments and models. In *Proceedings of the ninth WWW Conference*, 2000. Available at http://www9.org/w9cdrom/index.html.

[10] A. Buchsbaum, M. Goldwasser, S. Venkatasubramanian, and J. Westbrook. On external memory graph traversal. In *Proceedings of the ACM-SIAM Symposium on Discrete Algorithms*, pages 859–860, 2000.

[11] A. L. Buchsbaum and J. R. Westbrook. Maintaining hierarchical graph views. In *Proceedings of the Annual ACM-SIAM Symposium on Discrete Algorithms*, pages 566–575, 2000.

[12] Y.-J. Chiang, M. T. Goodrich, E. F. Grove, R. Tamassia, D. E. Vengroff, and J. S. Vitter. External-memory graph algorithms. In *Proceedings of the 6th Annual ACM-SIAM Symposium on Discrete Algorithms*, pages 139–149, 1995.

[13] F. Chin, J. Lam, and I. Chen. Efficient parallel algorithms for some graphs problems. *Communications of the ACM*, 25(9):659–665, 1982.

[14] E. Feuerstein and A. Marchetti-Spaccamela. Memory paging for connectivity and path problems in graphs. In *Proceedings of the International Symposium on Algorithms and Computation*, pages 416–425, 1993.

[15] T. Hagerup. Planar depth-first search in $O(\log n)$ parallel time. *SIAM Journal on Computing*, 19(4):678–704, 1990.

[16] D. Hutchinson, A. Maheshwari, and N. Zeh. An external memory data structure for shortest path queries. In *Proceedings of the 5th ACM-SIAM Computing and Combinatorics Conference*, volume 1627 of *Lecture Notes in Computer Science*, pages 51–60, 1999. To appear in Discrete Applied Mathematics.

[17] J. JáJá and R. Kosaraju. Parallel algorithms for planar graph isomorphism and related problems. *IEEE Transactions on Circuits and Systems*, 35(3):304–311, 1988.

[18] J. F. JáJá. *An introduction to parallel algorithms*, chapter 5, pages 222–227. Addison-Wesley, Reading, MA, 1992.

[19] V. Kumar and E. J. Schwabe. Improved algorithms and data structures for solving graph problems in external memory. In *Proceedings of the 8th IEEE Sumposium on Parallel and Distributed Computing*, pages 169–177, 1996.

[20] A. Maheshwari and N. Zeh. External memory algorithms for outerplanar graphs. In *Proceedings of the 10th International Symposium on Algorithms and Computation*, volume 1741 of *Lecture Notes in Computer Science*, pages 307–316. Springer Verlag, December 1999.

[21] A. Maheshwari and N. Zeh. I/O-efficient algorithms for graphs of bounded treewidth. In *Proc. ACM-SIAM Symp. on Discrete Algorithms*, pages 89–90, 2001.

[22] A. Maheshwari and N. Zeh. I/O-optimal algorithms for planar graphs using separators. In *Proceedings of the 13th Annual ACM-SIAM Symposium on Discrete Algorithms*, pages 372–381, 2002.

[23] K. Mehlhorn and U. Meyer. External-memory breadth-first search with sublinear I/O. In *Proceedings of the 10th Annual European Symposium on Algorithms*, volume 2461 of *Lecture Notes in Computer Science*, pages 723–735, 2002.

[24] U. Meyer. External memory BFS on undirected graphs with bounded degree. In *Proc. ACM-SIAM Symp. on Discrete Algorithms*, pages 87–88, 2001.

[25] G. L. Miller. Finding small simple cycle separators for 2-connected planar graphs. *Journal of Computer and System Sciences*, 32:265–279, 1986.

[26] K. Munagala and A. Ranade. I/O-complexity of graph algorithms. In *Proceedings of the 10th Annual ACM-SIAM Symposium on Discrete Algorithms*, pages 687–694, 1999.

[27] J. R. Smith. Parallel algorithms for depth-first searches I. planar graphs. *SIAM Journal on Computing*, 15(3):814–830, 1986.

L. Arge et al., *On External-Memory Planar DFS*, JGAA, 7(2) 105–129 (2003)129

[28] R. Tarjan. Depth-first search and linear graph algorithms. *SIAM Journal on Computing*, 1:146–159, 1972.

[29] R. Tarjan and U. Vishkin. Finding biconnected components and computing tree functions in logarithmic parallel time. *SIAM Journal on Computing*, 14(4):862–874, 1985.

[30] U. Ullman and M. Yannakakis. The input/output complexity of transitive closure. *Annals of Mathematics and Artificial Intellligence*, 3:331–360, 1991.

# Journal of Graph Algorithms and Applications

http://jgaa.info/

vol. 7, no. 2, pp. 131–140 (2003)


# Small Maximal Independent Sets and Faster Exact Graph Coloring


*David Eppstein*

Department of Information and Computer Science
University of California, Irvine
eppstein@ics.uci.edu



### Abstract

We show that, for any $n$-vertex graph $G$ and integer parameter $k$, there are at most $3^{4k-n}4^{n-3k}$ maximal independent sets $I \subset G$ with $|I| \leq k$, and that all such sets can be listed in time $\mathcal{O}(3^{4k-n}4^{n-3k})$. These bounds are tight when $n/4 \leq k \leq n/3$. As a consequence, we show how to compute the exact chromatic number of a graph in time $\mathcal{O}((4/3 + 3^{4/3}/4)^n) \approx 2.4150^n$, improving a previous $\mathcal{O}((1 + 3^{1/3})^n) \approx 2.4422^n$ algorithm of Lawler (1976).



Communicated by Giuseppe Liotta and Ioannis G. Tollis: submitted August 2001; revised April 2002.



A preliminary version of this paper appeared in the Proceedings of the 7th International Workshop on Algorithms and Data Structures, Lecture Notes in Computer Science 2125, Springer-Verlag, 2001, pp. 462–470.

D. Eppstein, *Faster Exact Graph Coloring*, JGAA, 7(2) 131–140 (2003)    132

# 1   Introduction

One of the earliest works in the area of worst-case analysis of NP-hard problems is a 1976 paper by Lawler [5] on graph coloring. It contains two results: an algorithm for finding a 3-coloring of a graph (if the graph is 3-chromatic) in time $\mathcal{O}(3^{n/3}) \approx 1.4422^n$, and an algorithm for finding the chromatic number of an arbitrary graph in time $\mathcal{O}((1 + 3^{1/3})^n) \approx 2.4422^n$. Since then, the area has grown, and there has been a sequence of papers improving Lawler's 3-coloring algorithm [1, 2, 4, 8], with the most recent algorithm taking time $\approx 1.3289^n$. However, there has been no improvement to Lawler's chromatic number algorithm.

Lawler's algorithm follows a simple dynamic programming approach, in which we compute the chromatic number not just of $G$ but of all its induced subgraphs. For each subgraph $S$, the chromatic number is found by listing all maximal independent subsets $I \subset S$, adding one to the chromatic number of $S \setminus I$, and taking the minimum of these values. The $\mathcal{O}((1 + 3^{1/3})^n)$ running time of this technique follows from an upper bound of $3^{n/3}$ on the number of maximal independent sets in any $n$-vertex graph, due to Moon and Moser [6]. This bound is tight in graphs formed by a disjoint union of triangles.

In this paper, we provide the first improvement to Lawler's algorithm, using the following ideas. First, instead of removing a maximal independent set from each induced subgraph $S$, and computing the chromatic number of $S$ from that of the resulting subset, we add a maximal independent set of $G \setminus S$ and compute the chromatic number of the resulting superset from that of $S$. This reversal does not itself affect the running time of the dynamic programming algorithm, but it allows us to constrain the size of the maximal independent sets we consider to at most $|S|/3$. We show that, with such a constraint, we can improve the Moon-Moser bound: for any $n$-vertex graph $G$ and integer parameter $k$, there are at most $3^{4k-n}4^{n-3k}$ maximal independent sets $I \subset G$ with $|I| \le k$. This bound then leads to a corresponding improvement in the running time of our chromatic number algorithm.

# 2   Preliminaries

We assume as given a graph $G$ with vertex set $V(G)$ and edge set $E(G)$. We let $n = |V(G)|$ and $m = |E(G)|$. A *proper coloring* of $G$ is an assignment of colors to vertices such that no two endpoints of any edge share the same color. We denote the chromatic number of $G$ (the minimum number of colors in any proper coloring) by $\chi(G)$.

If $V(G) = \{v_0, v_1, \ldots v_{n-1}\}$, then we can place subsets $S \subseteq V(G)$ in one-to-one correspondence with the integers $0, 1, \ldots 2^n - 1$:

$$S \leftrightarrow \sum_{v_i \in S} 2^i.$$

Subsets of vertices also correspond to induced subgraphs of $G$, in which we

D. Eppstein, *Faster Exact Graph Coloring*, JGAA, 7(2) 131–140 (2003)     133

include all edges between vertices in the subset. We make no distinction between these three equivalent views of a vertex subset, so e.g. we will write $\chi(S)$ to indicate the chromatic number of the subgraph induced by set $S$, and $X[S]$ to indicate a reference to an array element indexed by the number $\sum_{v_i \in S} 2^i$. We write $S < T$ to indicate the usual arithmetic comparison between two numbers, and $S \subset T$ to indicate the usual (proper) subset relation between two sets. Note that, if $S \subset T$, then also $S < T$, although the reverse implication does not hold.

A set $S$ is a *maximal $k$-chromatic subset* of $T$ if $S \subseteq T$, $\chi(S) = k$, and $\chi(S') > k$ for every $S \subset S' \subseteq T$. In particular, if $k = 1$, $S$ is a *maximal independent subset* of $T$.

For any vertex $v \in V(G)$, we let $N(v)$ denote the set of neighbors of $v$, including $v$ itself. If $S$ and $T$ are sets, $S \setminus T$ denotes the set-theoretic difference, consisting of elements of $S$ that are not also in $T$. $K_i$ denotes the complete graph on $i$ vertices. We write $\deg(v, S)$ to denote the degree of vertex $v$ in the subgraph induced by $S$.

We express our pseudocode in a syntax similar to that of C, C++, or Java. In particular this implies that array indexing is zero-based. We assume the usual RAM model of computation, in which a single cell is capable of storing an integer large enough to index the memory requirements of the program (thus, in our case, $n$-bit values are machine integers), and in which arithmetic and array indexing operations on these values are assumed to take constant time.

# 3  Small Maximal Independent Sets

**Theorem 1** *Let $G$ be an $n$-vertex graph, and $k$ be a nonnegative number. Then the number of maximal independent sets $I \subset V(G)$ for which $|I| \leq k$ is at most $3^{4k-n}4^{n-3k}$.*

**Proof:** We use induction on $n$; in the base case $n = 0$, there is one (empty) maximal independent set, and for any $k \geq 0$, $1 \leq 3^{4k}4^{-3k} = (81/64)^k$. Otherwise, we divide into cases according to the degrees of the vertices in $G$, as follows:

- If $G$ contains a vertex $v$ of degree three or more, then each maximal independent set $I$ either avoids $v$ (in which case $I$ itself is a maximal independent set of $G \setminus \{v\}$) or contains $v$ (in which case $I \setminus \{v\}$ is a maximal independent set of $G \setminus N(v)$). Thus, by induction, the number of maximal independent sets of cardinality at most $k$ is at most

$$3^{4k-(n-1)}4^{(n-1)-3k} + 3^{4(k-1)-(n-4)}4^{(n-4)-3(k-1)}$$

$$= (\frac{3}{4} + \frac{1}{4})3^{4k-n}4^{n-3k} = 3^{4k-n}4^{n-3k}$$

as was to be proved.

- If $G$ contains a degree-one vertex $v$, let its neighbor be $u$. Then each maximal independent set contains exactly one of $u$ or $v$, and removing

D. Eppstein, *Faster Exact Graph Coloring*, JGAA, 7(2) 131–140 (2003)     134

this vertex from the set produces a maximal independent set of either $G \setminus N(v)$ or $G \setminus N(u)$. If the degree of $u$ is $d$, this gives us by induction a bound of

$$3^{4(k-1)-(n-2)}4^{(n-2)-3(k-1)} + 3^{4(k-1)-(n-d-1)}4^{(n-d-1)-3(k-1)}$$
$$\leq \frac{8}{9}\, 3^{4k-n}4^{n-3k}$$

on the number of maximal independent sets of cardinality at most $k$.

- If $G$ contains an isolated vertex $v$, then each maximal independent set contains $v$, and the number of maximal independent sets of cardinality at most $k$ is at most

$$3^{4(k-1)-(n-1)}4^{(n-1)-3(k-1)} = \frac{16}{27}\, 3^{4k-n}4^{n-3k}.$$

- If $G$ contains a chain $u$-$v$-$w$-$x$ of degree two vertices, then each maximal independent set contains $u$, contains $v$, or does not contain $u$ and contains $w$. Thus in this case the number of maximal independent sets of cardinality at most $k$ is at most

$$2 \cdot 3^{4(k-1)-(n-3)}4^{(n-3)-3(k-1)} + 3^{4(k-1)-(n-4)}4^{(n-4)-3(k-1)}$$
$$= \frac{11}{12}\, 3^{4k-n}4^{n-3k}.$$

- In the remaining case, $G$ consists of a disjoint union of triangles, all maximal independent sets have exactly $n/3$ vertices, and there are exactly $3^{n/3}$ maximal independent sets. If $k \geq n/3$, then $3^{n/3} \leq 3^{4k-n}4^{n-3k}$. If $k < n/3$, there are no maximal independent sets of cardinality at most $k$.

Thus in all cases the number of maximal independent sets is within the claimed bound. $\qquad\square$

Croitoru [3] proved a similar bound with the stronger assumption that all maximal independent sets have $|I| \leq k$. When $n/4 \leq k \leq n/3$, our result is tight, as can be seen for a graph formed by the disjoint union of $4k - n$ triangles and $n - 3k$ $K_4$'s.

**Theorem 2** *There is an algorithm for listing all maximal independent sets of size at most $k$ in an $n$-vertex graph $G$, in time $\mathcal{O}(3^{4k-n}4^{n-3k})$.*

**Proof:** We use a recursive backtracking search, following the case analysis of Theorem 1: if there is a high-degree vertex, we try including it or not including it; if there is a degree-one vertex, we try including it or its neighbor; if there is a degree-zero vertex, we include it; and if all vertices form chains of degree-two vertices, we test whether the parameter $k$ allows any small maximal independent sets, and if so we try including each of a chain of three adjacent vertices. The

D. Eppstein, *Faster Exact Graph Coloring*, JGAA, 7(2) 131–140 (2003)     135

```
// List maximal independent subsets of S smaller than a given parameter.
//      S is a set of vertices forming an induced subgraph in G,
//      I is a set of vertices to be included in the MIS (initially zero), and
//      k bounds the number of vertices of S to add to I.
// We call processMIS(I) on each generated set. Some non-maximal sets may be
// generated along with the maximal ones, but all generated sets are independent.

void smallMIS (set S, set I, int k)
{
    if (S = 0 or k = 0) processMIS(I);
    else if (there exists v ∈ S with deg(v, S) ≥ 3)
    {
        smallMIS (S \ {v}, I, k);
        smallMIS (S \ N(v), I ∪ {v}, k − 1);
    }
    else if (there exists v ∈ S with deg(v, S) = 1)
    {
        let u be the neighbor of v;
        smallMIS (S \ N(u), I ∪ {u}, k − 1);
        smallMIS (S \ N(v), I ∪ {v}, k − 1);
    }
    else if (there exists v ∈ S with deg(v, S) = 0)
        smallMIS (S \ {v}, I ∪ {v}, k − 1);
    else if (some cycle in S is not a triangle or k ≥ |S|/3)
    {
        let u, v, and w be adjacent degree-two vertices,
            such that (if possible) u and w are nonadjacent;
        smallMIS (S \ N(u), I ∪ {u}, k − 1);
        smallMIS (S \ N(v), I ∪ {v}, k − 1);
        smallMIS (S \ ({u} ∪ N(w)), I ∪ {w}, k − 1);
    }
}
```

Figure 1: Algorithm for listing all small maximal independent sets.

D. Eppstein, *Faster Exact Graph Coloring*, JGAA, 7(2) 131–140 (2003)    136

same case analysis shows that this algorithm performs $\mathcal{O}(3^{4k-n}4^{n-3k})$ recursive calls.

Each recursive call can easily be implemented in time polynomial in the size of the graph passed to the recursive call. Since our $3^{4k-n}4^{n-3k}$ bound is exponential in $n$, even when $k = 0$, this polynomial overhead at the higher levels of the recursion is swamped by the time spent at lower levels of the recursion, and does not appear in our overall time bound. $\square$

A more detailed pseudocode description of the algorithm is shown in Figure 1. The given pseudocode may generate non-maximal as well as maximal independent sets, because (when we try not including a high degree vertex) we do not make sure that a neighbor is later included. This will not cause problems for our chromatic number algorithm, but if only maximal independent sets are desired one can easily test the generated sets and eliminate the non-maximal ones. The pseudocode also omits the data structures necessary to implement each recursive call in time polynomial in $|S|$ instead of polynomial in the number of vertices of the original graph.

## 4    Chromatic Number

We are now ready to describe our algorithm for computing the chromatic number of graph $G$. We use an array $X$, indexed by the $2^n$ subsets of $G$, which will (eventually) hold the chromatic numbers of certain of the subsets including $V(G)$ itself. We initialize this array by testing, for each subset $S$, whether $\chi(S) \le 3$; if so, we set $X[S]$ to $\chi(S)$, but otherwise we set $X[S]$ to $\infty$.

Next, we loop through the subsets $S$ of $V(G)$, in numerical order (or any other order such that all proper subsets of each set $S$ are visited before we visit $S$ itself). When we visit $S$, we first test whether $X[S] \ge 3$. If not, we skip over $S$ without doing anything. But if $X[S] \ge 3$, we loop through the small independent sets of $G\backslash S$, limiting the size of each such set to $|S|/X[S]$, using the algorithm of the previous section. For each independent set $I$, we set $X[S \cup I]$ to the minimum of its previous value and $X[S] + 1$.

Finally, after looping through all subsets, we return the value in $X[V(G)]$ as the chromatic number of $G$. Pseudocode for this algorithm is shown in Figure 2.

**Lemma 1** *Throughout the course of the algorithm, for any set $S$, $X[S] \ge \chi(S)$.*

**Proof:** Clearly this is true of the initial values of $X$. Then for any $S$ and any independent set $I$, we can color $S \cup I$ by using a coloring of $S$ and another color for each vertex in $I$, so $\chi(S \cup I) \le \chi(S) + 1 \le X[S] + 1$, and each step of our algorithm preserves the invariant. $\square$

**Lemma 2** *Let $M$ be a maximal $k + 1$-chromatic subset of $G$, and let $(S, I)$ be a partition of $M$ into a $k$-chromatic subset $S$ and an independent subset $I$, maximizing the cardinality of $S$ among all such partitions. Then $I$ is a maximal independent subset of $G \setminus S$ with $|I| \le |S|/k$, and $S$ is a maximal $k$-chromatic subset of $G$.*

D. Eppstein, *Faster Exact Graph Coloring*, JGAA, 7(2) 131–140 (2003)    137

```
int chromaticNumber (graph G)
{
    int X[2^n];
    for (S = 0; S ≤ 2^n; S++)
    {
        if (χ(S) ≤ 3) X[S] = χ[S];
        else X[S] = ∞;
    }
    for (S = 0; S ≤ 2^n; S++)
    {
        if (3 ≤ X[S] < ∞)
        {
            for (each maximal independent set I of G \ S with |I| ≤ |S|/X[S])
                X[S ∪ I] = min(X[S ∪ I], X[S] + 1);
        }
    }
    return X[V(G)];
}
```

Figure 2: Algorithm for computing the chromatic number of a graph.

**Proof:** If we have any $(k+1)$-coloring of $G$, then the partition formed by separating the largest $k$ color classes from the smallest color class satisfies the inequality $|I| \leq |S|/k$, so clearly this also is true when $(S, I)$ is the partition maximizing $|S|$. If $I$ were not maximal, due to the existence of another independent set $I \subset I' \subset G \setminus S$, then $S \cup I'$ would be a larger $(k+1)$-chromatic graph, violating the assumption of maximality of $M$.

Similarly, suppose there were another $k$-chromatic set $S \subset S' \subset G$. Then if $S' \cap I$ were empty, $S' \cup I$ would be a $(k+1)$-chromatic superset of $M$, violating the assumption of $M$'s maximality. But if $S' \cap I$ were nonempty, $(S', I \setminus S')$ would be a better partition than $(S, I)$, so in either case we get a contradiction. $\qquad\square$

**Lemma 3** *Let $M$ be a maximal $k+1$-chromatic subset of $G$. Then, when the outer loop of our algorithm reaches $M$, it will be the case that $X[M] = \chi(M)$.*

**Proof:** Clearly, the initialization phase of the algorithm causes this to be true when $\chi(M) \leq 3$. Otherwise, let $(S, I)$ be as in Lemma 2. By induction on $|M|$, $X[S] = \chi(S)$ at the time we visit $S$. Then $X[S] \geq 3$, and $|I| \leq |S|/X[S]$, so the inner loop for $S$ will visit $I$ and set $X[M]$ to $X[S] + 1 = \chi(M)$. $\qquad\square$

**Theorem 3** *We can compute the chromatic number of a graph $G$ in time $\mathcal{O}((4/3 + 3^{4/3}/4)^n)$ and space $\mathcal{O}(2^n)$.*

D. Eppstein, *Faster Exact Graph Coloring*, JGAA, 7(2) 131–140 (2003)     138

**Proof:** $V(G)$ is itself a maximal $\chi(G)$-chromatic subset of $G$, so Lemma 3 shows that the algorithm correctly computes $\chi(G) = X[V(G)]$. Clearly, the space is bounded by $\mathcal{O}(2^n)$. It remains to analyze the algorithm's time complexity.

First, we consider the time spent initializing $X$. Since we perform a 3-coloring algorithm on each subset of $G$, this time is

$$\sum_{S \subset V(G)} \mathcal{O}(1.3289^{|S|}) = \mathcal{O}\Big(\sum_{i=0}^{n} \binom{n}{i} 1.3289^i\Big) = \mathcal{O}(2.3289^i).$$

Finally, we bound the time in the main loop of the algorithm, which applies the algorithm of Theorem 2 to generate small independent subsets of each set $G \setminus S$. In the worst case, $X[S] = 3$ and we can only limit the size of the generated independent sets to $|S|/3$. We spend constant time adjusting the value of $X[S \cup I]$ for each generated set. Thus, the time can be bounded as

$$\sum_{S \subset V(G)} \mathcal{O}(3^{4\frac{|S|}{3} - |G \setminus S|} 4^{|G \setminus S| - 3\frac{|S|}{3}})$$

$$= \mathcal{O}\Big(\sum_{i=0}^{n} \binom{n}{i} 3^{\frac{7i}{3} - n} 4^{n - 2i}\Big)$$

$$= \mathcal{O}\Big(\big(\frac{4}{3} + \frac{3^{4/3}}{4}\big)^n\Big).$$

This final term dominates the overall time bound. $\qquad\square$

## 5  Finding a Coloring

Although the algorithm of the previous section finds the chromatic number of $G$, it is likely that an explicit coloring is desired, rather than just this number. The usual method of performing this sort of construction task in a dynamic programming algorithm is to augment the dynamic programming array with back pointers indicating the origin of each value computed in the array, but since storing $2^n$ chromatic numbers is likely to be the limiting factor in determining how large a graph this algorithm can be applied to, it is likely that also storing $2^n$ set indices will severely reduce its applicability.

Instead, we can simply search backwards from $V(G)$ until we find a subset $S$ that can be augmented by an independent set to form $V(G)$, and that has chromatic number $\chi(S) = \chi(G) - 1$ as indicated by the value of $X[S]$. We assign the first color to $G \setminus S$. Then, we continue searching for a similar subset $T \subset S$, etc., until we reach the empty set. Although not every set $S$ may necessarily have $X[S] = \chi(S)$, it is guaranteed that for any $S$ we can find $T \subset S$ with $S \setminus T$ independent and $X[T] = X[S] - 1$, so this search procedure always finds a correct coloring.

**Theorem 4** *After computing the array $X$ as in Theorem 3, we can compute an optimal coloring of $G$ in additional time $\mathcal{O}(2^n)$ and $\mathcal{O}(1)$ additional space.*

```
void color (graph G)
{
    compute array X as in Figure 2;
    S = V(G);
    for (T = 2^n - 1; T ≥ 0; T--)
    {
        if (T ⊂ S and X[S \ T] = 1 and X[T] = X[S] - 1)
        {
            color all vertices in S \ T with the same new color;
            S = T;
        }
    }
}
```

Figure 3: Algorithm for optimally coloring a graph.

**Proof:** Pseudocode for the coloring algorithm is shown in Figure 3. As we saw earlier, $X[S]$ can only be guaranteed equal to $\chi(S)$ when $\chi(S) \leq 3$ or when $S$ is maximal $k$-chromatic; however, $X[S]$ always provides a correct upper bound on $\chi(S)$. So, each iteration of the inner block of Figure 3 correctly decomposes the problem into a single independent color class and a remaining $(k-1)$-coloring problem. A subset $T$ satisfying the test will always be found, because the dynamic program must have used some $T$ to set the value of $X[S]$.

The time analysis follows since the algorithm consists of a simple loop over all subsets, performing simple subset tests and array lookups that can be executed in constant time each. $\qquad\square$


# 6   Conclusions

We have shown a bound on the number of small independent sets in a graph, shown how to list all small independent sets in time proportional to our bound, and used this algorithm in a new dynamic programming algorithm for computing the chromatic number of a graph as well as an optimal coloring of the graph.

Our bound on the number of small independent sets is tight for $n/4 \leq k \leq n/3$. Very recently, Nielsen [7] has shown similar tight bounds for all ranges of $k$. Although this extension of our results does not help our chromatic number algorithm, Nielsen was able to use it, together with algorithms for listing small maximal independent sets, as part of improved algorithms for four- and five-coloring. Both our algorithm and Nielsen's may take time proportional to the worst case bound, even for graphs with fewer maximal independent sets. It would be of interest to find an algorithm for listing all small maximal independent sets in time proportional to the number of generated sets rather than simply proportional to the worst case bound on this number.

D. Eppstein, *Faster Exact Graph Coloring*, JGAA, 7(2) 131–140 (2003)     140

Our worst case analysis of the chromatic number algorithm assumes that, every time we call the procedure for listing small maximal independent sets, this procedure achieves its worst case time bound. But is it really possible for all sets $G \setminus S$ to be worst case instances for this procedure? If not, perhaps the analysis of our coloring algorithm can be improved.

Can we prove a bound smaller than $\binom{n}{i}$ on the number of $i$-vertex maximal $k$-chromatic induced subgraphs of a graph $G$? If such a bound could be proven, even for $k = 3$, we could likely improve the algorithm presented here by only looping through the independent subgraphs of $G \setminus S$ when $S$ is maximal.

An alternative possibility for improving the present algorithm would be to find an algorithm for testing whether $\chi(G) \leq 4$ in time $o(1.415^n)$. Then we could test the four-colorability of all subsets of $G$ before applying the rest of our algorithm, and avoid looping over maximal independent subsets of $G \setminus S$ unless $X[S] \geq 4$. This would produce tighter limits on the independent set sizes and therefore reduce the number of independent sets examined. However such a result would be significantly better than the best known time bound, $\mathcal{O}(1.7504^n)$ for Nielsen's four-coloring algorithm [7].

# References


[1] R. Beigel and D. Eppstein.    3-coloring in time $\mathcal{O}(1.3446^n)$:    a no-MIS algorithm.    *Proc. 36th Symp. Foundations of Computer Science*, pp. 444–453. IEEE, October 1995, ftp://ftp.eccc.uni-trier.de/pub/eccc/reports/1995/TR95-033/index.html.

[2] R. Beigel and D. Eppstein. 3-coloring in time $\mathcal{O}(1.3289^n)$. ACM Computing Research Repository, June 2000, arXiv:cs.DS/0006046.

[3] C. Croitoru. On stables in graphs. *Proc. 3rd Coll. Operations Research*, pp. 55–60. Babes-Bolyai Univ., Cluj-Napoca, Romania, 1979.

[4] D. Eppstein. Improved algorithms for 3-coloring, 3-edge-coloring, and constraint satisfaction. *Proc. 12th Symp. Discrete Algorithms*, pp. 329–337. ACM and SIAM, January 2001, arXiv:cs.DS/0009006.

[5] E. L. Lawler. A note on the complexity of the chromatic number problem. *Inf. Proc. Lett.* 5(3):66–67, August 1976.

[6] J. W. Moon and L. Moser. On cliques in graphs. *Israel J. Math.* 3:23–28, 1965.

[7] J. M. Nielsen. On the number of maximal independent sets in a graph. Tech. Rep. RS-02-15, Center for Basic Research in Computer Science (BRICS), April 2002, http://www.brics.dk/RS/02/15/.

[8] I. Schiermeyer. Deciding 3-colourability in less than $\mathcal{O}(1.415^n)$ steps. *Proc. 19th Int. Worksh. Graph-Theoretic Concepts in Computer Science*, pp. 177–182. Springer-Verlag, Lecture Notes in Comp. Sci. 790, 1994.

# Journal of Graph Algorithms and Applications

http://jgaa.info/

vol. 7, no. 2, pp. 141–180 (2003)

# Deciding Clique-Width for Graphs of Bounded Tree-Width

*Wolfgang Espelage*    *Frank Gurski*    *Egon Wanke*

Department of Computer Science
D-40225 Düsseldorf, Germany
http://www.cs.uni-duesseldorf.de/
{espelage,gurski,wanke}@cs.uni-dueseldorf.de

### Abstract

We show that there exists a linear time algorithm for deciding whether a graph of bounded tree-width has clique-width $k$ for some fixed integer $k$.

Communicated by Giuseppe Liotta and Ioannis G. Tollis: submitted October 2001; revised July 2002 and February 2003.

---

The work of the second author was supported by the German Research Association (DFG) grant WA 674/9-2.

W. Espelage et al., *Deciding Clique-Width*, JGAA, 7(2) 141–180 (2003)     142

# 1   Introduction

The clique-width of a graph is defined by a composition mechanism for vertex-labeled graphs, see [CO00]. The operations are the vertex disjoint union $G \oplus H$ of two graphs $G$ and $H$, the addition of edges $\eta_{i,j}(G)$ between vertices labeled by $i$ and vertices labeled by $j$, and the relabeling $\rho_{i \to j}(G)$ of vertices labeled by $i$ into vertices labeled by $j$. The clique-width of a graph $G$ is the minimum number of labels needed to define $G$.

Graphs of bounded clique-width are interesting from an algorithmic point of view. A lot of NP-complete graph problems can be solved in polynomial time for graphs of bounded clique-width if the composition of the graph is explicitly given. For example, all graph properties which are expressible in monadic second order logic with quantifications over vertices and vertex sets ($MSO_1$-logic) are decidable in linear time on graphs of bounded clique-width, see [CMR00]. The $MSO_1$-logic has been extended by counting mechanisms which allow the expressibility of optimization problems concerning maximal or minimal vertex sets, see [CMR00]. All these graph problems expressible in extended $MSO_1$-logic can be solved in polynomial time on graphs of bounded clique-width. Furthermore, a lot of NP-complete graph problems which are not expressible in $MSO_1$-logic or extended $MSO_1$-logic like Hamiltonicity and a lot of partitioning problems can also be solved in polynomial time on graphs of bounded clique-width, see [EGW01, KR01, Wan94].

The following facts are already known about graphs of bounded clique-width. If a graph $G$ has clique-width at most $k$ then the edge complement $\overline{G}$ has clique-width at most $2k$, see [CO00]. Distance hereditary graphs have clique-width at most 3, see [GR00]. The set of all graphs of clique-width at most 2 is the set of all cographs. The clique-width of permutation graphs, interval graphs, grids and planar graphs is not bounded by some fixed integer, see [GR00]. An arbitrary graph with $n$ vertices has clique-width at most $n - r$, if $2^r < n - r$, see [Joh98].

One of the central open questions concerning clique-width is determining the complexity of recognizing and finding a decomposition with clique-width operations of graphs of clique-width at most $k$, for fixed $k \geq 4$. Clique-width of at most 2 is decidable in linear time, see [CPS85]. Clique-width of at most 3 is decidable in polynomial time, see [CHL$^+$00]. The recognition problem for graphs of clique-width at most $k$ is still open for $k \geq 4$. The complexity of the minimization problem where $k$ is additionally given to the input is also open, i.e., not known to be NP-complete nor known to be solvable in polynomial time.

A famous class of graphs for which a lot of NP-complete graph problems can be solved in polynomial time is the class of graphs of bounded tree-width, see Bodlaender [Bod98] for a survey. For every fixed integer $l$, it is decidable in linear time whether a given graph $G$ has tree-width $l$, see [Bod96]. All graph properties expressible in monadic second order logic with quantifications over vertex sets and edge sets ($MSO_2$-logic) are decidable in linear time for graphs of bounded tree-width by dynamic programming, see [Cou90]. The $MSO_2$-logic has also been extended by counting mechanisms to express optimization problems which can then be solved in polynomial time for graphs of bounded tree-width,

W. Espelage et al., *Deciding Clique-Width*, JGAA, 7(2) 141–180 (2003)     143

see [ALS91].

Clique-width seems to be "more powerful" than tree-width. Every graph of tree-width at most $l$ has clique-width at most $3 \cdot 2^{l-1}$, see [CR01]. Since the set of all cographs already contains all complete graphs, the set of all graphs of clique-width at most 2 does not have bounded tree-width. In [GW00], it is shown that every graph of clique-width at most $k$ which does not contain the complete bipartite graph $K_{n,n}$ for some $n > 1$ as a subgraph has tree-width at most $3k(n-1) - 1$.

An algorithm to decide a graph property on a graph of bounded tree-width can simply be obtained by partitioning the set of all so-called $l$-*terminal* graphs into a finite number of equivalence classes as follows. An $l$-terminal graph is a graph with a list of $l$ distinct vertices called terminals. Two $l$-terminal graphs $G$ and $H$ can be combined to a graph $G \circ H$ by taking the disjoint union of $G$ and $H$ and then identifying the $i$-th terminal of $G$ with the $i$-th terminal of $H$ for $1 \leq i \leq l$. They are called *replaceable* with respect to a graph property $\Pi$ if for all $l$-terminal graphs $J$ the answer to $\Pi$ is the same for $G \circ J$ and $H \circ J$. Replaceability is obviously an equivalence relation. A graph property $\Pi$ is decidable in linear time on a graph of bounded tree-width if there is a finite number of equivalence classes with respect to $\Pi$ for all $l$-terminal graphs and all $l \geq 0$. The linear time algorithm first computes a binary tree-decomposition $T_D$ for $G$ and then bottom-up the equivalence class for every $l$-terminal graph $G'$ represented by a complete subtree $T'_D$ of $T_D$. The equivalence class of $G'$ defined by subtree $T'_D$ with root $u'$ is computable in time $O(1)$ from the classes of the two $l$-terminal graphs defined by the two subtrees in $T'_D - \{u'\}$, see also [Arn85, ALS91, AP89, Bod97, Cou90, LW88, LW93].

In this paper, we prove that the graph property "clique-width at most $k$" divides the set of all $l$-terminal graphs into a finite number of equivalence classes. This implies that there exists a linear time algorithm for deciding "clique-width at most $k$" for graphs of bounded tree-width. Since every graph of tree-width $l$ has clique-width at most $3 \cdot 2^{l-1}$, there is also a linear time algorithm for computing the "exact clique-width" of a graph of bounded tree-width by testing "clique-width at most $k$" for $k = 1, \ldots, 3 \cdot 2^{l-1}$. Note that it remains still open whether the clique-width $k$ property is expressible in $MSO_2$-logic and whether "clique-width at most $k$" is decidable in polynomial time for arbitrary graphs.

The paper is organized as follows. In Section 2 we define the clique-width of vertex labeled graphs. Every graph of clique-width at most $k$ is defined by a $k$-expression $X$.

In Section 3, we define the $k$-expression tree of a $k$-expression. Every $k$-expression defines a unique $k$-expression tree and every $k$-expression tree defines a unique $k$-expression. We will mostly work with the expression tree instead of the expression, because many transformation steps are easier to explain for expression trees than for expressions.

In Section 4, we define a normal form for a $k$-expression. We show that for every $k$-expression there is an equivalent one in normal form.

In Section 5, we define $l$-terminal graphs, for some nonnegative integer $l$, and

W. Espelage et al., *Deciding Clique-Width*, JGAA, 7(2) 141–180 (2003)      144

an equivalence relation on the set of all $l$-terminal graphs, called replaceability. This equivalence relation is defined with respect to the graph property clique-width at most $k$. If the relation has a finite number of equivalence classes, then the graph property clique-width at most $k$ is decidable in linear time on graphs of tree-width at most $l$ by dynamic programming algorithms.

In Section 6, we give an overview about the proof of the main result.

In Section 7, we show that every combined graph $H \circ J$ of clique-width at most $k$ can be defined by a $k$-expression in normal form whose expression tree satisfies further special properties concerning the composition of $H$ and $J$ to $H \circ J$. Every of these special expression trees defines a connection tree for $H$. The set of all connection trees for $H$ is called the connection type of $H$. For fixed integers $k$ and $l$, there is only a fixed number of mutually different connection trees and thus a fixed number of connection types.

In Section 8, we show that if two $l$-terminal graphs $H_1$ and $H_2$ define the same connection type then they are replaceable with respect to property clique-width at most $k$. This shows that the equivalence relation defined in Section 5 has a finite number of equivalence classes, which implies that graph property clique-width at most $k$ is decidable in linear time for graphs of bounded tree-width.

## 2    Clique-width

We work with finite undirected *graphs* $G = (V_G, E_G)$, where $V_G$ is a finite set of *vertices* and $E_G \subseteq \{\{u, v\} \mid u, v \in V_G, \ u \neq v\}$ is a finite set of *edges*. A graph $J = (V_J, E_J)$ is a *subgraph* of $G$ if $V_J$ is a subset of $V_G$ and $E_J$ is a subset of $E_G \cap \{\{u, v\} \mid u, v \in V_J, \ u \neq v\}$. $J$ is an *induced subgraph* of $G$ if additionally $E_J = \{\{u, v\} \in E_G \mid u, v \in V_J\}$. $G$ and $J$ are *isomorphic* if there is a bijection $b : V_G \to V_J$ such that for every pair of vertices $u, v \in V_G$, $\{u, v\}$ is an edge of $G$ if and only if $\{b(u), b(v)\}$ is an edge of $J$. To distinguish between the vertices of (non-tree) graphs and trees, we simply call the vertices of the trees *nodes*.

The notion of clique-width for labeled graphs is first defined by Courcelle and Olariu in [CO00]. Let $[k] := \{1, \ldots, k\}$ be the set of all integers between 1 and $k$. A *$k$-labeled graph* $G = (V_G, E_G, \mathrm{lab}_G)$ is a graph $(V_G, E_G)$ whose vertices are labeled by a mapping $\mathrm{lab}_G : V_G \to [k]$. The $k$-labeled graph consisting of a single vertex labeled by some label $t \in [k]$ is denoted by $\bullet_t$. A $k$-labeled graph $J = (V_J, E_J, \mathrm{lab}_J)$ is a *$k$-labeled subgraph* of $G$ if $V_J \subseteq V_G$, $E_J \subseteq E_G \cap \{\{u, v\} \mid u, v \in V_J, \ u \neq v\}$ and $\mathrm{lab}_J(u) = \mathrm{lab}_G(u)$ for all $u \in V_J$. $G$ and $J$ are *isomorphic* if there is a bijection $b : V_G \to V_J$ such that $\{u, v\} \in E_G$ if and only if $\{b(u), b(v)\} \in E_J$, and for every vertex $u \in V_G$, $\mathrm{lab}_G(u) = \mathrm{lab}_J(b(u))$.

**Definition 2.1 (Clique-width, [CO00])** *Let $k$ be some positive integer. The class $CW_k$ of $k$-labeled graphs is recursively defined as follows.*

   *1. The $k$-labeled graphs $\bullet_t$ for $t \in [k]$ are in $CW_k$.*

W. Espelage et al., *Deciding Clique-Width*, JGAA, 7(2) 141–180 (2003)    145

2. Let $G = (V_G, E_G, lab_G) \in CW_k$ and $J = (V_J, E_J, lab_J) \in CW_k$ be two vertex disjoint $k$-labeled graphs. Then the $k$-labeled graph

$$G \oplus J := (V', E', lab')$$

defined by $V' := V_G \cup V_J$, $E' := E_G \cup E_J$, and

$$lab'(u) := \begin{cases} lab_G(u) & \text{if } u \in V_G \\ lab_J(u) & \text{if } u \in V_J \end{cases} , \forall u \in V'$$

is in $CW_k$.

3. Let $i, j \in [k]$, $i \neq j$, be two distinct integers and $G = (V_G, E_G, lab_G) \in CW_k$ be a $k$-labeled graph then

(a) the $k$-labeled graph $\eta_{i,j}(G) := (V_G, E', lab_G)$ defined by

$$E' := E_G \cup \{\{u, v\} \mid u, v \in V_G, \ u \neq v, \ lab(u) = i, \ lab(v) = j\}$$

is in $CW_k$ and

(b) the $k$-labeled graph $\rho_{i \to j}(G) := (V_G, E_G, lab')$ defined by

$$lab'(u) := \begin{cases} lab_G(u) & \text{if } lab_G(u) \neq i \\ j & \text{if } lab_G(u) = i \end{cases} , \forall u \in V_G$$

is in $CW_k$.

An expression $X$ built with the operations $\bullet_t, \oplus, \eta_{i,j}, \rho_{i \to j}$ for integers $t, i, j \in [k]$ is called a *$k$-expression*. To distinguish between the $k$-expression and the graph defined by the $k$-expression, we denote by $\text{val}(X)$ the graph defined by expression $X$. That is, $\text{CW}_k$ is the set of all graphs $\text{val}(X)$, where $X$ is a $k$-expression.

We say, a $k$-labeled graph $G$ has *clique-width at most $k$* if $G$ is contained in class $\text{CW}_k$, i.e., the set $\text{CW}_k$ is the set of all $k$-labeled graphs of *clique-width at most $k$*. The *clique-width* of a $k$-labeled graph $G$ is the smallest integer $k$ such that $G$ has clique-width at most $k$.

We sometimes use the simplified notions *labeled graph* and *expression* for a $k$-labeled graph and a $k$-expression, respectively. In these cases, however, either $k$ is known from the context, or $k$ is irrelevant for the discussion.

An *unlabeled* graph $G = (V_G, E_G)$ has *clique-width at most $k$* if there is some labeling $lab_G : V_G \to [k]$ of the vertices of $G$ such that the labeled graph $G' = (V_G, E_G, lab_G)$ has clique-width at most $k$. The *clique-width* of an *unlabeled* graph $G = (V_G, E_G)$ is the smallest integer $k$ such that there is some labeling $lab_G : V_G \to [k]$ of the vertices of $G$ such that the labeled graph $G' = (V_G, E_G, lab_G)$ has clique-width at most $k$.

If $X$ is a $k$-expression then obviously $\rho_{i \to 1}(X)$ is a $k$-expression for all $i \in [k], i > 1$. For the rest of this paper, we consider an unlabeled graph as a labeled graph in that all vertices are labeled by the same label, which is without loss of generality label 1. This allows us to use the notation "graph" without any confusion for labeled and unlabeled graphs.

W. Espelage et al., *Deciding Clique-Width*, JGAA, 7(2) 141–180 (2003)    146

# 3  Expression tree

Every $k$-expression $X$ has by its recursive definition a tree structure that is called the *$k$-expression tree $T$* for $X$. It is an ordered rooted tree whose nodes are labeled by the operations of the $k$-expression and whose arcs are directed from the leaves towards the root of $T$. The root of $T$ is labeled by the last operation of the $k$-expression.

**Definition 3.1 (Expression tree)** *The $k$-expression tree $T$ for $k$-expression $\bullet_t$ consists of a single node $r$ (the root of $T$) labeled by $\bullet_t$. The $k$-expression tree $T$ for $\eta_{i,j}(X)$ and $\rho_{i \to j}(X)$ consists of a copy $T'$ of the $k$-expression tree for $X$, an additional node $r$ (the root of $T$) labeled by $\eta_{i,j}$ or $\rho_{i \to j}$, respectively, and an additional arc from the root of $T'$ to node $r$. The $k$-expression tree $T$ for $X_1 \oplus X_2$ consists of a copy $T_1$ of the $k$-expression tree for $X_1$, a copy $T_2$ of the $k$-expression tree for $X_2$, an additional node $r$ (the root of $T$) labeled by $\oplus$ and two additional arcs from the roots of $T_1$ and $T_2$ to node $r$. The root of $T_1$ is the* left *child of $r$ and the root of $T_2$ is the* right *child of $r$.*

*A node of $T$ labeled by $\bullet_t$, $\eta_{i,j}$, $\rho_{i \to j}$, or $\oplus$ is called a* leaf, edge insertion node, relabeling node, *or* union node, *respectively.*

If integer $k$ is known from the context or irrelevant for the discussion, then we sometimes use the simplified notion *expression tree* for the notion $k$-expression tree. The leaves of expression tree $T$ for expression $X$ correspond to the vertices of graph $\mathrm{val}(X)$. For some node $u$ of expression tree $T$, let $T(u)$ be the subtree of $T$ induced by node $u$ and all nodes of $T$ from which there is a directed path to $u$. Note that $T(u)$ is always an expression tree. The expression $X(u)$ defined by $T(u)$ can simply be determined by traversing the tree starting from the root, where the left children are visited first. The vertices of $G'$ are the vertices of $G$ corresponding to the leaves of $T(u)$. The edges of $G'$ and the labels of the vertices of $G'$ are defined by expression $X(u)$. For two vertices $u, v$ of $G'$, every edge $\{u, v\}$ of $G'$ is also in $G$ but not necessarily vice versa. Two equal labeled vertices in $G'$ are also equal labeled in $G$ but not necessarily vice versa. The labeled graph $G'$ is denoted by $G(T, u)$ or simply $G(u)$, if tree $T$ is unique from the context. Figure 1 illustrates these notations.

# 4  Normal form

We next define a so-called *normal form* for a $k$-expression. This normal form does not restrict the class of $k$-labeled graphs that can be defined by $k$-expressions, but is very useful for the proof of our main result.

To keep the definition of our normal form as simple as possible, we enumerate the vertices in a graph $G = \mathrm{val}(X)$ defined by some $k$-expression $X$ as follows. The single vertex in $\mathrm{val}(\bullet_t)$ is the *first vertex* of $\mathrm{val}(\bullet_t)$. Let $G = \mathrm{val}(Y_1 \oplus Y_2)$. If $\mathrm{val}(Y_1)$ has $n$ vertices and $\mathrm{val}(Y_2)$ has $m$ vertices, then for $i = 1, \ldots, n$ the *$i$-th vertex* of $G$ is the $i$-th vertex of $\mathrm{val}(Y_1)$ and for $i = n+1, \ldots, n+m$ the *$i$-th vertex* of $G$ is the $(i-n)$-th vertex of $\mathrm{val}(Y_2)$. The *$i$-th vertex* of $\mathrm{val}(\eta_{i,j}(Y))$

W. Espelage et al., *Deciding Clique-Width*, JGAA, 7(2) 141–180 (2003)    147

$$X = \rho_{1\to 2}(\eta_{2,3}(((\eta_{1,2}(\bullet_1 \oplus \bullet_2)) \oplus (\eta_{1,2}(\bullet_1 \oplus \bullet_2))) \oplus \bullet_3))$$

$$X(u) = (\eta_{1,2}(\bullet_1 \oplus \bullet_2)) \oplus (\eta_{1,2}(\bullet_1 \oplus \bullet_2))$$

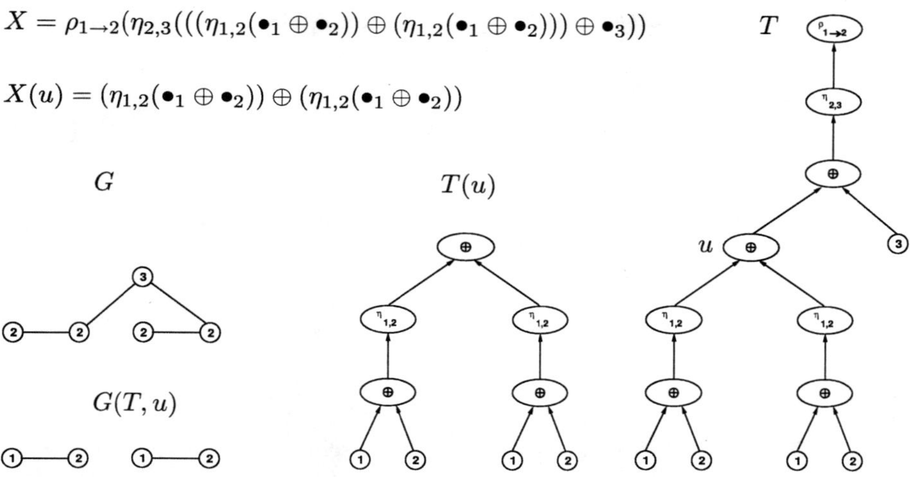

Figure 1: A 3-labeled graph $G$ defined by a 3-expression $X$ with 3-expression tree $T$, and a 3-labeled graph $G(T, u)$ defined by 3-expression $X(u)$ with 3-expression tree $T(u)$

and $\mathrm{val}(\rho_{i\to j}(Y))$ is the $i$-th vertex of $\mathrm{val}(Y)$. We say two expressions $X$ and $Y$ are *equivalent*, denoted by $X \equiv Y$, if $\mathrm{val}(X)$ and $\mathrm{val}(Y)$ are isomorphic in consideration of the order of the vertices, that is,

1. $\mathrm{val}(X)$ and $\mathrm{val}(Y)$ have the same number $n$ of vertices,

2. the $i$-th vertex in $\mathrm{val}(X)$, $1 \le i \le n$, has the same label as the $i$-th vertex in $\mathrm{val}(Y)$, and

3. there is an edge between the $i$-th and $j$-th vertex in $\mathrm{val}(X)$, $1 \le i, j \le n$, if and only if there is an edge between the $i$-th and $j$-th vertex in $\mathrm{val}(Y)$.

Otherwise $X$ and $Y$ are *not equivalent*, denoted by $X \not\equiv Y$.

If two $k$-expressions $X$ and $Y$ are equivalent then they do not need to be equal. If two $k$-labeled graphs $\mathrm{val}(X)$ and $\mathrm{val}(Y)$ defined by two $k$-expressions $X$ and $Y$ are isomorphic (see the second paragraph of Section 2) then $X$ and $Y$ do not need to be equivalent. Figure 2 shows an example.

**Definition 4.1 (Normal form)** *Our* normal form *for $k$-expressions is defined as follows.*

1. *The $k$-expression $\bullet_t$ for some $t \in [k]$ is in normal form.*

2. *If $Y_1$ and $Y_2$ are two $k$-expressions in normal form then the $k$-expression*

$$\rho_{i_n \to j_n}(\cdots \rho_{i_1 \to j_1}(\eta_{i'_{n'},j'_{n'}}(\cdots \eta_{i'_1,j'_1}(Y_1 \oplus Y_2)\cdots))\cdots)$$

*for $i_1, j_1, \ldots, i_n, j_n, i'_1, j'_1, \ldots, i'_{n'}, j'_{n'} \in [k]$ is in normal form if the following properties hold true.*

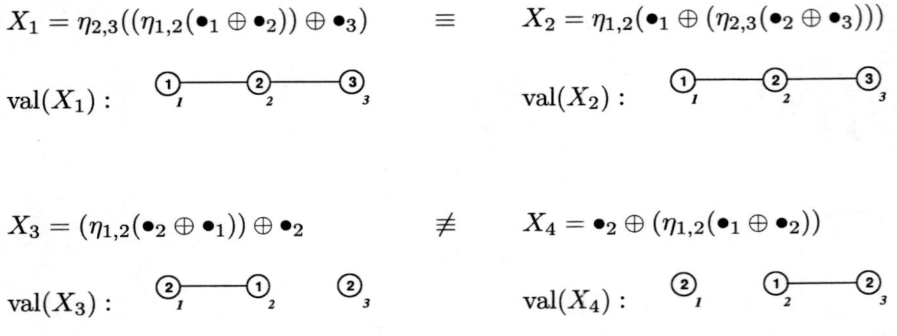

W. Espelage et al., *Deciding Clique-Width*, JGAA, 7(2) 141–180 (2003)    148

$$X_1 = \eta_{2,3}((\eta_{1,2}(\bullet_1 \oplus \bullet_2)) \oplus \bullet_3) \qquad \equiv \qquad X_2 = \eta_{1,2}(\bullet_1 \oplus (\eta_{2,3}(\bullet_2 \oplus \bullet_3)))$$

$$X_3 = (\eta_{1,2}(\bullet_2 \oplus \bullet_1)) \oplus \bullet_2 \qquad \not\equiv \qquad X_4 = \bullet_2 \oplus (\eta_{1,2}(\bullet_1 \oplus \bullet_2))$$

Figure 2: The small indices at the vertices represent their numbering with respect to the corresponding $k$-expression. The expressions $X_1$ and $X_2$ are equivalent but not equal. The labeled graphs $\mathrm{val}(X_3)$ and $\mathrm{val}(X_4)$ are isomorphic but the expressions $X_3$ and $X_4$ are not equivalent.

(a) *For every edge insertion operation* $\eta_{i'_{l'},j'_{l'}}$, $1 \leq l' \leq n'$,

$$\eta_{i'_{l'},j'_{l'}}(\eta_{i'_{l'-1},j'_{l'-1}}(\cdots \eta_{i'_1,j'_1}(Y_1 \oplus Y_2) \cdots)) \not\equiv \eta_{i'_{l'-1},j'_{l'-1}}(\cdots \eta_{i'_1,j'_1}(Y_1 \oplus Y_2) \cdots),$$

$$\eta_{i'_{l'},j'_{l'}}(Y_1) \equiv Y_1, \quad and \quad \eta_{i'_{l'},j'_{l'}}(Y_2) \equiv Y_2.$$

(b) *For every relabeling operation* $\rho_{i_l \to j_l}$, $1 \leq l \leq n$, *graph*

$$\mathrm{val}(\rho_{i_{l-1} \to j_{l-1}}(\cdots \rho_{i_1 \to j_1}(\eta_{i'_{n'},j'_{n'}}(\cdots \eta_{i'_1,j'_1}(Y_1 \oplus Y_2) \cdots)) \cdots))$$

*has a vertex labeled by* $i_l$ *and a vertex labeled by* $j_l$, *and*
$i_l \notin \{j_1, \ldots, j_{l-1}\}$.

(c) *For every pair of two distinct labels* $i, j \in [k], i \neq j$,

    i. *if* $\mathrm{val}(Y_1)$ *has a vertex labeled by* $i$ *and a vertex labeled by* $j$ *then*

$$\begin{aligned}
&\rho_{i_l \to j_l}(\cdots \rho_{i_1 \to j_1}(\eta_{i'_{n'},j'_{n'}}(\cdots \eta_{i'_1,j'_1}(\qquad Y_1 \quad \oplus \quad Y_2 \quad )\cdots))\cdots) \\
\not\equiv \; &\rho_{i_l \to j_l}(\cdots \rho_{i_1 \to j_1}(\eta_{i'_{n'},j'_{n'}}(\cdots \eta_{i'_1,j'_1}(\quad \rho_{i \to j}(Y_1) \quad \oplus \quad Y_2 \quad )\cdots))\cdots)
\end{aligned}$$

    *and*

    ii. *if* $\mathrm{val}(Y_2)$ *has a vertex labeled by* $i$ *and a vertex labeled by* $j$ *then*

$$\begin{aligned}
&\rho_{i_l \to j_l}(\cdots \rho_{i_1 \to j_1}(\eta_{i'_{n'},j'_{n'}}(\cdots \eta_{i'_1,j'_1}(\quad Y_1 \quad \oplus \qquad Y_2 \quad )\cdots))\cdots) \\
\not\equiv \; &\rho_{i_l \to j_l}(\cdots \rho_{i_1 \to j_1}(\eta_{i'_{n'},j'_{n'}}(\cdots \eta_{i'_1,j'_1}(\quad Y_1 \quad \oplus \quad \rho_{i \to j}(Y_2) \quad )\cdots))\cdots).
\end{aligned}$$

If $X$ is a $k$-expression in normal form then the operations between two union operations are ordered such that there are first the edge insertion operations and after that the relabeling operations. Edges are inserted and vertices are relabeled as soon as possible in the following sense. By Definition 4.1(2.(a)), every edge insertion operation $\eta_{i'_{l'},j'_{l'}}$ inserts at least one edge between a vertex of $\mathrm{val}(Y_1)$ and a vertex of $\mathrm{val}(Y_2)$ but no edge between two vertices of $\mathrm{val}(Y_1)$ or between two vertices of $\mathrm{val}(Y_2)$. By Definition 4.1(2.(b)), a relabeling operation $\rho_{i_l \to j_l}$ always relabels at least one vertex to some label already used by at least one other vertex. Thus every relabeling operation decreases the number of labels used by the vertices of the graph. Since none of the relabeling operations $\rho_{i_{l-1} \to j_{l-1}}, \ldots, \rho_{i_1 \to j_1}$ relabels anything to $i_l$, every vertex is relabeled at most once (between two union operations). The three properties, there is a vertex labeled by $i_l$, there is a vertex labeled by $j_l$, and $i_l \notin \{j_{l-1}, \ldots, j_1\}$ imply that $i_l \notin \{i_{l-1}, \ldots, i_1, j_{l-1}, \ldots, j_1\}$ and $j_l \notin \{i_{l-1}, \ldots, i_1\}$. Finally, by Definition 4.1(2.(c)), the number of labels used in the graphs defined by the subexpressions is always minimal.

The following observations are easy to verify. If $k$-expression $\rho_{i \to j}(Y)$ is in normal form then $k$-expression $Y$ is in normal form, if $k$-expression $\eta_{i',j'}(Y)$ is in normal form then $k$-expression $Y$ is in normal form, and if $k$-expression $Y_1 \oplus Y_2$ is in normal form then $k$-expression $Y_1$ and $k$-expression $Y_2$ are in normal form. That is, if an expression is in normal form, then every complete subexpression is in normal form.

**Theorem 4.2** *For every $k$-expression $X$ there is an equivalent $k$-expression in normal form.*

**Proof:** We show how to transform an arbitrary $k$-expression $X$ into an equivalent $k$-expression in normal form.

The following transformation steps can be used to transform a $k$-expression $X$ into an equivalent $k$-expression in that no edge insertion operation is applied directly after a relabeling operation.

Let $Z = \eta_{i',j'}(\rho_{i \to j}(Y))$ be a subexpression of $X$.

1. If $\{i,j\} \cap \{i',j'\} = \emptyset$, then $Z$ can be replaced by $\rho_{i \to j}(\eta_{i',j'}(Y))$, because the two operations do not affect each other.

2. If $i \in \{i',j'\}$, then we can omit the edge insertion operation $\eta_{i',j'}$, because it does not create an edge.

3. If $i \notin \{i',j'\}$ and $j \in \{i',j'\}$, then we distinguish between two cases. If $j = i'$ then $Z$ can be replaced by $\rho_{i \to j}(\eta_{i',j'}(\eta_{i,j'}(Y)))$, if $j = j'$ then $Z$ can be replaced by $\rho_{i \to j}(\eta_{i',j'}(\eta_{i',i}(Y)))$.

These transformation steps can be used to transform a $k$-expression $X$ into an equivalent $k$-expression such that all edge insertion and relabeling operations are in the right order with respect to Definition 4.1. The succeeding transformation steps will not change this *right* order.

Next we consider an induction on the number of union operations and the composition of $X$. The transformation steps do not change the number of union operations in the modified subexpressions.

Let

$$X = \rho_{i_n \to j_n}(\cdots \rho_{i_1 \to j_1}(\eta_{i'_{n'},j'_{n'}}(\cdots \eta_{i'_1,j'_1}(\bullet_t)\cdots))\cdots)$$

be a $k$-expression without any union operation. Then $X$ is equivalent to a $k$-expression $\bullet_j$ for some $j \in \{j_1, \ldots, j_n, t\}$.

Let

$$X = \eta_{i'_{l'},j'_{l'}}(\eta_{i'_{l'-1},j'_{l'-1}}(\cdots \eta_{i'_1,j'_1}(Y_1 \oplus Y_2)\cdots))$$

be a $k$-expression, where

$$X = \eta_{i'_{l'-1},j'_{l'-1}}(\cdots \eta_{i'_1,j'_1}(Y_1 \oplus Y_2)\cdots)$$

is in normal form.

1. If $val(Y_1)$ does not contain all edges between vertices labeled by $i'_{l'}$ and vertices labeled by $j'_{l'}$, then we transform the $k$-expression $\eta_{i'_{l'},j'_{l'}}(Y_1)$ into an equivalent $k$-expression $Y'_1$ in normal form and replace in $X$ the subexpression $Y_1$ by $Y'_1$. The transformation of $\eta_{i'_{l'},j'_{l'}}(Y_1)$ into normal form is possible by the inductive hypothesis. The same replacement is possible for $Y_2$, if necessary.

2. If

$$\eta_{i'_{l'},j'_{l'}}(\eta_{i'_{l'-1},j'_{l'-1}}(\cdots \eta_{i'_1,j'_1}(Y_1 \oplus Y_2)\cdots)) \equiv \eta_{i'_{l'-1},j'_{l'-1}}(\cdots \eta_{i'_1,j'_1}(Y_1 \oplus Y_2)\cdots)$$

then we omit operation $\eta_{i'_{l'},j'_{l'}}$ from $X$, because it does not create any edge.

The result is an equivalent $k$-expression in normal form.

Let

$$X = \rho_{i_n \to j_n}(\cdots \rho_{i_1 \to j_1}(\eta_{i'_{n'},j'_{n'}}(\cdots \eta_{i'_1,j'_1}(Y_1 \oplus Y_2)\cdots))\cdots)$$

be a $k$-expression, where subexpression

$$Z_0 := \eta_{i'_{n'},j'_{n'}}(\cdots \eta_{i'_1,j'_1}(Y_1 \oplus Y_2)\cdots)$$

is in normal form.

If graph $val(Y_1)$ has a vertex labeled by $i$ and a vertex labeled by $j$ for two distinct labels $i, j \in [k]$ such that

$$\begin{aligned}
&\rho_{i_l \to j_l}(\cdots \rho_{i_1 \to j_1}(\eta_{i'_{n'},j'_{n'}}(\cdots \eta_{i'_1,j'_1}(\quad\quad Y_1 \oplus Y_2 \quad)\cdots))\cdots) \\
\equiv\ &\rho_{i_l \to j_l}(\cdots \rho_{i_1 \to j_1}(\eta_{i'_{n'},j'_{n'}}(\cdots \eta_{i'_1,j'_1}(\quad \rho_{i \to j}(Y_1) \oplus Y_2 \quad)\cdots))\cdots),
\end{aligned}$$

then we transform the $k$-expression $\rho_{i \to j}(Y_1)$ into an equivalent $k$-expressions $Y'_1$ in normal form and replace in $k$-expression $X$ the subexpression $Y_1$ by $Y'_1$. The transformation of $\rho_{i_l \to j_l}(Y_1)$ into normal form is possible by the inductive hypothesis. After that we transform $Z_0$ with the new subexpression $Y_1$ into normal form, if necessary. This is possible by the inductive hypothesis and the transformation steps already defined above. The same replacement is possible for $Y_2$, if necessary. This procedure can be repeated at most $k - 1$ times for $Y_1$ and $Y_2$, because after every replacement the number of labels used by the vertices of $\mathrm{val}(Y_1)$ and $\mathrm{val}(Y_2)$ decreases by one.

We now compute for every label $l \in [k]$, the label $h(l)$ into which label $l$ is relabeled by performing the $n$ relabeling operations $\rho_{i_1 \to j_1}, \ldots, \rho_{i_n \to j_n}$ in this given order one after the other. Function $h : [k] \to [k]$ can be considered as a directed graph $H = (V_H, A_H)$ with vertex set $V_H = [k]$ and arc set $A_H = \{(l, h(l)) \mid l \in [k]\}$. Every vertex $l$ of $H$ has exactly one outgoing arc $(l, h(l))$.

We first remove all arcs $(l_1, l_2)$ from $H$ for which graph $\mathrm{val}(Z_0)$ has no vertex labeled by $l_1$ and all arcs $(l_1, l_2)$ for which $l_1 = l_2$, because these arcs do not represent any relabeling of vertices of $\mathrm{val}(Z_0)$. Next we consider every pair of two arcs $(l_1, l_2)$, $(l_2, l_3)$ of $H$ and simultaneously replace in expression $Z_0$ all labels $l_1$ by $l_2$ and all labels $l_2$ by $l_1$, and remove both arcs $(l_1, l_2)$ and $(l_2, l_3)$ from $H$. After that we insert a new arc $(l_1, l_3)$ into $H$ if $l_1 \neq l_3$. Note that $k$-expression $Z_0$ remains in normal form if two labels are exchanged in all operations of $Z_0$. Finally, we consider all arcs $(l_1, l_2)$ of $H$ for which graph $\mathrm{val}(Z_0)$ has no vertex labeled by $l_2$. We then simultaneously replace in expression $Z_0$ all labels $l_1$ by $l_2$ and all labels $l_2$ by $l_1$, and remove arc $(l_1, l_2)$ from $H$.

Now we can define the new relabeling by the remaining arcs of $H$. We remove step by step an arc $(l_1, l_2)$ from $H$ and apply the relabeling operation $\rho_{l_1 \to l_2}$ to the current $k$-expression $Z_i$ (which is initially $Z_0$) to get a new $k$-expression $Z_{i+1} = \rho_{l_1 \to l_2}(Z_i)$. This leads to a $k$-expression which is in normal form and equivalent to the original one. $\qquad\square$

The proof of Theorem 4.2 uses a simple relabeling trick to omit a relabeling operation $\rho_{i_l \to j_l}(X)$ if graph $\mathrm{val}(X)$ has no vertex labeled by $j_l$. This relabeling simultaneously replaces in expression $X$ all labels $i_l$ by $j_l$ and all labels $j_l$ by $i_l$. Let $X_{i_l \leftrightarrow j_l}$ be the resulting expression. If $X$ is in normal form then $X_{i_l \leftrightarrow j_l}$ is in normal form, and $X_{i_l \leftrightarrow j_l} \equiv \rho_{i_l \to j_l}(X)$.

## 5  Replaceability

Most of the bottom-up dynamic programming algorithms for deciding a graph property $\Pi$ on a tree-structured graph $G$ are based on the idea of substituting a subgraph of $G$ by a small so-called *replaceable* subgraph. The substitution is defined by a composition mechanism which is different for the various graph models. However, the notion of *replaceability* can be defined for every composition mechanism. Thus, the bottom-up dynamic programming techniques work in principle for all tree-structured graphs, more or less successfully.

For the analysis of *tree-width bounded graphs*, we need so-called *l-terminal graphs* and an operation denoted by $\circ$ which combines two $l$-terminal graphs by identifying vertices. Since we are mainly interested in labeled graphs, we use

W. Espelage et al., *Deciding Clique-Width*, JGAA, 7(2) 141–180 (2003)     152

a labeled version of $l$-terminal graphs. Terminal graphs are also called *sourced graphs*, see [ALS91].

**Definition 5.1 ($k$-labeled $l$-terminal graph)** *A $k$-labeled $l$-terminal graph is a system*

$$G = (V_G, E_G, P_G, lab_G)$$

*where $(V_G, E_G, lab_G)$ is a $k$-labeled graph and $P_G = (x_1, \ldots, x_l)$ is a sequence of $l \geq 0$ distinct vertices of $V_G$. The vertices in sequence $P_G$ are called terminal vertices or terminals for short. Vertex $x_i$, $1 \leq i \leq l$, is the $i$-th terminal of $G$. The other vertices in $V_G - P_G$ are called the inner vertices of $G$.*

*Let $H = (V_H, E_H, P_H, lab_H)$ and $J = (V_J, E_J, P_J, lab_J)$ be two vertex disjoint $k$-labeled $l$-terminal graphs such that the $i$-th terminal of $P_H$ has the same label as the $i$-th terminal of $P_J$ for $i = 1, \ldots, l$. Then the composition $H \circ J$ is the $k$-labeled graph obtained by taking the disjoint union of $(V_H, E_H, lab_H)$ and $(V_J, E_J, lab_J)$, and then identifying corresponding terminals, i.e., for $i = 1, \ldots, l$, identifying the $i$-th terminal of $H$ with the $i$-th terminal of $J$, and removing multiple edges.*

**Definition 5.2 (Replaceability of $k$-labeled $l$-terminal graphs)** *Let $\Pi$ be a graph property, i.e., $\Pi : \mathcal{G}_k \to \{true, false\}$, where $\mathcal{G}_k$ is the set of all $k$-labeled graphs. Two $k$-labeled $l$-terminal graphs $H_1$ and $H_2$ are called replaceable with respect to $\Pi$, denoted by $H_1 \sim_{\Pi,l} H_2$, if for every $k$-labeled $l$-terminal graph $J$,*

$$\Pi(H_1 \circ J) = \Pi(H_2 \circ J).$$

Figure 3 and 4 show three 4-labeled 3-terminal graphs $H_1, H_2, J$. The two 4-labeled 3-terminal graphs $H_1$ and $H_2$ are not replaceable, for example, with respect to Hamiltonicity, because $H_1 \circ J$ has a Hamilton cycle but $H_2 \circ J$ does not.

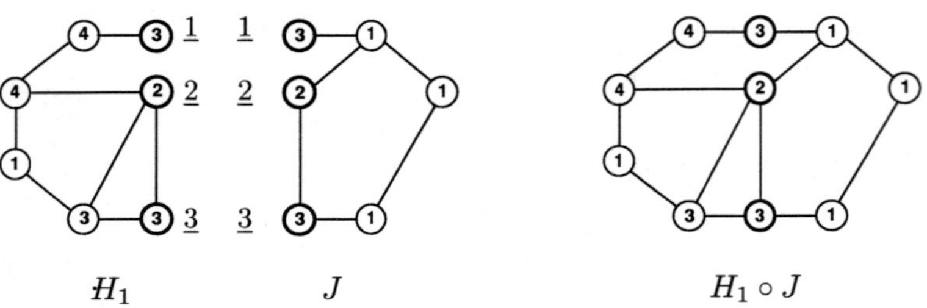


Figure 3: Two 4-labeled 3-terminal graphs $H_1, J$ and the composed graph $H_1 \circ J$. The underlined integers represent the numbering of the terminals.

It is well known that if $\sim_{\Pi,l}$ divides the set of all $k$-labeled $l$-terminal graphs into a finite number of equivalence classes then $\Pi$ is decidable in linear time for all $k$-labeled graphs of tree-width at most $l$. Note that linear time means under the assumption that integer $l$ is fixed and not part of the input.

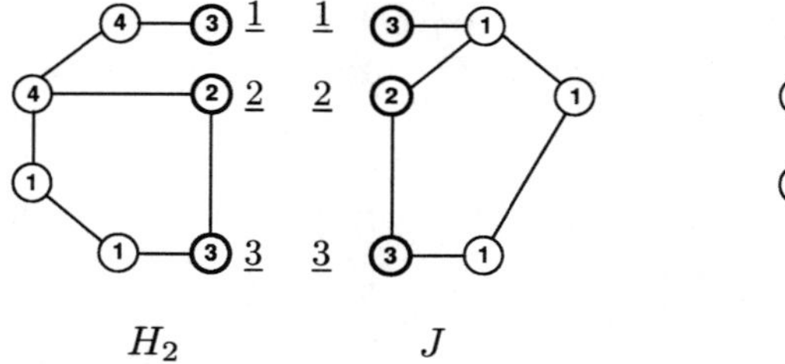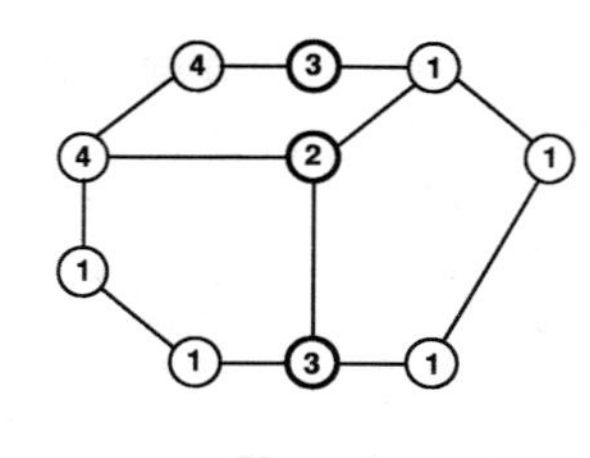

Figure 4: Two 4-labeled 3-terminal graphs $H_2, J$ and the composed graph $H_2 \circ J$

The same schema can be used to solve graph properties on graphs of bounded clique-width. In this case the composition of two vertex disjoint $k$-labeled graphs $H$ and $J$ is done by an operation $\times_S$, where $S \subseteq [k] \times [k]$. The composed $k$-labeled graph $H \times_S J$ is the disjoint union of $H$ and $J$ with all additional edges between vertices $u \in V_H$ and $v \in V_J$ for which $(\mathrm{lab}_H(u), \mathrm{lab}_J(v)) \in S$. Two $k$-labeled graphs $H_1$ and $H_2$ are replaceable with respect to some graph property $\Pi$, denoted by $\sim_{\Pi,k}$, if for all $k$-labeled graphs $J$ and all $S \subseteq [k] \times [k]$,

$$\Pi(H_1 \times_S J) \;=\; \Pi(H_2 \times_S J).$$

If this equivalence relation $\sim_{\Pi,k}$ has a finite number of equivalence classes then property $\Pi$ is decidable in linear time for all $k$-labeled graphs $\mathrm{val}(X)$ of clique-width at most $k$ if the $k$-expression $X$ is given to the input (integer $k$ is assumed to be fixed and not part of the input).

For all who are interested in the details how to solve a graph property on a tree-width or clique-width bounded graph with the bottom-up techniques mentioned above, we refer to [Arn85, AP89, ALS91, Bod97, Bod98, CMR00, Cou90, EGW01, KR01, LW93, Wan94]. These details are not necessary for this paper.

## 6   Overview

In this section, we intuitively explain how the proof of our main result is running. Let $\Pi_k$ be the graph property clique-width at most $k$. Let $\sim_{\Pi_k, l}$ be the equivalence relation defined for $k$-labeled $l$-terminal graphs as in Definition 5.2. Our aim is to show that $\sim_{\Pi_k, l}$ divides the set of all $k$-labeled $l$-terminal graphs into a finite number of equivalence classes, for every fixed $k \geq 1$ and every fixed $l \geq 0$. This would imply that the graph property clique-width at most $k$ is decidable in linear time for graphs of tree-width at most $l$.

For every $k$-labeled $l$-terminal graph $G$ we will define a so-called *connection type* consisting of a set of so-called *connection trees*. We will show that two $k$-labeled $l$-terminal graphs are replaceable with respect to the graph property clique-width at most $k$ if they are of the same connection type, but not necessarily vice versa. Thus, the number of equivalence classes of $\sim_{\Pi_k, l}$ can be

W. Espelage et al., *Deciding Clique-Width*, JGAA, 7(2) 141–180 (2003)    154

bounded by the number of mutually different connection types for all $k$-labeled $l$-terminal graphs. If there is a finite number of mutually different connection types for every fixed $k \geq 1$ and every fixed $l \geq 0$, then $\sim_{\Pi_k,l}$ has a finite number of equivalence classes, and our main result follows.

The outline of the proof can easily be explained more precisely, but still intuitively, with a simplified version of the connection tree. To distinguish between the real connection tree and the simplified one, we will call the simplified version the *strong connection tree*. Let $H$ and $J$ be two $k$-labeled $l$-terminal graphs such that the $k$-labeled graph $H \circ J$ has clique-width at most $k$, see also Figure 5. Let $X$ be a $k$-expression for $H \circ J$ and let $T$ be the $k$-expression tree of $X$. The $k$-expression tree $T$ can be decomposed into two subtrees, say $T_H$ and $T_J$, as follows. Subtree $T_H$ describes the $k$-labeled subgraph of $H \circ J$ induced by the vertices of $H$. That is, $T_H$ consists of the leaves of $T$ representing the vertices of $H$ and of all nodes of $T$ on the paths from these leaves to the root of $T$. Subtree $T_J$ is defined in the same way with respect to the vertices of $J$. Note that $T_H$ and $T_J$ are not necessarily expression tress. Every node of $T$ is in at least one of these two subtrees $T_H$ and $T_J$. Some of the nodes of $T$ are contained in both subtrees. More precisely, the root of $T$ is in both subtrees and at least the leaves of $T$ representing the identified terminals of $H$ and $J$, and all nodes of $T$ on the paths of these leaves to the root of $T$ are in both subtrees.

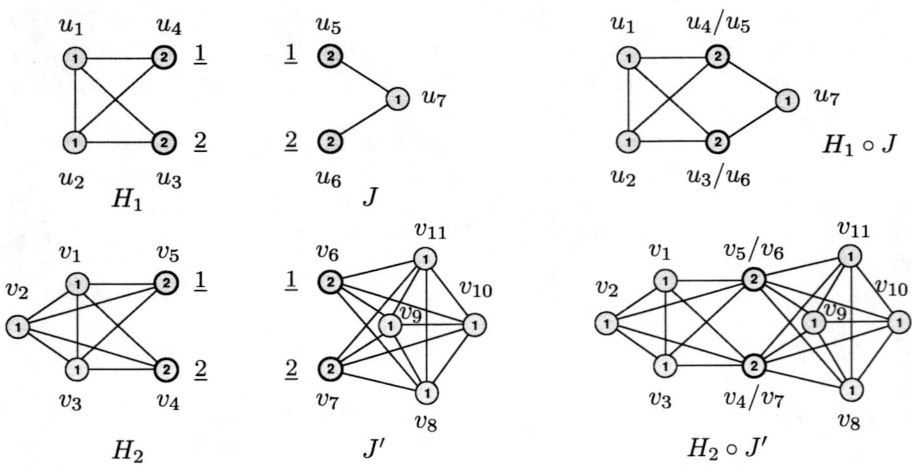


Figure 5: Four 2-labeled 2-terminal graphs $H_1$, $H_2$, $J$, and $J'$, and the two 2-labeled graphs $H_1 \circ J$ and $H_2 \circ J'$

The common part of both subtrees $T_H$ and $T_J$, denoted by $C$, defines a *strong connection tree* for $H$. The leaves in the common part $C$ are either leaves of $T$ or union nodes. If a leaf $u$ represents a vertex of $H \circ J$ obtained by identifying the $i$-th terminal of $H$ with the $i$-th terminal of $J$, then $u$ will additionally be labeled by index $i$. Let $u$ be a union node of the $k$-expression

W. Espelage et al., *Deciding Clique-Width*, JGAA, 7(2) 141–180 (2003)     155

tree $T$ and let $u_l$ and $u_r$ be the left and right child of $u$ in $T$. If $u$ is in the common part $C$ but $u_l$ or $u_r$ is not, then we add a left child $v_l$ to $C$ or a right child $v_r$ to $C$, respectively, such that we get an ordered tree in that every union node has a left child and a right child. If $u_l$ ($u_r$) is a node of $T_H$ but not a node of $T_J$, then the inserted leaf $v_l$ ($v_r$, respectively) is labeled by the set $L$ of all labels of the vertices in the $k$-labeled graph defined by the $k$-expression subtree $T(u_l)$ ($k$-expression subtree $T(u_r)$, respectively) of $T$. Figure 6 illustrates such a labeling of the inserted leaves by an example. The existence of the non-empty labeling $L$ indicates that the leaf represents a subtree of $T_H$ and not a subtree of $T_J$. These leaves are called *internal* leaves, the other leaves are called *external* leaves. The notions *internal* and *external* refer to the association that the left argument $H$ is the internal graph for which we compute the connection tree, and the right argument $J$ is the external graph, i.e., the environment to which $H$ is attached. The resulting structure $C$ is called a *strong connection tree* for $H$. To get all strong connection trees for $H$, we have to consider all $k$-labeled $l$-terminal graphs $J$ such that $H \circ J$ has clique-width at most $k$, and all possible $k$-expressions for $H \circ J$. The set of all strong connection trees for $H$ is the *strong connection type* of $H$.

Let us next explain why two $k$-labeled $l$-terminal graphs $H_1$ and $H_2$ of the same strong connection type are replaceable with respect to clique-width at most $k$. After that we consider the size of the strong connection tees. Assume $H_1 \circ J$ has clique-width at most $k$ for some $k$-labeled $l$-terminal graph $J$. Let $X$ be a $k$-expression for $H_1 \circ J$. Let $T$ be the $k$-expression tree of $X$ and let $T_{H_1}$ and $T_J$ be the two subtrees for $H_1$ and $J$, respectively. The common part of $T_{H_1}$ and $T_J$ defines a strong connection tree $C$ for $H_1$ which is, by our assumption, also a strong connection tree for $H_2$. That is, there has to be at least one $k$-labeled $l$-terminal graph $J'$ such that $H_2 \circ J'$ has clique-width at most $k$. Furthermore, there has to be a $k$-expression $X'$ with a $k$-expression tree $T'$ for $H_2 \circ J'$ such that the common part of the two subtrees $T'_{H_2}$ and $T'_{J'}$ defines the same strong connection tree $C$ for $H_2$. Now we can replace in $k$-expression tree $T$ subtree $T_{H_1}$ by subtree $T'_{H_2}$, see Figure 7. This can easily be done by substituting the corresponding subtrees represented by the internal leaves. Let $T''$ be the resulting $k$-expression tree we get after this replacement.

It is easy to verify that the $k$-expression tree $T''$ defines the $k$-labeled graph $H_2 \circ J$. The vertices from $H_2$ and $J$ are labeled in the $k$-labeled graph defined by $T''$ as in the $k$-labeled graphs $H_2 \circ J'$ and $H_1 \circ J$, respectively. Two vertices from $H_2$ or two vertices from $J$ are connected by an edge if and only if they are connected by an edge in $H_2 \circ J'$ or $H_1 \circ J$, respectively. This is, because the subtrees defined by the paths from the involved leaves to the roots are equal in both $k$-expression trees $T''$ and $T$ or in both $k$-expression trees $T''$ and $T'$, respectively.

The additional $L$-labeling of the internal leaves in the strong connection tree $C$ is necessary to ensure that $T''$ defines no forbidden edge between an inner vertex $u_1$ of $H_2$ and an inner vertex $u_2$ of $J$. If the graph defined by $T''$ has such a forbidden edge then $H_1 \circ J$ would also have at least one such forbidden edge, because the corresponding subgraph of $H_1$ would have at least one vertex

W. Espelage et al., *Deciding Clique-Width*, JGAA, 7(2) 141–180 (2003)     156

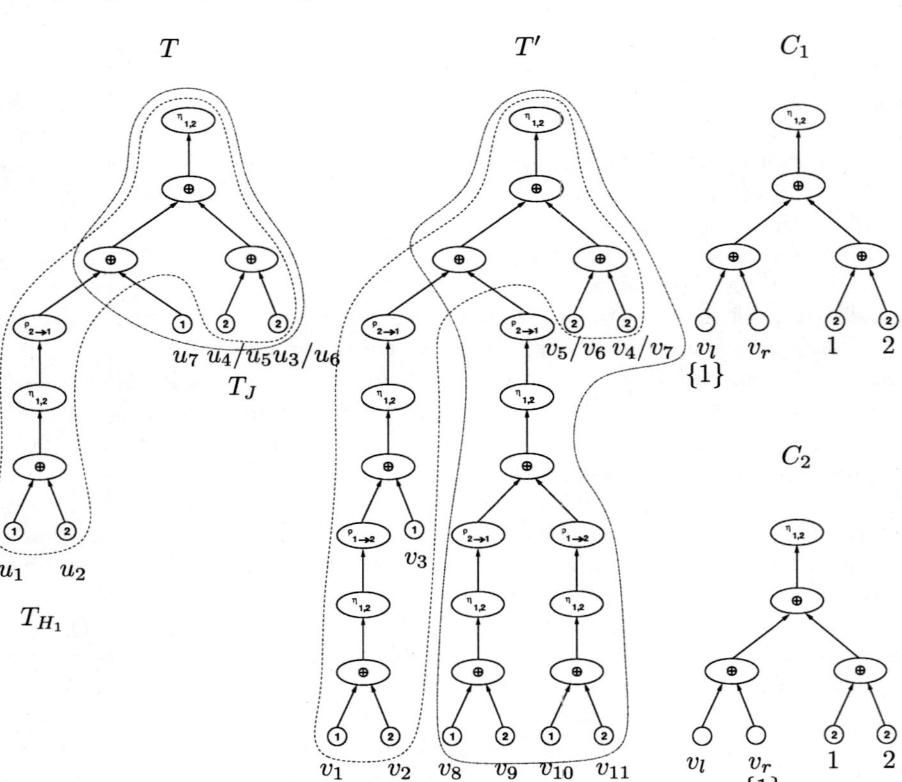

Figure 6: A 2-expression tree $T$ for the 2-labeled graph $H_1 \circ J$ and a 2-expression tree $T'$ for the 2-labeled graph $H_2 \circ J'$. $C_1$ is a strong connection tree for $H_1$ and $H_2$. $C_2$ is a strong connection tree for $J$ and $J'$.

labeled as $u_1$ of $H_2$. If for every $k$-labeled $l$-terminal graph $J$, graph $H_1 \circ J$ has clique-width at most $k$ if and only if $H_2 \circ J$ has clique-width at most $k$, then obviously $H_1$ and $H_2$ are replaceable with respect to clique-width at most $k$.

Finally, we have to consider the size of the strong connection trees. The size of the common part of the two subtrees $T_H$ and $T_J$ can, unfortunately, not be bounded by some constant depending only on $k$ and $l$. However, the main part of the next section is the proof that for every $k$-labeled graph $H \circ J$ of clique-width at most $k$ there is at least one $k$-expression tree $T$ such that the information we really need from the common part of the two subtrees $T_H$ and $T_J$ can be bounded. This information is still tree-structured and will be defined in the next section as the real connection tree.

We will show step by step that there is a $k$-expression tree for $H \circ J$ in that the paths in the common part of $T_H$ and $T_J$ have the following structure. We divide the operations of the nodes of $T$ into $H$-operations and $J$-operations. An

W. Espelage et al., *Deciding Clique-Width*, JGAA, 7(2) 141–180 (2003)     157

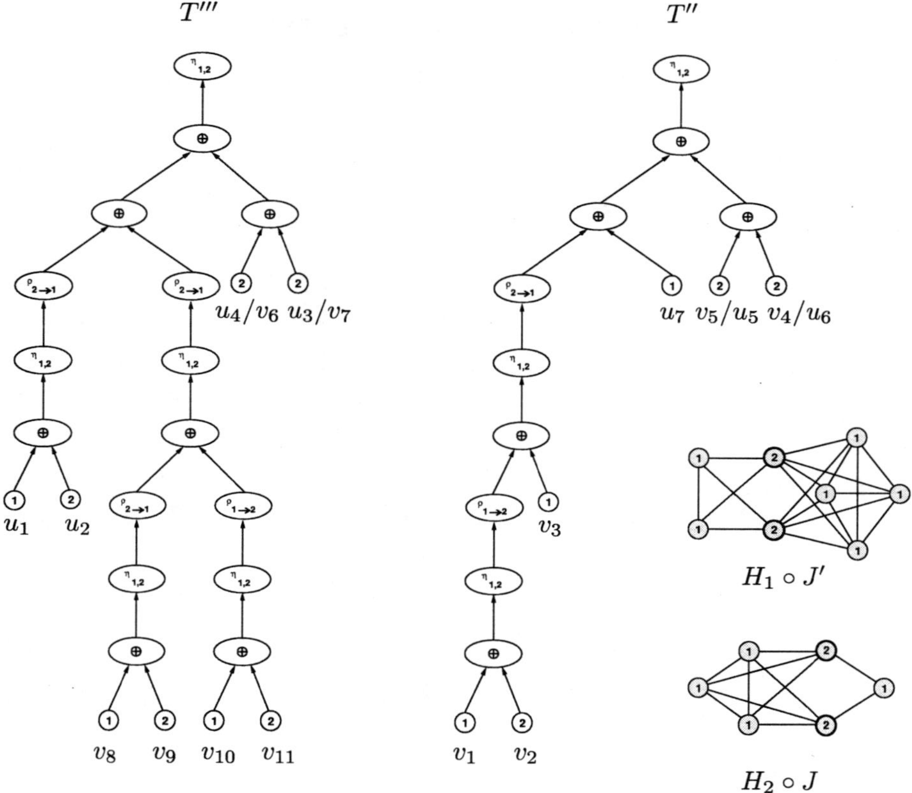

Figure 7: A 2-expression tree $T'''$ for the 2-labeled graph $H_1 \circ J'$ and a 2-expression tree $T''$ for the 2-labeled graph $H_2 \circ J$.

$H$-operation changes a label of a vertex from $H$ or inserts an edge incident to a vertex from $H$. A $J$-operation does anything concerning the vertices from $J$. Some of the operations could even be $H$-operations and an $J$-operations. In the next section, we will prove that there is always a $k$-expression tree $T$ for $H \circ J$ such that in the common part of $T_H$ and $T_J$ the number of times the classification into $H$- and $J$-operations changes along a path from a leaf to the root can be bounded by some constant depending only on $k$ and $l$. This property finally allows us to define a connection structure of bounded size, which we call the *connection tree* for $H$. The main idea is to replace the unbounded subpaths with certain operations of the same type by single so-called *bridge* nodes.

W. Espelage et al., *Deciding Clique-Width*, JGAA, 7(2) 141–180 (2003)    158

# 7  Determining the connection type

We consider the case where we have a $k$-labeled $l$-terminal graph

$$H = (V_H, E_H, P_H, \mathrm{lab}_H)$$

and a $k$-labeled $l$-terminal graph $J = (V_J, E_J, P_J, \mathrm{lab}_J)$ such that $H$ and $J$ are vertex disjoint and the combined graph

$$G = (V_G, E_G, \mathrm{lab}_G) = H \circ J$$

has clique-width at most $k$.

We partition the vertex set $V_G$ of $G$ into three disjoint sets $U_H, U_J, U_P$ such that $U_H \cup U_J \cup U_P = V_G$. Vertex set $U_H = V_H - P_H$ contains the inner vertices from $H$, vertex set $U_J = V_J - P_J$ contains the inner vertices from $J$, and vertex set $U_P$ contains the joined terminals from $H$ and $J$. Vertex set $U_P$ has exactly $l$ vertices, because the $l$ terminals of $H$ are identified with the $l$ terminals of $J$. Note that graph $G$ does not have any edge between a vertex of $U_H$ and a vertex of $U_J$.

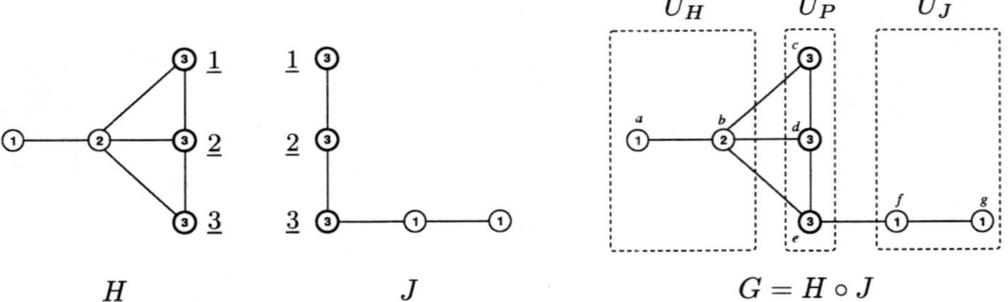


Figure 8: Two 3-labeled 3-terminal graphs $H$ and $J$, the 3-labeled graph $G = H \circ J$, and the partition of its vertices into $U_H$, $U_P$, and $U_J$

Let $T$ be a $k$-expression tree for $G = H \circ J$. The subtree $T_P$ of $T$ is defined by the $l$ leaves of $T$ that correspond to the $l$ vertices of $U_P$ and by all nodes of $T$ on the paths from these leaves to the root of $T$, see Figure 9. Thus the root of $T_P$ is the root of $T$. Tree $T_P$ is in general not an expression tree. It is only an expression tree if neither $H$ nor $J$ has inner vertices. In this case, $T_P$ and $T$ are equal.

Our intention is to show that for each such pair $H$, $J$ as above there is always at least one $k$-expression tree $T$ for $G$ such that $T_P$ has a very special form. This special form represents the necessary information how $H$ and $J$ are combined. We will see that the size of this connection information will depend only on $k$ and $l$ but not on the size of $H$ or $J$.

The following four subsections start with a lemma that allows us to consider a more restricted $k$-expression tree $T$ than before. The restrictions are expressed

W. Espelage et al., *Deciding Clique-Width*, JGAA, 7(2) 141–180 (2003)    159

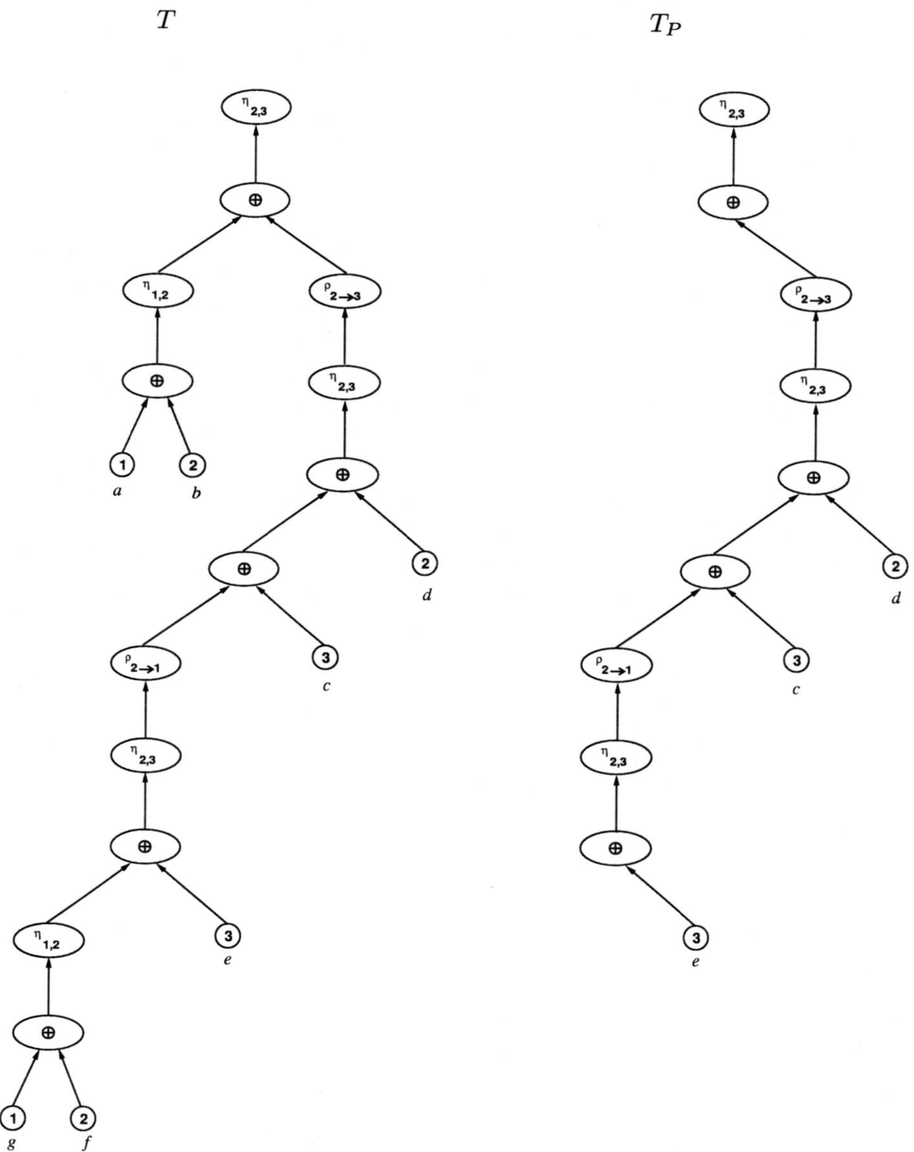

Figure 9: A 3-expression tree $T$ for the 3-labeled graph $G$ of Figure 8 and the subtree $T_P$ of $T$

by certain properties that have to be satisfied. The lemmas show that this is always possible without loss of generality.

## Partition into paths of type 1, 1.*a*, and 1.*b*

**Lemma 7.1** *There is always a k-expression tree $T$ for $G$ that satisfies the following property.*

**Property 7.2** *Let $u_1$ be a union node of $T$ such that one of its children $u_0$ is in $T_P$ and the other child $u_0'$ is not in $T_P$. Then the vertices of $G(u_0')$ are either all from $U_H$ or all from $U_J$.*

**Proof:** Since $G(u_0')$ does not contain vertices from $U_P$, we know that the vertices of $G(u_0')$ are all from $U_H \cup U_J$. If the vertices of $G(u_0')$ are not all from $U_H$ or not all from $U_J$ then let $T_H$ and $T_J$ be the two $k$-expression trees that define the subgraphs of $G(u_0')$ induced by the vertices of $U_H$ and $U_J$, respectively. $T_H$ and $T_J$ can easily be constructed from $T(u_0')$ by removing subtrees whose leaves represent only vertices from $U_J$ or $U_H$, respectively. A union node that loses one of its children can be omitted by making the remaining child to the child of its parent node. Then we replace subtree $T(u_0')$ by $T_H$ and $T_J$ as follows. We insert a new union node $v_0$ between $u_1$ and $u_0$, and make the roots of $T_H$ and $T_J$ to the second child of $u_1$ and $v_0$, respectively. The expression of the resulting tree obviously defines the same graph as before but $u_1$ now satisfies Property 7.2. This can be done for all union nodes which do not satisfy Property 7.2. See also Figure 10. $\qquad\square$

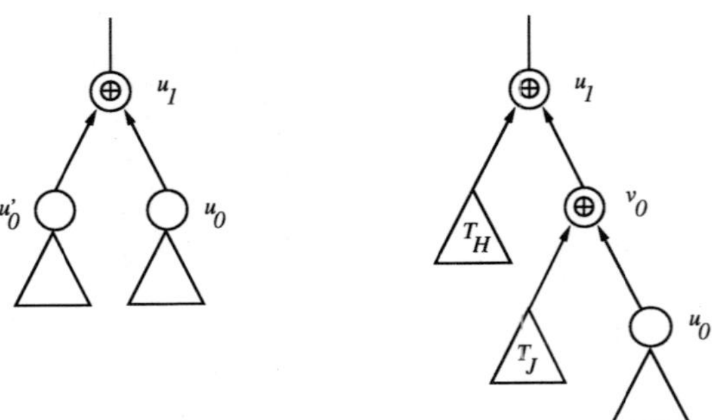


Figure 10: A transformation step used in the proof of Lemma 7.1

Let $X$ be the $k$-expression of $k$-expression tree $T$ which satisfies Property 7.2. Then we can apply the transformation steps of the proof of Theorem 4.2 to get a $k$-expression in normal form equivalent to $X$. This is possible because the transformation steps of the transformation into normal form only rearrange some relabeling and edge insertion operations. They do not change Property 7.2 of $T$. ¿From now on we will assume that $T$ satisfies Property 7.2 and that $X$ is in normal form.

Let $u_1$ be a union node of $T$ such that one of its children $u_0$ is in $T_P$ and the other child $u_0'$ is not in $T_P$. We define $\xi(u_1) := 0$ or $\xi(u_1) := 1$ if the vertices of $G(u_0')$ are all from $U_H$ or all from $U_J$, respectively. In all other cases and in

W. Espelage et al., *Deciding Clique-Width*, JGAA, 7(2) 141–180 (2003)    161

the case where $u_1$ is not a union node, we say $\xi(u_1)$ is undefined. For better readability we write $\xi(u_1) = H$ instead of $\xi(u_1) = 0$ and $\xi(u_1) = J$ instead of $\xi(u_1) = 1$. This does not mean that $\xi(u_1)$ is the graph $H$ or $J$, but only that all vertices of $G(u_0')$ are from $U_H$ or $H_J$, respectively. By Lemma 7.1, we can now assume that $\xi(u_1)$ is well defined for all union nodes $u_1$ of $T$ for which exactly one of their children is not in $T_P$.

The tree $T_P$ with $l$ leaves now consists of at most $2l - 1$ maximal paths $p = (u_1, \ldots, u_{s'})$, $s' \geq 1$, such that $u_1$ is a union node with two children in $T_P$ or $u_1$ has only one child in $T_P$ which is a leaf. The last node $u_{s'}$ of such a path $p$ is either the root of $T_P$ or a child of some union node whose children are both in $T_P$. All the graphs $G(u_s)$ for $s = 1, \ldots, s'$ contain the same vertices of $U_P$. Such a path of $T_P$ is called a 1-*path* or *path of type* 1. Every non-leaf node of $T_P$ is in exactly one of these paths of type 1.

A maximal subpath $(u_1, \ldots, u_{r'}, \ldots, u_{s'})$ of $T_P$ such that $u_1$ is a union node, $u_2, \ldots, u_{r'}$ are edge insertion nodes, and $u_{r'+1}, \ldots, u_{s'}$ are relabeling nodes, is called a *frame* of $T_P$. Every frame has at most $\binom{k}{2} + k$ nodes, because there is exactly one union node $u_1$, there are at most $\binom{k}{2}$ edge insertion nodes $u_2, \ldots, u_{r'}$, and at most $k - 1$ relabeling nodes $u_{r'+1}, \ldots, u_{s'}$. Figure 11 shows the general structure of a frame.

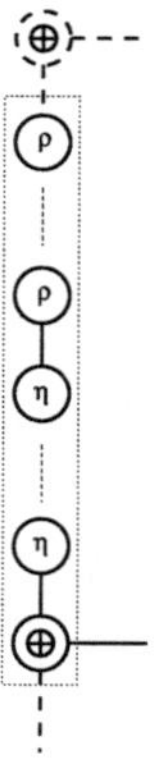

Figure 11: A frame always starts with a union node followed by edge insertion nodes and relabeling nodes.

The first frame of every 1-path is called a *path of type* 1.*a*. The remaining part, if not empty, is called a *path of type* 1.*b*. There are at most $2l - 1$ paths of type 1.*a* and at most $2l - 1$ paths of type 1.*b*. Every 1.*a* path has at most $\binom{k}{2} + k$ nodes, because it is a frame. For every union node $u_1$ of a 1.*b*-path there is either $\xi(u_1) = H$ or $\xi(u_1) = J$.

## Partition into paths of type 2.*a* and 2.*b*

For some node $u_s$ of the $k$-expression tree $T$, let $L_H(u_s)$, $L_J(u_s)$, and $L_P(u_s)$ be the label sets of the vertices of $G(u_s)$ which are from $U_H$, $U_J$, and $U_P$,

W. Espelage et al., *Deciding Clique-Width*, JGAA, 7(2) 141–180 (2003)      162

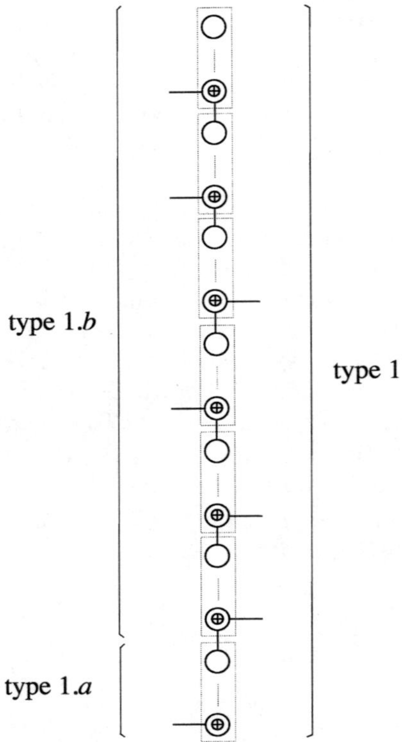


Figure 12: Every 1-path $p$ is divided into a path of type 1.$a$ and a path of type 1.$b$. The path of type 1.$a$ is the first frame of $p$. The path of type 1.$b$ is the remaining part of $p$, which can also be empty.

respectively. The intersection sets

$$L_H(u_s) \cap L_J(u_s), \quad L_P(u_s) \cap L_H(u_s), \quad L_P(u_s) \cap L_J(u_s),$$

and

$$L_P(u_s) \cap L_H(u_s) \cap L_J(u_s)$$

are abbreviated by

$$L_{H \cap J}(u_s), \quad L_{P \cap H}(u_s), \quad L_{P \cap J}(u_s)$$

and

$$L_{P \cap H \cap J}(u_s),$$

respectively.

**Lemma 7.3** *There is always a k-expression tree $T$ for $G$ such that the k-expression $X$ of $T$ is in normal form and $T$ satisfies Property 7.2 and additionally Property 7.4.*

W. Espelage et al., *Deciding Clique-Width*, JGAA, 7(2) 141–180 (2003)     163

**Property 7.4** *Let $u_s$ be a relabeling node of $T_P$ labeled by $\rho_{i \to j}$ and let $u_{s-1}$ be the child of $u_s$ in $T_P$. If $i \in L_P(u_{s-1})$ then $j \in L_P(u_{s-1})$.*

**Proof:** By induction on the height of $T_P$. Assume $X$ is in normal form and $T_P$ satisfies Property 7.2. Let $q = (u_1, \ldots, u_{r'}, \ldots, u_{s'})$ be a frame of $T_P$ and $u_s$, $r' < s \le s'$, be a relabeling node labeled by $\rho_{i \to j}$. Let $i \in L_P(u_{s-1})$. By the inductive hypothesis, we assume that $T(u_{s-1})$ already satisfies Property 7.4.

If $j \notin L_P(u_{s-1})$ then we simultaneously replace in the expression of subtree $T(u_{r'})$ every label $i$ by label $j$ and every label $j$ by label $i$. The expression of the resulting subtree $T(u_{r'})$ is still in normal form and $T(u_{r'})$ satisfies Property 7.2 and 7.4. Since $i$ is not involved in the relabeling operations of the nodes $u_{r'+1}, \ldots, u_{s-1}$, the resulting expression $X$ is obviously in normal form and defines the same graph as before, and $T(u_s)$ satisfies Property 7.2 and Property 7.4. $\qquad\square$

Let $u_{s-1}$ be the child of some relabeling node $u_s$ of $T_P$. By Lemma 7.3, we can now assume that

$$L_P(u_{s-1}) \quad \supseteq \quad L_P(u_s).$$

If $L_P(u_{s-1}) = L_P(u_s)$, then the reverse inclusion holds true for the sets $L_{P \cap H}(u_s)$ and $L_{P \cap J}(u_s)$, i.e.,

$$L_{P \cap H}(u_{s-1}) \quad \subseteq \quad L_{P \cap H}(u_s) \quad \text{and} \quad L_{P \cap J}(u_{s-1}) \quad \subseteq \quad L_{P \cap J}(u_s),$$

because a relabeling of a label from $L_{P \cap H}(u_{s-1})$ or $L_{P \cap J}(u_{s-1})$ is always a relabeling of a label from $L_P(u_{s-1})$.

This allows us to divide every $1.b$-path $p$ into *paths of type 2.a* and *paths of type 2.b* as follows. The $2.a$-paths are the frames $q = (u_1, \ldots, u_{r'}, \ldots, u_{s'})$ of $p$ for which at least one of the following two properties holds true.

1. There is some relabeling node $u_s$, $r' < s \le s'$, such that

$$L_P(u_{s-1}) \supsetneq L_P(u_s), \quad L_{P \cap H}(u_{s-1}) \subsetneq L_{P \cap H}(u_s),$$

   or

$$L_{P \cap J}(u_{s-1}) \subsetneq L_{P \cap J}(u_s).$$

2.

$$L_{P \cap H}(u_0) \subsetneq L_{P \cap H}(u_1) \quad \text{or} \quad L_{P \cap J}(u_0) \subsetneq L_{P \cap J}(u_1),$$

   where $u_0$ is the child of union node $u_1$ which is in $T_P$.

It is easy to verify that this is equivalent to property

$$L_P(u_0) \supsetneq L_P(u_{s'}), \quad L_{P \cap H}(u_0) \subsetneq L_{P \cap H}(u_{s'}), \quad \text{or} \quad L_{P \cap J}(u_0) \subsetneq L_{P \cap J}(u_{s'}).$$

where $u_0$ is the child of union node $u_1$ which is in $T_P$.

The $2.a$-paths are the frames $q$ of the $1.b$-paths for which either the number of labels in $L_P$ decreases or the number of labels in $L_{P \cap H}$ or $L_{P \cap J}$ increases.

W. Espelage et al., *Deciding Clique-Width*, JGAA, 7(2) 141–180 (2003)      164

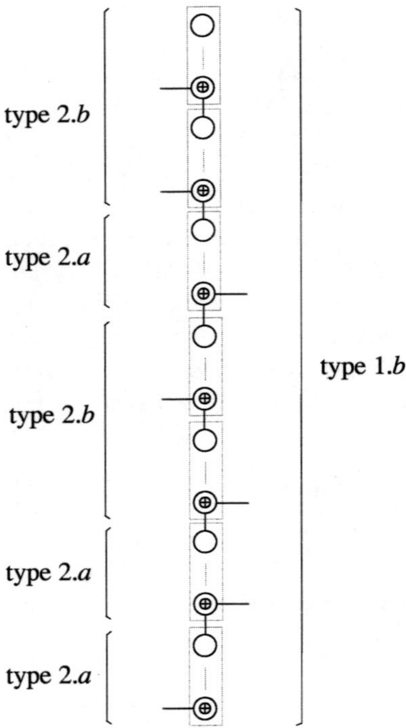

Figure 13: Every 1.$b$-path is divided into paths of type 2.$a$ and paths of type 2.$b$. The 2.$a$-paths are the frames $q$ of the 1.$b$-paths for which the number of labels in $L_P$ decreases or the number of labels in $L_{P \cap H}$ or $L_{P \cap J}$ increases.

The 2.$b$-paths are the remaining parts of the 1.$b$-paths. In a 2.$b$-path $p$ all the sets $L_P(u_s)$ are equal, all the sets $L_{P \cap H}(u_s)$ are equal, all the sets $L_{P \cap J}(u_s)$ are equal, and thus also all the sets $L_{P \cap H \cap J}(u_s)$ are equal, for all nodes $u_s$ of $p$ including the child $u_0$ of the first node $u_1$ which is in $T_P$. See also Figure 13.

For a frame $q = (u_1, \ldots, u_{r'}, \ldots, u_{s'})$ of a 2.$b$-path let

$$L_P(q) = L_P(u_{s'}), \quad L_{P \cap H}(q) = L_{P \cap H}(u_{s'}),$$

$$L_{P \cap J}(q) = L_{P \cap J}(u_{s'}), \quad \text{and} \quad L_{P \cap H \cap J}(q) = L_{P \cap H \cap J}(u_{s'}).$$

We use $q$ as the argument instead of some node of $q$ to emphasize that the sets above are equal for all nodes of $q$ including the child $u_0$ of the first node of $q$ which is in $T_P$. It is easy to count that for every 1.$b$-path $p$ there are at most $3k - 1$ paths of type 2.$a$ and thus at most $3k$ paths of type 2.$b$. A worst case example for $k = 3$ is shown in the following table. The $j$-th row shows the labeling for the last node $u_i$ of the $j$-th 2.$a$-frame.

W. Espelage et al., *Deciding Clique-Width*, JGAA, 7(2) 141–180 (2003)     165

| $j$ | $L_P(u_i)$ | $L_{P \cap H}(u_i)$ | $L_{P \cap J}(u_i)$ |
|---|---|---|---|
| 1 | $\{1,\,2,\,3\}$ | $\{1\}$ | $\emptyset$ |
| 2 | $\{1,\,2,\,3\}$ | $\{1,\,2\}$ | $\emptyset$ |
| 3 | $\{1,\,2,\,3\}$ | $\{1,\,2,\,3\}$ | $\emptyset$ |
| 4 | $\{1,\,2,\,3\}$ | $\{1,\,2,\,3\}$ | $\{1\}$ |
| 5 | $\{1,\,2,\,3\}$ | $\{1,\,2,\,3\}$ | $\{1,\,2\}$ |
| 6 | $\{1,\,2,\,3\}$ | $\{1,\,2,\,3\}$ | $\{1,\,2,\,3\}$ |
| 7 | $\{1,\,2\}$ | $\{1,\,2\}$ | $\{1,\,2\}$ |
| 8 | $\{1\}$ | $\{1\}$ | $\{1\}$ |

## Partition into paths of type 3.*a* and 3.*b*

**Lemma 7.5** *There is always a k-expression tree $T$ for $G$ such that the k-expression $X$ of $T$ is in normal form and $T$ satisfies Property 7.2, Property 7.4, and additionally Property 7.6.*

**Property 7.6** *Let $p$ be a 2.b-path of $T_P$, and $q = (u_1, \ldots, u_{r'}, \ldots, u_{s'})$ be a frame of $p$ such that node $u_s$, $r' < s \leq s'$, is a relabeling node labeled by $\rho_{i \to j}$. If $i \in L_{H \cap J}(u_{s-1})$ then $j \in L_{H \cap J}(u_1)$.*

Before we prove Lemma 7.5 let us emphasize that label $j$ will even be from $L_{H \cap J}(u_1)$ and not only from $L_{H \cap J}(u_{s-1})$.

**Proof:** Assume $X$ is in normal form, $T$ satisfies Property 7.2 and Property 7.4, and $T(u_{s-1})$ satisfies additionally Property 7.6 for some $s$, $r' < s \leq s'$. Let $i \in L_{H \cap J}(u_{s-1})$.

If $j \in L_P(q)$ then the assumption $i \in L_{H \cap J}(u_{s-1})$ and the relabeling $\rho_{i \to j}$ at node $u_s$ imply $j \in L_{P \cap H \cap J}(u_s) = L_{P \cap H \cap J}(q)$ and thus $j \in L_{H \cap J}(u_1)$.

If $j \notin L_P(q)$ and $j \notin L_{H \cap J}(u_1)$ then we simultaneously replace in the expression of subtree $T(u_{r'})$ every label $i$ by label $j$ and every label $j$ by label $i$. The new expression of the resulting subtree $T(u_{r'})$ is still in normal form and subtree $T(u_{r'})$ still satisfies the Properties 7.2, 7.4, and 7.6. Let $\rho_{i_1 \to j_1}, \ldots, \rho_{i_{l-1} \to j_{l-1}}$ be the relabeling operations of the nodes $u_{r'+1}, \ldots, u_{s-1}$. Label $i$ is not involved in these relabeling operations, i.e., $i \notin \{i_1, \ldots, i_{l-1}, j_1, \ldots, j_{l-1}\}$. Label $j$ is not relabeled by these relabeling operations, i.e., $j \notin \{i_1, \ldots, i_{l-1}\}$, and none of these relabeling operations relabels some label of $L_{H \cap J}(u_{r'})$ to $j$ in the original expression, because the original tree $T(u_{r'})$ satisfies Property 7.6 and $j \notin L_{H \cap J}(u_1)$. Thus the new expression of the resulting tree $T(u_s)$ is in normal form and defines the same graph as before, and tree $T(u_s)$ now satisfies the Properties 7.2, 7.4, and 7.6. $\qquad\square$

For some node $u_s$ of $T_P$ and some label $j \in [k]$ let $\text{forb}_P(u_s, j)$ be the set of all labels $i \in L_P(u_s)$ such that graph $G(u_s)$ has two non adjacent vertices, one labeled by $i$ and one labeled by $j$. If the set $\text{forb}_P(u_s, j)$ is empty then either graph $G(u_s)$ has no vertex labeled by $j$ or every vertex of $G(u_s)$ labeled by $j$ is adjacent to every vertex of $G(u_s)$ labeled by some label of $L_P(u_s)$. (Remember that $L_P(u_s)$ is always non-empty for the nodes $u_s$ of $T_P$).

W. Espelage et al., *Deciding Clique-Width*, JGAA, 7(2) 141–180 (2003)     166

Let $q = (u_1, \ldots, u_{r'}, \ldots, u_{s'})$ be a frame of $T_P$ such that $u_1$ is a union node. Let $u_0$ be the child of $u_1$ which is in $T_P$. If one of the edge insertion nodes $u_r$, $1 < r \le r'$, is labeled by $\eta_{i,j}$ then $i \notin \mathrm{forb}_P(u_0, j)$ and $j \notin \mathrm{forb}_P(u_0, i)$, because otherwise $\eta_{i,j}$ would create a forbidden edge between two vertices from $G(u_0)$.

Let $u_s$ be a relabeling node of $T_P$ labeled by $\rho_{i \to j}$. If $i \notin L_P(u_{s-1})$ then obviously

$$\mathrm{forb}_P(u_s, j) = \mathrm{forb}_P(u_{s-1}, j) \cup \mathrm{forb}_P(u_{s-1}, i).$$

Intuitively speaking, a vertex labeled by some label of $L_P(u_{s-1})$ is not adjacent in $G(u_s)$ to some vertex labeled by $j$ if and only if it is not adjacent in $G(u_{s-1})$ to some vertex labeled by $j$ or $i$.

**Lemma 7.7** *Assume expression tree $T$ satisfies Property 7.2, Property 7.4, and Property 7.6 and the $k$-expression $X$ of $T$ is in normal form. Let $p$ be a 2.b-path of $T_P$, let $q = (u_1, \ldots, u_{r'}, \ldots, u_{s'})$ be a frame of $p$, and let $u_0$ be the child of $u_1$ which is in $T_P$. If a node $u_s$, $r' < s \le s'$, is a relabeling node labeled by $\rho_{i \to j}$ and if $i \in L_{H \cap J}(u_{s-1})$ then*

$$\mathrm{forb}_P(u_0, i) \subsetneq \mathrm{forb}_P(u_{s'}, j) \ \text{ and } \ \mathrm{forb}_P(u_0, j) \subsetneq \mathrm{forb}_P(u_{s'}, j).$$

**Proof:** Since $i$ is not involved in the relabeling operations of the nodes $u_{r'+1}$, $\ldots, u_{s-1}$, label $i$ is also in $L_{H \cap J}(u_1)$. By Property 7.6, we know that $j \in L_{H \cap J}(u_1)$ and thus $i, j \in L_{H \cap J}(u_1)$. Let $u_0'$ be the other child of $u_1$ which is not in $T_P$. Without loss of generality, let $\xi(u_1) = H$. Since $i$ and $j$ are both in $L_J(u_1)$ and since the vertices of $G(u_0')$ are all from $U_H$, graph $G(u_0)$ has at least one vertex labeled by $i$ and at least one vertex labeled by $j$.

If label $i$ or label $j$ is involved in an edge insertion operation $\eta_{i',j'}$ of the nodes $u_2, \ldots, u_{r'}$ then the other label of $\{i', j'\}$ has to be in $L_P(q) - L_{H \cup J}(u_1)$, i.e., is not in $L_{H \cup J}(u_1)$, where $L_{H \cup J}(u_1)$ is defined by $L_H(u_1) \cup L_J(u_1)$. Otherwise, a forbidden edge between a vertex from $U_H$ and a vertex from $U_J$ is created, because $i$ and $j$ are both in $L_H(u_1)$ and both in $L_J(u_1)$. By our normal form Property 2.(a), we know that all these edge insertion operations do not create a new edge between two vertices from $G(u_0')$ or two vertices from $G(u_0)$. Thus every of these edge insertion operations in that label $i$ or $j$ is involved creates an edge between a vertex from $G(u_0')$ labeled by $i$ or $j$, respectively, and a vertex from $G(u_0)$ labeled by some label of $L_P(q) - L_{H \cup J}(u_1)$.

If every label of $\mathrm{forb}_P(u_0, i)$ is also in $\mathrm{forb}_P(u_0, j)$ then an additional relabeling $\rho_{i \to j}$ applied to the expression represented by $T(u_0)$ does not change the graph $G(u_{s'})$. This contradicts normal form Property 2.(c). On the other hand, if every label of $\mathrm{forb}_P(u_0, j)$ is also in $\mathrm{forb}_P(u_0, i)$ then an additional relabeling $\rho_{j \to i}$ applied to the expression represented by $T(u_0)$ does not change the graph $G(u_{s'})$. This also contradicts normal form Property 2.(c).

So there has to be at least one label in $\mathrm{forb}_P(u_0, i)$ which is not in $\mathrm{forb}_P(u_0, j)$ and one label in $\mathrm{forb}_P(u_0, j)$ which is not in $\mathrm{forb}_P(u_0, i)$. Since $\mathrm{forb}_P(u_0, i) \subseteq \mathrm{forb}_P(u_{s-1}, i)$ and $\mathrm{forb}_P(u_0, j) \subseteq \mathrm{forb}_P(u_{s-1}, j)$, the result follows. $\qquad\square$

W. Espelage et al., *Deciding Clique-Width*, JGAA, 7(2) 141–180 (2003)    167

Next we divide every 2.$b$-path $p$ into *paths of type 3.a* and *paths of type 3.b* as follows. The 3.$a$-paths are the frames $q = (u_1, \ldots, u_{r'}, \ldots, u_{s'})$ of $p$ for which

$$L_{H \cap J}(u_0) \subsetneqq L_{H \cap J}(u_{s'}) \quad \text{or} \quad \mathrm{forb}_P(u_0, j) \subsetneqq \mathrm{forb}_P(u_{s'}, j)$$

for some $j \in L_P(u_0) \cup L_{H \cap J}(u_0)$, where $u_0$ is the child of $u_1$ in $T_P$.

The sets above can change their size in a frame of a 2.$b$-path as follows.

1. $L_{H \cap J}(u_0) \subseteq L_{H \cap J}(u_1)$ and $\mathrm{forb}_P(u_0, j) \subseteq \mathrm{forb}_P(u_1, j)$, because a union operation can not remove labels from $L_{H \cap J}(u_0)$ or $\mathrm{forb}_P(u_0, j)$, respectively.

2. $L_{H \cap J}(u_{r-1}) = L_{H \cap J}(u_r)$ and $\mathrm{forb}_P(u_0, j) \subseteq \mathrm{forb}_P(u_r, j)$ for $r = 2, \ldots, r'$, because the edge insertion operations do not change the labels, and do not create edges between two vertices from $G(u_0)$, respectively. Note that they can not remove labels from $\mathrm{forb}_P(u_0, j)$, although they can remove labels from $\mathrm{forb}_P(u_1, j)$.

3. Let $u_s$, $r' < s \leq s'$, be a relabeling node labeled by $\rho_{j \to j'}$.

   (a) If $j \notin L_P(u_{s-1}) \cup L_{H \cap J}(u_{s-1})$, then $L_{H \cap J}(u_{s-1}) \subseteq L_{H \cap J}(u_s)$.

   (b) If $j \in L_P(u_{s-1}) \cup L_{H \cap J}(u_{s-1})$ then $j \in L_{H \cap J}(u_{s-1})$, because we consider a 2.$b$-path. By Lemma 7.5, $j' \in L_{H \cap J}(u_{s-1})$ and thus $L_{H \cap J}(u_{s-1}) \supsetneqq L_{H \cap J}(u_s)$. By Lemma 7.7, $\mathrm{forb}_P(u_s, j) = \emptyset$, $\mathrm{forb}_P(u_0, j) \subsetneqq \mathrm{forb}_P(u_s, j')$ and $\mathrm{forb}_P(u_0, j') \subsetneqq \mathrm{forb}_P(u_s, j')$.

The size of $\mathrm{forb}_P(u_{s-1}, j)$ for some $j \in L_P(u_{s-1}) \cup L_{H \cap J}(u_{s-1})$ can only become smaller in case 3.($b$), where $j \in L_{H \cap J}(u_{s-1})$ is relabeled into another label $j' \in L_{H \cap J}(u_{s-1})$. In this case $\mathrm{forb}_P(u_s, j) = \emptyset$, because $G(u_s)$ has no vertex labeled by $j$.

A simple idea shows that the number of 3.$a$-paths (3.$a$-frames) can be bounded by $(k + 1)^{k+1}$. For a node $u_s$ let $\alpha(u_s) = (z_0, \ldots, z_{k'})$ be the vector, where $k' = |L_P(u_s)|$ and $z_t$, $0 \leq t \leq k'$, is the number of sets $\mathrm{forb}_P(u_s, j)$, $j \in L_P(u_s) \cup L_{H \cap J}(u_s)$, of size $t$. We say vector $(z'_0, \ldots, z'_{k'})$ is larger than vector $(z_0, \ldots, z_{k'})$, denoted by

$$(z'_0, \ldots, z'_{k'}) > (z_0, \ldots, z_{k'}),$$

if there is some $t$, $0 \leq t \leq k'$, such that $z'_t > z_t$ and $z'_{t'} = z_{t'}$ for $t' = t+1, \ldots, k'$. For every 3.$a$-path $q = (u_1, \ldots, u_{r'}, \ldots, u_{s'})$, we have $\alpha(u_{s'}) > \alpha(u_0)$, where $u_0$ is the child of $u_1$ which is in $T_P$. This bounds the number of 3.$a$-paths by $(k + 1)^{k+1}$. Note that this bound is not really tight.

The remaining parts of $p$ are the 3.$b$-paths. In a 3.$b$-path $p$ all the sets $L_{H \cap J}(u_s)$ are equal for all nodes $u_s$ of $p$ including the child of the first node which is in $T_P$. To emphasizes this we define $L_{H \cap J}(q) = L_{H \cap J}(u_{s'})$ for the frames $q = (u_1, \ldots, u_{r'}, \ldots, u_{s'})$ of a 3.$b$-path $p$. The sets $\mathrm{forb}_P(u_s, j)$ for $j \in L_P(u_0) \cup L_{H \cap J}(u_0)$ do not need to be equal for all nodes $u_s$ of $p$. In a frame $q = (u_1, \ldots, u_{r'}, \ldots, u_{s'})$ of a 3.$b$-path $p$, it could be that there is some $r$, $1 \leq r < r'$, such that set $\mathrm{forb}_P(u_r, j)$ has a label which is not in $\mathrm{forb}_P(u_0, j)$. However, we know that for $s = r', \ldots, s'$, $\mathrm{forb}_P(u_0, j) = \mathrm{forb}_P(u_s, j)$.

W. Espelage et al., *Deciding Clique-Width*, JGAA, 7(2) 141–180 (2003)    168

## Partition into paths of type $4$

To partition the paths of type $3.b$ into paths of type $4$, we need three more lemmas. The first lemma already holds for paths of type $1.b$, but we use it only for paths of type $3.b$.

**Lemma 7.8** *Let $q = (u_1, \ldots, u_{r'}, \ldots, u_{s'})$ be a frame of a 1.b-path $p$ such that $u_r$, $1 < r \leq r'$, is an edge insertion node labeled by $\eta_{i',j'}$. Let $u_0$ and $u_0'$ be the two children of $u_1$, where $u_0$ is in $T_P$. If $\xi(u_1) = H$ (if $\xi(u_1) = J$) then operation $\eta_{i',j'}$ only inserts edges between vertices from graph $G(u_0')$ labeled by labels of $L_H(u_1)$ (of $L_J(u_1)$) and vertices from $G(u_0)$ not labeled by labels of $L_J(u_1)$ (of $L_H(u_1)$, respectively).*

**Proof:** If $\xi(u_1) = H$ (if $\xi(u_1) = J$) then all vertices of $G(u_0')$ are from $U_H$ (from $U_J$, respectively). Since $\eta_{i',j'}$ creates at least one edge between a vertex from $G(u_0')$ and a vertex from $G(u_0)$ and since there is no edge between a vertex from $U_H$ and a vertex from $U_J$, one label of $\{i', j'\}$ has to be in $L_H(u_1)$ (in $L_J(u_1)$) and the other label of $\{i', j'\}$ can not be in $L_J(u_1)$ (in $L_H(u_1)$, respectively). $\square$

The next lemma shows that the relabeling operations of a frame

$$q = (u_1, \ldots, u_{r'}, \ldots, u_{s'})$$

from a 3.b-path with $\xi(u_1) = H$ relabels only inner vertices from $H$.

**Lemma 7.9** *Let $q = (u_1, \ldots, u_{r'}, \ldots, u_{s'})$ be a frame of a 3.b-path $p$ such that node $u_s$, $r' < s \leq s'$, is a relabeling node labeled by $\rho_{i \to j}$.*

1. *If $\xi(u_1) = H$ then $i \in L_H(u_{s-1}) - L_P(q) - L_J(u_{s-1})$ and $j \in L_H(u_{s-1})$.*

2. *If $\xi(u_1) = J$ then $i \in L_J(u_{s-1}) - L_P(q) - L_H(u_{s-1})$ and $j \in L_J(u_{s-1})$.*

**Proof:** Since in a 3.b-path $p$, the labels of $L_P(q)$ and $L_{H \cap J}(q)$ are not relabeled, label $i$ can only be in $L_H(u_{s-1}) - L_P(q) - L_J(u_{s-1})$ or $L_J(u_{s-1}) - L_P(q) - L_H(u_{s-1})$.

If $i \in L_H(u_{s-1}) - L_P(q) - L_J(u_{s-1})$ then we get $j \in L_H(u_{s-1})$, otherwise $L_{P \cap H}(u_{s-1}) \subsetneq L_{P \cap H}(u_s)$ or $L_{H \cap J}(u_{s-1}) \subsetneq L_{H \cap J}(u_s)$. Both are not possible in a 3.b-path. On the other hand, if $i \in L_J(u_{s-1}) - L_P(q) - L_H(u_{s-1})$ then we get $j \in L_J(u_{s-1})$, otherwise $L_{P \cap J}(u_{s-1}) \subsetneq L_{P \cap J}(u_s)$ or $L_{H \cap J}(u_{s-1}) \subsetneq L_{H \cap J}(u_s)$.

Let $u_0$ be the child of $u_1$ which is in $T_P$. Assume first that $\xi(u_1) = H$, $i \in L_J(u_{s-1}) - L_P(q) - L_H(u_{s-1})$, and $j \in L_J(u_{s-1})$. Then $G(u_0)$ has a vertex labeled by $i$ and a vertex labeled by $j$.

1. If $j$ is not involved in an edge insertion operation of the nodes $u_2, \ldots, u_r$ then in $G(u_0)$ label $i$ can be relabeled into $j$, without changing $G(u_{s'})$. This contradicts our normal form Property 2.(c).

W. Espelage et al., *Deciding Clique-Width*, JGAA, 7(2) 141–180 (2003)     169

2. If $j$ is involved in some edge insertion operation of the nodes $u_2, \ldots, u_r$ then $j$ is contained in $L_{H \cap J}(u_1) = L_{H \cap J}(q) = L_{H \cap J}(u_0)$ and thus $\mathrm{forb}_P(u_{s'}, j) = \mathrm{forb}_P(u_0, j)$, and we can also relabel $i$ into $j$ in graph $G(u_0)$ without changing the resulting graph $G(u_{s'})$. This also contradicts our normal form Property 2.(c).

Thus, we get $i \in L_H(u_{s-1}) - L_P(q) - L_J(u_{s-1})$ and $j \in L_H(u_{s-1})$. For $\xi(u_1) = J$, we get $i \in L_J(u_{s-1}) - L_P(q) - L_H(u_{s-1})$ and $j \in L_J(u_{s-1})$.     $\square$

In the proof of the next lemma, we will frequently rearrange frames in a path of type 3.*b*. Assume a path $p$ consists of two consecutive frames, i.e.,

$$p = (u_1, \ldots, u_{r'}, \ldots, u_{s'},\ u_{s'+1}, \ldots, u_{r''}, \ldots, u_{s''}),$$

where $u_1$ and $u_{s'+1}$ are union nodes, $u_2, \ldots, u_{r'}$ and $u_{s'+2}, \ldots, u_{r''}$ are edge insertion nodes, and $u_{r'+1}, \ldots, u_{s'}$ and $u_{r''+1}, \ldots, u_{s''}$ are relabeling nodes. Let $u_0'$ and $u_0$ be the two children of $u_1$, where $u_0$ is in $T_P$, and let $u_0''$ be the child of $u_{s'+1}$ which is not in $T_P$.

If we exchange the two frames of $p$ then we get the new path

$$p' = (u_{s'+1}, \ldots, u_{r''}, \ldots, u_{s''}, u_1, \ldots, u_{r'}, \ldots, u_{s'}).$$

In the resulting expression tree, union node $u_{s'+1}$ has the two children $u_0''$ and $u_0$, and union node $u_1$ has the two children $u_0'$ and $u_{s''}$. The left-right order of the children of $u_1$ and $u_{s'+1}$ is not changed. That is, if $u_0'$ is the left child (right child) of $u_1$ in the original expression tree then $u_0'$ is the left child (right child, respectively) of $u_1$ in the new expression tree, and if $u_0''$ is the left child (right child) of $u_{s'+1}$ in the original expression tree then $u_0''$ is the left child (right child, respectively) of $u_{s'+1}$ in the new expression tree.

This rearrangement changes the expression defined by the original expression tree $T(u_{s''})$ as follows, see also Figure 14. Let $X_1, X_2, X_3$ be the expressions defined by the expression trees $T(u_0')$, $T(u_0)$, and $T(u_0'')$, respectively. Without loss of generality, let $u_0'$ be the left child of $u_1$ and $u_0''$ be the right child of $u_{s'+1}$. Let $\eta_{i_2, j_2}, \ldots, \eta_{i_{r'}, j_{r'}}$ be the edge insertion operations of the nodes $u_2, \ldots, u_{r'}$, let $\rho_{i_{r'+1} \to j_{r'+1}}, \ldots, \rho_{i_{s'} \to j_{s'}}$ be the relabeling operations of the nodes $u_{r'+1}, \ldots, u_{s'}$, let $\eta_{i_{s'+2}, j_{s'+2}}, \ldots, \eta_{i_{r''}, j_{r''}}$ be the edge insertion operations of the nodes $u_{s'+2}, \ldots, u_{r''}$, and let $\rho_{i_{r''+1} \to j_{r''+1}}, \ldots, \rho_{i_{s''} \to j_{s''}}$ be the relabeling operations of the nodes $u_{r''+1}, \ldots, u_{s''}$. Then the original expression defined by $T(u_{s''})$ is

$$\rho_{i_{s''} \to j_{s''}}(\cdots \rho_{i_{r''+1} \to j_{r''+1}}(\eta_{i_{r''}, j_{r''}}(\cdots \eta_{i_{s'+2}, j_{s'+2}}(Y \oplus X_3) \cdots)) \cdots),$$

where

$$Y = \rho_{i_{s'} \to j_{s'}}(\cdots \rho_{i_{r'+1} \to j_{r'+1}}(\eta_{i_{r'}, j_{r'}}(\cdots \eta_{i_2, j_2}(X_1 \oplus X_2) \cdots)) \cdots).$$

The expression of the new expression tree $T(u_{s'})$ which we get after exchanging the two frames is

$$\rho_{i_{s'} \to j_{s'}}(\cdots \rho_{i_{r'+1} \to j_{r'+1}}(\eta_{i_{r'}, j_{r'}}(\cdots \eta_{i_2, j_2}(X_1 \oplus Y') \cdots)) \cdots),$$

W. Espelage et al., *Deciding Clique-Width*, JGAA, 7(2) 141–180 (2003)     170

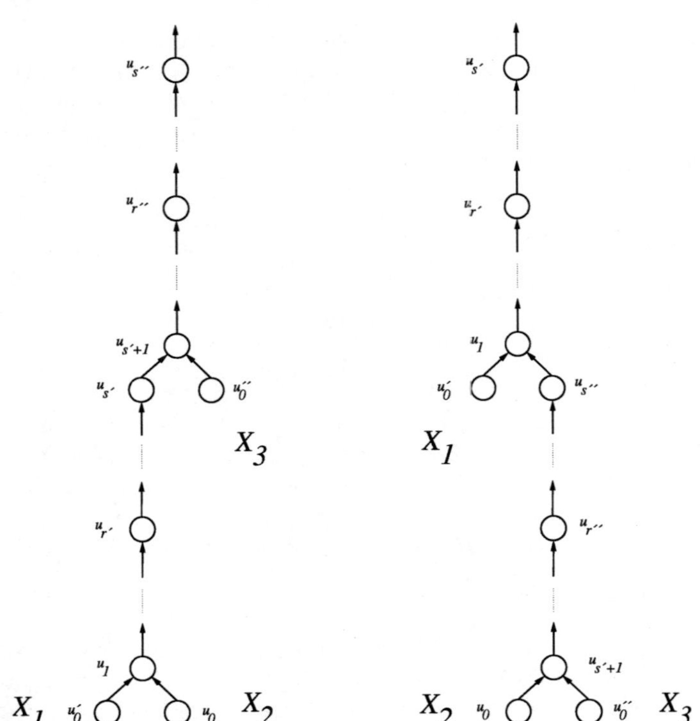

Figure 14: The rearrangement of two frames.

where

$$Y' = \rho_{i_{s''} \to j_{s''}} (\cdots \rho_{i_{r''+1} \to j_{r''+1}} (\eta_{i_{r''}, j_{r''}} (\cdots \eta_{i_{s'+2}, j_{s'+2}} (X_2 \oplus X_3) \cdots )) \cdots ).$$

Note that the new expression and the original expression do not need to be equivalent, but the order of the leaves in the tree is not changed.

We need the following additional notation. Let $u_s$ be a node of $T$. The labels of $L_H(u_s) - L_P(u_s) - L_J(u_s)$ and $L_J(u_s) - L_P(u_s) - L_H(u_s)$ are called the *unfixed labels* of $G(u_s)$. Within a path of type 3.b, only unfixed labels are relabeled, see also Lemma 7.9. The vertices labeled by unfixed labels of $G(u_s)$ are called *unfixed vertices* of $G(u_s)$.

**Lemma 7.10** *There is always a $k$-expression $X$ in normal form such that tree $T$ satisfies Property 7.2, Property 7.4, Property 7.6, and additionally Property 7.11.*

**Property 7.11** *Every 3.b-path $p$ is divided into at most $3 \cdot 2^{2(k+1)}$ paths $p'$ such that either for all frames $q = (u_1, \ldots, u_{r'}, \ldots, u_{s'})$ of $p'$ $\xi(u_1) = H$ or for all frames $q = (u_1, \ldots, u_{r'}, \ldots, u_{s'})$ of $p'$ $\xi(u_1) = J$.*

W. Espelage et al., *Deciding Clique-Width*, JGAA, 7(2) 141–180 (2003)     171

**Proof:** Let $q_1, \ldots, q_t$ be the frames of a 3.$b$-path $p$, i.e., $p = q_1 \odot \cdots \odot q_t$, where operation $\odot$ is the concatenation of paths.

For a frame $q = (u_1, \ldots, u_{s'})$ of $p$, let $\xi(q) := \xi(u_1)$, $\mathrm{unfix}_H(q) = |L_H(u_{s'}) - L_P(u_{s'}) - L_J(u_{s'})|$, $\mathrm{unfix}_J(q) = |L_J(u_{s'}) - L_P(u_{s'}) - L_H(u_{s'})|$ and

$$\begin{aligned}
\min_H(p) &= \min\{\mathrm{unfix}_H(q_1), \ldots, \mathrm{unfix}_H(q_t)\} \quad \text{and} \\
\min_J(p) &= \min\{\mathrm{unfix}_J(q_1), \ldots, \mathrm{unfix}_J(q_t)\}.
\end{aligned}$$

Let $r_1, r_2$, $1 \le r_1 \le r_2 \le t$, such that $r_2 - r_1$ is maximal and either

$$\begin{aligned}
\mathrm{unfix}_H(q_{r_1}) &= \min_H(p) & \text{and} & & \mathrm{unfix}_J(q_{r_2}) &= \min_J(p) \quad \text{or} \\
\mathrm{unfix}_J(q_{r_1}) &= \min_J(p) & \text{and} & & \mathrm{unfix}_H(q_{r_2}) &= \min_H(p).
\end{aligned}$$

If $q_{i_1}, \ldots, q_{i_n}$, $1 \le i_1 < i_2 < \cdots < i_n \le t$, are the frames for which

$$\mathrm{unfix}_H(q_{i_1}) = \mathrm{unfix}_H(q_{i_2}) = \cdots = \mathrm{unfix}_H(q_{i_n}) = \min_H(p)$$

and if $q_{j_1}, \ldots, q_{j_m}$, $1 \le j_1 < j_2 < \cdots < i_m \le t$, are the frames for which

$$\mathrm{unfix}_J(q_{j_1}) = \mathrm{unfix}_J(q_{j_2}) = \cdots = \mathrm{unfix}_J(q_{j_m}) = \min_J(p),$$

then either $r_1 = i_1$ and $r_2 = j_m$ or $r_1 = j_1$ and $r_2 = i_n$.

We divide the path $p$ into three parts $p_{\text{first}}, p_{\text{middle}}, p_{\text{last}}$ such that $p = p_{\text{first}} \odot p_{\text{middle}} \odot p_{\text{last}}$. Subpath $p_{\text{middle}}$ starts with frame $q_{r_1}$ and ends with frame $q_{r_2}$, see also Figure 15.

If the first part $p_{\text{first}}$ or the last part $p_{\text{last}}$ of $p$ are not empty then they will be partitioned in the same way as $p$. Since

$$\begin{aligned}
\min_J(p_{\text{first}}) &> \min_J(p) & \text{and} & & \min_H(p_{\text{last}}) &> \min_H(p) \quad \text{or} \\
\min_H(p_{\text{first}}) &> \min_H(p) & \text{and} & & \min_J(p_{\text{last}}) &> \min_J(p),
\end{aligned}$$

the partition procedure yields at most $2^{2(k+1)}$ such paths $p_{\text{middle}}$.

Assume $\mathrm{unfix}_H(q_{r_1}) = \min_H(p)$ and $\mathrm{unfix}_J(q_{r_2}) = \min_J(p)$. The second case where $\mathrm{unfix}_J(q_{r_1}) = \min_J(p)$ and $\mathrm{unfix}_H(q_{r_2}) = \min_H(p)$ runs analogously. Then $\xi(q_{r_1}) = H$ and $\xi(q_{r_2}) = J$ and we rearrange the frames in path $p_{\text{middle}}$ such that in the new path there is first frame $q_{r_1}$, then all frames $q$ with $\xi(q) = J$ and then all remaining frames $q$ with $\xi(q) = H$. If we move all frames $q$ of $p_{\text{middle}}$ where $\xi(q) = J$ to the front, then the remaining frames $q$ of $p_{\text{middle}}$ where $\xi(q) = H$ (except frame $q_{r_1}$) will automatically move to the end. This rearrangement will yield at most $3 \cdot 2^{2(k+1)}$ paths such that for all frames $q$ of every path all $\xi(q)$ are equal.

The order of the frames $q$ with $\xi(q) = J$ or $\xi(q) = H$ is not changed, i.e, it is the same order as in the original path $p_{\text{middle}}$. The order of the nodes in the frames is also not changed when moving frames. To ensure that the resulting expression is really equivalent to the original one, we perform a relabeling of the unfixed labels as follows.

For every node $u_s$ of the new path $p_{\text{middle}}$, we define step by step a bijection $b_{u_s} : [k] \to [k]$. The idea is to use for the operations on subgraph $G(u_s)$ label

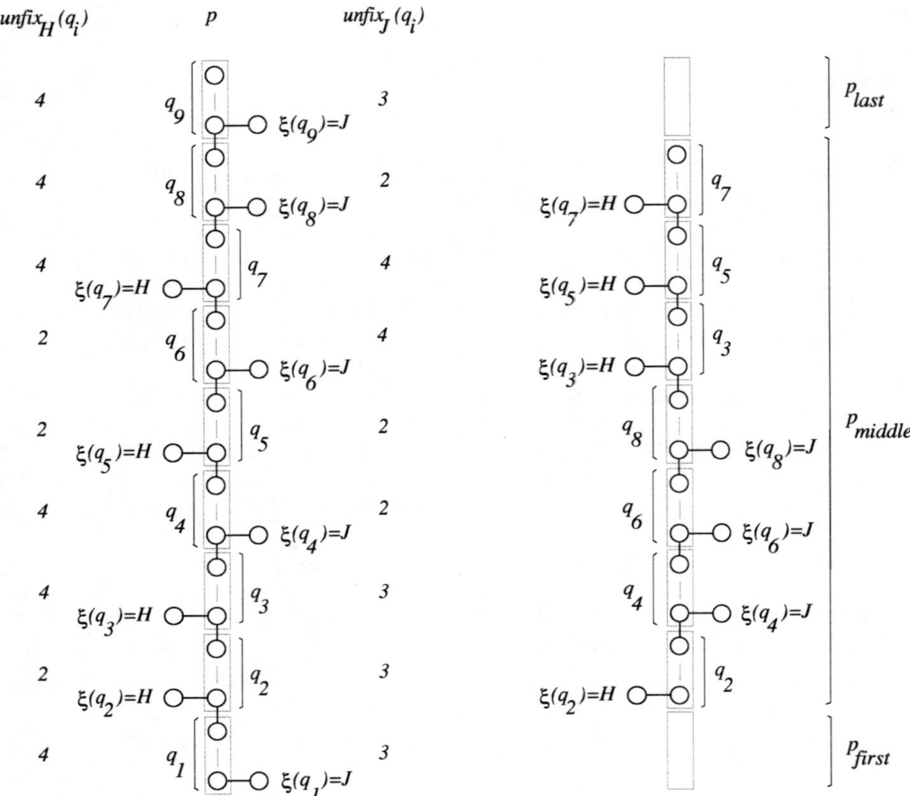

Figure 15: The partition of path $p$ into three parts $p_{\text{first}}, p_{\text{middle}}, p_{\text{last}}$, where $\min_H(p) = 2$, $\min_J(p) = 2$, $r_1 = 2$, and $r_2 = 8$

$b_{u_s}(i)$ instead of label $i$. For all labels $i \in L_P(u_s) \cup L_{H \cap J}(u_s)$, we will have $b_{u_s}(i) = i$, because these labels are not relabeled along a path of type 3.$b$.

The bijections for the nodes of $q_{r_1}$ are identities. The other bijections are defined step by step depending on the operations of the nodes along the new path $p_{\text{middle}}$.

We consider the nodes $u_s$ of the new path $p_{\text{middle}}$ in the given order starting with the parent node of the last node of frame $q_{r_1}$.

1. If $u_s$ is a union node then let $u'_{s-1}$ and $u_{s-1}$ be the two children of $u_s$, where $u'_{s-1}$ is not in $T_P$. The bijection $b_{u_s}$ of $u_s$ is initially the bijection $b_{u_{s-1}}$ of the child $u_{s-1}$.

   We then simultaneously replace in the expression of subtree $T(u'_{s-1})$ every label $i$ by $b_{u_s}(i)$. After that, we verify whether there is a vertex $w$ of $G(u'_{s-1})$ which is an unfixed vertex in the original tree $T(u_s)$ but not an unfixed vertex in the new tree $T(u_s)$.

If there is such a vertex $w$ originally labeled by $i$ and now labeled by $b_{u_s}(i)$, then we choose an arbitrary label $j$ such that $b_{u_s}(j)$ is not used up to now, i.e., $b_{u_s}(j) \notin L_P(u_s) \cup L_H(u_s) \cup L_J(u_s)$ with respect to the expression defined up to now by the new tree $T(u_s)$. We define $b_{u_s}(i) := b_{u_s}(j)$ and $b_{u_s}(j) := b_{u_s}(i)$, and simultaneously exchange in the expression of subtree $T(u'_{s-1})$ every label $b_{u_s}(i)$ by $b_{u_s}(j)$ and every label $b_{u_s}(j)$ by $b_{u_s}(i)$.

2. If $u_s$ is an edge insertion node labeled by $\eta_{i,j}$, then bijection $b_{u_s}$ is the bijection $b_{u_{s-1}}$ and node $u_s$ will be labeled by operation $\eta_{b_{u_s}(i),b_{u_s}(j)}$.

3. If $u_s$ is a relabeling node labeled by $\rho_{i \to j}$, then bijection $b_{u_s}$ is the bijection $b_{u_{s-1}}$ and node $u_s$ will be labeled by operation $\rho_{b_{u_s}(i) \to b_{u_s}(j)}$.

For the final node $u_s$ of the new path $p_{\text{middle}}$ we perform one additional relabeling of the resulting expression defined by $T(u_s)$ such that the unfixed vertices of $G(u_s)$ are labeled as in the graph defined by the final node of the original path $p_{\text{middle}}$.

All these relabeling steps are possible, because for all nodes $u_s$ in the first part of $p_{\text{middle}}$ $\text{unfix}_H(u_s) = \min_H(p)$, and for all nodes $u_s$ of the last part of $p_{\text{middle}}$ $\text{unfix}_J(u_s) = \min_J(p)$. Thus, there are always enough unused labels to relabel the unfixed vertices. Note that the labels of the sets $L_P(u_s)$ and $L_{H \cap J}(u_s)$ are unchanged along the nodes of $p_{\text{middle}}$.

It remains to show that the new expression is equivalent to the original expression. Let $q = (u_1, \ldots, u_{r'}, \ldots, u_{s'})$ be a frame of the original path $p_{\text{middle}}$ where $\xi(u_1) = H$. The other case where $\xi(u_1) = J$ runs analogously and is even less complicated. Frame $q$ can be moved by the rearrangement toward the end of $p_{\text{middle}}$. Let $u'_0$ and $u_0$ be the two children of $u_1$, where $u_0$ is in $T_P$. Node $u'_0$ is also a child of $u_1$ in the new expression tree, because the children of the union nodes which are not in $T_P$ are not changed by the rearrangement of the frames.

Consider now an edge insertion operation $\eta_{i,j}$ of some node $u_r$, $1 < r \le r'$, of frame $q$ in the original expression. By Lemma 7.8, we know that the edge insertion operation $\eta_{i,j}$ of node $u_r$ creates only edges between vertices from $G(u'_0)$ and vertices from $G(u_0)$. We also know that one of the two labels $i, j$ is from $L_H(u_1)$ and the other is not from $L_J(u_1)$. Without loss of generality, let $j \in L_H(u_1)$ and $i \notin L_J(u_1)$.

The rearrangement of the frames does not change the order of the leaves in the expression tree, see Figure 14. It also does not change the order of the frames $q$ with the same $\xi(q)$ on path $p_{\text{middle}}$. Since $\eta_{i,j}$ creates only edges between vertices from $U_H$ and vertices from $U_H \cup U_P$, all these edges are also created by the corresponding edge insertion operation $\eta_{b_{u_r}(i),b_{u_r}(j)}$ in the new expression.

Assume edge insertion operation $\eta_{b_{u_r}(i),b_{u_r}(j)}$ in the new expression creates an edge which is not in the graph defined by the original expression. Then one of the nodes of this edge has to be in $U_J$. This node can only be labeled by $b_{u_r}(j)$, because $i \notin L_J(u_1)$ for $u_1$ from the original expression tree, and by our

W. Espelage et al., *Deciding Clique-Width*, JGAA, 7(2) 141–180 (2003)     174

relabeling procedure $b_{u_r}(i) \notin L_J(u_1)$ for $u_1$ from the new expression tree. Now we get $b_{u_r}(j) \in L_{H \cap J}(u_1)$, $b_{u_r}(i) \in L_P(u_1)$, and $b_{u_r}(i) \notin \mathrm{forb}_P(u_{s'}, b_{u_r}(j))$ for $u_1$ from the new expression tree. Since all sets $L_P(u_s)$ are equal for all nodes $u_s$ of $p_{\mathrm{middle}}$, all sets $L_{H \cap J}(u_s)$ are equal for all nodes $u_s$ of $p_{\mathrm{middle}}$, and all sets $\mathrm{forb}_P(u_s, j)$ are equal for all the last nodes $u_s$ of all frames of $p_{\mathrm{middle}}$, and since these labels are not relabeled by our relabeling procedure, we get that all edges created by $\eta_{b_{u_r}(i), b_{u_r}(j)}$ have to be in the graph defined by the original expression. This contradicts our assumption.

Thus the original expression and the new expression are equivalent. Note that the normal form property and the Properties 7.2, 7.4, and 7.6 are also not changed by the rearrangement of the frames. $\square$

Lemma 7.10 allows us to divide every 3.$b$-path into at most $3 \cdot 2^{2(k+1)}$ *paths of type* 4. The paths of type 4 are the those parts of the paths $p_{\mathrm{middle}}$ in that for all frames $q$ all $\xi(q)$ are equal.

## The connection type of $H$

Let us summarize how the paths of tree $T_P$ are partitioned now. Tree $T_P$ consists of

1. at most $2l - 1$ paths of type 1.$a$,

2. at most $(2l - 1) \cdot (3k - 1)$ paths of type 2.$a$,

3. at most $(2l - 1) \cdot 3k \cdot (k + 1)^{k+1}$ paths of type 3.$a$, and

4. at most $(2l - 1) \cdot 3k \cdot ((k + 1)^{k+1} + 1) \cdot 3 \cdot 2^{2(k+1)}$ paths of type 4.

Every non-leaf node of $T_P$ is in exactly one of these paths of type 1.$a$, 2.$a$, 3.$a$, or 4. Every path of type 1.$a$, 2.$a$, or 3.$a$ has at most $\binom{k}{2} + k$ nodes, because these paths are frames. For all frames $q = (u_1, \ldots, u_{s'})$ in a path of type 4 all $\xi(u_1)$ are equal, all sets $L_P(q)$ and $L_{H \cap J}(q)$ are equal, and all sets $\mathrm{forb}_P(u_{s'}, j)$ are equal for all $j \in L_P(q) \cup L_{H \cap J}(q)$. For a node $u_s$ of $T_P$, let $L_P(u_s)$ be the set of all *terminal labels*, $L_H(u_s)$ be the set of all *internal labels*, and $L_J(u_s)$ be the set of all *external labels* for node $u_s$.

We now replace every 4-path $p = (u_1, \ldots, u_{s'})$ of $T_P$ which consists of more than one frame by some so-called *bridge node* node $v$. Let $u_0$ be the child of $u_1$ which is in $T_P$ and $u_{s'+1}$ be the parent node of $u_{s'}$ in $T_P$. Then the path $(u_0, u_1, \ldots, u_{s'}, u_{s'+1})$ is replaced by path $(u_0, v, u_{s'+1})$. Node $v$ is called an *internal* bridge node if $\xi(u_1) = H$ and an *external bridge node* if $\xi(u_1) = J$. Every bridge node represents a 4-path with more than one frame. Note that in a succeeding replacement the nodes $u_0$ and $u_{s'+1}$ can also be bridge nodes. At every bridge node we store the information whether it is internal or external, and the set of all terminal labels $L_P(u_{s'})$, the set of all internal labels $L_H(u_{s'})$, the set of all external labels $L_J(u_{s'})$, and the pairs $(\mathrm{forb}_P(u_{s'}, j), j)$ for all $j \in L_P(u_{s'}) \cup L_{H \cap J}(u_{s'})$.

W. Espelage et al., *Deciding Clique-Width*, JGAA, 7(2) 141–180 (2003)     175

A union node $u_1$ is called an *internal union node* if $\xi(u_1) = H$ and an *external union node* if $\xi(u_1) = J$. At every union node $u_s$ of $T_P$ for which $\xi(u_s)$ is defined, we store the information whether $u_s$ is internal or external.

At every non-bridge node $u_s$ of $T_P$ we store additionally to the clique-width operation the set of all terminal labels $L_P(u_s)$, the set of all internal labels $L_H(u_s)$, the set of all external labels $L_J(u_s)$, and all pairs $(\mathrm{forb}_P(u_s, j), j)$ for all $j \in L_P(u_s) \cup L_{H \cap J}(u_s)$. If a leaf $u_s$ of $T_P$ represents a vertex of $G$ obtained by joining the $i$-th terminal vertex from $H$ with the $i$-th terminal vertex from $J$, then leaf $u_s$ is additionally labeled by index $i$. The result $C$ is called a *connection tree* for the $k$-labeled $l$-terminal graph $H$.

The set of all mutually different connection trees of $H$ with respect to all $k$-labeled $l$-terminal graphs $J$ is called the *connection type* of $H$. Two connection trees $C_1$, $C_2$ for $H$ are *equivalent* if there is a bijection $b$ between the nodes of $C_1$ and $C_2$ such that

1. node $u_{s-1}$ is a child (left child, right child) of node $u_s$ in $C_1$ if and only if node $b(u_{s-1})$ is a child (left child, right child, respectively) of node $b(u_s)$ in $C_2$,

2. node $u_s$ of $C_1$ is an external or internal union node if and only if node $b(u_s)$ of $C_2$ is an external or internal union node, respectively,

3. node $u_s$ of $C_1$ is an external or internal bridge node if and only if node $b(u_s)$ of $C_2$ is an external or internal bridge node, respectively,

4. node $u_s$ of $C_1$ and node $b(u_s)$ of $C_2$ store the same terminal label sets, internal label sets, external label sets, the same pairs $(\mathrm{forb}_P(u_s, j), j)$, and the same clique-width operation,

5. node $u_s$ of $C_1$ and node $b(u_s)$ of $C_2$ store the same index if $u_s$ and $b(u_s)$ are leaves representing a vertex obtained by joining to terminal vertices.

Note that the two notions *connection tree* and *connection type* are always defined with respect to graph property clique-width at most $k$. For better readability, we will sometimes omit this extension.

# 8 Main result

The following theorem implies the main result of this paper.

**Theorem 8.1** *If two $k$-labeled $l$-terminal graphs are of the same connection type with respect to graph property clique-width at most $k$, then they are replaceable with respect to graph property clique-width at most $k$.*

**Proof:** Let $H_1$ and $H_2$ be two $k$-labeled $l$-terminal graphs such that $H_1$ and $H_2$ are of the same connection type. Let $J$ be any $k$-labeled $l$-terminal graph such that $H_1 \circ J$ has clique-width at most $k$. We will show that $H_2 \circ J$ has

W. Espelage et al., *Deciding Clique-Width*, JGAA, 7(2) 141–180 (2003)    176

also clique-width at most $k$. This implies that $H_1$ and $H_2$ are replaceable with respect to graph property clique-width at most $k$.

Let $T_1$ be a $k$-expression tree for $H_1 \circ J$ which defines connection tree $C$. Let $T_{1,P}$ be the subtree of $T_1$ defined by the leaves of $T_1$ which represent the joined terminal vertices of $H_1$ and $J$, and by the nodes on the paths from these leaves to the root of $T_1$.

Since $H_1$ and $H_2$ are of the same connection type, $C$ is also a connection tree for $H_2$ with respect to some $k$-labeled $l$-terminal graph $J'$. Let $T'$ be a $k$-expression tree for $H_2 \circ J'$ which defines connection tree $C$. Let $T'_P$ be the subtree of $T'$ defined by the leaves of $T'$ which represent the joined terminal vertices of $H_2$ and $J'$, and by the nodes on the paths from these leaves to the root of $T'$.

Since $T_1$ and $T'$ define the same connection tree $C$, there is a one-to-one correspondence between some nodes of $T_1$, $T'$, and $C$. For a node $u$ of $C$, we write $u^C$ to indicate that $u$ is a node of $C$. If $v$ is the corresponding node of $T_1$, then we write $u^{T_1}$ for $v$. The corresponding node in $T'$ is denoted by $u^{T'}$. We use this notation also for frames and paths.

Our aim is to define a new $k$-expression tree $T_2$ from $T_1$ and $T'$ such that $T_2$ defines $H_2 \circ J$. We start with a copy $T'_1$ of the $k$-expression tree $T_1$. Let $T'_{1,P}$ be defined for $T'_1$ in the same way as $T_{1,P}$ is defined for $T_1$. Let $u^C_1$ be an internal union node, let $u^{T'_1}_0$ be the child of $u^{T'_1}_1$ which is not in $T'_{1,P}$, and let $u^{T'}_0$ be the child of $u^{T'}_1$ which is not in $T'_P$. By our notation, it is clear from which trees these nodes are. For every such node $u^C_0$ we replace in the copy $T'_1$ of $T_1$ the subtrees $T'_1(u^{T'_1}_0)$ by the subtrees $T'(u^{T'}_0)$. Let $u^C$ be an internal bridge node, let $p^{T'_1}$ be the corresponding 4-path in $T'_1$ and $p^{T'}$ be the corresponding 4-path in $T'$. For every such node we substitute in the current tree $T'_1$ the 4-path $p^{T'_1}$ by the 4-path $p^{T'}$. This substitution includes all the subtrees at the children of the union nodes of $p^{T'_1}$ and $p^{T'}$ which are not in $T'_{1,P}$ and $T'_P$, respectively. The resulting tree is denoted by $T_2$. It is clear that $T_2$ is a $k$-expression tree.

It remains to show that $k$-expression tree $T_2$ defines $H_2 \circ J$. Let $T_{2,P}$ be the subtree of $T_2$ defined by the leaves of $T_2$ which represent the joined terminal vertices of $H_2$ and $J$, and by the nodes on the paths from these leaves to the root of $T_2$.

We first show that the vertices in the $k$-labeled graph $H_2 \circ J$ are labeled as in the $k$-labeled graph defined by $k$-expression tree $T_2$. There is obviously a one-to-one correspondence between the vertices of $H_2 \circ J$ and the vertices of the graph defined by $T_2$, because $T_2$ is constructed from $T_1$ and $T'$ which define $J$ and $H_2$. Let $T_{2,H_2}$ and $T_{2,J}$ be the subtrees of $T_2$ defined by the leaves which represent vertices of $H_2$ and $J$, respectively, and by the nodes on the paths from these leaves to the roots. The vertices of $H_2$ are labeled in $H_2 \circ J$ as in the graph defined by $k$-expression tree $T_2$, because these vertices are only relabeled by relabeling operations of $T_{2,H_2}$ which do not belong to the external 4-paths of $T_{2,P}$. (Here external and internal 4-path means that the path is copied from $T_1$ and $T'$, respectively.) The same holds for the vertices of $J$, because these vertices are only relabeled by relabeling operations of $T_{2,J}$ which do not belong

W. Espelage et al., *Deciding Clique-Width*, JGAA, 7(2) 141–180 (2003)    177

to the internal 4-paths of $T_{2,P}$.

Next we show that all edges of $H_2 \circ J$ are also in the graph defined by $T_2$. Let $T'_{H_2}$ and $T_{1,J}$ be the subtrees of $T'$ and $T_1$, respectively, defined by the leaves which represent vertices of $H_2$ and $J$, respectively, and by the nodes on the paths from these leaves to the roots. Let $e$ be an edge of $H_2 \circ J$. If the end vertices of $e$ are both from $H_2$ or both from $J$ then $e$ is created by an edge insertion node $u_s^{T'}$ or $u_s^{T_1}$ which is also in $T'_{H_2}$ or $T_{1,J}$, respectively. The composition of $T_2$ now implies that node $u_s^{T_2}$ exists in $T_2$ and the corresponding edge is also contained in the graph defined by $T_2$. Thus, all edges of $H_2 \circ J$ are in the graph defined by $T_2$.

The most interesting part is to show that the edge insertion operations of $T_2$ do not create any edge which is not in $H_2 \circ J$. An edge insertion node $u_s^{T_2}$ of $T_2$ which does not belong to $T_{2,P}$ creates only edges which are also in $H_2 \circ J$, because the corresponding subtree defined by $T_2(u_s^{T_2})$ is either completely copied from $T'$ or completely copied from $T_1$.

Assume now the edge insertion node $u_s^{T_2}$ belongs also to $T_{2,P}$. Let $\eta_{i,j}$ be the edge insertion operation of $u_s^{T_2}$. If $u_s^{T_2}$ is not from a 4-path of $T_{2,P}$ which consists of more than one frame, then $u_s^{T_2}$ is also in $C$. Then there is an equivalent edge insertion node $u_s^{T_1}$ or $u_s^{T'}$ in $T_1$ or $T'$ which is labeled as $u_s^{T_2}$ in $T_2$. This equal labeling ensures that the edge insertion operation $\eta_{i,j}$ defines a new edge between a vertex labeled by $i$ and a vertex labeled by $j$ if it defines at least one such edge in $T_1$ or $T'$.

If $u_s^{T_2}$ is from a 4-path $p^{T_2}$ which is also in $T_{2,P}$ and which consists of more than one frame, then without loss of generality, let $p^{T_2}$ be copied from $T'$, i.e., let $p^{T_2}$ be an internal 4-path for which $\xi(q^{T_2}) = H_2$ for all frames $q^{T_2}$ of $p^{T_2}$. Let $u^C$ be the corresponding internal bridge node for $p^{T_2}$ and let $u'^C$ be the child of $u^C$ in $C$. The child $u'^C$ can be a bridge node or a usual node. If it is a usual node then the equal labeling of $u'^{T_2}$ and $u'^{T'}$ ensures that $\eta_{i,j}$ defines a new edge between a vertex labeled by $i$ and a vertex labeled by $j$ if it defines at least one such edge in $T'$.

If child $u'^C$ is a bridge node then let $v^{T_2}$ be the last node of the path of $T_2$ which is represented by $u'^C$ in $C$. If $u'^C$ is an internal (external) bridge node then the equal labeling of $v^{T_2}$ and $v^{T'}$ (and $v^{T_1}$, respectively) ensures that $\eta_{i,j}$ defines a new edge between a vertex labeled by $i$ and a vertex labeled by $j$ if it defines such an edge in $T'$. Thus, every edge in the graph defined by $T_2$ is also in $H_2 \circ J$, and vice versa. $\qquad\square$

By Theorem 8.1 and the fact that there is only a finite number of connection types for fixed integers $l$ and $k$ it follows that the equivalence relation $\sim_{\Pi_k,l}$ has a finite number of equivalence classes, where $\Pi_k$ is the graph property clique-width at most $k$. This implies the following corollary.

**Corollary 8.2** *For every integer $k$, there exists a linear time algorithm for deciding clique-width at most $k$ of a graph of bounded tree-width.*

Since the clique-width of a graph of tree-width $l$ is bounded by $3 \cdot 2^{l-1}$, see [CR01], there is also an algorithm which minimizes the clique-width of a graph of bounded tree-width in linear time.

**Corollary 8.3** *There exists a linear time algorithm for computing the clique-width of a graph of bounded tree-width.*

The corollary above only states that such a linear-time algorithm for deciding clique-width $k$ for graphs of bounded tree-width exists. Although the proof is constructive, the resulting algorithm seems to be only interesting from a theoretical point of view.

Note that our result does not imply that the clique-width $k$ property is expressible in counting $MSO_2$-logic. The equivalence between a finite number of equivalence classes of $\sim_{\Pi_k,l}$ and monadic second-order definability is only given for special graph classes as for example for graphs of bounded tree-width [Lap98], but not for the class of all graphs.

# Acknowledgments

The authors wish to thank the anonymous referees for several useful suggestions.

# References

[ALS91]   S. Arnborg, J. Lagergren, and D. Seese. Easy problems for tree-decomposable graphs. *Journal of Algorithms*, 12:308–340, 1991.

[AP89]    S. Arnborg and A. Proskurowski. Linear time algorithms for NP-hard problems on graphs embedded in $k$-trees. *Discrete Applied Mathematics*, 23:11–24, 1989.

[Arn85]   S. Arnborg. Efficient algorithms for combinatorial problems on graphs with bounded decomposability – A survey. *BIT*, 25:2–23, 1985.

[Bod96]   H.L. Bodlaender. A linear-time algorithm for finding tree-decompositions of small treewidth. *SIAM Journal on Computing*, 25(6):1305–1317, 1996.

[Bod97]   H.L. Bodlaender. Treewidth: Algorithmic techniques and results. In *Proceedings of Mathematical Foundations of Computer Science*, volume 1295 of *LNCS*, pages 29–36. Springer-Verlag, 1997.

[Bod98]   H.L. Bodlaender. A partial $k$-arboretum of graphs with bounded treewidth. *Theoretical Computer Science*, 209:1–45, 1998.

[CHL$^+$00] D.G. Corneil, M. Habib, J.M. Lanlignel, B. Reed, and U. Rotics. Polynomial time recognition of clique-width at most three graphs. In *Proceedings of Latin American Symposium on Theoretical Informatics (LATIN '2000)*, volume 1776 of *LNCS*. Springer-Verlag, 2000.

[CMR00]   B. Courcelle, J.A. Makowsky, and U. Rotics. Linear time solvable optimization problems on graphs of bounded clique-width. *Theory of Computing Systems*, 33(2):125–150, 2000.

[CO00]   B. Courcelle and S. Olariu. Upper bounds to the clique width of graphs. *Discrete Applied Mathematics*, 101:77–114, 2000.

[Cou90]   B. Courcelle. The monadic second-order logic of graphs I: Recognizable sets of finite graphs. *Information and Computation*, 85:12–75, 1990.

[CPS85]   D.G. Corneil, Y. Perl, and L.K. Stewart. A linear recognition algorithm for cographs. *SIAM Journal on Computing*, 14(4):926–934, 1985.

[CR01]   D.G. Corneil and U. Rotics. On the relationship between clique-width and treewidth. In *Proceedings of Graph-Theoretical Concepts in Computer Science*, volume 2204 of *LNCS*, pages 78–90. Springer-Verlag, 2001.

[EGW01]   W. Espelage, F. Gurski, and E. Wanke. How to solve NP-hard graph problems on clique-width bounded graphs in polynomial time. In *Proceedings of Graph-Theoretical Concepts in Computer Science*, volume 2204 of *LNCS*, pages 117–128. Springer-Verlag, 2001.

[GR00]   M.C. Golumbic and U. Rotics. On the clique-width of some perfect graph classes. *IJFCS: International Journal of Foundations of Computer Science*, 11(3):423–443, 2000.

[GW00]   F. Gurski and E. Wanke. The tree-width of clique-width bounded graphs without $K_{n,n}$. In *Proceedings of Graph-Theoretical Concepts in Computer Science*, volume 1938 of *LNCS*, pages 196–205. Springer-Verlag, 2000.

[Joh98]   Ö. Johansson. Clique-decomposition, NLC-decomposition, and modular decomposition - relationships and results for random graphs. *Congressus Numerantium*, 132:39–60, 1998.

[KR01]   D. Kobler and U. Rotics. Polynomial algorithms for partitioning problems on graphs with fixed clique-width. In *Proceedings of the ACM-SIAM Symposium on Discrete Algorithms*, pages 468–476. ACM-SIAM, 2001.

[Lap98]   D. Lapoire. Recognizability equals definability, for every set of graphs of bounded tree-width. In *Proceedings 15th Annual Symposium on Theoretical Aspects of Computer Science*, volume 1373 of *LNCS*, pages 618–628. Springer-Verlag, 1998.

W. Espelage et al., *Deciding Clique-Width*, JGAA, 7(2) 141–180 (2003)     180

[LW88]    T. Lengauer and E. Wanke. Efficient solution of connectivity prob-
          lems on hierarchically defined graphs. *SIAM Journal on Computing*,
          17(6):1063–1080, 1988.

[LW93]    T. Lengauer and E. Wanke. Efficient analysis of graph properties on
          context-free graph languages. *Journal of the ACM*, 40(2):368–393,
          1993.

[Wan94]   E. Wanke. $k$-NLC graphs and polynomial algorithms. *Discrete Ap-
          plied Mathematics*, 54:251–266, 1994,
          revised version: `http://www.cs.uni-duesseldorf.de/~wanke`.

# Journal of Graph Algorithms and Applications

http://jgaa.info/

vol. 7, no. 2, pp. 181–201 (2003)


# Visual Ranking of Link Structures


*Ulrik Brandes*

Department of Mathematics & Computer Science
University of Passau
http://algo.fmi.uni-passau.de/~brandes/
brandes@algo.fmi.uni-passau.de

*Sabine Cornelsen*

Department of Computer & Information Science
University of Konstanz
http://www.inf.uni-konstanz.de/~cornelse/
cornelse@inf.uni-konstanz.de


### Abstract


Methods for ranking World Wide Web resources according to their position in the link structure of the Web are receiving considerable attention, because they provide the first effective means for search engines to cope with the explosive growth and diversification of the Web. Closely related methods have been used in other disciplines for quite some time.

We propose a visualization method that supports the simultaneous exploration of a link structure and a ranking of its nodes by showing the result of the ranking algorithm in one dimension and using graph drawing techniques in the remaining one or two dimensions to show the underlying structure. We suggest to use a simple spectral layout algorithm, because it does not add to the complexity of an implementation already used for ranking, but nevertheless produces meaningful layouts. The effectiveness of our visualizations is demonstrated with example applications, in which they provide valuable insight into the link structure and the ranking mechanism alike. We consider them useful for the analysis of query results, maintenance of search engines, and evaluation of Web graph models.



Communicated by Giuseppe Liotta and Ioannis G. Tollis: submitted October 2001; revised December 2002.


---


Research supported in part by the Deutsche Forschungsgemeinschaft (DFG) under grant Br 2158/1-1 and the European Commission within FET Open Project COSIN (IST-2001-33555).

U. Brandes and S. Cornelsen, *Visual Ranking*, JGAA, 7(2) 181–201 (2003) 182

# 1 Introduction

The directed graph induced by the hyperlink structure of the Web has been recognized as a rich source of information. Understanding and exploiting this structure has a proven potential to help dealing with the explosive growth and diversification of the Web. Probably the most widely recognized example of this kind is the PageRank index employed by the Google search engine [9].

PageRank is but one of many models and algorithms to rank Web resources according to their position in a hyperlink structure (see, e.g., [36, 29, 13, 1, 8, 12]). We propose a method to complement rankings with a meaningful visualization of the graph they are computed on.

While graph visualization is an active area of research as well [14, 28], its integration with quantitative network analyses is only beginning to receive attention. It is, however, rather difficult to understand the determinants of, say, a particular ranking if its results do not influence the way in which the structure is visualized.

A design for graph visualizations showing a vertex valuation in its structural context is introduced in [6, 5]. In two-dimensional diagrams of social networks, the vertical dimension of the layout area is used to represent exactly the value assigned to each actor (a constraint), and a layout of the horizontal dimension is determined to make the diagram readable (an objective). Since the networks in question are relatively small (no more than a hundred vertices), an adaptation of the Sugiyama framework for layered graph drawing [38] is used for horizontal layout.

The guiding principle in the above design is axis separation: in one dimension the most important information is conveyed precisely, and in another the perception of its basis is eased. To facilitate visual exploration of ranking methods on larger link structures such as Web graphs, we propose to apply the same principle, but with a very different layout algorithm that is more appropriate for the specific type of data.

Standard rankings are based on spectral methods and iterative computation, but the same methods can also be used for graph layout. In the present application they are particularly well-suited, because densely connected subgraphs are clustered. On the Web, such subgraphs correspond to related resources and graphical clustering is therefore highly desirable. By using the axis separation principle and spectral layout techniques, a uniform approach to visual ranking of link structures is achieved.

The paper is organized as follows. In Section 2, we recall some fundamental spectral properties of graphs. Link-based ranking is surveyed in Section 3, and formally and computationally similar layout techniques are described in Section 4. Applications in which our visualization approach may be useful are discussed in Section 5 and examples with generated and real-world data are provided. We conclude in Section 6.

U. Brandes and S. Cornelsen, *Visual Ranking*, JGAA, 7(2) 181–201 (2003)  183

# 2  Preliminaries

The structural features of the Web can be captured in a directed graph $G = (V, E)$, where the set $V$ of vertices represents the set of resources on the Web, and there is a directed edge $(u, v) \in E$ from a resource $u$ to a resource $v$, if $u$ contains a hyperlink to $v$. All graphs considered in this paper are assumed to be connected. We do not allow parallel edges, but self-loops and a positive real weight $\omega_{uv}$ for every edge. Let $A(G) = A = (A_{uv})_{u,v \in V}$ be the *adjacency matrix* of a graph, i.e. $A_{uv} = \omega_{uv}$ if $(u, v) \in E$, and $A_{uv} = 0$ otherwise. The *indegree* (*outdegree*), $d_v^+$ ($d_v^-$), of a vertex $v \in V$ is $\sum_{u:(u,v) \in E} A_{uv}$ ($\sum_{w:(v,w) \in E} A_{vw}$).

We will make extensive use of algebraic properties of graphs. If $A$ is a square matrix and $Ap = \lambda p$, $\lambda$ is called an *eigenvalue* of $A$ and $p$ an associated *eigenvector*. Note that, if $p$ is an eigenvector associated with $\lambda$, then $cp$, $c \in \mathrm{R}$, is also. The *multiplicity* of an eigenvalue is the number of distinct eigenvectors associated with it. Counting multiplicities, an $n \times n$ matrix has $n$ eigenvalues. The multiset $\Lambda(A) = \{\lambda_1, \ldots, \lambda_n\}$ of its eigenvalues with their respective multiplicity is called the *spectrum* of $A$.

We recall some important properties of spectra. The following lemma applies in particular to adjacency matrices of undirected graphs.

**Lemma 1** *Let $A$ be a real symmetric $n \times n$ matrix.*

1. *All eigenvalues of $A$ are real.*

2. *Any two eigenvectors of $A$ with distinct eigenvalues are orthogonal.*

3. *Let $\Lambda(A) = \{\lambda_1, \ldots, \lambda_n\}$, then*

   (a) $\Lambda(cA) = \{c\lambda_1, \ldots, c\lambda_n\}$ *for all $c \in \mathrm{R}$,*
   *in particular $\Lambda(-A) = \{-\lambda_1, \ldots, -\lambda_n\}$, and*

   (b) $\Lambda(I + A) = \{1 + \lambda_1, \ldots, 1 + \lambda_n\}$.

For directed graphs, we have the following theorem, which is a version of the fundamental Perron-Frobenius Theorem reformulated for our purposes.

**Theorem 2** *If $A$ is the adjacency matrix of a strongly connected graph $G$, then there is an ordering $\lambda_1 \geq |\lambda_2| \geq \cdots \geq |\lambda_n|$ of its eigenvalues such that $\lambda_1$ is real and simple, and $-\lambda_1$ is an eigenvalue of $A$ if and only if $G$ is bipartite. Moreover, the entries of a non-zero eigenvector associated with $\lambda_1$ are either all negative or all positive real numbers.*

For further background on matrix computations and algebraic properties of graphs we refer to [21] and [20].

# 3  Structural Ranking of Web Resources

Any real-valued vector $p = (p_v)_{v \in V}$ defined on the vertices of a graph is called a *prominence index*, where $p_v$ is the *prominence* of vertex $v$. A *ranking* is obtained

U. Brandes and S. Cornelsen, *Visual Ranking*, JGAA, 7(2) 181–201 (2003) 184

from a prominence index by ordering the vertices according to non-increasing prominence.

Many models have been proposed to capture an explicitly or implicitly defined notion of a vertex's prominence in a graph [27, 25, 16, 4, 17, 36, 29, 1, 13, 12, and many more]. Though in general only defined for undirected graphs, we first outline eigenvector centrality, because it nicely illustrates some important commonalities of the popular ranking methods that we discuss below.

Assume that the prominence of a vertex is understood to be proportional to the combined prominence of its neighbors, $\lambda p_v = \sum_{u:\{u,v\}\in E} \omega_{uv} p_u$, $v \in V$, where the constant $\lambda$ is introduced so that the system of equations has a non-zero solution. This definition yields the eigensystem of the (transposed) adjacency matrix,

$$\lambda p = A^T p = Ap, \qquad (eigenvector\ centrality\ [3])$$

and every eigenvector of $A = A(G)$ gives a ranking of the vertices for the above notion of prominence, although the *principal eigenvector*, i.e. the one associated with the eigenvalue of largest magnitude, is generally preferred [4, 17]. The principal eigenvector can be obtained by power iteration, which starts with any non-zero vector and iteratively multiplies the matrix with the current solution, e.g. $p^{(0)} \leftarrow 1$ and

$$p^{(k+1)} \quad \leftarrow \quad A \cdot p^{(k)}.$$

Since the matrices considered here originate from large and sparse graphs, multiplication is carried out by computing $p_v^{(k+1)} \leftarrow \sum_{u:\{u,v\}\in E} \omega_{uv} p_u^{(k)}$ for every $v \in V$. To prevent the entries of the iterates from growing out of range, each vector is normalized such that the magnitude of the largest entry equals the number of vertices in the graph (recall that multiples of eigenvectors are eigenvectors as well). This normalization scheme is applied in all subsequently described iterative computations without explicit mentioning.

More elaborate indices defined on directed graphs are discussed below. In Figure 1 they are illustrated on an acyclic grid. The grid is placed in a plane and each grid point is then lifted according to its prominence.

## 3.1 Hubs and authorities

A natural notion of prominence for a Web resource is the extent to which it is referred to by other Web pages, in particular by those pages that specialize in listing useful resources. In turn, the property of being such a list of useful resources is a notion of prominence in itself. In these complementary and mutually reinforcing notions prominent resources are called *authorities* (resources with useful information) and *hubs* (pages with useful links).

The hub score of a page is proportional to the combined authority of the resources it links to, and the authority of a resource is proportional to the combined hub score of the pages linking to it. In practice, hub and authority

U. Brandes and S. Cornelsen, *Visual Ranking*, JGAA, 7(2) 181–201 (2003)  185

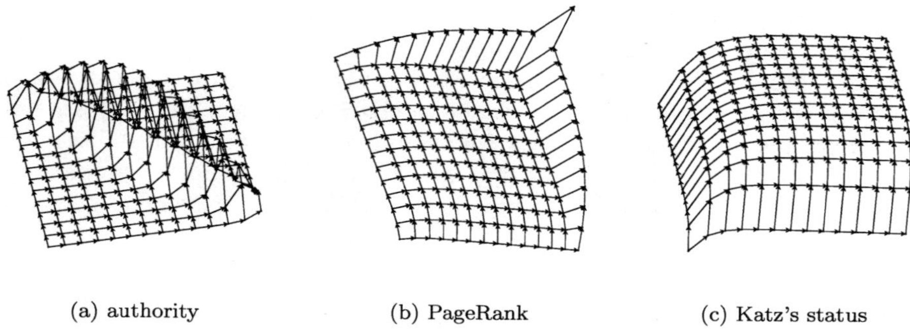

(a) authority        (b) PageRank        (c) Katz's status

Figure 1: Comparison of prominence indices on a directed grid

scores are thus computed by iterating $p^{(0)} \leftarrow 1$ and

$$p^{(2k+1)} \leftarrow A^T \cdot p^{(2k)}$$
$$p^{(2k+2)} \leftarrow A \cdot p^{(2k+1)}.$$

For $h^{(k)} = p^{(2k)}$ and $a^{(k)} = p^{(2k+1)}$, the alternating iteration can be written as

$$h^{(k+1)} \leftarrow AA^T \cdot h^{(k)} \qquad\qquad (hubs\ [29])$$
$$a^{(k+1)} \leftarrow A^T A \cdot a^{(k)}. \qquad\qquad (authorities\ [29])$$

In this formulation, it is easy to see that the hub and authority indices in a graph with adjacency matrix $A$ correspond to eigenvector centrality in the weighted undirected graphs with symmetric adjacency matrix $AA^T$ and $A^T A$, respectively.

As can be seen in Figure 1(a), vertices on and above the falling diagonal of the grid have the highest authority, because they are in the midst of the undirected graph induced by $A^T A$. Compare this to the undirected graph induced by $AA^T$, indicating why the best hubs are found on and below this diagonal.

## 3.2 PageRank

In another variant of eigenvector centrality the contribution of each vertex to another vertex's prominence is weighted by its outdegree, $p_v = \sum_{u:(u,v)\in E} \frac{\omega_{uv} p_u}{d_u^+}$ (see e.g. [35, 11]). If we require $p$ to be a probability distribution over the set of vertices, this notion has a nice interpretation as the stationary distribution of the simple random walk on the graph (or random surfer on the Web, if you will), in which each edge leaving a vertex is chosen with equal probability.

Let $M = D^{-1}A$ be the adjacency matrix normalized so that the rows sum to one, where $D$ is the diagonal matrix with the outdegrees on the diagonal. Then, $M$ is a stochastic matrix of transition probabilities, and a stationary

U. Brandes and S. Cornelsen, *Visual Ranking*, JGAA, 7(2) 181–201 (2003)  186

distribution $p = M^T \cdot p$ satisfies the above notion of prominence. However, if a vertex has outdegree zero, the computation breaks down, and strongly connected components may cause an overdue increase of the prominence of their vertices. This so-called "sink problem" can be avoided by introducing an escape mechanism. Let $\hat{p}$ be an a-priori probability distribution over the vertices (e.g., user preferences or general popularity of a resource), then with probability $\omega$ the random walk picks an edge of the graph whereas with the remaining probability, it jumps to any other vertex according to $\hat{p}$. The index is thus defined by

$$p = \omega M^T p + (1 - \omega)\hat{p} \qquad\qquad (PageRank\ [8])$$
$$= \left(\omega M^T + (1 - \omega)\hat{p} \cdot \mathbf{1}^T\right) \cdot p.$$

The second equality holds because $p$ is a probability distribution. From the second expression it can be seen that PageRank is the eigenvector centrality of a weighted graph with a complete set of additional escape edges. This modified matrix is irreducible and aperiodic so that the iteration $p^{(0)} \leftarrow \frac{1}{n}\mathbf{1}$ and

$$p^{(k+1)} \leftarrow \left(\omega M + (1 - \omega)\mathbf{1} \cdot \hat{p}^T\right)^T \cdot p^{(k)}$$

converges to a unique prominence vector. On the grid in Figure 1(b), the random surfer may jump to any vertex, but is most likely to walk towards the upper and right side of the grid, from where the only continuation is towards the upper right corner.

## 3.3 Katz's status index

As a generalization of simply using indegrees to measure 'status' in social networks, the prominence of a vertex is determined by the number of directed paths of arbitrary length ending in the vertex, where the influence of longer paths is attenuated by a decay factor. Recall that the entries of the $k$-th power of the adjacency matrix of an unweighted graph give the number of paths of length $k$ between every pair of vertices. Therefore, this notion of prominence is determined by

$$p = \left(\sum_{k=1}^{\infty}(\alpha A^T)^k\right) \cdot \mathbf{1}, \qquad\qquad (Katz's\ status\ [27])$$

where parameter $\alpha$ corresponds to the fraction of status that is passed along a directed edge. For sufficiently small values of $\alpha$ (a convenient choice is $\frac{1}{\Delta+1}$, where $\Delta$ is the minimum of the maximum in- or outdegree of any vertex in the graph), the sum converges to $(I - \alpha A^T)^{-1} - I$. Therefore, the status vector can be obtained by solving $\left(\alpha^{-1}I - A^T\right) \cdot p = d$, where $d$ is the vector of indegrees. Solving this system of linear equations directly is prohibitive for large graphs. Standard sparse matrix approaches approximate a solution iteratively. The update step in Jacobi iteration, for instance, yields $p^{(k+1)} \leftarrow \alpha A^T \cdot p^{(k)} + \alpha d$. This iteration nicely reflects the underlying notion of adding contributions from

U. Brandes and S. Cornelsen, *Visual Ranking*, JGAA, 7(2) 181–201 (2003)  187

vertices farther and farther away. The same can be observed in Figure 1(c), where the attenuated influence from vertices in the lower left does not suffice to discriminate the prominence of vertices in the upper right any more.

In a sense, the above definitions of prominence are contained in the following generic formulation of status in networks. It puts a twist on eigenvector centrality through the addition of an a-priori prominence vector $\hat{p}$,

$$p = A^T p + \hat{p}. \qquad (\textit{Hubbel's status } [25])$$

By choosing appropriate weights and a-priori prominences, we obtain eigenvector centrality and PageRank. Reordering, we have $p = (I - A^T)^{-1} \cdot \hat{p}$, provided the inverse exists. If it does, it equals $\sum_{k=0}^{\infty}(A^T)^k$, and therefore $p = \left(\sum_{k=0}^{\infty}(A^T)^k\right) \cdot \hat{p} = \left(I + \sum_{k=1}^{\infty}(A^T)^k\right) \cdot \hat{p}$. With uniform edge weights and $\hat{p} = 1$ we obtain a prominence index in which every component is by one larger than Katz's status index.

# 4  Spectral Graph Layout

In the previous section we emphasized formal similarities in the definition of popular prominence indices. In practice, all of them are computed by some variant of sparse matrix power iteration, i.e. by iterating over all vertices, and, for each vertex, combining the current scores of its neighbors. Implementation of these algorithms is thus trivial.

In this section, we introduce a layout algorithm that produces meaningful layouts using the same principles as the ranking methods. It is therefore a simple matter to complement an existing system for ranking vertices to compute a layout of the graph on the fly, since both parts of the system can operate synchronously on the same data

## 4.1  Layout with eigenvectors

For layout, we consider the unweighted, undirected, simple graph obtained by omitting weights, directions, self-loops, and multiple edges. Note that edge directions are sufficiently represented in the prominence dimension.

Let $A$ be the adjacency matrix of a simple undirected graph $G$ and $D = D(G)$ its diagonal *degree matrix*. We consider the *Laplacian matrix* $L = D - A$, which has interesting applications in many areas (see, e.g., [33]). Its usefulness for drawing graphs was first described in [22] and is based on the observation that minimizing the associated quadratic form

$$x^T L x = \sum_{\{u,v\} \in E} (x_u - x_v)^2, \qquad (1)$$

corresponds to minimizing the squared distance between pairs of adjacent vertices if $x$ is interpreted as a vector of vertex positions. This objective functions

U. Brandes and S. Cornelsen, *Visual Ranking*, JGAA, 7(2) 181–201 (2003)  188

is closely related to standard graph drawing methods, since it can be interpreted as the energy of a physical system consisting of rings (the vertices) that are tied together by springs (the edges) of natural length zero. In other words, we have a spring-embedder with zero-length springs and no repelling forces.

The energy-minimum state of the above system is obtained by assigning the same position to all vertices (recall that we assume connectedness of the graph). These undesirable single-point solutions can be avoided by fixing some selected vertices at distinct positions. Minimization subject to these boundary conditions yields the well-known barycentric layout model of Tutte [39]. However, the final layout is contingent on the fixed vertices and their position. While placing a face of a triconnected planar graph on a convex polygon yields a planar layout of the graph, there are no general rules on which vertices to place where in more general graphs.

Other alternatives include the addition of repulsive forces between nodes [15, 18] and the use of springs with non-zero length [26]. Although these methods have been extended to be applicable on graphs with thousands of vertices [19, 23, 40], there implementation is far from trivial.

Spectral methods take a different approach and yield a trivial, parameter-free algorithm working toward a globally optimal solution with respect to the above quadratic objective function. Note that the undesired minima $x = c\mathbf{1}$ are the eigenvectors associated with eigenvalue zero, i.e. $Lx = 0$. More generally, if $(\lambda, x)$ is any eigenpair of $L$, then $\lambda = \frac{x^T L x}{x^T x}$. We therefore want to minimize

$$\frac{x^T L x}{x^T x} \text{ subject to } \mathbf{0} \neq x \perp \mathbf{1},$$

since the eigenvectors of a symmetric matrix are orthogonal. Hence, the desired solution is an eigenvector associated with the second-smallest eigenvalue of $L$. This vector is called the *Fiedler vector* and, because of its distance minimization property, frequently used in graph partitioning (see, e.g., [37]). For the same reason, it yields a useful one-dimensional layout of a graph, because edges are short and hence dense subgraphs are clustered. Another argument in favor of using the Fiedler vector for horizontal layout is its successful application in drawing bipartite graphs in two-layers with few crossing edges [34].

If rankings ought to be visualized in three dimensions (cf. Figure 1), a reasonable choice for the second free dimension is the normalized eigenvector minimizing the objective function subject to being orthogonal to $\mathbf{1}$ and the first solution.

An example of two-dimensional layouts obtained from barycentric layout, a typical spring embedder, and two orthogonal eigenvectors of $L$ is provided in Figure 2. While the spring embedder produces more uniform edge lengths, the eigenvectors emphasize structural clustering of vertices.

## 4.2   Computing the layout

Eigenvectors associated with the smallest eigenvalues of large sparse matrices are usually computed using Lanczos' method. However, all popular prominence

U. Brandes and S. Cornelsen, *Visual Ranking*, JGAA, 7(2) 181–201 (2003)  189

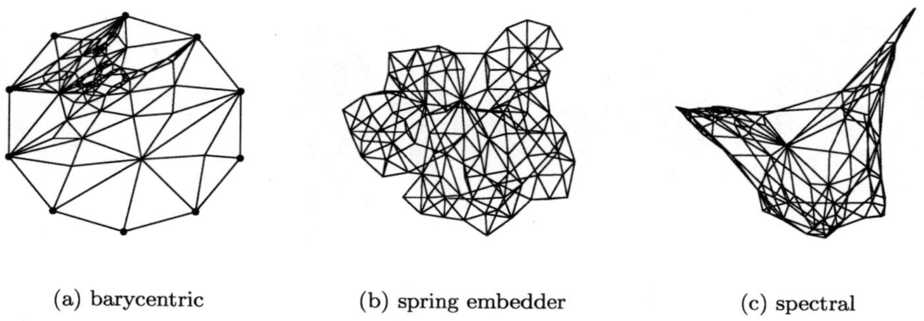

(a) barycentric     (b) spring embedder     (c) spectral

Figure 2: Two-dimensional layouts of a random planar triconnected graph

indices are computed using a variant of the much simpler power iteration, which only gives an eigenvector associated with the eigenvalue of largest magnitude. To be able to apply the same simple algorithm and thus synchronize layout and prominence computation, we reverse the eigenvalues of the Laplacian.

By Lemma 1, all eigenvalues of $L$ are real, and since $L$ is positive-semidefinite they are non-negative. By Gershgorin's Theorem, the largest eigenvalue is no more than twice the maximum vertex degree $\Delta$ of the graph, so that again by Lemma 1 the matrix $L' = 2\Delta \cdot I - L$ has the same eigenvectors as $L$, but the order of the corresponding eigenvalues is reversed.

Straightforward application of power iteration on $L'$ returns the principal eigenvector of $L'$, which is the trivial eigenvector $\mathbf{1}$ associated with the smallest eigenvalue of $L$. Power iteration on a vector that is orthogonal to the principal eigenvector yields an eigenvector of the second-largest eigenvalue of $L'$, and hence the desired layout for the first dimension. If needed, iterating on a vector that is orthogonal to both the trivial eigenvector and the approximate solution for the first dimension yields the second dimension.

A vector $y$ is orthogonalized with respect to another vector $x$ by setting $y \leftarrow y - \frac{x^T \cdot y}{x^T \cdot x} x$. Orthogonalization with respect to the trivial eigenvector $\mathbf{1}$ is even easier, since it corresponds to subtracting, from each entry of $y$, the mean of all its entries. To obtain vectors $x$ and $y$ for a two-dimensional layout we thus carry out the following augmented power iteration on random starting vectors $x^{(0)}, y^{(0)}$ that are repeatedly orthogonalized with respect to $\mathbf{1}$ and to one another

$$x^{(k+1)} \leftarrow L' \cdot x^{(k)}; \qquad x^{(k+1)} \leftarrow x^{(k+1)} - \frac{1}{n} \sum_{v \in V} x_v^{(k+1)}$$

$$y^{(k+1)} \leftarrow L' \cdot y^{(k)}; \qquad y^{(k+1)} \leftarrow y^{(k+1)} - \frac{1}{n} \sum_{v \in V} y_v^{(k+1)}$$

$$y^{(k+1)} \leftarrow y^{(k+1)} - \frac{x^{(k+1)T} \cdot y^{(k+1)}}{x^{(k+1)T} \cdot x^{(k+1)}} x^{(k+1)}$$

U. Brandes and S. Cornelsen, *Visual Ranking*, JGAA, 7(2) 181–201 (2003)  190

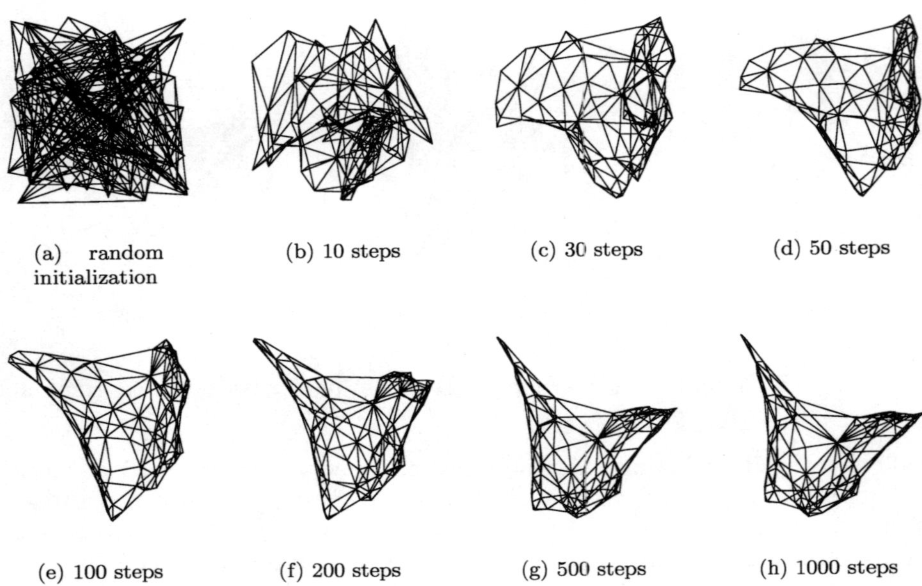

(a) random initialization

(b) 10 steps

(c) 30 steps

(d) 50 steps

(e) 100 steps

(f) 200 steps

(g) 500 steps

(h) 1000 steps

Figure 3: Typical convergence behavior of the power iteration for layout. Note that $x$-coordinates do not change significantly after 30 iterations

Intuitively, the layout is centered, rectified, and (due to the normalization) zoomed after each multiplication with $L'$. The last two lines are omitted if only one dimension needs to be determined for the layout.

Note that in our setting the potentially slow convergence of power iteration is of minor importance since all we are looking for is a vector that approximately minimizes the quadratic objective function (1). Though overall convergence depends on the ratio of the largest eigenvalues, the iterate quickly moves toward a subspace spanned by eigenvectors associated with large eigenvalues. Only then, when the largest eigenvalues need to be separated, does the slow-down take effect. Figure 3 gives a typical, qualitative example.

As a quantitative measure of convergence we use the *residual* $r(x) = \| L'x - \frac{x^T L' x}{x^T x} x \|^2$, that is the squared distance of $x$ from being an eigenvector associated with the current eigenvalue estimate $\frac{x^T L' x}{x^T x}$. Recall that we normalize after each iteration such that the magnitude of the largest entry (the largest coordinate) equals the number of vertices. We consider a layout to be of sufficient quality, if the residual is of the same order, i.e. if on the average each vertex is one unit off its optimal position. The entire one-dimensional layout algorithm is given in Algorithm 1. Note that it requires no external parameters, and is trivial to implement along with a ranking algorithm.

We compared the number of iterations needed for layout to that needed for ranking. Since ranking is the important information to be conveyed, it is

U. Brandes and S. Cornelsen, *Visual Ranking*, JGAA, 7(2) 181–201 (2003) 191

---

**Algorithm 1:** One-dimensional spectral layout

> **Input:** simple, connected, undirected graph $G = (V, E)$, $n = |V|$
>
> **Output:** one-dimensional layout $x = (x_v)_{v \in V}$
>
> $r = \infty$;
>
> $x \leftarrow n \cdot \mathbf{1}$;
>
> **while** $\frac{r}{n} > 1$ **do**
>
> $\quad x' \leftarrow L'x$;
>
> $\quad x' \leftarrow x' - \frac{\sum_{v \in V} x'_v}{n} \cdot \mathbf{1}$;
>
> $\quad r \leftarrow \| x' - \frac{x^T x'}{x^T x} x \|^2$;
>
> $\quad x \leftarrow \frac{n}{\max\limits_{v \in V} x'_v} \cdot x'$;

---

required to be precise. Convergence of ranking iterations is assumed when the corresponding residual is below 1 (rather than the number of vertices). Our experience suggests that the number of iterations needed for the layout is larger than that for ranking, but not by much. In Figure 4, convergence of ranking and layout is compared on example graphs.

For larger graphs with tens of thousands of nodes, the simple algorithm nevertheless becomes to slow, especially when compared with the fastest-converging ranking algorithms. A much more sophisticated multiscale algorithm [30] to obtain the Fiedler vector is available, though.

# 5  Application Examples

We demonstrate our visualization approach on three different kinds of data: random Web graphs generated from popular models, a search engine example constructed from an AltaVista query, and a bibliographic data set. Our C++-implementations use LEDA, the *Library of Efficient Data Types and Algorithms* [32].

## 5.1  Web graph models

In the *linear growth model* [31], a graph grows one vertex at a time. At each time step, a prototype is chosen among already existing vertices, and a new vertex is generated. This new vertex is then assigned a fixed number of outgoing edges. With some fixed probability, the $i$th of these edges points to a randomly selected vertex among those already existing (creation case), and with the remaining probability it points to the same vertex as the $i$th outgoing edge of the prototype vertex (copying case). Our generator does not introduce multiple edges, and if a prototype happens to not have enough outgoing edges, no edge is introduced in the copying case. Clearly, all graphs evolving like this are acyclic.

In the *exponential growth model* [31], a graph grows by a fixed fraction of its current size at each time step. New vertices receive a fixed number of loops,

U. Brandes and S. Cornelsen, *Visual Ranking*, JGAA, 7(2) 181–201 (2003)  192

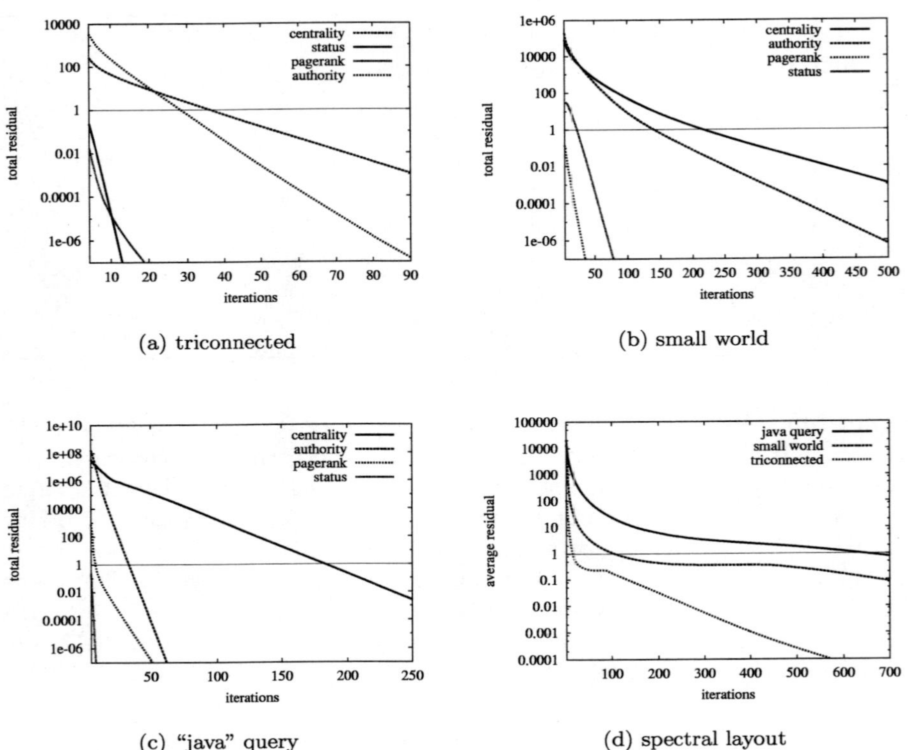



Figure 4: Convergence of ranking and layout compared (logarithmic scale). The three graphs are the triconnected planar graph of Figure 2, the random small world of Figure 5(c) (Section 5.1), and the Web graph of Figure 6 (Section 5.2)

and for each already existing edge, its target receives a new incoming edge for which, with some fixed probability, the source is chosen uniformly at random from the new vertices, and otherwise from the existing vertices with probability proportional to their current outdegree. We used a simpler model in which existing vertices are chosen uniformly at random as well.

For the *small-world model* [41], we initially generate a cyclic sequence of vertices and let a vertex link to a fixed number of predecessors and successors. Then, each edge is rewired with some small probability by choosing a new destination uniformly at random.

Figure 5 shows spectral layouts of graphs generated according to these models and rankings replacing the vertical dimension with PageRank as an example of a prominence index. The linear model graph has about 750 vertices and was generated with desired outdegree 7 and copying probability 0.3, where some of the vertices created last where trimmed because of poor connectivity. The expo-

U. Brandes and S. Cornelsen, *Visual Ranking*, JGAA, 7(2) 181–201 (2003)  193

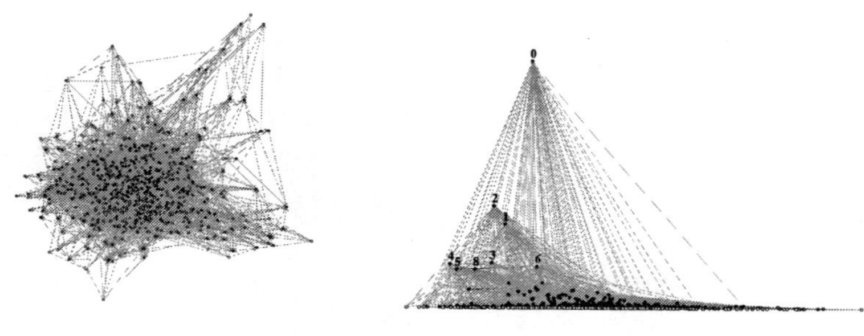

(a) Linear growth evolving copying model [31]

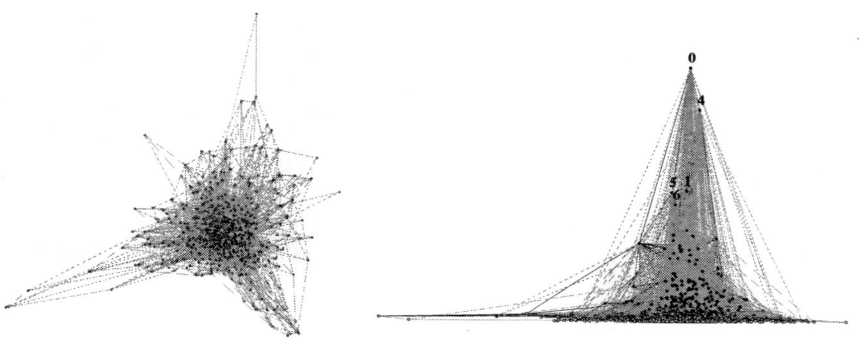

(b) Exponential growth evolving copying model [31]

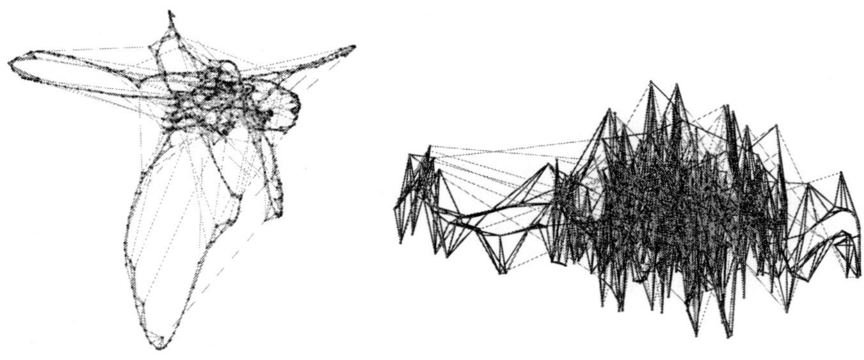

(c) Small-world model [41]

Figure 5: Web models (2D spectral layout and 1D spectral layout vs. PageRank)

U. Brandes and S. Cornelsen, *Visual Ranking*, JGAA, 7(2) 181–201 (2003)  194

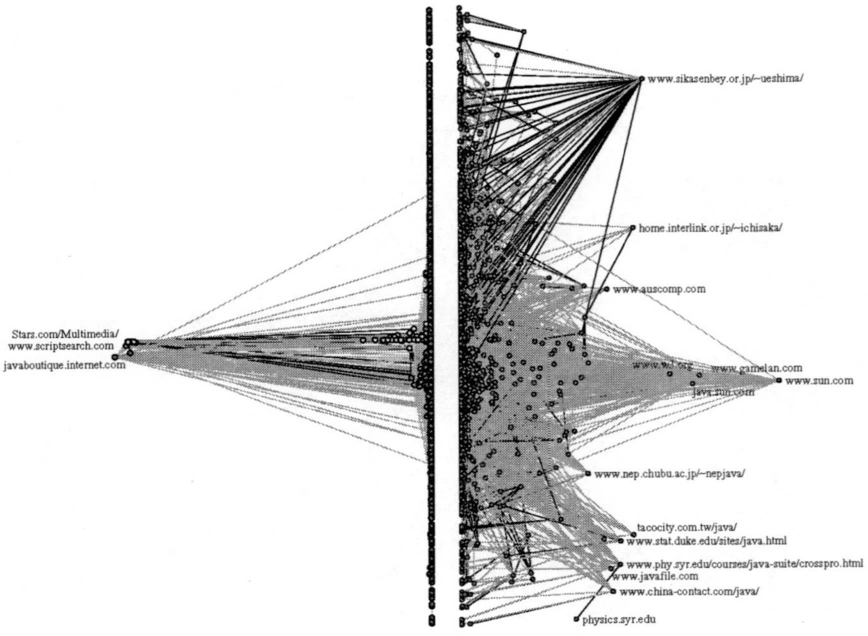


Figure 6: Authority and PageRank visualization of "java" query result

nential model graph was generated with 10% growth rate, desired outdegree 7, 7 initial loops, and probability 0.5 for choosing the origin among new vertices. It originally had about 1000 vertices, but again roughly a quarter of the vertices last created were removed to achieve more robust structure. The 750 vertices in the small-world graph originally linked to their six nearest cyclic neighbors and edges were rewired with probability 0.05. In all rankings, edge directions are indicated by color (gray edges point upwards, black edges point downwards).

There is no visible clustering in the evolving copying models. Moreover, the prominence of resources appears to be correlated with their age (also with the other indices outlined in Section 3). The figures thus graphically support the conclusion of [31] that *death processes*, i.e. the occasional deletion of vertices and edges, might be necessary for the evolving copying models to be realistic. In the small-world model, the spectral layout reveals a cycle crumpled by chords, and the ranking shows that the model yields a rather egalitarian structure.

Our generators are slightly simplified versions of the original ones and our samples are not representative. Their sole purpose is to demonstrate the potential utility of ranked visualizations in the exploration and comparison of different models and parameterizations.

U. Brandes and S. Cornelsen, *Visual Ranking*, JGAA, 7(2) 181–201 (2003)  195

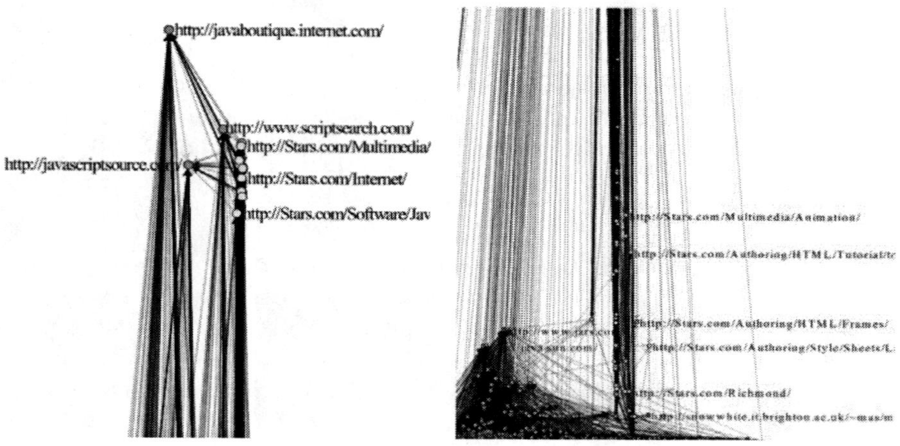

Figure 7: Detecting a data preparation flaw by visual inspection

## 5.2  Search engine query results

The data for this example was compiled in a way similar to the HITS algorithm [29]. We asked the AltaVista search engine for pages containing the word "java" and used the first 200 URLs it returned as the root set. It was then expanded by asking AltaVista for pages containing links to resources in the root set (backward extension), and adding resources linked to by pages in the root set (forward extension). The graph was completed by adding edges for all links between pages in the resulting set of vertices. The computations were carried out on the only large component of this graph from which some poorly connected vertices were removed to prevent extreme clustering. The graph has more than 5000 vertices and 15000 edges.

In Figure 6, this graph is shown twice, with vertices positioned vertically according to the Fiedler vector, and horizontally according to one of two prominence indices. Again, links from more to less prominent resources are colored black.

The most prominent resources under the PageRank measure match our expectations, but there are some surprising recommendations as well. It is clearly visible that some of these serve distinct user groups, like the Japanese directory in the upper right. Note that, without zooming into the image, we may not conclude that vertically close vertices are closely connected. However, it is safe to assume that vertically separated vertices are relatively distant in the structure. This feature can serve to distinguish query results which contain a keyword that is used in different contexts (see the "jaguar"-query example in [29]).

Figure 6 also shows that the top authorities are surprisingly distinguished from the rest of the graph, and quite different from our expectations. Most of them are located at `Stars.com`, a large repository for developers ("Web Developer's Virtual Library"). Since they are well connected among each other,

U. Brandes and S. Cornelsen, *Visual Ranking*, JGAA, 7(2) 181–201 (2003)  196

it is by virtue of our layout approach that their vertical position is similar, and thus this phenomenon could be detected by visual exploration. In Figure 7, resources at this site are colored lighter. Not surprisingly, vertices with high hub scores are from this site as well. This simple example graphically explains why the original HITS algorithm does not consider links within a site.

## 5.3  Bibliographic networks

Web graphs may be viewed as citation networks, and there exist many other bibliographic relations between publications and authors of written works. A discussion of techniques to analyze bibliographic networks is beyond the scope of this paper, but there is evidence that carefully applied network analytic methods can provide insight into the structure of a research area by identifying, e.g., prominent works or thematic clusters. We refer the reader to [42] for an introduction to bibliographic analysis and to [10] for an example of a system constructing and visualizing graphs from various bibliographic relations.

The application of our visual ranking approach to bibliographic networks is illustrated by citation data made available for the 2001 Graph Drawing Contest [2], held in conjunction with the 9th International Symposium on Graph Drawing. Since bibliographic networks typically contain loosely connected subgraphs which are difficult for spectral approaches to draw properly, it is proposed in [7] to weaken the diagonal of the Laplacian matrix. This modification serves to spread vertices more uniformly.

Each vertex in Figure 8 represents a paper that appears in one of the proceedings of the symposia from 1994 and 2000. While the color indicates the year of the symposium, height and width represent the number of citations received and made, respectively. As noted in Section 3, Kleinberg's hub and authority scores correspond to eigenvector centrality scores in the undirected graphs $AA^T$ and $A^TA$. In bibliometrics, these are known as the bibliographic coupling and co-citation graphs. A hub is thus a potential survey, while an authority is an influential paper. Note, however, that the specific data at hand is certainly not sufficient to draw valid conclusions about the role and importance of a publication.

We have chosen this data set because the emerging patterns even for this small data set happen to resemble at least some of our intuition about the field. In particular, the horizontal clustering produced by the spectral layout algorithm does indeed correspond to a thematic clustering. The small cluster in the far right, for instance, are the Graph Drawing Contest Reports, connected only to the mainstream papers that form the adjacent dense cluster. Moving to the left, topics change via orthogonal drawing and 3D to the less intensely studied visibility representations and proximity drawings.

U. Brandes and S. Cornelsen, *Visual Ranking*, JGAA, 7(2) 181–201 (2003)  197

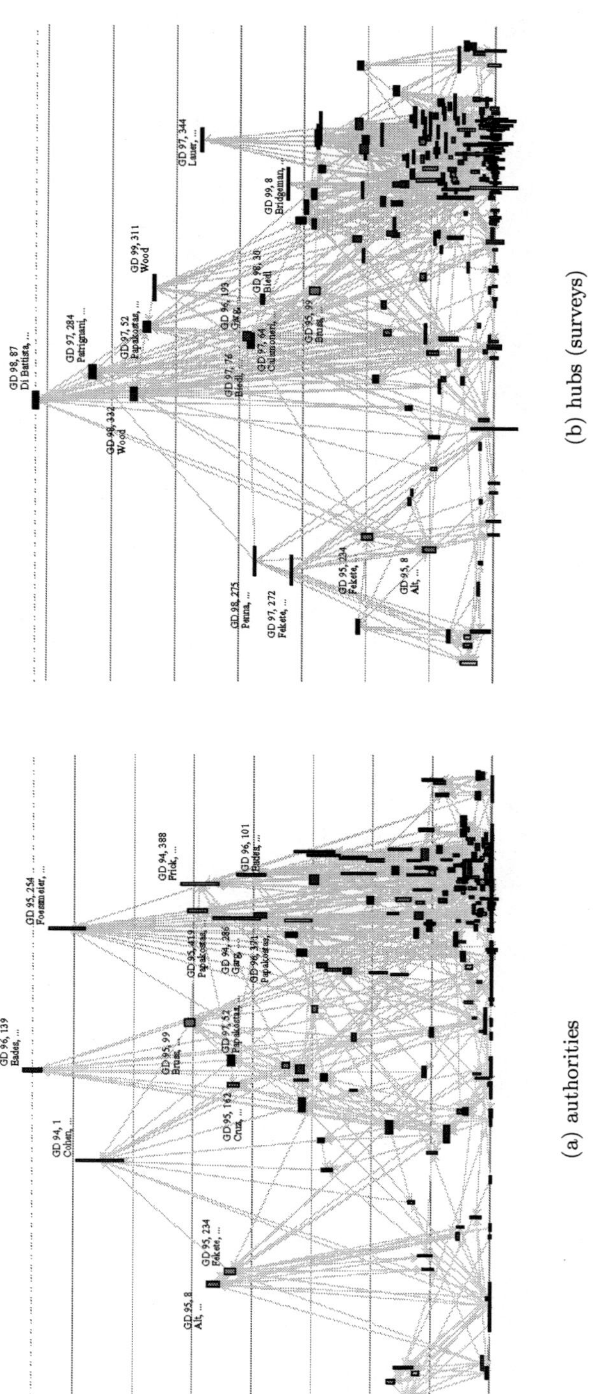


Figure 8: Citations between papers in proceedings of symposia on Graph Drawing (data from the 2001 GD Contest [2])

U. Brandes and S. Cornelsen, *Visual Ranking*, JGAA, 7(2) 181–201 (2003) 198

## 6 Conclusions

We have proposed a method for Web graph visualization that provides unambiguous identification of prominent resources while showing the entire graph and its clustering. In the simplest approach, the layout for our visualizations can be computed synchronously with common link-based rankings.

We expect the proposed visualization design to be particularly useful for visual exploration of ranked structures, for teaching and experimenting with ranking procedures, and for evaluation and illustration of stochastic models of the Web graph.

For graphs with tens of thousands of vertices, power iteration becomes costly because of its slow convergence. While speed-up techniques that reorganize storage to reduce external memory access [24] carry over to the layout algorithm, more sophisticated layout algorithms are available. Several recently introduced methods [19, 40, 23] produce layouts similar to the spectral approach. With a new multiscale technique for eigenvector layout computation [30], however, our approach extends directly.

The main advantage of spectral graph layout, its correspondence with distance minimization and hence with clustering, becomes a drawback in cases where the underlying undirected graph is poorly connected, since denser subgraphs will be clustered in a very small interval. Experiments with modifications of the Laplacian matrix [7] suggest that this problem can be addressed without changing the iteration significantly.

**Acknowledgments.** We thank Marco Gaertler for collecting the "java"-query data used in Section 5.2.

## References

[1] Krishna Bharat and Monika R. Henzinger. Improved algorithms for topic distillation in a hyperlinked environment. In *Proc. 21st Ann. Intl. ACM SIGIR Conf. Research and Development in Information Retrieval*, pages 104–111, 1998.

[2] Therese C. Biedl and Franz J. Brandenburg. Graph-drawing contest report. In *Proc. 9th Intl. Symp. Graph Drawing (GD '01)*, Springer LNCS 2265, pages 513–522, 2002.

[3] Phillip Bonacich. Factoring and weighting approaches to status scores and clique identification. *Journal of Mathematical Sociology*, 2:113–120, 1972.

[4] Phillip Bonacich. Power and centrality: A family of measures. *American Journal of Sociology*, 92:1170–1182, 1987.

[5] Ulrik Brandes, Jörg Raab, and Dorothea Wagner. Exploratory network visualization: Simultaneous display of actor status and connections. *Journal of Social Structure*, 2(4), 2001.

[6] Ulrik Brandes and Dorothea Wagner. Contextual visualization of actor status in social networks. In *Data Visualization 2000. Proc. 2nd Joint Eurographics/IEEE TCVG Symp. Visualization (VisSym '00)*, pages 13–22. Springer, 2000.

[7] Ulrik Brandes and Thomas Willhalm. Visualization of bibliographic networks with a reshaped landscape metaphor. In *Proc. 4th Joint Eurographics/IEEE TCVG Symp. Visualization (VisSym '02)*, pages 159–164. ACM Press, 2002.

[8] Sergey Brin, Rajeev Motwani, Lawrence Page, and Terry Winograd. What can you do with a web in your pocket? *IEEE Bulletin of the Technical Committee on Data Engineering*, 21(2):37–47, 1998.

[9] Sergey Brin and Lawrence Page. The anatomy of a large-scale hypertextual Web search engine. *Computer Networks and ISDN Systems*, 30(1–7):107–117, 1998.

[10] Anne Büggemann-Klein, Rolf Klein, and Britta Landgraf. BibRelEx: Exploring bibiliographic databases by visualization of annotated content-based relations. *D-Lib Magazine*, 5(11), 1999.

[11] Ronald S. Burt. *Toward a Structural Theory of Action: Network Models of Social Structure, Perception, and Action.* Academic Press, 1982.

[12] Soumen Chakrabarti, Byron E. Dom, Ravi Kumar, Prabhakar Raghavan, Sridhar Rajagopalan, Andrew S. Tomkins, David Gibson, and Jon M. Kleinberg. Mining the Web's link structure. *IEEE Computer*, 32(8):60–67, 1999.

[13] Soumen Chakrabarti, Byron E. Dom, Prabhakar Raghavan, Sridhar Rajagopalan, David Gibson, and Jon M. Kleinberg. Automatic resource compilation by analyzing hyperlink structure and associated text. *Computer Networks and ISDN Systems*, 30(1–7):65–74, 1998.

[14] Giuseppe Di Battista, Peter Eades, Roberto Tamassia, and Ioannis G. Tollis. *Graph Drawing: Algorithms for the Visualization of Graphs.* Prentice Hall, 1999.

[15] Peter Eades. A heuristic for graph drawing. *Congressus Numerantium*, 42:149–160, 1984.

[16] Linton C. Freeman. Centrality in social networks: Conceptual clarification. *Social Networks*, 1:215–239, 1979.

[17] Noah E. Friedkin. Theoretical foundations for centrality measures. *American Journal of Sociology*, 96(6):1478–1504, May 1991.

[18] Thomas M.J. Fruchterman and Edward M. Reingold. Graph-drawing by force-directed placement. *Software—Practice and Experience*, 21(11):1129–1164, 1991.

U. Brandes and S. Cornelsen, *Visual Ranking*, JGAA, 7(2) 181–201 (2003) 200

[19] Pawel Gajer, Michael T. Goodrich, and Stephen G. Kobourov. A fast multi-dimensional algorithm for drawing large graphs. In *Proc. 8th Intl. Symp. Graph Drawing (GD 2000)*, Springer LNCS 1984, pages 211–221, 2001.

[20] Chris Godsil and Gordon Royle. *Algebraic Graph Theory*, volume 207 of *Graduate Texts in Mathematics*. Springer, 2001.

[21] Gene H. Golub and Charles F. van Loan. *Matrix Computations*. Johns Hopkins University Press, 3rd edition, 1996.

[22] Kenneth M. Hall. An $r$-dimensional quadratic placement algorithm. *Management Science*, 17(3):219–229, 1970.

[23] David Harel and Yehuda Koren. A fast multi-scale method for drawing large graphs. In *Proc. 8th Intl. Symp. Graph Drawing (GD 2000)*, Springer LNCS 1984, pages 183–196, 2001.

[24] Taher H. Haveliwala. Efficient computation of pagerank. Technical Report 1999-31, Database Group, Stanford University, 1999.

[25] Charles H. Hubbell. An input-output approach to clique identification. *Sociometry*, 28:377–399, 1965.

[26] Tomihisa Kamada and Satoru Kawai. An algorithm for drawing general undirected graphs. *Information Processing Letters*, 31:7–15, 1989.

[27] Leo Katz. A new status index derived from sociometric analysis. *Psychometrika*, 18:39–43, 1953.

[28] Michael Kaufmann and Dorothea Wagner, editors. *Drawing Graphs: Methods and Models*, volume 2025 of *Lecture Notes in Computer Science*. Springer, 2001.

[29] Jon M. Kleinberg. Authoritative sources in a hyperlinked environment. *Journal of the Association for Computing Machinery*, 46(5):604–632, September 1999.

[30] Yehuda Koren, Liran Carmel, and David Harel. ACE: A fast multiscale eigenvectors computation for drawing huge graphs. In *Proc. IEEE Symp. Information Visualization 2002 (InfoVis '02)*, pages 137–144, 2002.

[31] Ravi Kumar, Prabhakar Raghavan, Sridhar Rajagopalan, D. Sivakumar, Andrew S. Tomkins, and Eli Upfal. Stochastic models for the Web graph. In *Proc. 41st Ann. IEEE Symp. Foundations of Computer Science (FOCS 2000)*, pages 57–65, 2000.

[32] Kurt Mehlhorn and Stefan Näher. *The LEDA Platform of Combinatorial and Geometric Computing*. Cambridge University Press, 1999.

U. Brandes and S. Cornelsen, *Visual Ranking*, JGAA, 7(2) 181–201 (2003) 201

[33] Bojan Mohar. Some applications of Laplace eigenvalues of graphs. In Gena Hahn and Gert Sabidussi, editors, *Graph Symmetry: Algebraic Methods and Applications*, NATO ASI Series C 497, pages 225–275. Kluwer, 1997.

[34] Matthew Newton, Ondrej Sýkora, and Imrich Vrťo. Two new heuristics for two-sided bipartite graph drawings. In Michael T. Goodrich and Stephen G. Kobourov, editors, *Proceedings of the 10th International Symposium on Graph Drawing (GD '02)*, volume 2528 of *Lecture Notes in Computer Science*. Springer, 2002.

[35] Gabriel Pinski and Francis Narin. Citation influence for journal aggregates of scientific publications: Theory, with applications to the literature of physics. *Information Processing and Management*, 12(5):297–312, 1976.

[36] Peter Pirolli, James Pitkow, and Ramana Rao. Silk from a sow's ear: Extracting usable structures from the Web. In *Proc. ACM Conf. Human Factors in Computing Systems (CHI '96)*, pages 118–125, 1996.

[37] Daniel A. Spielman and Shang-Hua Teng. Spectral graph partitioning works: Planar graphs and finite element meshes. In *Proc. 37th Ann. IEEE Symp. Foundations of Computer Science (FOCS '96)*, pages 96–105, 1996.

[38] Kozo Sugiyama, Shojiro Tagawa, and Mitsuhiko Toda. Methods for visual understanding of hierarchical system structures. *IEEE Trans. Systems, Man and Cybernetics*, 11(2):109–125, 1981.

[39] William T. Tutte. How to draw a graph. *Proceedings of the London Mathematical Society, Third Series*, 13:743–768, 1963.

[40] Christopher Walshaw. A multilevel algorithm for force-directed graph drawing. In *Proc. 8th Intl. Symp. Graph Drawing (GD 2000)*, Springer LNCS 1984, pages 171–182, 2001.

[41] Duncan J. Watts and Steven H. Strogatz. Collective dynamics of "small-world" networks. *Nature*, 393:440–442, 1998.

[42] Howard D. White and Katherine W. McCain. Bibliometrics. *Annual Review of Information Science and Technology*, 24:119–186, 1989.

# Journal of Graph Algorithms and Applications

`http://jgaa.info/`

vol. 7, no. 2, pp. 203–220 (2003)

# An Approach for Mixed Upward Planarization

*Markus Eiglsperger*    *Frank Eppinger*    *Michael Kaufmann*

Wilhelm-Schickard-Institut für Informatik
Universität Tübingen,
Sand 13, 72076 Tübingen, Germany,
`eiglsper@informatik.uni-tuebingen.de`
`eppinger@informatik.uni-tuebingen.de`
`mk@informatik.uni-tuebingen.de`

### Abstract

In this paper, we consider the problem of finding a mixed upward planarization of a mixed graph, i.e., a graph with directed and undirected edges. The problem is a generalization of the planarization problem for undirected graphs and is motivated by several applications in graph drawing. We present a heuristic approach for this problem which provides good quality and reasonable running time in practice, even for large graphs. This planarization method combined with a graph drawing algorithm for upward planar graphs can be seen as a real alternative to the well known Sugiyama algorithm.

Communicated by Giuseppe Liotta and Ioannis G. Tollis: submitted October 2001;
revised December 2002.

Research supported in part by DFG-Grant Ka812/8-1

M. Eiglsperger et al., *Mixed Planarization*, JGAA, 7(2) 203–220 (2003)    204

# 1    Introduction

Research for upward drawings of digraphs has been studied extensively in the last years. One reason is that such drawings have many applications in areas like workflow, project management and data flow.

An *upward drawing* of a digraph is a drawing such that all the edges are represented by curves monotonically increasing in the vertical direction. Note that such a drawing exists only if the digraph is acyclic.

A straightforward generalization of upward drawings are *mixed upward drawings*. In mixed upward drawings, only a part of the edges in the graph are directed and must point upward. Note that such a drawing exists only if the directed part of the graph is acyclic.

Mixed drawings arise in applications where the edges of the graph can be partitioned into a set which denotes structural information and a another set which does not carry structural information. An example is UML class diagrams[4] arising in software engineering. In these diagrams, the vertices of the graph represent classes in an object-oriented software system, and edges represent relations between these classes. There are two main types of relations: *generalizations* and *associations*. The generalization relations describe structural information and form a directed acyclic subgraph in the diagram. It is an often employed convention to draw generalizations upward, whereas associations can have arbitrary directions[21].

The most popular approach for creating upward drawings of digraphs is probably the Sugiyama algorithm [22]. The main idea of the Sugiyama algorithm is to assign layers to the vertices of the graph, such that edges point in ascending layer order. In a next step, the number of crossings are minimized by ordering the nodes in the layer. In the last step the nodes are assigned coordinates. For a fixed layer assignment, we call a graph *level planar* if it has a drawing which respects the layering and has no crossings. Several heuristics have been proposed for this step and used in practice, but there are also efficient algorithms to solve the level planarity problem [14][13]. There have been several attempts to apply the Sugiyama algorithm also to mixed graphs, i.e., in [21] the approach is used for UML class diagrams.

The principal step of the Sugiyama algorithm, the layer assignment, is also its most severe drawback. The layer assignment restricts the freedom of choice for the crossing minimization algorithm drastically, and there may be large differences between the number of crossings for different layer assignments of the same graph. Also, the generalizations of the Sugiyama algorithm for the mixed case have to assign layers to nodes with no directed adjacent edges. This only works when there is a low number of them, but if the directed part of the mixed graph is only small, the results are not satisfying and the layer assignment to the nodes seems artificial.

In this work we propose a new drawing strategy for upward drawings of directed graphs which is based on the concept of *upward planarity*. A directed graph is *upward planar* if it can be drawn upward without edge crossings. Our strategy consists of two phases. In the first phase, we make the input graph

M. Eiglsperger et al., *Mixed Planarization*, JGAA, 7(2) 203–220 (2003)    205

upward planar by replacing edge crossings by dummy nodes. We call the result of this phase *upward planarization*. In the second phase, an upward planar drawing of the upward planarization is generated and the dummy nodes are discarded.

A similar strategy has been applied very successfully in the area of drawing undirected graphs. The most popular algorithms based on this strategy are perhaps the graph drawing algorithms descending from the topology-shape-metrics approach [2][23] for orthogonal drawings. Recently, we showed in [9] how to extend the topology-shape-metrics approach to mixed upward planar drawings of *mixed upward planar graphs*, i.e., mixed graphs that have a planar drawing in which the directed part of the graph is drawn upward. The above strategy can also be applied to this algorithm, the first phase then consisting of finding a mixed upward planarization. An alternative way to draw directed graphs using the topology-shape-metrics approach is based on the concept of *quasi-upward planarity*, which is introduced in [19]. However the algorithm does not guarantee that all edges of a directed graph point upward when the graph is not upward planar.

In the remainder of this paper, we concentrate on the first phase of the topology-shape-metrics approach, see [6] for a survey on graph drawing algorithms for upward planar graphs. We give an efficient heuristics that computes a high quality upward planarization of a directed graph. We concentrate on a heuristic approach, since the upward planarity test problem is already NP-complete. To our knowledge, this is the first time that this problem has been studied; work on planarization has been restricted to the undirected case until now. We also give a generalization of our algorithm for mixed graphs.

We want to emphasize that the topology-shape-metrics approach above is only one possible application of the new algorithm. Our approach also deserves attention as a stand-alone product which might be applicable in other environments.

The rest of the paper is organized as follows. Section 2 gives the formal definitions of the upward and the mixed upward planarization problem. In Section 3, we present an algorithm which solves the upward planarization problem. In Section 4, we generalize the results of Section 3 to the mixed case. Section 5 contains the results of empirical experiments performed with our algorithm. Finally, Section 6 concludes this work.

## 2    Upward and Mixed Upward Planarization Problem

A drawing of a graph (digraph) is a mapping of its nodes to points in the plane and of its edges to open jordan curves. A graph (digraph) is *planar* if it has a drawing where no two edges have a common point. An *upward drawing* of a digraph is a drawing such that all the edges are represented by curves monotonically increasing in the vertical direction. A digraph is *upward planar*

M. Eiglsperger et al., *Mixed Planarization*, JGAA, 7(2) 203–220 (2003)      206

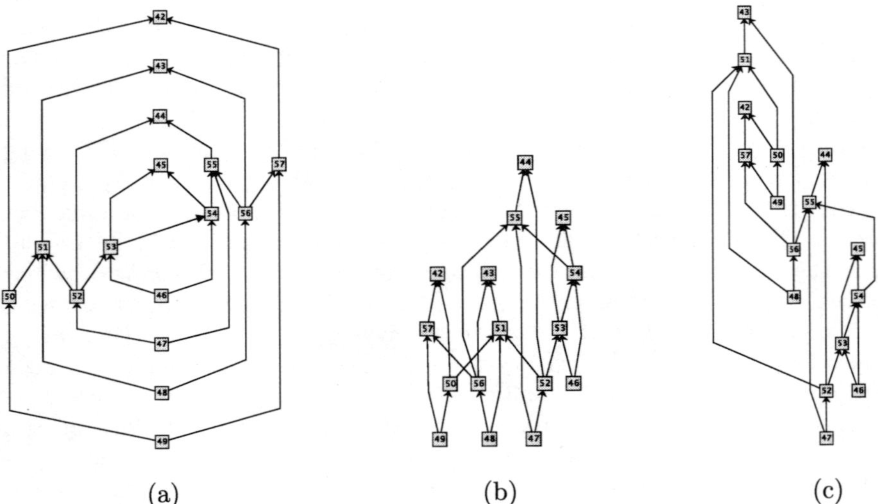

Figure 1: Three upward drawings of a directed graph. The graph is upward planar and can therefore be drawn without crossings (a). The Sugiyama-approach produces seven crossings (b) whereas our new method produces only two crossings (c).

if it has a drawing which is upward and planar at the same time. Please note that there are graphs which have an upward drawing and also have a planar drawing but do not have an upward planar drawing. An *embedding* of a graph is defined as a *cyclic ordering* of the adjacent edges of each vertex of the graph. An embedding is *planar* if there is a planar drawing of the graph which preserves this ordering. An *upward embedding* of a graph is a *linear ordering* of the adjacent edges of each vertex of the graph in which the incoming and outgoing edges form an interval. An upward embedding is *planar* if there is an upward planar drawing of the graph which preserves the corresponding ordering. Preserving the ordering means that the linear ordering is equivalent to the ordering that can be obtained by ordering the edges according to the angle they form with a ray leaving the vertex in direction of the negative x-axis. We assume in the remainder of the paper that graphs have no multiple edges and selfloops.

Given a directed graph $G = (V, E)$, the graph $G' = (V \cup V', E')$ is an *upward planarization* of $G$ with crossing number $|V'|$ if and only if

- $G'$ is upward planar,

- $deg(v) = 4$ for all $v \in V'$, and

- $\forall e = (v, w) \in E$, there is a path $p(e) = (v_0, v_1), (v_1, v_2), \ldots, (v_{n-1}, v_n)$ in $G'$ with $v = v_0$, $w = v_n$ and $v_i \in V', 0 < i < n$. Every edge in $E'$ is contained in such a path, and two paths have no edge in common.

A mixed graph is a three-tuple $G = (V, E_d, E_u) \subseteq (V, V \times V, V \times V)$, where

M. Eiglsperger et al., *Mixed Planarization*, JGAA, 7(2) 203–220 (2003)    207

$V$ is the set of vertices, $E_d$ is the set of directed edges and $E_u$ is the set of undirected edges.

The mixed graph $G = (V, E_d, E_u)$ is *mixed upward planar* if there is a planar drawing of $G$ where each edge in $E_d$ is represented by a curve monotonically increasing in the vertical direction.

A mixed upward embedding is *planar* if there is a mixed upward planar drawing of the graph which preserves the corresponding ordering.

Determining for a (mixed) graph $G$ a (mixed) upward planarized graph is called the *(mixed) upward planarization problem*. Determining for a (mixed) graph $G$ a (mixed) upward planarized graph with minimal crossing number is called the *(mixed) upward crossing minimization problem*.

Because the decision problem whether a graph has an upward embedding is a special case of the upward crossing minimization problem and the directed case is a special case of the mixed case, it follows that:

**Corollary 1 ([10])** *The upward crossing minimization problem and the mixed upward crossing minimization problem are both NP-hard.*

# 3 Upward Planarization

In this section, we propose an algorithmic framework for the upward crossing minimization problem. This framework is derived from techniques for the planarization of undirected graphs, see i.e. [6].

The framework consists of three parts:

1. Construct an upward planar subgraph.

2. Determine an upward embedding of this subgraph.

3. Insert the edges not contained in the subgraph, one by one.

In the first step, a subgraph of the input graph is calculated which is upward planar. For this subgraph, an upward embedding is determined in the second step. Of course, these two steps are only conceptually separated and can be combined to one step. Note that finding a maximum upward planar subgraph, i.e., finding an upward planar subgraph with the maximum number of edges, is NP-hard. In the third step, the edges which are not part of the upward planar subgraph are inserted incrementally into the embedding. Additionally, we can perform some local optimizations on the resulting planarization to improve the quality of it.

## 3.1 Maximum Upward Planar Subgraph

The maximum upward planar subgraph problem can be stated as follows: Given a directed graph $G = (V, E)$. Find $E' \subseteq E$ such that the directed graph $G = (V, E)$ is upward planar with maximum number of edges.

M. Eiglsperger et al., *Mixed Planarization*, JGAA, 7(2) 203–220 (2003)    208

The maximum planar subgraph problem is a related problem and a lot of algorithms have been proposed for its solution [11],[12],[17],[15],[20]. All of them, except [15], which can also compute the optimal solution when no time limit is specified, are heuristics, since the problem is NP-complete. Cimikowski [5] compared some of them empirically. In his comparison, the algorithm of Jünger and Mutzel(JM)[15] performed best in solution quality, followed by the algorithm of Goldschmidt and Takvorian (GT)[11]. The fastest algorithm was the one based on PQ-trees[11], but its performance in terms of the solution quality was significantly lower than JM and GT. In [20] Resende and Ribero give a randomized formulation of GT and show on the same test set as [5] that their formulation achieves better results with the same running time performance, except for one family of graphs where JM performs better.

However, the algorithm of GT is much easier to implement in contrast to the algorithm of JM. JM is a branch-and-cut algorithm and is, therefore, based on sophisticated algorithms for linear programming.

Because of its performance and its implementation issues, we use GT as a starting point. In the next section, we review the GT algorithm and show in the following section how it can be modified to calculate upward planar embeddings.

## 3.2   The Goldschmidt/Takvorian Planarization Algorithm

In this section, we review the main components of GT, the two-phase heuristics of Goldschmidt and Takvorian[11]. Our description follows the one in [20]. The first phase of GT consists in devising an ordering $\Pi$ of the set of vertices of $V$ of the input graph $G$. This ordering should possibly infer a Hamiltonian path. The vertices of $G$ are placed on a vertical line according to the ordering $\Pi$ obtained in the first phase, such that as many edges as possible between adjacent vertices can also be placed on the line. All other edges are drawn as arcs either right or left of the line.

The second phase of GT partitions the edge set $E$ of $G$ into subsets $\mathcal{L}$ (left of the line), $\mathcal{R}$ (right of the line), and $\mathcal{B}$ (the remainder) in such a way that $|\mathcal{L} + \mathcal{R}|$ is large (ideally maximum) and that no two edges both in $\mathcal{L}$ or both in $\mathcal{R}$ cross with respect to the sequence $\Pi$ devised in the first phase.

Let $\pi(v)$ denote the relative position of vertex $v \in V$ within vertex sequence $\Pi$. Furthermore, let $e_1 = (a, b)$ and $e_2 = (c, d)$ be two edges of $G$, such that, without loss of generality, $\pi(a) < \pi(b)$ and $\pi(c) < \pi(d)$. These edges are said to *cross* if, with respect to sequence $\Pi$, $\pi(a) < \pi(c) < \pi(b) < \pi(d)$ or $\pi(c) < \pi(a) < \pi(d) < \pi(b)$.

The *conflict graph* has a vertex for every edge in $G$ and two vertices are adjacent if the corresponding edges cross with respect to $\Pi$. It follows directly from its definition that the conflict graph is an *overlap graph*, i.e. a graph whose vertices can be represented as intervals, and two vertices are adjacent if and only if the corresponding intervals intersect but none of the two is contained by the other.

An induced bipartite subgraph of the conflict graph represents a valid assignment of the edges in $G$ to the sets $\mathcal{L}, \mathcal{R}$ and $\mathcal{B}$. Since finding a maximal

M. Eiglsperger et al., *Mixed Planarization*, JGAA, 7(2) 203–220 (2003)    209

induced bipartite subgraph is NP-complete, even for overlap graphs, GT uses a heuristics. This heuristics calculates two disjoint independent sets of the conflict graph which, together, are a bipartite subgraph of the conflict graph.

A maximum independent set of an overlap graph can be calculated in time $O(NM)$, where $N$ is the number of different interval endpoints and $M$ is the number of edges in the overlap graph with the algorithm of Asano, Imai and Mukaiyama[1]. In our setting, $N \leq n$ and $M = m$, which leads to a running time of $O(nm)$.

## 3.3 The directed version of the GT algorithm

We now present our variant of the GT algorithm for planar upward subgraph calculation. In order to change the GT algorithm to get an upward planar subgraph, we have to modify the first step of GT, the construction of the vertex order. The vertex order must ensure that no directed edge has a target vertex which is in the order before the source vertex. This is achieved by using algorithm *vertex order* as a first phase of GT. We call this variant *directed GT* or, shorter, DGT to distinguish it from the original formulation. Algorithm 1, *directed vertex order*, is a modification of the algorithm [11]. It is a variation of a standard topological sorting algorithm and tries to maximize the number of edges between consecutive vertices in the ordering. It has been shown in [20] that this improves the quality of the result of GT. The ordering is constructed incrementally. Assume that vertex $v$ is the vertex chosen in the previous step. The algorithm chooses a vertex in the next step which is adjacent to $v$, but which is not the successor of an unchosen vertex. If this is not possible, it takes a vertex of minimal degree which, additionally, is not the successor of an unchosen vertex. As the first vertex, it chooses a vertex with no incoming edge with minimal degree.

**Lemma 1** *Let $G$ be a directed graph. If the vertex order $\Pi$ in the first phase of the GT algorithm is a topological order of $G$, the result of GT is an upward planar subgraph of $G$.*

**Proof:** Placing the vertices on a vertical line according to the ordering used by GT and drawing the edges in $\mathcal{L}$ as arcs on the left side of the line and the edges in $\mathcal{R}$ on the right side of the line yields an upward planar drawing of the subgraph calculated by GT.                                           $\square$

**Lemma 2** *The vertex order calculated by algorithm* vertex order *is a topological order of $G$.*

**Proof:** The algorithm *vertex order* increments in each iteration the current ordering by a vertex with indegree zero. This is similar to a folklore topological sorting algorithm, see, i.e., [18] for details.                                           $\square$

From the sets $\mathcal{L}$ and $\mathcal{R}$ and the permutation $\Pi$, we can now easily obtain the upward planar embedding: For each node $v \in V$ we sort the edges with source

M. Eiglsperger et al., *Mixed Planarization*, JGAA, 7(2) 203–220 (2003)      210

---

**Algorithm 1:** vertex order

---

**Input:** A directed graph $G = (V, E)$
**Output:** A permutation $\Pi$ on the vertices
Select $v_1$ from $G$ with zero indegree and minimal outdegree;
$\mathcal{V} = V \setminus \{v_1\}$;
$G_1 = $ directed graph induced on $G$ by $\mathcal{V}$;
**for** $k = 2, \ldots, |V|$ **do**
$\quad \mathcal{U} = \{v \in \mathcal{V} | \text{indegree}(v) = 0 \text{ in } G_k\}$;
$\quad$ **if** $v_{k-1}$ *is connected to a vertex in* $\mathcal{U}$ **then**
$\quad\quad$ select $v_k$ as vertex in $\mathcal{U}$ adjacent to $v_{k-1}$ with min. degree in $G_{k-1}$
$\quad$ **else**
$\quad\quad$ select $v_k$ as vertex in $\mathcal{U}$ with min. degree in $G_{k-1}$
$\quad$ **end**
$\quad \mathcal{V} = \mathcal{V} \setminus v_k$;
$\quad G_k = $ directed graph induced on $G$ by $\mathcal{V}$;
**end**
**return** $\Pi = (v_1, v_2, \ldots, v_{|V|})$

---

$v$ in $\mathcal{L}$ decreasing according to $\Pi$ and the edges with source $v$ in $\mathcal{R}$ increasing according to $\Pi$ and concatenate these two ordered list to one. For the incoming edges, we first sort the edges with target $v$ in $\mathcal{R}$ decreasing according to $\Pi$ and the edges with source $v$ in $\mathcal{L}$ increasing according to $\Pi$ and append the result to the list of outgoing edges.

We conclude the section with the following theorem:

**Theorem 1** *Algorithm DGT computes an upward planar subgraph, together with an upward planar embedding of this subgraph, in time $O(nm)$.*

## 3.4   Edge Insertion

There is an interesting difference between the insertion of directed and undirected edges. In the undirected case, the edges which are not part of the planar subgraph in the first step can be inserted independently of each other. This is different in the directed case. Here, we cannot insert an edge into the drawing without looking at the remaining edges which have to be inserted later. The reason for this is that introducing dummy nodes in the graph introduces changes in the ordering of the vertices of the graph. This may introduce directed cycles if an edge is added later. (See Figure 2).

Assume that the dashed edges have to be inserted in Fig. 2(a), and we start by inserting edge $(5, 9)$. When we do not work carefully and insert edge $(5, 9)$ as in Fig. 2(b), we produce a crossing $C$ with edge $(1, 3)$ and some new edges, where $C$ is involved. Then, it is no longer possible to introduce edge $(3, 4)$ without destroying the upwardness property because of the new directed cycle $5 - C - 3 - 4 - 5$.

We call a vertex with indegree 0 a *source*, and a vertex with outdegree 0 a

M. Eiglsperger et al., *Mixed Planarization*, JGAA, 7(2) 203–220 (2003)    211

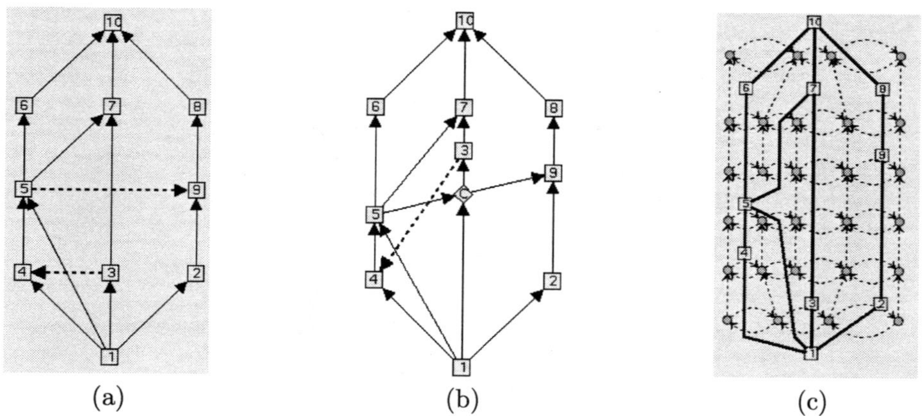

(a)    (b)    (c)

Figure 2: Edge insertion: Critical configuration and the routing graph

*sink.* A directed acyclic graph is called an *s-t graph* if it has exactly one sink and one source. We first restrict ourselves to s-t graphs. We show later how we can remove this restriction.

As shown above, we have to avoid cycles when we insert edges. We avoid this by *layering* the graph. A valid layering $l$ of a directed graph $G = (V, E)$ is a mapping of $V$ to integers such that $l(v) > l(u)$ for each edge $(u, v) \in E$. We then construct a routing graph $R$. The routing graph contains, for each face $f$ and for each layer that $f$ spans, a vertex. Two vertices lying in neighboring layers and representing the same face are connected by a directed edge of weight 0 in increasing layer order. Additionally, two vertices at the same layer $i$ of adjacent faces are connected by an edge of weight 1 if the source vertex of an edge separating these two faces is less than or equal to $i$ and the layer of the target node is greater than $i$.

In this graph, there are no edges in decreasing layer order. Each edge of weight 1 represents one crossing. A shortest path in the routing graph represents, therefore, an insertion of an edge with minimal number of crossings with respect to the given layering. Figure 2(c) shows an example for a routing graph.

Let $s(f)$, resp. $t(f)$, denote the source, resp. sink, of a face $f$. Note that in an s-t graph, every face has exactly one source and one sink. Furthermore, $lf(e)$, resp. $rf(e)$, denotes the face on the left, resp., right side of $e$. We consider the outer face as two faces, the *left-outer face*, resp. the *right-outer face*, which denote the left, resp. right part, of the outer face. Algorithm 2, *directed edge insertion*, summarizes the construction. It takes as input our current upward planar graph $G$, the set of remaining edges $F$ and one edge $e \in F$. The output $G'$ is a planarization of $G$ and $e$. It uses the subroutine subdivide$(G, e)$ which splits an edge $e = (a, b)$ into two edges $(a, w), (w, b)$, adds the vertex $w$ to $G$ and returns the created vertex $w$.

**Lemma 3** *The graph $G'$ calculated in* directed edge insertion *is an s-t graph and upward planar.*

M. Eiglsperger et al., *Mixed Planarization*, JGAA, 7(2) 203–220 (2003)     212

---

**Algorithm 2:** directed edge insertion

**Input**: Embedded upward planar s-t graph $G = (V, E)$, $F \subseteq V \times V$,
$\quad\quad e = (a, b) \in F$

**Output**: Embedded upward planarized s-t graph $G'$ of $G = (V, E \cup e)$

calculate faces from embedding;
determine valid layering $l$ of $(V, E \cup F)$;
Let $R$ be an empty directed graph;
**for** *every face $f$ of $G$* **do**
$\quad$ create vertices $v(f, i)$ for $l(s(f)) \leq i < l(t(f))$ in $R$;
$\quad$ **for** $l(s(f)) < i < l(t(f))$ **do**
$\quad\quad$ create edge of weight 0 from $v(f, i - 1)$ to $v(f, i)$ in $R$;
$\quad$ **end**
**end**
**for** *every edge $e' = (c, d)$ of $E$* **do**
$\quad$ **for** $l(c) \leq i < l(d)$ **do**
$\quad\quad$ create edge of weight 1 from $v(rf(e'), i)$ to $v(lf(e'), i)$ in $R$;
$\quad\quad$ create edge of weight 1 from $v(lf(e'), i)$ to $v(rf(e'), i)$ in $R$;
$\quad$ **end**
**end**
create vertex $v(a)$ and $v(b)$ in $R$ representing $a$ resp. $b$;
Insert edge of weight 0 from $v(a)$ to $v(f, l(a))$ in $R$ if $f$ is adjacent to $a$
and such a vertex exists;
Insert edge of weight 0 from $v(b)$ to $v(f, l(b) - 1)$ in $R$ if $f$ is adjacent to
$b$ and such a vertex exists;
Calculate shortest path $p$ from $v(a)$ to $v(b)$ in $R$;
$E' = E, V' = V, G' = (V', E')$;
Let $e_0, \ldots, e_n$ be the edges of weight 1 in $p$;
**for** $0 \leq i \leq n$ **do**
$\quad$ $w_i =$ subdivide$(G', e_i)$;
**end**
Add an edge between $a$ and $w_0$, $w_n$ and $b$, and $w_i$ and $w_{i+1}$ in $E'$;
**return** $G'$

---

**Proof:** In the edge insertion step, we do not decrease the indegree or the outdegree of any vertex existing already in the input graph. Therefore, we only have to show that none of the inserted vertices is a sink or a source. But this is true, since each of these vertices has indegree two and outdegree two. $G'$ is upward planar, since there are no crossings and the layering is observed. $\quad\square$

**Lemma 4** *The graph $(V', E' \cup F \setminus e)$ is acyclic.*

**Proof:** Assume that there is a directed cycle. Each inserted vertex $w_i$ is induced by an edge of weight 1 in the routing graph which connects two face vertices lying in the same layer. Assign this layer to node $w_i$. Then, each vertex has a layer assigned, and there are no edges which point in decreasing layer order. Thus, the cycle can only contain vertices in the same layer. These can only be

M. Eiglsperger et al., *Mixed Planarization*, JGAA, 7(2) 203–220 (2003)    213

vertices $w_i$ by the construction of the layering. But, from this fact it follows that the shortest path had a directed cycle which is a contradiction.    □

**Lemma 5** *Algorithm* directed edge insertion *has time complexity* $O(|V|^2)$,

**Proof:** The faces of the graph can be computed in linear time from the embedding. A valid layering with a minimal number of layers can also be computed in linear time using a topological sorting. The maximum number of layers is linear, since a topological sorting is an upper bound for the number of layers. Hence, the number of vertices in the routing graph is $O(|V|^2)$ and, since each vertex has constant degree, the total size of the routing graph is $O(|V|^2)$. Because the maximum cost of an edge is 1, we can use Dial's shortest path algorithm[8] which has linear running time in this case. The insertion of the edge can clearly be done in linear time.    □

The following theorem summarizes the lemmas above.

**Theorem 2** *Algorithm* directed edge insertion *inserts one edge in an embedded upward planar s-t graph* $G = (V, E)$ *in time* $O(|V|^2)$ *without introducing cycles with a set of not yet inserted edges. The planarized graph is an s-t graph.*

## 3.5   The Complete Algorithm

Algorithm 3, *upward planarization*, contains a description of the algorithm. Note that if the input graph for the second phase is not an s-t graph, we augment it to a planar upward s-t graph, see [3] for a linear time algorithm. Edges in the routing graph representing an edge added in the augmenting step are assigned weight 0, because they do not introduce a real crossing. The removed edges are inserted in random order in the second phase. After the routing, the augmenting edges are removed. Note that the augmentation does not affect the worst-case running time of the algorithm, since the number of edges in the graph remains linear in the number of nodes.

---

**Algorithm 3:** upward-planarization

---
    calculate embedded mixed upward planar subgraph with DGT;
    augment subgraph to an s-t graph;
    **for** *Each removed directed edge* **do**
        call algorithm directed edge insertion;
    **end**
    remove edges inserted in augmentation process;

---

From the discussion above, we derive the following theorem:

**Theorem 3** *Let* $G = (V, E)$ *be a directed graph. Algorithm 3,* upward-planarization, *creates an embedded upward planarized graph of* $G$ *in time* $O(|V||E| + (|V| + c)^2|E|)$, *where* $c$ *is the number of crossings of the planarized graph. When* $G$ *is sparse, i.e.* $|E| = O(|V|)$, *the algorithm* upward-planarization *runs in time* $O((|V| + c)^2|V|)$.

M. Eiglsperger et al., *Mixed Planarization*, JGAA, 7(2) 203–220 (2003)    214

However, the time bound in the theorem above is very pessimistic. In our experiments, the running time of the algorithm is reasonable, even for larger graphs.

## 3.6  Rerouting

In this section, we present a local optimization method for an upward planarization. One step of the method removes a path representing an edge from the planarization, and tries to reinsert it with fewer crossings. To test whether it can be reinserted with fewer crossings we first augment the graph to an s-t-graph after the removal of the path. Then we construct from this s-t graph the routing graph. Testing whether the edge can be inserted with fewer crossings reduces again to a shortest path problem in the routing graph. If we succeed, we change the planarization according to the new routing, otherwise, we do not change the planarization. In any case, the augmented edges are removed. We iterate this local optimization until we either do not make any further improvements, i.e., there is no edge for which we can find a routing with less crossings. This is realized by defining a set of edges $Cand$ which contains all edges of the original graph which have crossings in the planarization. In each iteration we randomly choose one edge and perform the local optimization step for the path defined by this edge. If the planarization had been improved we recalculate $Cand$ and start again. We stop when $Cand$ is empty. Since the local optimization is time consuming, the total number of local optimizations steps can be bounded by a constant.

## 4  Mixed Upward Planarization

In this section we show how the concepts in the previous sections can be extended to the mixed case, i.e., the input graph is a mixed graph.

For the mixed planar subgraph calculation, we also use the GT algorithm. As in the upward case, we only have to take care of the vertex ordering. We use Algorithm 4, *mixed vertex order*, a modified version of Algorithm *vertex-order* for this, which ignores the direction of the undirected edges.

The modifications follow the intuition that the undirected edges allow more freedom since we can choose the direction. So, the idea is to prefer the directed edges when computing the planar subgraph. One variant of the planar subgraph algorithm that takes this aspect into account, is to extend the GT approach by assigning different weights to the directed and undirected edges and then optimize over the weighted sum of the edges [1]. The actual choice of the weights depends on the application as well as on the class of graphs to consider and is the subject of further research.

The upward edge insertion algorithm can be extended to the mixed case by directing the undirected edges in $G$ temporarily according to the ordering in GT. We then insert the removed directed edges iteratively in the graph as described in section 3.4. Next we undirect the temporarily directed edges. Finally we

---

**Algorithm 4:** mixed vertex order

---

   **Input**: A mixed graph $G = (V, E, F)$

   **Output**: A permutation $\Pi$ on the vertices

   Select $v_1$ from $G$ with zero indegree and minimal degree;

   $\mathcal{V} = V \setminus \{v_1\}$;

   $G_1 = $ mixed graph induced on $G$ by $\mathcal{V}$;

   **for** $k = 2, \ldots, |V|$ **do**

      $\mathcal{U} = \{v \in \mathcal{V} | \text{indegree}(v) = 0 \text{ in } G_k\}$;

      **if** $v_{k-1}$ *is connected to a vertex in* $\mathcal{U}$ **then**

         select $v_k$ as vertex in $\mathcal{U}$ adjacent to $v_{k-1}$ with min. degree in $G_{k-1}$

      **else**

         select $v_k$ as vertex in $\mathcal{U}$ with min. degree in $G_{k-1}$

      **end**

      $\mathcal{V} = \mathcal{V} \setminus v_k$;

      $G_k = $ mixed graph induced on $G$ by $\mathcal{V}$;

   **end**

   **return** $\Pi = (v_1, v_2, \ldots, v_{|V|})$

---

insert the removed undirected edges by an standard edge insertion algorithm for undirected graphs [6]. Algorithm 5, *mixed-upward-planarization*, summarizes this.

---

**Algorithm 5:** mixed-upward-planarization

---

   calculate embedded mixed upward planar subgraph with GT;

   direct undirected edges in the subgraph temporarly;

   augment subgraph to an s-t graph;

   **for** *Each removed directed edge* **do**

      call algorithm directed edge insertion;

   **end**

   remove edges inserted in augmentation process;

   undirect edges which have been directed;

   **for** *Each removed undirected edge* **do**

      call algorithm undirected edge insertion;

   **end**

---

**Theorem 4** *Let $G = (V, E_d, E_u)$ be a mixed graph. Algorithm* mixed-upward-planarization *creates an embedded mixed upward planarized graph of $G$ in time* $O(|V|(|E_d| + |E_u|) + (|V| + c)^2 |E_d| + (|V| + c)|E_u|)$, *where $c$ is the number of crossings in the planarized graph.*

## 5   Experiments

In this section we present the results of an experimental comparison of our algorithm with the Sugiyama algorithm for directed acyclic graphs.

M. Eiglsperger et al., *Mixed Planarization*, JGAA, 7(2) 203–220 (2003)    216

All experiments have been performed on a Pentium IV System with 1.8 GHz and 512 Megabyte main-memory running Linux. We implemented our algorithm in pure JAVA based on the yFiles library [24]. For the experiments we used a randomized version of GT which takes the largest subgraph from 150 different node orderings. In the experiments we did not use rerouting. We compare our algorithm to the Sugiyama implementation in yFiles [24]. This implementation uses a randomized version of the iterated barycenter method. This method was the clear winner of an experimental comparison of heuristics for the crossing minimization problem of layered graph [16]. All experiments have been performed using JDK 1.4.1.

We performed our experiments on three test sets:

**Rome Graphs.** The Rome-graphs test suite[7] contains about 10.000 undirected connected graphs. The number of nodes in the test suite ranges from 10 to 100, the average density of the graphs ranging from 1 to 2 with average value of 1.3. We transformed the undirected graphs to directed acyclic graphs by directing the edges according to an ordering of the nodes of the graph. As ordering we chose the implicit ordering of the nodes as defined in the file.

**Upward Planar Graphs.** There are two test sets consisting of connected upward planar graphs, the first contains graphs with density 1.3, the second graphs with density 2.6. Both test sets contain 910 upward planar graphs, the number of nodes ranging in each form 10 to 100, containing 10 graphs for each node count. The graphs were generated the following way: First a random set of points in a triangle was generated. For this point set a Delaunay triangulation was performed which yields a planar triangulated graph. We deleted edges randomly until we reached the desired density. To assure that the generated graphs were connected we computed a spanning tree of the triangulated graph by randomized DFS and ensured that edges in the spanning tree are not deleted in the previous step. Finally we directed the edges according to the coordinates of their endpoints.

**Graphs With Limited Height.** There are two test sets which contain connected directed graphs with maximum height three, the first with density 1.3, the second with density 2.6. Maximum height three means in this setting that they have a layer assignment with at most three layers. Both test sets contain 910 upward planar graphs, the number of nodes ranging in each form 10 to 100, containing 10 graphs for each node count. The graphs were generated the following way: First we distributed randomly the nodes in three layers. To assure that a generated graph was connected we generated a spanning tree for it. Then we inserted edges randomly between nodes in neighbored layers until the desired density was reached.

Figure 3 shows the results of our experiments. The left diagrams show the relation between the average number of crossings and the number of vertices. The right diagrams show the relation between average running time and the

M. Eiglsperger et al., *Mixed Planarization*, JGAA, 7(2) 203–220 (2003)      217

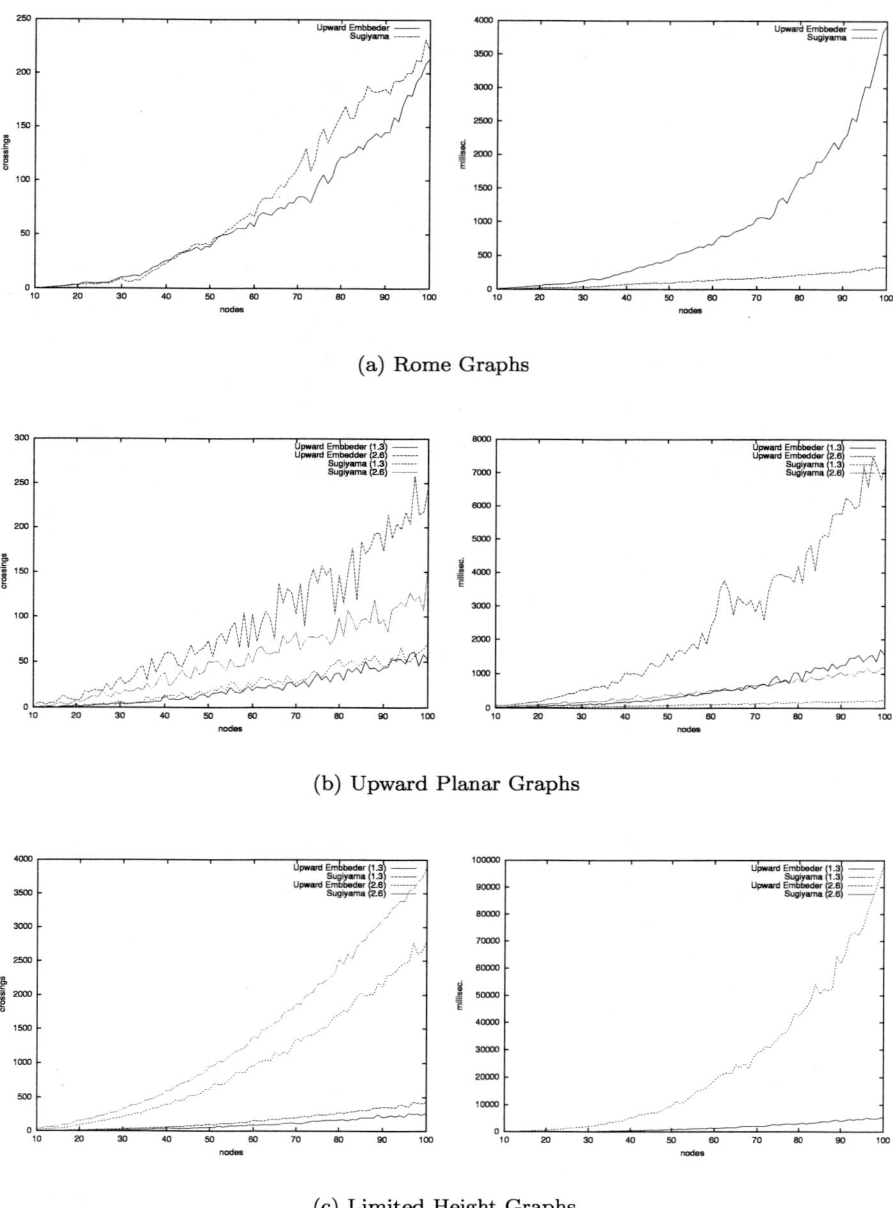


(a) Rome Graphs

(b) Upward Planar Graphs

(c) Limited Height Graphs

Figure 3: Results of experiments: Number Jof crossings and running time in milliseconds.

number of vertices. For the test sets with density 1.3 our algorithm yields better results than the Sugiyama approach. In the case for limited height graphs, the improvements are drastic. For the test sets with density 2.6 the Sugiyama approach is the clear winner. In terms of running time, the Sugiyama approach clearly outperforms our approach, however the running time of our algorithm is still acceptable for interactive use.

# 6    Conclusion

In this paper, we gave a new algorithm for the problem of finding a upward planarization for graphs with directed and undirected edges as well. Our approach generalizes the related problem for undirected graph and emphasizes on the practical needs for such methods, namely practical efficiency and good quality, even for large graphs. The concept is designed so flexible that many additional requirements like constraints or interactivity might be incorporated. Hence, together with a graph drawing algorithm for upward planar graphs it can be viewed as a reasonable alternative to the well known Sugiyama algorithm.

# Acknowledgments

The authors wish to thank the referees for their useful suggestions.

# References


[1] T. Asano, H. Imai, and A. Mukaiyama. Finding a maximum weight independent set of a circle graph. *IEICE Tranactions*, E74:681–683, 1991.

[2] C. Batini, E Nardelli, and R. Tamassia. A layout algorithm for data-flow diagrams. *IEEE Trans. Softw. Eng.*, SE-12(4):538–546, 1986.

[3] P. Bertolazzi, G. Di Battista, G. Liotta, and C. Mannino. Upward drawings of triconnected digraphs. *Algorithmica*, 6(12):476–497, 1996.

[4] G. Booch, J. Rumbaugh, and I. Jacobson. *The Unified Modeling Language.* Addison-Weslay, 1999.

[5] R. Cimikowski. An analysis of heuristics for the maximum planar subgraph problem. In *Proceedings of the 6th ACM-SIAM Symposium of Discrete Algorithms*, pages 322–331, 1995.

[6] G. Di Battista, P. Eades, R. Tamassia, and I. G. Tollis. *Graph Drawing: Algorithms for the Visualization of Graphs.* Prentice Hall, 1999.

[7] G. Di Battista, A. Garg, G. Liotta, R. Tamassia, E. Tassinari, and F. Vargiu. An experimental comparison of four graph drawing algortihms. *Comput. Geom. Theory Appl.*, 7:303–325, 1997.

M. Eiglsperger et al., *Mixed Planarization*, JGAA, 7(2) 203–220 (2003)    219

[8] R. Dial. Algorithm 360: Shortest path forest with topological ordering. *Communications of ACM*, 12:632–633, 1969.

[9] M. Eiglsperger, U. Foessmeier, and M .Kaufmann. Orthogonal graph drawing with constraints. In *Proc. 11th ACM-SIAM Symposium on Discrete Algorithms*, pages 3–11, 2000.

[10] A. Garg and R. Tamassia. On the complexity of upward and rectilinear planarity testing. In *Proceedings of the 2nd International Symposium on Graph Drawing (GD'94)*, volume 894 of *LNCS*, pages 286–297, 1995.

[11] O. Goldschmidt and A. Takvorian. An efficient graph planarization two-phase heuristic. *Networks*, 24:69–73, 1994.

[12] R. Jayakumar, K. Thulasiraman, and M. N. S. Swamy. $o(n^2)$ algorithms for graph planarization. In *IEEE Trans. on CAD*, volume 8 (3), pages 257–267, 1989.

[13] M. Jünger and S. Leipert. Level planar embedding in linear time. In J. Kratochvil, editor, *Proceedings of the 7th International Symposium on Graph Drawing*, volume 1547, pages 72–81. Springer Verlag, 2000.

[14] M. Jünger, S. Leipert, and P. Mutzel. Level planarity testing in linear time. In *Proceedings of the 6th International Symposium on Graph Drawing (GD'98)*, volume 1547 of *LNCS*, pages 224–237, 1998.

[15] M. Jünger and P. Mutzel. Solving the maximum weight planar subgraph problem by branch & cut. In *Proc. of the 3rd conference on integer programming and combinatorial optimization (IPCO)*, pages 479–492, 1993.

[16] M. Jünger and P. Mutzel. 2-layer straightline crossing minimization: Performance of exact and heuristic algorithms. *Journal of Graph Algorithms and Applications (JGAA)*, 1(1):1–25, 1997.

[17] G. Kant. An $O(n^2)$ maximal planarization algorithm based on PQ-trees. Technical Report RUU-CS-92-03, CS Dept., Univ. Utrecht, Netherlands, 1992.

[18] K. Mehlhorn and S. Näher. *The* LEDA *Platform of Combinatorial and Geometric Computing.* Cambridge University Press, 1999.

[19] W. Didimo P. Bertolazzi, G. Di Battista. Quasi-upward planarity. *Algorithmica*, 32(3):474–506, 2002.

[20] M. Resende and C. Ribeiro. A grasp for graph planarization. *Networks*, 29:173–189, 1997.

[21] J. Seemann. Extending the Sugiyama algorithm for drawing UML class diagrams: Towards automatic layout of object-oriented software diagrams. In *Proceedings of the 5th International Symposium on Graph Drawing (GD'97)*, volume 1353 of *LNCS*, pages 415–424, 1997.

[22] K. Sugiyama, S. Tagawa, and M. Toda. Methods for visual understanding of hierarchical system structures. *IEEE Transactions on Systems, Man, and Cybernetics*, 11(2):109–125, February 1981.

[23] R. Tamassia. On embedding a graph in the grid with the minimum number of bends. *SIAM Journal on Computing*, 16(3):421–444, 1987.

[24] R. Wiese, M. Eiglsperger, and M. Kaufmann. yfiles: Visualization and automatic layout of graphs. In *Proceedings of the 9th International Symposium on Graph Drawing (GD'01)*, LNCS, pages 453–454. Springer, 2001.

# Journal of Graph Algorithms and Applications

http://jgaa.info/

vol. 7, no. 2, pp. 221–241 (2003)

# Upward Embeddings and Orientations of Undirected Planar Graphs

*Walter Didimo*

Dipartimento di Ingegneria Elettronica e dell'Informazione
Università di Perugia
via G. Duranti 93, 06125 Perugia, Italy.
http://www.diei.unipg.it/~didimo/
didimo@diei.unipg.it

*Maurizio Pizzonia*

Dipartimento di Informatica e Automazione
Università di Roma Tre
via della Vasca Navale 79, 00146 Roma, Italy.
http://www.dia.uniroma3.it/~pizzonia/
pizzonia@dia.uniroma3.it

## Abstract

An *upward embedding* of an embedded planar graph specifies, for each vertex $v$, which edges are incident on $v$ "above" or "below" and, in turn, induces an *upward orientation* of the edges from bottom to top. In this paper we characterize the set of all upward embeddings and orientations of an embedded planar graph by using a simple flow model, which is related to that described by Bousset [3] to characterize bipolar orientations. We take advantage of such a flow model to compute upward orientations with the minimum number of sources and sinks of 1-connected embedded planar graphs. We finally devise a new algorithm for computing visibility representations of 1-connected planar graphs using our theoretic results.

Communicated by Giuseppe Liotta and Ioannis G. Tollis: submitted October 2001; revised April 2002.

Research partially supported by "Progetto ALINWEB: Algoritmica per Internet e per il Web", MIUR Programmi di Ricerca Scientifica di Rilevante Interesse Nazionale.

W. Didimo and M. Pizzonia, *Upward Embeddings*, JGAA, 7(2) 221–241 (2003)222

# 1  Introduction

Let $G$ be an undirected planar graph with a given planar embedding. Loosely speaking, an *upward embedding* (also called an *upward representation*) of $G$ is specified by splitting, for each vertex $v$ of $G$, the ordered circular list of the edges that are incident on $v$ into two linear lists (from left to right) $E_{above}(v)$ and $E_{below}(v)$, in such a way that there exists a planar drawing $\Gamma$ of $G$ with the following properties: (i) all the edges are monotone in vertical direction; (ii) for each vertex $v$ the edges in $E_{above}(v)$ ($E_{below}(v)$) are incident on $v$ above (below) the horizontal line through $v$.

A drawing $\Gamma$ that verifies properties (i) and (ii) is said to be an *upward drawing* of $G$. An orientation of all edges of $\Gamma$ from bottom to top defines an orientation of $G$, which we call an *upward orientation* of $G$. Hence, each upward embedding of $G$ induces an upward orientation of $G$. Figure 1 shows an upward embedding of an embedded planar graph and the upward orientation induced by it.

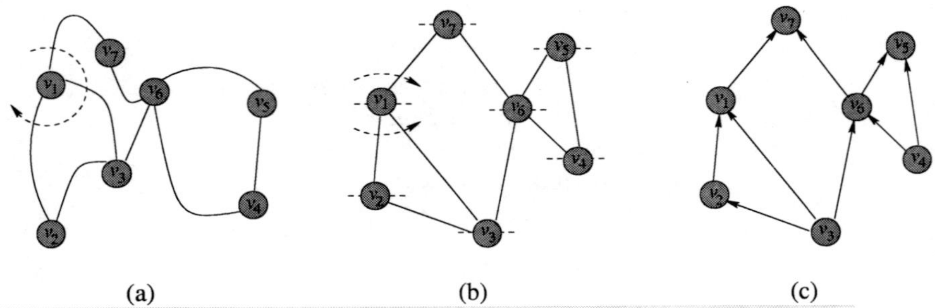

(a)                                          (b)                                          (c)

Figure 1: (a) An embedded planar graph. (b) An upward embedding of the embedded planar graph. For each vertex $v_i$ of the graph the edges in $E_{below}(v_i)$ and $E_{above}(v_i)$ are drawn incident below and above the horizontal line through $v_i$, respectively. (c) The upward orientation induced by the upward embedding.

An embedded planar graph has in general many upward embeddings and upward orientations within the given embedding. Although upward embeddings and orientations have been widely studied within specific theoretic and application domains, as far as we know no complete combinatorial characterizations have been provided in the case of general embedded planar graphs. In the present paper we investigate this problem and we show how our theoretic results have interesting applicability in graph drawing.

An important class of upward orientations, deeply studied in the literature, is represented by the so called *bipolar orientations* (or *st-orientations*). A bipolar orientation of an undirected planar graph $G$ is an upward orientation of $G$ with exactly one source $s$ (vertex without in-edges) and one sink $t$ (vertex without out-edges). A bipolar orientation of $G$ with source $s$ and sink $t$ exists if and only if $G \cup \{(s,t)\}$ is biconnected. Finding a bipolar orientation of a planar

W. Didimo and M. Pizzonia, *Upward Embeddings*, JGAA, 7(2) 221–241 (2003)223

graph is the first step of many algorithms in graph theory and graph drawing. A complete and elegant study of the properties of bipolar orientations has been provided by de Fraysseix et. al. [5], and a characterization of bipolar orientations in terms of a network flow model has been described by Bousset [3].

Czyzowicz, Kelly and Rival [14, 13, 4, 16] provide several theoretic results about upward orientations and upward drawings of ordered set and planar lattices, that is, special classes of combinatorial structures.

Several results on upward embeddings of digraphs have been also provided in the literature. In this case, the orientation of the edges of the graph is given, and a classical problem consists of finding an upward (planar) embedding that preserves such an orientation. Clearly, an upward embedding of a digraph might not exist. Bertolazzi et al. [1] describe a polynomial time algorithm for testing the existence of upward embeddings of a digraph within a given planar embedding. The algorithm is also able to construct an upward embedding if there exists one. In the variable embedding setting the upward planarity testing problem is NP-complete [9], but it can be solved in polynomial time for digraphs with a single source [2].

The main contributions of this paper are listed below:

- Starting from the properties on upward planarity of digraphs given in [1], we provide a complete characterization of the set of all upward embeddings and orientations of any embedded planar graph (Section 3.1). It is based on a network flow model, which is a generalization of that used by Bousset [3] for characterizing bipolar orientations. In particular, if the graph is biconnected, our flow model also captures all bipolar orientations of the graph.

- We describe flow based polynomial time algorithms for computing upward embeddings of the input graph. Such algorithms allow us to handle partial specifications of the upward embedding (Section 3.1). Further, we provide a polynomial time algorithm to compute upward orientations with the minimum number of sources and sinks (Section 3.2). Upward orientations with the minimum number of sources and sinks can be viewed as a natural extension of the concept of bipolar orientations to 1-connected graphs.

- We describe a simple technique to compute visibility representations of 1-connected planar graphs (Section 4), which can be of practical interest for graph drawing applications. It is based on the computation of an upward embedding of the graph, and does not require running any augmentation algorithm to initially make the graph biconnected. Compared to a standard technique that uses the good approximation algorithm described by Fialko and Mutzel [8] to make the graph biconnected, the algorithm we propose is theoretically faster, simpler to implement, and achieves similar results in terms of area of the visibility representation.

In Section 2 we recall formal definitions and known results on upward embeddings and orientations of undirected planar graphs.

W. Didimo and M. Pizzonia, *Upward Embeddings*, JGAA, 7(2) 221–241 (2003)224

## 2 Basic Definitions and Results on Upward Embeddings

A graph is 1-*connected* (or *connected*) if there exists a path between any pair of its vertices. A vertex of the graph whose removal disconnects the graph is called a *cutvertex*. A connected graph is 2-*connected* (or *biconnected*) if it has no cutvertex. Given a 1-connected graph $G$, a *biconnected component* (or *block*) of $G$ is a maximal biconnected subgraph of $G$. Observe that each cutvertex of $G$ belongs to at least two distinct blocks of $G$, and that each edge of $G$ belongs to exactly one block of $G$. The decomposition of a graph into its blocks can be easily done in linear time [18].

A *drawing* $\Gamma$ of a graph $G$ maps each vertex $u$ of $G$ into a point $p_u$ of the plane and each edge $(u, v)$ of $G$ into a Jordan curve between $p_u$ and $p_v$. $\Gamma$ is *planar* if two distinct edges never intersect except at common end-points. $G$ is *planar* if it admits a planar drawing. A planar drawing $\Gamma$ of $G$ divides the plane into topologically connected regions called *faces*. Exactly one of these faces is unbounded, and it is said to be *external*; the others are called *internal* faces. Also, for each vertex $v$ of $G$, $\Gamma$ induces a circular clockwise ordering of the edges incident on $v$. The choice $\phi$ of such an ordering for each vertex of $G$ and of an external face is called a *planar embedding* of $G$. A planar graph $G$ with a given planar embedding $\phi$ is called an *embedded planar graph* and denoted by $G_\phi$. A *drawing* of $G_\phi$ is a planar drawing of $G$ that induces $\phi$ as the planar embedding.

Let $G_\phi$ be an (undirected) embedded planar graph. An *upward embedding* $\mathcal{E}_\phi$ of $G_\phi$ is a splitting of the adjacency lists of all vertices of $G_\phi$ such that:

**(E1)** For each vertex $v$ of $G_\phi$ the circular clockwise list $L(v)$ of the edges incident on $v$ is split into two linear lists (from left to right), $E_{below}(v)$ and $E_{above}(v)$, so that the circular list obtained by concatenating $E_{above}(v)$ and the reverse of $E_{below}(v)$ is equal to $L(v)$.

**(E2)** There exists a planar drawing $\Gamma(\mathcal{E}_\phi)$ of $G_\phi$ such that all the edges are monotone in vertical direction and for each vertex $v$ of $G_\phi$ the edges of $E_{below}(v)$ and $E_{above}(v)$ are incident on $v$ below and above the horizontal line through $v$, respectively. We say that $\Gamma(\mathcal{E}_\phi)$ is a *drawing* of $\mathcal{E}_\phi$ and an *upward drawing* of $G_\phi$.

From (E2) the following is immediate.

**Property 1** *Given an upward embedding of $G_\phi$, for each edge $e = (u, v)$ of $G_\phi$ either $e \in E_{above}(u) \cap E_{below}(v)$ or $e \in E_{below}(u) \cap E_{above}(v)$.*

An upward embedding $\mathcal{E}_\phi$ of $G_\phi$ uniquely induces an *upward orientation* $\mathcal{O}_\phi$ of $G_\phi$. Namely, for each edge $e = (u, v)$ such that $e \in E_{above}(u)$ and $e \in E_{below}(v)$, we orient $e$ from $u$ to $v$ (see Figure 1). Conversely, an upward orientation defines in general a class of possible upward embeddings inducing it (see Figure 2). A *source* of $\mathcal{E}_\phi$ is a vertex $v$ of $G_\phi$ such that $E_{below}(v)$ is empty. A source has only out-edges with respect to orientation $\mathcal{O}_\phi$. A *sink* of $\mathcal{E}_\phi$ is

W. Didimo and M. Pizzonia, *Upward Embeddings*, JGAA, 7(2) 221–241 (2003)225

a vertex $v$ of $G_\phi$ such that $E_{above}(v)$ is empty. A sink has only in-edges with respect to $\mathcal{O}_\phi$.

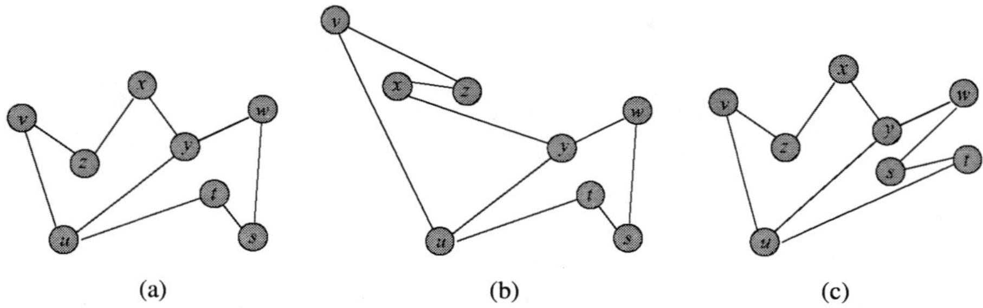

(a)         (b)         (c)

Figure 2: Three different upward embeddings that induce the same upward orientation.

Given a vertex $v$ of $G_\phi$, we denote by $deg(v)$ the number of edges incident on $v$. An *angle* of $G_\phi$ at vertex $v$ is a pair of clockwise consecutive edges incident on $v$. In particular, if $deg(v) = 1$, and if we denote by $e$ the edge incident on $v$, $(e, e)$ is an angle. Given a splitting of the adjacency lists of $G_\phi$ that verifies (E1), an angle $(e_1, e_2)$ at vertex $v$ of $G_\phi$ can be of three different types (see Figure 3 for an example):

- *large*: (i) both $e_1$ and $e_2$ belong to $E_{below}(v)$ $(E_{above}(v))$, and (ii) $e_1$ and $e_2$ are the first (last) edge and the last (first) edge of $E_{below}(v)$ $(E_{above}(v))$, respectively. We associate a label L with a large angle.

- *flat*: if: (i) $e_1 \in E_{below}(v)$ and $e_2 \in E_{above}(v)$ or, (ii) $e_1 \in E_{above}(v)$ and $e_2 \in E_{below}(v)$. We associate a label F with a flat angle.

- *small*: in all the other cases. We associate a label S with a small angle.

Figure 4 shows the labeling of the angles of an embedded planar graph $G_\phi$ determined by an upward embedding $\mathcal{E}_\phi$. Each drawing of $\mathcal{E}_\phi$ maps the angles of $G_\phi$ to geometric angles such that large and small angles always correspond to geometric angles larger and smaller than 180 degrees, respectively. Both the two edges that form a large or a small angle at vertex $v$ are incident on $v$ either above or below the horizontal line through $v$. Instead, a flat angle at vertex $v$ corresponds to a geometric angle that can be either larger or smaller than 180 degrees, but in any case one of its edges is incident on $v$ above the horizontal line through $v$ while the other edge is incident on $v$ below the same line.

Let $f$ be a face of $G_\phi$. We call *border* of $f$ the alternating circular list of vertices and edges that form the boundary of $f$. Note that, if the graph is not biconnected an edge or a vertex may appear more than once in the border of $f$. We say that an angle $(e_1, e_2)$ at vertex $v$ *belongs* to face $f$ if $e_1$, $e_2$, and $v$ belong

W. Didimo and M. Pizzonia, *Upward Embeddings*, JGAA, 7(2) 221–241 (2003)226

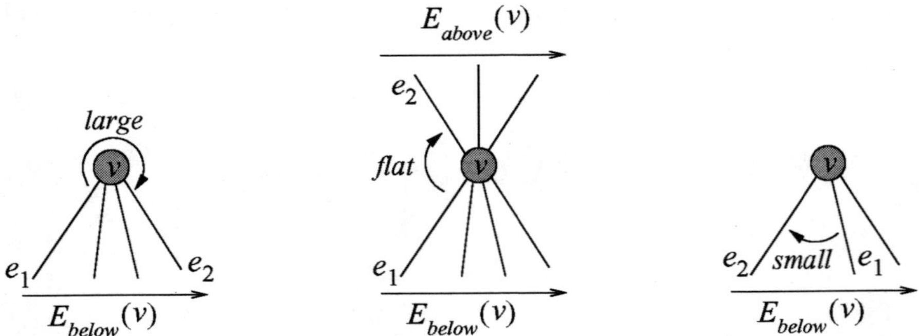

Figure 3: Examples of large, flat, and small angles.

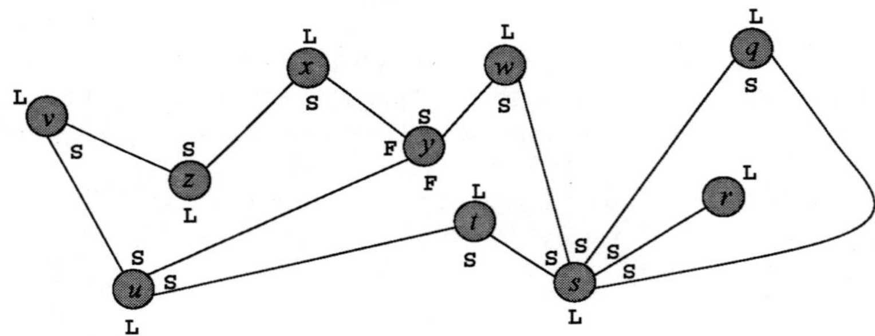

Figure 4: The labeling of the angles of an embedded planar graph determined by an upward embedding of the graph.

to the border of $f$. The *degree* of $f$, denoted by $deg(f)$, is the number of edges in the border of $f$. Observe that, $deg(f)$ is equal to the number of angles of $f$.

Consider now any labeling of the angles of $G_\phi$ with labels L, S, and F. For each face $f$ of $G_\phi$ denote by $L(f)$, $S(f)$, and $F(f)$ the number of angles that belong to $f$ with label L, S, and F, respectively. Also, for each vertex $v$ of $G_\phi$ denote by $L(v)$, $S(v)$, and $F(v)$ the number of angles at vertex $v$ with label L, S, and F, respectively. The following lemma is a direct consequence of a known result on upward planarity [1].

**Lemma 1** *Let $\mathcal{E}_\phi$ be a splitting of the adjacency lists of $G_\phi$ that verifies (E1), and consider the labeling of the angles of $G_\phi$ determined by it. $\mathcal{E}_\phi$ is an upward embedding of $G_\phi$ if and only if the following properties hold:*

(FIN) $S(f) = L(f) + 2$, *for each internal face $f$ of $G_\phi$.*

(FEX) $S(f) = L(f) - 2$, *for the external face $f$ of $G_\phi$.*

W. Didimo and M. Pizzonia, *Upward Embeddings*, JGAA, 7(2) 221–241 (2003)227

(VL0) $F(v) = 2$, $S(v) = deg(v) - 2$, and $L(v) = 0$, for each vertex $v$ of $G_\phi$ such that both $E_{above}(v)$ and $E_{below}(v)$ are not empty.

(VL1) $F(v) = 0$, $S(v) = deg(v) - 1$, and $L(v) = 1$, for each vertex $v$ of $G_\phi$ such that either $E_{above}(v)$ or $E_{below}(v)$ is empty.

Properties (VL0) and (VL1) of Lemma 1 state that if $\mathcal{E}_\phi$ is an upward embedding of $G_\phi$, each source or sink of $\mathcal{E}_\phi$ has exactly one large angle and no flat angle, while each vertex that is neither a source nor a sink has exactly two flat angles and no large angle. The next lemma provides a different formulation for properties (FIN) and (FEX).

**Lemma 2** *Properties* (FIN) *and* (FEX) *of* Lemma 1 *are equivalent to the following properties:*

(FIN') $deg(f) - 2 = 2L(f) + F(f)$, for each internal face $f$ of $G_\phi$.

(FEX') $deg(f) + 2 = 2L(f) + F(f)$, for the external face $f$ of $G_\phi$.

**Proof:**
Property (FIN) is equivalent to property (FIN') since $deg(f) = L(f) + S(f) + F(f)$. The equivalence between properties (FEX) and (FEX') can be proved analogously.

*q.e.d.*

Let $G_\phi$ be an embedded planar graph, $\mathcal{E}_\phi$ be an upward embedding of $G_\phi$, and $\mathcal{O}_\phi$ be the upward orientation induced by $\mathcal{E}_\phi$. Also, denote by $D_\phi$ the directed graph obtained by $G_\phi$ orienting its edges according to $\mathcal{O}_\phi$.

In Section 4 we need to compute a super-digraph of $D_\phi$ with only one source and one sink (*st-digraph*) and preserving the upward embedding $\mathcal{E}_\phi$ when restricted to $D_\phi$. In the following we recall an algorithm for this purpose. Further details can be found in [1].

Given a face $f$ of $D_\phi$, a vertex $v$ of $f$ with consecutive incident edges $e_1$ and $e_2$ on the boundary of $f$ is a *switch* of $f$ if $e_1$ and $e_2$ are both incoming or both outgoing $v$ (note that $e_1$ and $e_2$ may coincide if the graph is not biconnected). In the former case $v$ is a *sink-switch*, in the latter a *source-switch*. Observe that a source (sink) of $D_\phi$ is source-switch (sink-switch) of all its incident faces; a vertex of $D_\phi$ that is not a source or a sink is a switch of all its incident faces except two.

Consider the labeling of the angles of $D_\phi$ induced by its upward embedding. Let $v$ be a switch of a face $f$ of $D_\phi$, and let $e_1$ and $e_2$ be two consecutive edges on the boundary of $f$ that are incident on $v$. Clearly, $(e_1, e_2)$ is an angle of $f$. We call $v$ an $s_S$-*switch* ($s_L$-*switch*) of $f$ if $v$ is a source-switch of $f$ and if $(e_1, e_2)$ is labeled S (L). We call $v$ a $t_S$-*switch* ($t_L$-*switch*) of $f$ if $v$ is a sink-switch of $f$ and if $(e_1, e_2)$ is labeled S (L). Note that each S or L labels of a face is associated with a switch.

A *complete saturator* of $D_\phi$ is a set of vertices and edges, not belonging to $D_\phi$, with which we augment $D_\phi$. More precisely, a complete saturator consists of

W. Didimo and M. Pizzonia, *Upward Embeddings*, JGAA, 7(2) 221–241 (2003)228

two vertices $s$ and $t$, edge $(s,t)$, and a set of edges $(u,v)$ (each edge a *saturating edge*), such that (see Figure 5 (a)):

- vertices $u$ and $v$ are switches of the same face, or $u = s$ and $v$ is an $s_S$-switch of the external face, or $u$ is a $t_L$-switch of the external face and $v = t$,

- if $u, v \neq s, t$, either $u$ is an $s_S$-switch and $v$ is an $s_L$-switch or $u$ is a $t_L$-switch and $v$ is a $t_S$-switch; in the former case we say that $u$ *saturates* $v$ and in the latter case we say that $v$ *saturates* $u$,

- the graph $D_\phi$ augmented with the vertices and the edges of the complete saturator is an upward embedded graph with an *st*-orientation (st-digraph); the upward embedding of such a digraph restricted to the vertices and the edges of $D_\phi$ coincides with $\mathcal{E}_\phi$.

A simple linear time algorithm for computing a complete saturator of $D_\phi$ is given in [1]. This algorithm works in two main steps:

In the first step it recursively decomposes each face $f$ of $G_\phi$ adding a suitable number of saturating edges that split $f$. After this step, there are no more $s_L$-switches and $t_L$-switches in the internal faces of the digraph. Also, the $s_L$-switches and $t_L$-switches of the external face $f$ are not alternated in the border of $f$.

In the second step the algorithm further decomposes the external face $f$, adding the vertices $s$, $t$ and connecting $s$ to every $s_L$-switch of $f$, and $t$ to every $t_L$-switch of $f$.

In the following we briefly recall the algorithm for decomposing a face $f$ of $D_\phi$. More details can be found in [1]. We denote by $\sigma_f$ the sequence of labels of the angles of $f$ encountered in clockwise order while moving on the boundary of $f$. Also, we denote by $s_L$ an L label of $\sigma_f$ with associated a source-switch of $f$ and by $t_L$ an L label of $\sigma_f$ with associated a sink-switch of $f$. Similarly, we use symbols $s_S$ and $t_S$ to denote S-labels with associated a source-switch of $f$ and a sink-switch of $f$, respectively.

**Algorithm** *Saturate-Face(f)*

1. If $f$ has exactly one source-switch and one sink-switch then return.

2. Find a subsequence $(x, y, z)$ in $\sigma_f$ such that $x$ is an L label, and $y$ and $z$ are S labels. Let $v_x$, $v_y$, and $v_z$ be the switches of $f$ associated with $x$, $y$, and $z$, respectively.

3. Split $f$ into two faces $f'$ and $f''$ by inserting one edge; after the split, $f''$ always consists of the part of $f$ containing $v_x$, $v_y$, and $v_z$ plus the new edge; $f''$ has only one source and only one sink. Two cases are possible for the new edge:

W. Didimo and M. Pizzonia, *Upward Embeddings*, JGAA, 7(2) 221–241 (2003)229

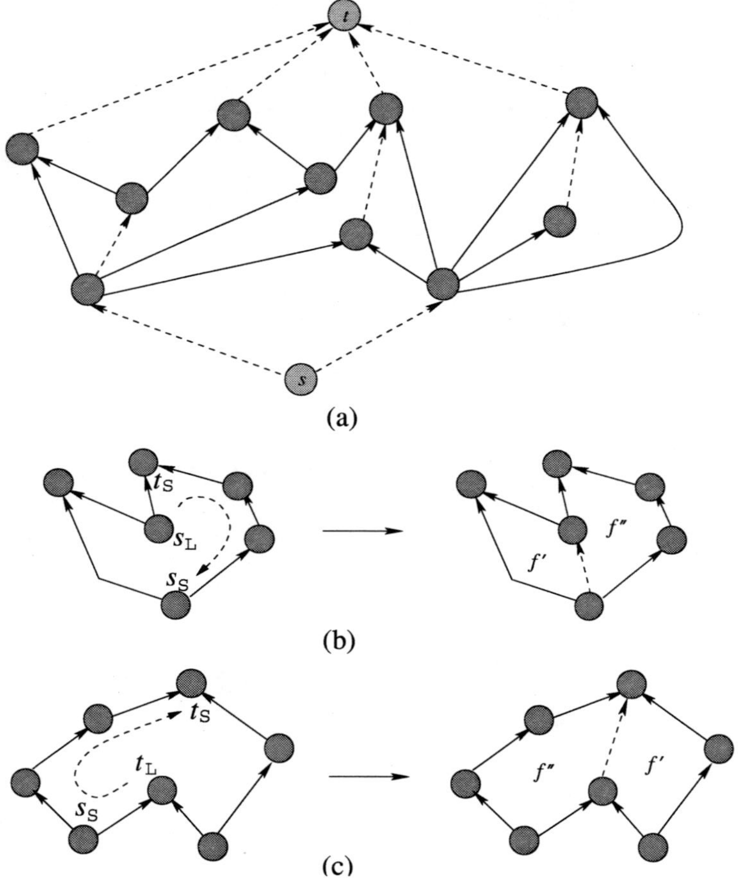

Figure 5: (a) An upward embedded digraph with a complete saturator. The edges of the saturator are dashed. (b) Illustration of Case 1 of algorithm *Saturate-Face(f)*. (c) Illustration of Case 2 of algorithm *Saturate-Face(f)*

Case 1 $(x, y, z) = (s_{\mathrm{L}}, t_{\mathrm{S}}, s_{\mathrm{S}})$: Add edge $(v_z, v_x)$; $f'$ consists of the part of $f$ that does not contain $v_y$ plus the new edge. Also, $\sigma_{f'}$ is obtained from $\sigma_f$ by replacing the subsequence $(x, y, z)$ with an $s_{\mathrm{S}}$. (see Figure 5 (b)).

Case 2 $(x, y, z) = (t_{\mathrm{L}}, s_{\mathrm{S}}, t_{\mathrm{S}})$: Add edge $(v_x, v_z)$; $f'$ consists of the part of $f$ that does not contain $v_y$ plus the new edge. Also, $\sigma_{f'}$ is obtained from $\sigma_f$ by replacing the subsequence $(x, y, z)$ with a $t_{\mathrm{S}}$. (see Figure 5 (c)).

4. Apply *Saturate-Face(f')*.

W. Didimo and M. Pizzonia, *Upward Embeddings*, JGAA, 7(2) 221–241 (2003)230

# 3 Characterizing Upward Embeddings

In this section we provide a complete characterization of the set of all upward embeddings of a general embedded planar graph (Section 3.1); it also implies a characterization of the upward orientations of the given graph. We model such a set of upward embeddings by using a simple network flow technique, which extends and generalizes that described by Bousset [3] for characterizing bipolar orientations. Also, we show how it is possible to add costs to our flow model in order to compute in polynomial time an upward orientation with the minimum number of sources and sinks (Section 3.2).

## 3.1 A Flow Model Characterizing Upward Embeddings

The following theorem characterizes the class of labelings that can be determined by any upward embedding of an embedded planar graph. It is important to observe that the characterization of such a class of labelings does not depend either on the choice of a splitting of the adjacency lists of the graph, in contrast to the result given in Lemma 1, or on the choice of an orientation of the graph.

**Theorem 1** *Let $\mathcal{L}$ be any labeling of the angles of an embedded planar graph $G_\phi$ with labels L, S, and F. $\mathcal{L}$ is the labeling determined by an upward embedding of $G_\phi$ if and only if the following properties hold:*

(FIN') $deg(f) - 2 = 2L(f) + F(f)$, *for each internal face $f$ of $G_\phi$.*

(FEX') $deg(f) + 2 = 2L(f) + F(f)$, *for the external face $f$ of $G_\phi$.*

(VL) *For each vertex $v$ either $F(v) = 2$ and $L(v) = 0$ or $F(v) = 0$ and $L(v) = 1$.*

**Proof:**
The necessary condition is an immediate consequence of Lemma 1 and Lemma 2. In fact, if $\mathcal{L}$ is determined by an upward embedding, then properties (FIN), (FEX), (VL0), and (VL1) of Lemma 1 hold. From Lemma 2 properties (FIN) and (FEX) are equivalent to properties (FIN') and (FEX'); further, properties (VL0) and (VL1) imply that one of the two cases of property (VL) holds, for each vertex of $G_\phi$.

To prove the sufficiency of the condition we consider a labeling $\mathcal{L}$ that verifies properties (FIN'), (FEX'), and (VL), and construct an upward embedding of $G_\phi$ that determines $\mathcal{L}$. From $\mathcal{L}$, we construct a splitting $\mathcal{E}_\phi$ of the adjacency lists of $G_\phi$ as follows:

- We observe that there exists at least two distinct vertices $s$ and $t$ on the external face $f$ having an angle labeled with L. In fact, from property (FEX') (that is equivalent to property (FEX) of Lemma 1) we must have that $L(f) = S(f) + 2$. We assign all the edges incident on $s$ to the list $E_{above}(s)$ (we set $E_{below}(s)$ empty). Namely, if $(e_1, e_2)$ is the angle with label L at vertex $s$, $e_2$ and $e_1$ will be the first edge and the last edge of $E_{above}(s)$, respectively.

W. Didimo and M. Pizzonia, *Upward Embeddings*, JGAA, 7(2) 221–241 (2003)231

- We execute a breadth first search starting from $s$. At each step we visit a different vertex $v$ and split the list of the edges that are incident on $v$. In a breadth first search all the edges (and hence all the angles) incident on a vertex $v$ are explored when $v$ is visited. We chose to scan these edges in clockwise order. Namely, suppose that $v$ is visited by moving from vertex $u$ through edge $e_0 = (u, v)$ ($e_0$ is the parent edge of $v$ in the breadth first search). If $e_0$ is in $E_{above}(u)$ we put $e_0$ in $E_{below}(v)$, while if $e_0$ is in $E_{below}(u)$ we put $e_0$ in $E_{above}(v)$. Suppose that $e_0, e_1, \ldots, e_k$ are the edges incident on $v$ in this clockwise ordering. For each $e_i$ ($i = 0, \ldots, k - 1$) we consider the label $l$ of angle $(e_i, e_{i+1})$, and we decide if $e_{i+1}$ has to be assigned to $E_{above}(v)$ or to $E_{below}(v)$. Note that, at this point, $e_i$ has been already assigned to one of the two lists. The following cases are possible: (1) If $l = $ L and $e_i \in E_{below}(v)$ then $e_{i+1}$ is put at the end of $E_{below}(v)$. (2) If $l = $ L and $e_i \in E_{above}(v)$ then $e_{i+1}$ is put at the start of $E_{above}(v)$. (3) If $l = $ S and $e_i \in E_{below}(v)$ then $e_{i+1}$ is put immediately before $e_i$ in $E_{below}(v)$. (4) If $l = $ S and $e_i \in E_{above}(v)$ then $e_{i+1}$ is put immediately after $e_i$ in $E_{above}(v)$. (5) If $l = $ F and $e_i \in E_{below}(v)$ then $e_{i+1}$ is put at the start of $E_{above}(v)$. (6) If $l = $ F and $e_i \in E_{above}(v)$ then $e_{i+1}$ is put at the end of $E_{below}(v)$.

It is easy to see that $\mathcal{E}_\phi$ verifies (E1). To prove that $\mathcal{E}_\phi$ is an upward embedding of $G_\phi$ we need only to prove that properties (VL0) and (VL1) of Lemma 1 are verified (since properties (FIN) and (FEX) are equivalent to properties (FIN') and (FEX')). From property (VL) we only have two possible cases for the labels of the angles at each vertex $v$ of $G_\phi$.

- $F(v) = 2$ and $L(v) = 0$. This implies that, for splitting the edges incident on $v$ cases (1) and (2) are never applied, cases (5) and (6) are applied twice in the total, and cases (3) and (4) are applied $deg(v) - 2$ times in the total. Also, cases (5) and (6) imply that neither $E_{above}(v)$ nor $E_{below}(v)$ will be empty. This matches property (VL0).

- $F(v) = 0$ and $L(v) = 1$. This implies that, for splitting the edges incident on $v$, either case (1) or case (2) is applied once, cases (5) and (6) are never applied, and either case (3) or case (4) is applied $deg(v) - 1$ times. Also, observe that each of the cases (1), (2), (3), and (4) always puts $e_{i+1}$ in the same list as $e_i$, and that either (1) and (3) or (2) and (4) are applied. This guarantees that exactly one of the two lists $E_{above}(v)$ and $E_{below}(v)$ will be empty. This matches property (VL1).

Finally, since no other cases are possible, properties (VL0) and (VL1) of Lemma 2 hold.

*q.e.d.*

We call *upward labeling* of $G_\phi$ a labeling of the angles of $G_\phi$ that verifies properties (FIN'), (FEX'), and (VL) of Theorem 1. The result of Theorem 1 allows the description of all upward embeddings of $G_\phi$ in terms of upward labelings

W. Didimo and M. Pizzonia, *Upward Embeddings*, JGAA, 7(2) 221–241 (2003)232

of $G_\phi$. Note that, the proof of the theorem provides a procedure to construct the upward embedding associated with an upward labeling. Actually, for each upward labeling, there are exactly two "symmetric" upward embeddings that determine it; they are obtained one from the other by simply exchanging list $E_{above}(v)$ with list $E_{below}(v)$ for each vertex $v$ and then reversing such lists (see Figure 7 (b) for an example).

We now provide a network flow model that characterizes all the upward labelings of $G_\phi$. Because of the above considerations, this flow model provides a characterization of all upward embeddings of $G_\phi$. We associate with $G_\phi$ a flow network $\mathcal{N}_\phi$, such that the integer feasible flows on $\mathcal{N}_\phi$ are in one-to-one correspondence with the upward labelings of $G_\phi$. Flow network $\mathcal{N}_\phi$ is a directed graph defined as follows (see Figure 6):

- The nodes of $\mathcal{N}_\phi$ are the vertices (*vertex-nodes*) and the faces (*face-nodes*) of $G_\phi$. Each vertex-node supplies flow 2 and each face-node associated with face $f$ of $G_\phi$ demands a flow equal to $deg(f) - 2$ if $f$ is internal and $deg(f) + 2$ if $f$ is external.

- With each angle of $G_\phi$ at vertex $v$ in face $f$ there is an associated arc $(v, f)$ of $\mathcal{N}_\phi$ with lower capacity 0 and upper capacity 2.

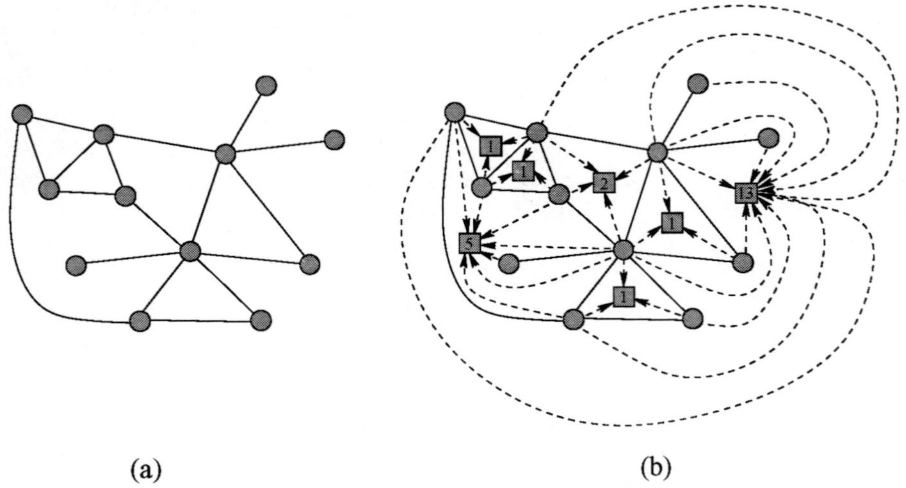

(a)                                      (b)

Figure 6: (a) An embedded planar graph $G_\phi$. (b) Flow network $\mathcal{N}_\phi$ associated with $G_\phi$. The vertex-nodes are circles and the face-nodes are squares. Each face-node is labeled with its demand. The arcs of the networks are dashed.

Observe that in $\mathcal{N}_\phi$ the total demand is equal to the total supply. In fact:

$$\sum_{f \in F} (deg(f) - 2) + 4 = \sum_{f \in F} deg(f) - 2|F| + 4 = 2|E| - 2|F| + 4 = 2|V|.$$

W. Didimo and M. Pizzonia, *Upward Embeddings*, JGAA, 7(2) 221–241 (2003)233

The intuitive interpretation of the flow model in terms of upward embedding is as follows: (i) Each unit of flow represents a flat angle, with the convention that a large angle counts as two flat angles; an arc $a$ of $\mathcal{N}_\phi$ has flow $0, 1$, or $2$, depending on the fact that its associated angle is small, flat, or large, respectively. (ii) The demand of each face-node and the supply of each vertex-node reflect the balancing properties (FIN'), (FEX') and (VL). Figure 7 shows a feasible flow on the network associated with an embedded planar graph, the corresponding upward labeling, and the two "symmetric" upward embeddings associated with the labeling. Theorem 2 formally proves the correctness of the intuitive interpretation described above .

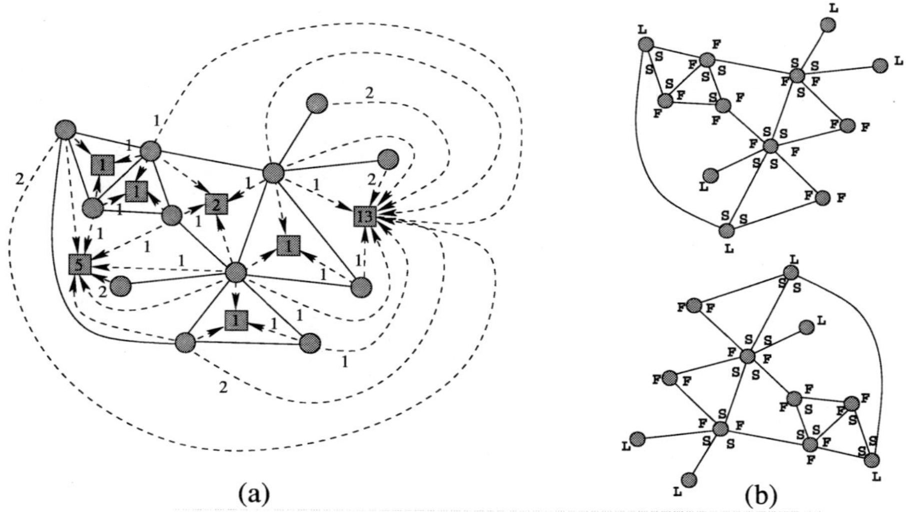

Figure 7: (a) A feasible flow on the network associated with an embedded planar graph. Only the flow values different from zero are shown. (b) The upward labeling $\mathcal{L}$ corresponding to the flow and the two "symmetric" upward embeddings associated with $\mathcal{L}$.

We remark that network $\mathcal{N}_\phi$ is related to the flow model used by Bousset for describing bipolar orientations of biconnected embedded planar graphs. The flow values in such a model do not allow to represent large angles (the allowed flow values are only $0$ or $1$), and the source and the sink of the final orientation are prescribed. Our flow model extends and generalizes the model of Bousset to 1-connected planar graphs, by allowing the representation of any kind of upward orientations and embeddings, including the bipolar orientations for biconnected graphs.

**Theorem 2** *Let $G_\phi$ be an embedded planar graph and let $\mathcal{N}_\phi$ be the flow network associated with $G_\phi$. There is a one-to-one correspondence between the set of the upward labelings of $G_\phi$ and the set of the integer feasible flows on $\mathcal{N}_\phi$.*

W. Didimo and M. Pizzonia, *Upward Embeddings*, JGAA, 7(2) 221–241 (2003)234

**Proof:**
Consider an upward labeling $\mathcal{L}$ of $G_\phi$. From it we construct an integer feasible flow $x$ of $\mathcal{N}_\phi$ as follows. For each angle $\alpha$ of $G_\phi$ let $a$ be the arc of $\mathcal{N}_\phi$ associated with $\alpha$. We set $x(a) = 2$ if $\alpha$ is labeled L, $x(a) = 1$ if $\alpha$ is labeled F, and $x(a) = 0$ if $\alpha$ is labeled S. The above construction is clearly an injective transformation. In fact, there is a one-to-one correspondence between angles of $G_\phi$ and arcs of $\mathcal{N}_\phi$ and hence, different labelings of the same angle of $G_\phi$ produces different values of flow on the corresponding arc of $\mathcal{N}_\phi$. We now prove that flow $x$ is feasible. From the construction of $x$ and from property (VL) of $\mathcal{L}$, it follows that every vertex-node of $\mathcal{N}_\phi$ supplies flow 2 (and demands flow 0). Hence, the balance property of $x$ on every vertex-node of $\mathcal{N}_\phi$ is verified. Let $f$ be an internal face of $G_\phi$, and consider the face-node of $\mathcal{N}_\phi$ associated with $f$. From the construction of $x$, such a face-node receives a flow equal to $2L(f) + F(f)$ and supplies flow 0 ; hence, from property (FIN') of $\mathcal{L}$, it demands a flow equal to $deg(f) - 2$. The same reasoning applies for the external face, using property (FEX'). Hence, also the balance property of $x$ on every face-node is verified. Finally, since on each arc of $\mathcal{N}_\phi$ we assign an integer amount of flow in the range $[0, 2]$, the lower and upper capacities on the arcs of $\mathcal{N}_\phi$ are respected by $x$.

Conversely, consider an integer feasible flow $x$ of $\mathcal{N}_\phi$, and construct from $x$ a labeling $\mathcal{L}$ of $G_\phi$, by applying a transformation that is the reverse of that described above . Namely, for each arc $a$ of $\mathcal{N}_\phi$ denote by $\alpha$ the corresponding angle of $G_\phi$. Labeling $\mathcal{L}$ is constructed by assigning label L, F, and S to $\alpha$, depending on the case that $x(a) = 2$, $x(a) = 1$, and $x(a) = 0$, respectively. By using the properties of $x$ and the same reasoning applied above, it is easy to prove that $\mathcal{L}$ is an upward labeling of $G_\phi$. $\hfill q.e.d.$

Theorem 1 and Theorem 2 allow us to compute an upward embedding of an embedded planar graph $G_\phi$ by computing an integer feasible flow on network $\mathcal{N}_\phi$. We now analyze the running time complexity of computing an upward embedding by means of a flow technique.

Network $\mathcal{N}_\phi$ has $O(n)$ vertices and edges, where $n$ denotes the number of vertices of $G_\phi$. Both $\mathcal{N}_\phi$ and an upward embedding associated with a feasible flow on $\mathcal{N}_\phi$ can be constructed in linear time. We now observe that $\mathcal{N}_\phi$ can be easily reduced to an equivalent unit capacity network $\mathcal{N}_\phi^*$ with a single source $s$ and a single sink $t$ and with $O(n)$ nodes and arcs. On $\mathcal{N}_\phi^*$ we can apply Dinic's algorithm to compute in $O(n^{3/2})$ time a feasible (maximum) flow [7]. Namely, $\mathcal{N}_\phi^*$ is obtained from $\mathcal{N}_\phi$ by replacing each arc $a$ with two unit capacity arcs having the same direction as $a$, by connecting $s$ to each vertex-node with two unit capacity arcs, by connecting each internal face-node $f$ to $t$ with $deg(f) - 2$ unit capacity arcs, and by connecting the external face-node $h$ to $t$ with $deg(h) + 2$ unit capacity arcs. Finally, node $s$ supplies flow $2|V|$ and node $t$ demands flow $2|V|$, while all the other nodes demand and supply flow 0. The following theorem summarizes the complexity analysis.

**Theorem 3** *There exists a flow technique for computing an upward embedding*

W. Didimo and M. Pizzonia, *Upward Embeddings*, JGAA, 7(2) 221–241 (2003)235

*of an undirected embedded planar graph in $O(n^{3/2})$ time and $O(n)$ space, where n denotes the number of vertices of the graph.*

There are two main advantages of computing upward embeddings of a general planar graph $G_\phi$ by using the flow model described so far. First, no augmentation algorithm has to be used to make the input graph biconnected (we just apply a standard flow algorithm). Second, it is possible to deal with partially specified embeddings. In particular it is possible to constrain an angle to be large by fixing flow 2 on the corresponding arc of the network and to constrain a vertex to be neither a source nor a sink by reducing to 1 the upper capacity of its leaving arcs in the network. Also observe that in the presence of constraints a feasible solution might not exist, and in this case a feasible flow is not found.

In the next section we describe how to compute upward embeddings with the minimum number of sources and sinks, by adding costs to our network.

## 3.2   Minimizing Sources and Sinks

Computing an upward embedding of $G_\phi$ with the minimum number of sources and sinks (which we call *optimal upward embedding* for simplicity) is equivalent to computing an upward embedding with the minimum number of large angles. Clearly, if the graph is biconnected, the problem is reduced to the computation of a bipolar orientation. For this reason, we regard the concept of optimal upward embedding as the natural extension of the definition of bipolar orientation to the case of general connected graphs.

The flow model we use to compute an optimal upward orientation of $G_\phi$ is a simple variation of the one described for characterizing upward embeddings (see Section 3.1). We add a linear number of arcs to network $\mathcal{N}_\phi$ and we equip the arcs of the new network with costs. Each unit of cost represents a large angle. We also reduce the upper capacity of all the arcs of the network. More in detail, the new network $\widetilde{\mathcal{N}_\phi}$ is derived from $\mathcal{N}_\phi$ as follows: for each angle of $G_\phi$ at vertex $v$ in face $f$ we substitute its associated arc in $\mathcal{N}_\phi$ with a pair of directed arcs $a_v = (v, f), a'_v = (v, f)$. Both the new arcs have lower capacity 0 and upper capacity 1. Also, arc $a_v$ has cost 0 while arc $a'_v$ has cost 1.

Let $x$ be a minimum cost flow on $\widetilde{\mathcal{N}_\phi}$. The interpretation of the flow in terms of upward labeling is similar to the one given for $\mathcal{N}_\phi$, with a slight variation due to the additional arcs and costs. We first observe that for each pair of arcs $a_v$, $a'_v$ it never happens $x(a_v) = 0$ and $x(a'_v) = 1$, due to the fact that the cost of $a_v$ is 0 and that the cost of $a'_v$ is 1. In fact, if $x(a_v) = 0$ and $x(a'_v) = 1$, then there would exist a negative cost cycle represented by the two arcs $a'_v, a_v$, and it would be possible to derive a new flow $x'$ from $x$ by simply exchanging one unit of flow between $a'_v$ and $a_v$ (i.e., $x'(a_v) = 1$ and $x'(a'_v) = 0$). This would imply that $x'$ has a cost smaller than the cost of $x$, in contrast to the assumption that $x$ has the minimum cost. Hence, the only possibilities for the flow on arcs $a_v, a'_v$ are: (i) $x(a_v) = x(a'_v) = 0$, the angle associated with arcs $a_v, a'_v$ is small.

W. Didimo and M. Pizzonia, *Upward Embeddings*, JGAA, 7(2) 221–241 (2003)236

(ii) $x(a_v) = 1$ and $x(a_v') = 0$, the angle associated with arcs $a_v, a_v'$ is flat. (iii) $x(a_v) = x(a_v') = 1$, the angle associated with arcs $a_v, a_v'$ is large.

Note that, only in the third case we have cost 1 on arcs $a_v, a_v'$, while in the other two cases we have cost 0. This implies that the total cost of flow $x$ on $\widetilde{\mathcal{N}_\phi}$ represents the total number of large angles of the corresponding upward embedding of $G_\phi$. Hence, since $x$ has the minimum cost, the corresponding upward embedding has the minimum number of large angles.

Let $n$ be the number of vertices of $G_\phi$. Since network $\widetilde{\mathcal{N}_\phi}$ is planar and has $O(n)$ vertices, and since its total demand (supply) is $O(n)$, a minimum cost flow on $\widetilde{\mathcal{N}_\phi}$ can be computed in $O(n^{\frac{7}{4}} \log n)$ time by the algorithm described in [10]. The following theorem summarizes the main contribution of this section.

**Theorem 4** *There exists an $O(n^{\frac{7}{4}} \log n)$ time algorithm that computes an upward embedding of an embedded 1-connected planar graph with the minimum number of sources and sinks.*

We conclude this section by giving an upper bound on the number of sources and sinks of an optimal upward embedding.

**Lemma 3** *An optimal upward embedding of an embedded planar graph $G_\phi$ has at most $B + 1$ sources and sinks, where $B$ is the number of blocks of $G_\phi$.*

**Proof:** We prove the lemma by induction on $B$. If $B = 1$, the graph is biconnected and an optimal upward embedding of it has exactly one source and one sink. Suppose that the lemma is true for each graph with $B \geq 1$ blocks, and consider a graph $G_\phi$ with $B + 1$ blocks. We select any block $C$ of $G_\phi$ such that $C$ contains exactly one cutvertex of $G_\phi$ and there is no block nested into $C$. Note that such a block always exists. Let $G'_{\phi'}$ be the graph obtained from $G_\phi$ by removing $C$ and let $\mathcal{E}'_{\phi'}$ be an optimal upward embedding of $G'_{\phi'}$. From the inductive hypothesis, $\mathcal{E}'_{\phi'}$ has at most $B + 1$ sources and sinks. From $\mathcal{E}'_{\phi'}$ we construct an upward embedding of $G_\phi$. Such an upward embedding coincides with $\mathcal{E}'_{\phi'}$ for the subgraph $G'_{\phi'}$ and it is determined on $C$ as follows. We always embed $C$ above or below its cutvertex $v$, according to $\mathcal{E}'_{\phi'}$ and according to the planar embedding of $G_\phi$. Namely, let $e_1$ and $e_2$ be the two edges (not necessarily distinct) of $G_\phi$ encountered immediately before and after $C$ in the clockwise ordering around $v$. Three distinct cases are possible for $\mathcal{E}'_{\phi'}$:

- If both $e_1$ and $e_2$ belong to $E_{above}(v)$, we compute an upward embedding of $C$ with exactly one source and one sink, where the source is $v$, and we embed it above $v$ in $\mathcal{E}'_{\phi'}$ (see Figure 8 (a)).

- If both $e_1$ and $e_2$ belong to $E_{below}(v)$, we compute an upward embedding of $C$ with exactly one source and one sink, where the sink is $v$, and we embed it below $v$ in $\mathcal{E}'_{\phi'}$ (see Figure 8 (b)).

- If one between $e_1$ and $e_2$ belongs to $E_{above}(v)$ while the other edge belongs to $E_{below}(v)$, we arbitrarily choose to compute an upward embedding of $C$ with exactly one source and one sink, where the source is $v$, and we embed it above $v$ in $\mathcal{E}'_{\phi'}$ (see Figure 8 (c)).

W. Didimo and M. Pizzonia, *Upward Embeddings*, JGAA, 7(2) 221–241 (2003)237

The obtained upward embedding has at most one source or one sink more than $\mathcal{E}'_{\phi'}$, since vertex $v$ is in common between $C$ and $G'_{\phi'}$. Therefore, an optimal upward embedding of $G_\phi$ has at most $B + 2$ sources and sinks.

*q.e.d.*

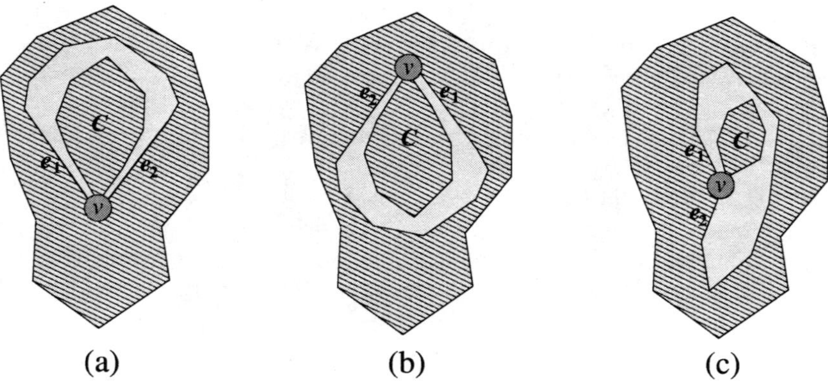

(a)        (b)        (c)

Figure 8: Illustration of the proof of Lemma 3.

The bound of Lemma 3 is strict and a class of plane graphs whose upward embeddings have $B + 1$ sources and sinks can be obtained by nesting each block into another, as shown by the example of Figure 9.

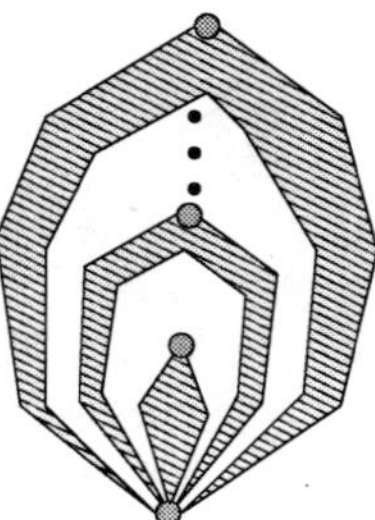

Figure 9: A class of embedded planar graphs whose optimal upward embeddings have $B + 1$ sources and sinks (circles).

# 4   Algorithms for Visibility Representations

We use the above results on upward embeddings to compute drawings of general connected planar graphs. Namely, we focus on graph drawing algorithms which require the computation of a (*weak-*)*visibility representation* of the input graph

W. Didimo and M. Pizzonia, *Upward Embeddings*, JGAA, 7(2) 221–241 (2003)238

as a preliminary step [6]. In a visibility representation (see Figure 10), each vertex is mapped to a horizontal segment and each edge $(u, v)$ is mapped to a vertical segment between the segments associated with $u$ and $v$; horizontal segments do not overlap, and each vertical segment only intersects its extreme horizontal segments.

A standard technique [6] for constructing a visibility representation of a planar graph $G$ first computes a bipolar orientation of $G$ and then computes the coordinates of the drawing from this orientation. If $G$ is not biconnected the technique needs to augment the graph to a biconnected planar one, in order to compute a bipolar orientation of it. The augmentation algorithm adds to $G$ a suitable number of dummy edges, which will be removed in the final drawing. However, this technique has several drawbacks: (i) Adding too many dummy edges may lead to a final drawing with area much bigger than necessary. On the other side, the problem of adding the minimum number of edges to make a planar graph biconnected and still planar is NP-hard [12]. (ii) Although a good approximation algorithm for the above augmentation problem exists [8] (which reaches the optimal solution in many cases), implementing it efficiently is quite difficult, because it requires us to deal with the *block cutvertex tree* [11] of the graph and with an efficient incremental planarity testing algorithm. In fact, such an approximation algorithm has $O(n^2 T)$ running time, where $T$ is the amortized time bound per query or insertion operation of the incremental planarity testing algorithm. (iii) The presence of dummy edges in the graph makes difficult to handle with partial assignments of the upward embedding.

Tamassia and Tollis [17] provide a different linear time algorithm for computing visibility representations of general connected graphs. At each step of the algorithm a visibility representation of a new distinct block of the graph is computed and suitably merged to the current drawing. However, merging operations require the execution of scaling down geometric operations, which may lead to a final drawing with a big area on an integer grid. Also, the algorithm has many degrees of freedom about how to perform some topological operations and about the choice of the ordering in which the blocks are considered; different decisions may lead to very different results.

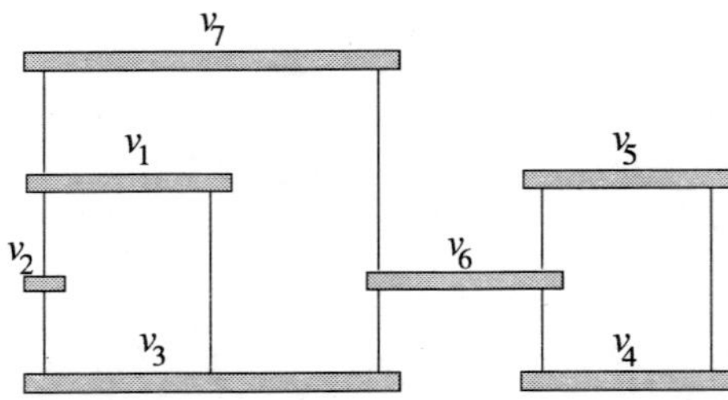

Figure 10: A visibility representation of the upward embedded graph shown in Figure 1(b).

W. Didimo and M. Pizzonia, *Upward Embeddings*, JGAA, 7(2) 221–241 (2003)239

We propose the following algorithm for computing a visibility representation of a 1-connected embedded planar graph $G_\phi$.

**Algorithm** *Visibility-Upward-Embedding*

1. Compute an upward embedding $\mathcal{E}_\phi$ of $G_\phi$ by calculating a feasible flow on network $\mathcal{N}_\phi$.

2. Compute an upward embedded $st$-digraph $S_\phi$ including $G_\phi$ and preserving $\mathcal{E}_\phi$ on $G_\phi$, by using the linear time saturation procedure described at the end of Section 2.

3. Compute a visibility representation of $S_\phi$ (within its upward embedding) by using any known linear time algorithm [6], and then remove the edges introduced by the saturation procedure.

Algorithm *Visibility-Upward-Embedding* has $O(n^{3/2})$ running time, because its time complexity is dominated by the cost of computing a feasible flow on $\mathcal{N}_\phi$. We experimentally observed that the area of the visibility representations produced by this algorithm can be dramatically improved by computing upward embeddings with the minimum number of sources and sinks. To do that we just apply a min-cost-flow algorithm in Step 1. Clearly, in this case, the running time of the whole algorithm grows to $O(n^{\frac{7}{4}} \log n)$.

We have also slightly refined Algorithm *Visibility Upward Embedding* aiming to get a certain control over the width and the height of visibility representations of 1-connected planar graphs. After we have computed an upward embedding with the minimum number of switches we rearrange the blocks around the cutvertices in the upward embedding. Namely, if $v$ is a cutvertex we place all the blocks of $v$ either above or below. This often leads to a reduction of the height and to an increase in the width. Such a rearrangement is performed in linear time by exploiting the flow network associated with the embedded planar graph. We experimented such an approach on a randomly generated test suite of 1820 graphs whose number $n$ of vertices ranges from 10 to 100 (20 instances for each value of $n$). A detailed description of the procedure used to generate the graphs can be found in [15]. We averaged the width and the height on all the graphs having the same number of vertices. Charts in Figure 11 graphically show the results of the experimentation for the maximum number of cutvertices $k(k = 0 \ldots 8)$ whose blocks have been rearranged.

Also, Figure 12 compares the area of the drawings computed with this strategy, where $k$ is chosen equal to the total number of cutvertices of the graph, against the area of the drawings computed with a standard technique which uses the approximation algorithm in [8] to initially make the graph biconnected. In the two strategies we use the same algorithm for constructing the visibility representation from the $st$-digraph. Experimentally, for the considered test suite, the running time of the two algorithms is comparable (less than one second for the largest graphs).

W. Didimo and M. Pizzonia, *Upward Embeddings*, JGAA, 7(2) 221–241 (2003)240

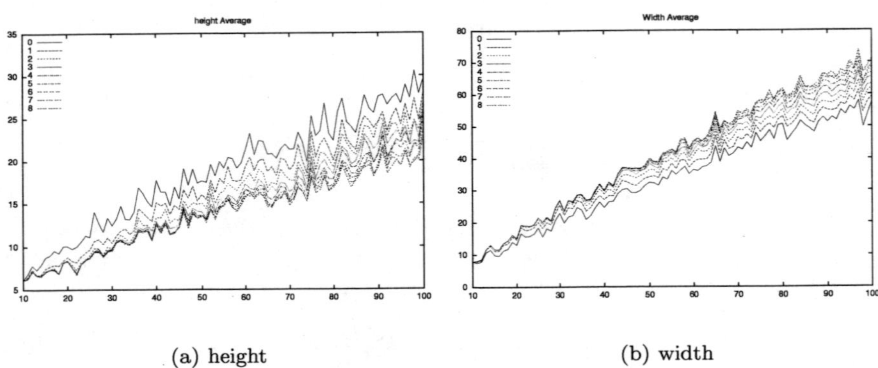

(a) height            (b) width

Figure 11: The charts show how rearranging the blocks around cutvertices affects the width and the height of the visibility representation.

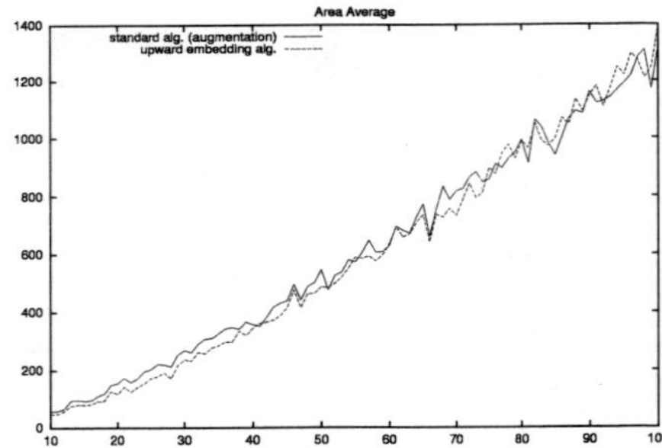

Figure 12: Area of the drawings computed with our strategy against the area of the drawings computed with a standard technique based on a sophisticated augmentation algorithm (average values). The $x$-axis represents the number of vertices.

# 5    Open Problems

There are several open problems that we plan to study in the near future. For example, we are interested in an algorithm for counting and enumerating all upward embeddings of an embedded planar graph without repetitions. Also, is it possible to pass from an upward embedding to any other in linear time? Is there a linear time algorithm to compute optimal upward embeddings of embedded planar graphs? What about non-embedded planar graphs? Finally, from an applications point of view we believe that the techniques shown in this paper may be successfully refined to compute drawings that approximate a given width/height ratio.

W. Didimo and M. Pizzonia, *Upward Embeddings*, JGAA, 7(2) 221–241 (2003)241

# References


[1] P. Bertolazzi, G. Di Battista, G. Liotta, and C. Mannino. Upward drawings of triconnected digraphs. *Algorithmica*, 6(12):476–497, 1994.

[2] P. Bertolazzi, G. Di Battista, C. Mannino, and R. Tamassia. Optimal upward planarity testing of single-source digraphs. *SIAM J. Comput.*, 27(1):132–169, 1998.

[3] M. Bousset. A flow model of low complexity for twisting a layout. In *Workshop on Graph Drawing (GD'93)*, pages 43–44, 1993.

[4] J. Czyzowicz, A. Pelc, and I. Rival. Drawing orders with few slopes. Technical Report TR-87-12, Department of Computer Science, University of Ottawa, 1987.

[5] H. de Fraysseix, P. O. de Mendez, and P. Rosenstiehl. Bipolar orientations revisited. *Discrete Appl. Math.*, 56:157–179, 1995.

[6] G. Di Battista, P. Eades, R. Tamassia, and I. G. Tollis. *Graph Drawing*. Prentice Hall, Upper Saddle River, NJ, 1999.

[7] S. Even and R. E. Tarjan. Network flow and testing graph connectivity. *SIAM J. Comput.*, 4:507–518, 1975.

[8] S. Fialko and P. Mutzel. A new approximation algorithm for the planar augmentation problem. In *Symposium on Discrete Algorithms (SODA'98)*, pages 260–269, 1998.

[9] A. Garg and R. Tamassia. On the computational complexity of upward and rectilinear planarity testing. In R. Tamassia and I. G. Tollis, editors, *Graph Drawing (Proc. GD '94)*, volume 894 of *Lecture Notes Comput. Sci.*, pages 286–297. Springer-Verlag, 1995.

[10] A. Garg and R. Tamassia. A new minimum cost flow algorithm with applications to graph drawing. In S. C. North, editor, *Graph Drawing (Proc. GD '96)*, volume 1190 of *Lecture Notes Comput. Sci.*, pages 201–216. Springer-Verlag, 1997.

[11] F. Harary. *Graph Theory*. Addison-Wesley, Reading, MA, 1972.

[12] G. Kant and H. L. Bodlaender. Planar graph augmentation problems. In *Proc. 2nd Workshop Algorithms Data Struct.*, volume 519 of *Lecture Notes Comput. Sci.*, pages 286–298. Springer-Verlag, 1991.

[13] D. Kelly. Fundamentals of planar ordered sets. *Discrete Math.*, 63:197–216, 1987.

[14] D. Kelly and I. Rival. Planar lattices. *Canad. J. Math.*, 27(3):636–665, 1975.

[15] M. Pizzonia. Engineering of graph drawing algorithms for applications. *PhD thesis*, 2001. Dipartimento di Informatica e Sistemistica, Università "La Sapienza" di Roma.

[16] I. Rival. Reading, drawing, and order. In I. G. Rosenberg and G. Sabidussi, editors, *Algebras and Orders*, pages 359–404. Kluwer Academic Publishers, 1993.

[17] R. Tamassia and I. G. Tollis. A unified approach to visibility representations of planar graphs. *Discrete Comput. Geom.*, 1(4):321–341, 1986.

[18] R. E. Tarjan. Depth-first search and linear graph algorithms. *SIAM J. Comput.*, 1(2):146–160, 1972.

Volume 7:3 (2003)

## Journal of Graph Algorithms and Applications

http://jgaa.info/

vol. 7, no. 3, pp. 245–251 (2003)

# Crossing Numbers and Cutwidths

*Hristo N. Djidjev*

Department of Computer Science
Warwick University
hristo@dcs.warwick.ac.uk

*Imrich Vrťo*

Department of Informatics
Institute of Mathematics, Slovak Academy of Science
http://www.ifi.savba.sk/~imrich
vrto@savba.sk

### Abstract

The crossing number of a graph $G = (V, E)$, denoted by $\mathrm{cr}(G)$, is the smallest number of edge crossings in any drawing of $G$ in the plane. Wee assume that the drawing is good, i.e., incident edges do not cross, two edges cross at most once and at most two edges cross in a point of the plane. Leighton [13] proved that for any $n$-vertex graph $G$ of bounded degree, its crossing number satisfies $\mathrm{cr}(G) + n = \Omega(\mathrm{bw}^2(G))$, where $\mathrm{bw}(G)$ is the bisection width of $G$. The lower bound method was extended for graphs of arbitrary vertex degrees to $\mathrm{cr}(G) + \frac{1}{16} \sum_{v \in G} d_v^2 = \Omega(\mathrm{bw}^2(G))$ in [16, 20], where $d_v$ is the degree of any vertex $v$. We improve this bound by showing that the bisection width can be replaced by a larger parameter - the cutwidth of the graph. Our result also yields an upper bound for the path-width of $G$ in terms of its crossing number.

Communicated by Giuseppe Liotta: submitted September 2002; revised June 2003.

Supported by the VEGA grant No. 02/3164/23 and the DFG grant No. Hr14/5-1.

Djidjev and Vrťo, *Crossings and Cutwidths*, JGAA, 7(3) 245–251 (2003)    246

# 1   Introduction

The crossing number of a graph $G = (V, E)$, denoted by $\mathrm{cr}(G)$, is the smallest number of edge crossings in any drawing of $G$ in the plane. It represents a fundamental measure of non-planarity of graphs but is attractive from practical point of view too. It is known that the aesthetics and readability of graph-like structures (information diagrams, class hierarchies, flowcharts...) heavily depends on the number of crossings [4, 17], when the structures are visualized on a 2-dimensional medium. Another natural appearance of the problem is in the design of printed circuit boards and VLSI circuits [13]. The area of a VLSI circuit is strongly related to the crossing number of the underlying graph. The problem is NP-hard [6] and the best theoretical exact and approximation algorithms are in [5, 9]. A survey on heuristics is in [3]. Concerning crossing numbers of standard graphs, there are only a few infinite classes of graphs for which exact or tight bounds are known [12]. The main problem is the lack of efficient lower bound methods for estimating the crossing numbers of explicitly given graphs. The survey on known methods is in [19]. One of the powerful methods is based on the bisection width concept. The bisection width of a graph $G$ is the minimum number of edges whose removal divides $G$ into two parts having at most $2|V|/3$ vertices each. Leighton [13] proved that in any $n$-vertex graph $G$ of bounded degree, the crossing number satisfies $\mathrm{cr}(G) + n = \Omega(\mathrm{bw}^2(G))$. The lower bound was extended to

$$\mathrm{cr}(G) + \frac{1}{16} \sum_{v \in V} d_v^2 \geq \frac{1}{40} \mathrm{bw}^2(G)$$

in [16, 20], where $d_v$ is the degree of any vertex $v$. We improve this bound by showing that the bisection width can be replaced by a larger parameter - the cutwidth of the graph, denoted by $\mathrm{cw}(G)$ and defined as follows. Let $G = (V, E)$ be a graph. Let $\phi : V \to \{1, 2, 3, ..., |V|\}$ be an injection. Then

$$\mathrm{cw}(G) = \min_{\phi} \max_{i} |\{uv \in E : \phi(u) \leq i < \phi(v)\}|.$$

Note that the cutwidth is a standard graph invariant appearing e.g. in the linear VLSI layouts [21], and is related to such a classical topic like the discrete isoperimetric problem [1]. We prove

$$\mathrm{cr}(G) + \frac{1}{16} \sum_{v \in G} d_v^2 \geq \frac{1}{1176} \mathrm{cw}^2(G). \tag{1}$$

Ignoring the constant factors, the improvement is evident as $\mathrm{cw}(G) \geq \mathrm{bw}(G)$ and there are connected graphs with $\mathrm{bw}(G) = 1$ but with arbitrarily large cutwidth. For example, let $G$ be a graph obtained by joining two $K_{\frac{n}{2}}$'s by an edge. Then clearly $\mathrm{cw}(G) = \Omega(n^2)$. If $\mathrm{cw}(G) \approx \mathrm{bw}(G)$ then the bisection lower bound is better up to a constant factor because of the small constant in our estimation. An improvement on it remains an open problem. Anyway, the

Djidjev and Vrťo, *Crossings and Cutwidths*, JGAA, 7(3) 245–251 (2003)    247

aim of this note is to show that from the asymptotic point of view, the graph invariant that essentially influences the crossing number is not the bisection width but the cutwidth. The new crossing number lower bound is tight up to a constant factor for a large class of graphs. Following Pach and Tóth [15], for almost all $n$-vertex and $m$-edge graphs $G$, $\mathrm{bw}(G) \geq m/10$, where $m \geq 10n$. As $\mathrm{cw}(G) \geq \mathrm{bw}(G)$, the lower bound (1) implies $\mathrm{cr}(G) = \Omega(\mathrm{cw}^2(G)) = \Omega(m^2)$. On the other hand, trivially $\mathrm{cr}(G) = O(m^2) = O(\mathrm{cw}^2(G))$.

Moreover, the additive term $\sum_{v \in G} d_v^2$ cannot be removed, since the crossing number of any planar graph is 0 and there exists a planar graph $P$ (e.g. the star) such that $\mathrm{cw}^2(P) = \Omega(\sum_{v \in P} d_v^2)$.

As a byproduct we obtain the following contribution to topological graph theory. The path-decomposition of a graph $G$ is a sequence $D = X_1, X_2, ..., X_r$ of vertex subsets of $G$, such that every edge of $G$ has both ends in some set $X_i$ and if a vertex of $G$ occurs in some sets $X_i$ and $X_j$ with $i < j$, then the same vertex occurs in all sets $X_k$ with $i < k < j$. The width of $D$ is the maximum number of vertices in any $X_i$ minus 1. The path-width of $G$, $\mathrm{pw}(G)$, is the minimum width over all path-decompositions of $G$.

A graph $G = (V, E)$ is $k$-crossing critical if $\mathrm{cr}(G) = k$ and $\mathrm{cr}(G - e) < \mathrm{cr}(G)$, for all edges $e \in E$. Hliněný [10] proved that $\mathrm{pw}(G) \leq 2^{f(k)}$, where $f(k) = O(k^3 \log k)$. This answers an open question of Geelen et al. [8] whether crossing critical graphs with bounded crossing numbers have bounded path-widths.

Our result implies another relation between pathwidths and crossing numbers. If $\mathrm{cr}(G) = k$, then $\mathrm{pw}(G) \leq 9\sqrt{k + \sum_{v \in V} d_v^2}$, without the crossing-criticality assumption.

## 2  A New Lower Bound

We will make use of the following theorem [7].

**Theorem 1** *Let $G = (V, E)$ be a planar graph with non-negative weights on its vertices that sum up to one and every weight is at most $\frac{2}{3}$. Let $d_v$ denote the degree of any vertex $v$. Then there exists at most $\frac{\sqrt{3} + \sqrt{2}}{2}\sqrt{\sum_{v \in V} d_v^2}$ edges whose removal divides $G$ into disjoint subgraphs $G_1 = (V_1, E_1)$ and $G_2 = (V_2, E_2)$ such that the weight of each is at most $\frac{2}{3}$.*

Theorem 2 implies an upper bound for the cutwidth of planar graphs which deserves an independent interest.

**Theorem 2** *For any planar graph $G = (V, E)$*

$$\mathrm{cw}(G) \leq \frac{6\sqrt{2} + 5\sqrt{3}}{2}\sqrt{\sum_{v \in V} d_v^2},$$

*where $d_v$ is the degree of any vertex $v$.*

Djidjev and Vrťo, *Crossings and Cutwidths*, JGAA, 7(3) 245–251 (2003)     248

**Proof:** Apply Theorem 1 to $G$. Assign weights to vertices:

$$\text{weight}(u) = \frac{d_u^2}{\sum_{v \in V} d_v^2}.$$

1. Let $\text{weight}(u) \le \frac{2}{3}$ for all $u$. By deleting $\frac{\sqrt{3}+\sqrt{2}}{2}\sqrt{\sum_{v \in V} d_v^2}$ edges we get graphs $G_1 = (V_1, E_1)$ and $G_2 = (V_2, E_2)$ such that for $i = 1, 2$ $\text{weight}(V_i) \le \frac{2}{3}$, which implies

$$\sum_{v \in V_i} d_v^2 \le \frac{2}{3} \sum_{v \in V} d_v^2.$$

2. Assume there exists a vertex $u$ such that $\text{weight}(u) > \frac{2}{3}$. By deleting edges adjacent to $u$ we get disjoint subgraphs $G_1 = (V_1, E_1)$ and $G_2 = (V_2, E_2)$, where $G_2$ is a one vertex graph. We have $\text{weight}(V_1) = 1 - \text{weight}(u) < \frac{2}{3}$ and

$$\sum_{v \in V_1} d_v^2 < \frac{2}{3} \sum_{v \in V} d_v^2.$$

The number of edges between $G_1$ and $G_2$ is

$$d_u \le \frac{\sqrt{3} + \sqrt{2}}{2} \sqrt{\sum_{v \in V} d_v^2}.$$

Placing the graphs $G_1$ and $G_2$ consecutively on the line and adding the deleted edges we obtain the estimation

$$\text{cw}(G) \le \max\{\text{cw}(G_1), \text{cw}(G_2)\} + \frac{(\sqrt{3} + \sqrt{2})}{2} \sqrt{\sum_{v \in V} d_v^2}.$$

Solving the recurrence we find

$$\text{cw}(G) \le \frac{\sqrt{3} + \sqrt{2}}{2} \sum_{i=0}^{\infty} \left(\frac{2}{3}\right)^{i/2} \sqrt{\sum_{v \in V} d_v^2} = \frac{6\sqrt{2} + 5\sqrt{3}}{2} \sqrt{\sum_{v \in V} d_v^2}.$$

$\square$

Our main result is

**Theorem 3** *Let $G = (V, E)$ be a graph. Let $d_v$ denote the degree of any vertex $v$. Then the crossing number of $G$ satisfies*

$$\text{cr}(G) + \frac{1}{16} \sum_{v \in V} d_v^2 \ge \frac{1}{1176} \text{cw}^2(G).$$

Djidjev and Vrťo, *Crossings and Cutwidths*, JGAA, 7(3) 245–251 (2003)    249

**Proof:** Consider a drawing of $G$ with $\mathrm{cr}(G)$ crossings. Introducing a new vertex at each crossing results in a plane graph $H$ with $\mathrm{cr}(G)+n$ vertices. By Theorem 2 we have

$$\mathrm{cw}(H) \leq \frac{6\sqrt{2}+5\sqrt{3}}{2}\sqrt{\sum_{v\in H} d_v^2} = \frac{6\sqrt{2}+5\sqrt{3}}{2}\sqrt{\sum_{v\in G} d_v^2 + 16\mathrm{cr}(G)}.$$

Finally, note that $\mathrm{cw}(G) \leq \mathrm{cw}(H)$, which proves the claim. $\qquad\square$

This result immediately gives an upper bound for the path-width of $G$ in terms of its crossing number as the result of Kinnersley [11] implies that $\mathrm{pw}(G) \leq \mathrm{cw}(G)$.

**Corollary 1** *Let $G = (V, E)$ be a graph. Then*

$$\mathrm{pw}(G) < \frac{6\sqrt{2}+5\sqrt{3}}{2}\sqrt{16\mathrm{cr}(G) + \sum_{v\in V} d_v^2}.$$

# 3  Final Remarks

We proved a new lower bound formula for estimating the crossing numbers of graphs. The former method was based on the bisection width of graphs. Our method replaces the bisection width by a stronger parameter - the cutwidth. While the bisection width of a connected graph can be just one edge, which implies a trivial lower bound only, the cutwidth based method gives nontrivial lower bounds in most cases. A drawback of the method is the big constant factor in the formula.

A natural question arises how to find or estimate the cutwidth of a graph. The most frequent approach so far was its estimation from below by the bisection width. This of course degrades the cutwidth method to the bisection method. Provided that $\mathrm{cw}(H)$ in known or estimated from below, for some graph $H$, we can use a well-known relation $\mathrm{cw}(G) \geq \mathrm{cw}(H)/\mathrm{cg}(H,G)$, see [18], where is the congestion of $G$ in $H$ defined as follows. Consider an injective mapping of vertices of $H$ into the vertices of $G$ and a mapping of edges of $H$ into paths in $G$. Take the maximal number of paths traversing an edge. Minimizing this maximum over all possible mappings gives $\mathrm{cg}(H,G)$. Another possibility is to use the strong relation of the cutwidth problem to the so called discrete edge isoperimetric problem [1]. Informally, the problem is to find, for a given $k$, a $k$-vertex subset of a graph with the smallest "edge boundary". A good solution to the isoperimetric problem provides a good lower bound for the cutwidth.

Recently Pach and Tardos [14] proved another relation between crossing numbers and a special edge cut of a graph (Corollary 5), which resembles our Theorem 2.3. But neither of the two statements implies the other.

# Acknowledgement

We thank both referees for insightful comments and suggestions.

# References

[1] S.L. Bezrukov. Edge isoperimetric problems on graphs. In L. Lovász, A. Gyarfás, G.O.H. Katona, A. Recski, A., L.A. Székely, editors, *Graph Theory and Combinatorial Biology*, volume 7 of Bolyai Society Mathematical Studies, pages 157-197, Akadémia Kiadó, Budapest, 1999.

[2] S. Bezrukov, J.D. Chavez, L.H. Harper, M. Röttger, U.-P. Schroeder. The congestion of $n$-cube layout on a rectangular grid. *Discrete Mathematics* 213:13-19, 2000.

[3] R. Cimikowski, Algorithms for the fixed linear crossing number problem, *Discrete Applied Mathematics* 122:93-115, 2002.

[4] G. Di Battista, P. Eades, R. Tamassia, I.G. Tollis. *Graph Drawing: Algorithms for Visualization of Graphs*. Prentice Hall, 1999.

[5] G. Even, S. Guha, B. Schieber. Improved approximations of crossings in graph drawing and VLSI layout area. In *Proc. 32th Annual Symposium on Theory of Computing*, pages 296-305, 2000.

[6] M.R. Garey, D.S. Johnson. Crossing number is NP-complete. *SIAM J. Algebraic and Discrete Methods* 4:312-316, (1983).

[7] H. Gazit, G.L. Miller. Planar separators and the Euclidean norm. In *Intl. Symposium on Algorithms*, volume 450 of Lecture Notes in Computer Science, pages 338-347, Springer Verlag, 1990.

[8] J.F. Geelen, R.B. Richter, G. Salazar. Embedding graphs on surfaces. Submitted to *J. Combinatorial Theory-B*.

[9] M. Grohe. Computing crossing numbers in quadratic time. In *33rd Annual ACM Symposium on Theory of Computing*, pages 231-236, 2001.

[10] P. Hliněný. Crossing critical graphs and path-width. In *Proc. 9th Intl. Symposium on Graph Drawing*, volume 2265 of Lecture Notes in Computer Science, pages 102-114, Springer Verlag, 2001.

[11] N. Kinnersley. The vertex separation number of a graph equals its path-width, *Information Processing Letters* 142:345-350, 1992.

[12] A. Liebers. Methods for planarizing graphs - a survey and annotated bibliography, *J. of Graph Algorithms and Applications* 5:1-74, 2001.

[13] F.T. Leighton. *Complexity Issues in VLSI*. M.I.T. Press, Cambridge, 1983.

[14] J. Pach, G. Tardos. Untangling a polygon. *Discrete and Computational Geometry* 28:585-592, 2002.

[15] J. Pach, G. Tóth. Thirteen problems on crossing numbers. *Geombinatorics* 9:194-207, 2000.

Djidjev and Vrťo, *Crossings and Cutwidths*, JGAA, 7(3) 245–251 (2003)    251

[16] J. Pach, F. Shahrokhi, M. Szegedy. Applications of crossing numbers. *Algorithmica* 16:11-117, 1996.

[17] H. Purchase. Which aestethic has the greatest effect on human understanding? In *Proc. 5th Intl. Symposium on Graph Drawing*, volume 1353 of Lecture Notes in Computer Science, pages 248-261, Springer Verlag, 1997.

[18] A. Raspaud, O. Sýkora, I. Vrťo. Congestion and Dilation, Similarities and Differences - a Survey. In *Proc. 7th Intl. Colloquium on Structural Information and Communication Complexity*, pages 269-280, Carleton Scientific, 2000.

[19] F. Shahrokhi, O. Sýkora, L.A. Székely, I. Vrťo, Crossing numbers: bounds and applications. In I. Bárány, K. Böröczky, editors, *Intuitive Geometry*, vol. 6 of Bolyai Society Mathematical Studies, pages 179-206, Akadémia Kiadó, Budapest, 1997.

[20] O. Sýkora, I. Vrťo. On VLSI layouts of the star graph and related networks. *Integration, The VLSI Journal* 17:83-93, 1994.

[21] Wei-Liang Lin, A. H. Farrahi, M. Sarrafzadeh. On the power of logic resynthesis. *SIAM J. Computing* 29:1257-1289, 2000.

## Journal of Graph Algorithms and Applications

http://jgaa.info/

vol. 7, no. 3, pp. 253–285 (2003)

# A Multilevel Algorithm for Force-Directed Graph-Drawing

*Chris Walshaw*

School of Computing and Mathematical Sciences,
University of Greenwich,
Old Royal Naval College, Greenwich,
London, SE10 9LS, UK.
http://www.gre.ac.uk/~c.walshaw
C.Walshaw@gre.ac.uk

### Abstract

We describe a heuristic method for drawing graphs which uses a multilevel framework combined with a force-directed placement algorithm. The multilevel technique matches and coalesces pairs of adjacent vertices to define a new graph and is repeated recursively to create a hierarchy of increasingly coarse graphs, $G_0, G_1, \ldots, G_L$. The coarsest graph, $G_L$, is then given an initial layout and the layout is refined and extended to all the graphs starting with the coarsest and ending with the original. At each successive change of level, $l$, the initial layout for $G_l$ is taken from its coarser and smaller child graph, $G_{l+1}$, and refined using force-directed placement. In this way the multilevel framework both accelerates and appears to give a more global quality to the drawing. The algorithm can compute both 2 & 3 dimensional layouts and we demonstrate it on examples ranging in size from 10 to 225,000 vertices. It is also very fast and can compute a 2D layout of a sparse graph in around 12 seconds for a 10,000 vertex graph to around 5-7 minutes for the largest graphs. This is an order of magnitude faster than recent implementations of force-directed placement algorithms.

**Keywords:** graph-drawing, multilevel optimisation, force-directed placement.

Communicated by Michael Kaufmann: submitted March 2002; revised August 2003.

C. Walshaw, *Multilevel Force-Directed Drawing*, JGAA, 7(3) 253–285 (2003)254

# 1  Introduction

Graph-drawing algorithms form a basic enabling technology which can be used to help with the understanding of large sets of inter-related data. By presenting data in a visual form it can often be more easily digested by the user and both regular patterns and anomalies can be identified. However most data sets do not contain any explicit information on how they should be laid out for easy viewing, although normally such a layout will depend on the relationships between pieces of data. Thus if we model the data points with the vertices of a graph and the relationships with the edges we can use graph-based technology and, in particular, graph-drawing algorithms to infer a 'good' layout from an arbitrary data set based on the relationships.

There has been considerable research into graph-drawing in recent years and a comprehensive survey can be found in [2]. Many such algorithms are based on physical models and the vertices are placed so as to minimise the 'energy' in the physical system (see below, §2.3). Typically such algorithms are able to display structures and symmetries in the graph but their computational cost in terms of CPU time is very high.

## 1.1  Motivation

The motivation behind our approach to graph-drawing arises from our work in the field of graph partitioning and the multilevel paradigm, e.g. [20, 21]. In recent years it has been recognised that an effective way of enhancing partitioning algorithms is to use multilevel techniques and this strategy has been successfully developed to overcome the localised nature of the Kernighan-Lin and other partition optimisation algorithms, e.g. [12]. The multilevel process has also recently been successfully applied to the travelling salesman and graph colouring problems and appears to work (for combinatorial problems at least) by sampling and smoothing the objective function, [20], thus imparting a more global perspective to the optimisation.

This is an important consideration for graph-drawing; the localised positioning of a vertex relative to fixed neighbours is actually fairly easy and it is the global untangling of the graph which is more difficult or time consuming. We therefore aim to use the multilevel ideas to both enhance the layout and accelerate the graph-drawing process.

In this paper (and an earlier version, [19]) we apply multilevel ideas to force-directed placement (FDP) algorithms. In fact such ideas have been previously suggested in the graph-drawing literature and for example in 1991 Fruchterman & Reingold, [7], suggested the possible use of 'a multigrid technique that allows whole portions of the graph to be moved', whilst Davidson & Harel, [1], suggest a multilevel approach to 'expedite the SA [simulated annealing] process'. More recently Hadany & Harel, [9], and in particular Harel & Koren, [10], have actually used multilevel ideas (or as they refer to them, *multiscale*) and are able to robustly handle graphs of up to 15,000 vertices. However their algorithm uses the placement scheme of Kamada & Kawai, [13], which requires the graph the-

C. Walshaw, *Multilevel Force-Directed Drawing*, JGAA, 7(3) 253–285 (2003)255

oretic distances (path lengths) between pairs of vertices, and hence the overall complexity of the method contains an $O(N^2)$ term. Gajer *et al.*, who subsequently developed a similar scheme, [8], managed to reduce this complexity by calculating these distances dynamically and also enhanced the scheme by computing the layout in higher dimensions. All three of these approaches ([8, 9, 10]) share many features with the algorithm outlined here (although derived independently) and confirm that the multilevel paradigm can be a powerful tool for force-directed placement irrespective of the specific FDP algorithm used.

A related but somewhat different idea is that of multilevel drawings, e.g. [3, 6]. Rather than using the multilevel process to create a good layout of the original graph, a multilevel graph is created, either by natural clustering which exists in the graph or by artificial means similar to those applied here. Each level is drawn on a plane at a different height and the entire structure can then be used to aid understanding of the graph at multiple abstraction levels, [5].

Finally, although not strictly related to the multilevel ideas described here, it is worth mentioning that Koren *et al.* have recently developed a number of other graph-drawing schemes which can work even faster than multilevel force-directed placement (although the layout quality is often somewhat inferior). In particular, these include the use of the algebraic multigrid techniques, [14], and (building on the ideas due to Gajer *et al.*, [8]) the development of higher-dimensional embeddings, [11].

## 2 A multilevel algorithm for graph-drawing

In this section we describe how we combine the multilevel optimisation ideas with our variant of a force-directed placement algorithm.

### 2.1 Notation and Definitions

Let $G = G(V, E)$ be an undirected graph of vertices $V$, with edges $E$ and which we will assume is connected. For any vertex $v$ let $\Gamma_v$ be the neighbourhood of, or set of vertices adjacent to, $v$, i.e., $\Gamma_v = \{u \in V : (u, v) \in E\}$. We use the $|.|$ operator to denote the size of a set so that $|V|$ is the number of vertices in the graph and $|\Gamma_v|$ is the number of vertices adjacent to $v$ (the *degree* of $v$). We also use $|.|$ to denote the weight of a vertex; since weighted vertices in the coarsened graphs represent sets of vertices from the original graph, the weight of a coarsened vertex is just equivalent to the number of original vertices in the set it represents. We then use $||.||$ to denote Euclidean distance in either 2D or 3D.

### 2.2 The multilevel framework

As stated above, the inspiration behind our graph-drawing scheme is the multilevel paradigm, e.g. [20]. The idea is to coalesce *clusters* of vertices to define

C. Walshaw, *Multilevel Force-Directed Drawing*, JGAA, 7(3) 253–285 (2003)256

a new graph and recursively iterate this procedure to create a hierarchy of increasingly coarse graphs, $G_0, G_1, \ldots$ and until the size of the coarsest graph falls below some threshold. The coarsest graph, $G_L$, is then given an initial layout and the layout is refined and extended to all the graphs starting with the coarsest and ending with the original. At each successive change of level, $l$, the initial layout for $G_l$ is taken from its coarser and smaller child graph, $G_{l+1}$, and refined using force-directed placement. Thus the algorithm does not actually operate simultaneously on multiple levels of the graph (as, for example, a multigrid algorithm might) but instead refines the layout at each level and then extends the result to the next level down.

### 2.2.1  Graph coarsening

There are many ways to create a coarser graph $G_{l+1}(V_{l+1}, E_{l+1})$ from $G_l(V_l, E_l)$ and clustering algorithms are an active area of research within the field of graph-drawing amongst others, e.g. [3, 17]. Usually such clustering algorithms seek to retain the more important structural features of the graph in order that the visualisation of each level is meaningful in itself. However, here we are only interested in the drawing of the original graph and as such we seek a fast and efficient (i.e., not necessarily optimal) algorithm that judiciously reduces the size of the graph. Thus, if too many vertices are clustered together in one step it may depreciate the benefits of the multilevel paradigm and in particular inhibit the force-directed placement algorithm, as applied to $G_l$, from making use of the positioning obtained for $G_{l+1}$. Conversely, if each clustering only shrinks the graph by a small fraction, the multilevel scheme may be significantly slowed by having to compute the layout for a multitude of fairly similar coarse graphs. To suit these requirements we choose (as is typical for partitioning) a coarsening approach known as *matching* in which each vertex is matched with at most one neighbour, so that clusters are thus formed of at most two vertices and the number of vertices in the coarsened graph $G_{l+1}$ is no less than half the number in $G_l$.

Computing a matching is equivalent to finding a maximal independent subset of graph edges which are then collapsed to create the coarser graph. The set is independent if no two edges in the set are incident on the same vertex (so no two edges in the set are adjacent), and maximal if no more edges can be added to the set without breaking the independence criterion. Having found such a set, each selected edge is collapsed and the vertices, $u_1, u_2 \in V_l$ say, at either end of it are merged to form a new vertex $v \in V_{l+1}$ with weight $|v| = |u_1| + |u_2|$.

The problem of computing a matching of the vertices is known as the maximum cardinality matching problem. Although there are optimal algorithms to solve this problem, they are of at least $O(|V|^{2.5})$, e.g. [15]. Unfortunately this is too slow for our purposes and, since it is not essential for the multilevel process to solve the problem optimally, we use a variant of the edge contraction heuristic proposed by Hendrickson & Leland, [12]. Their method of constructing a maximal independent subset of edges is to create a randomly ordered list of the vertices and visit them in turn, matching each unmatched vertex with

C. Walshaw, *Multilevel Force-Directed Drawing*, JGAA, 7(3) 253–285 (2003)257

an unmatched neighbouring vertex (or with itself if no unmatched neighbours exist). Matched vertices are removed from the list.

If there are several unmatched neighbours the choice of which to match with can be random, but in order to keep the weight of the vertices in the coarser graphs as uniform as possible, we choose to match with the neighbouring vertex with the smallest weight (note that even if the original graph $G_0$ is unweighted, $G_l$ for $l = 1, 2, \ldots$ will be weighted). In the case of several such minimally weighted neighbours a random choice is made from amongst them. Other matching heuristics were tested (e.g. such as one that prefers to match across heavily weighted edges) but did not reveal any noticeable benefits and in the end the choice was based purely on empirical evidence (not presented here).

### 2.2.2 The initial layout

Having constructed the series of graphs until the number of vertices in the coarsest graph, $G_L$, is smaller than some threshold, we need to compute an initial layout for $G_L$. However, if the graph is coarsened down to 2 vertices (which because of the mechanisms of the coarsening will be connected by a single weighted edge) we can simply place these vertices at random with no loss of generality.

Note that contraction down to 2 vertices should always be possible provided the graph is connected (assumed, §2.1). To see this consider that every connected graph of $|V|$ vertices must have at least $|V| - 1$ edges and that the collapsing of an edge results in a connected graph. Thus, if $|V| > 2$ there must be at least one edge which can be collapsed to create a graph with $|V| - 1$ vertices and so on by induction.

### 2.2.3 Uncoarsening

At each level $l$ the layout on graph $G_l(V_l, E_l)$ is refined and then extended to its parent $G_{l-1}(V_{l-1}, E_{l-1})$. This uncoarsening step is a trivial matter and matched pairs of vertices, $v_1, v_2 \in V_{l-1}$, are placed at the same position as the cluster, $v \in V_l$, which represents them.

## 2.3 The force-directed placement algorithm

We use a standard drawing algorithm to refine the layout on the graph, $G_l$, at each level $l$. There has been considerable research into graph-drawing paradigms, [2], and here we are interested in *straight-line* drawing schemes and, in particular, *spring-embedder* or *force-directed placement* algorithms. The original concept came from a paper by Eades, [4], and is based on the idea of replacing vertices by rings or hinges and edges by springs. The vertices are given initial positions, usually random, and the system is released so that the springs move the vertices to a minimal energy state (i.e., so that the springs are compressed or extended as little as possible).

C. Walshaw, *Multilevel Force-Directed Drawing*, JGAA, 7(3) 253–285 (2003)258

Unfortunately the local spring forces are insufficient to globally untangle a graph and so such algorithms also employ global repulsive forces, calculated between every pair of vertices in the graph, and thus the system resembles an $n$-body problem. Such repulsive forces between non-adjacent vertices do not have an analogue in the spring system but are a crucial part of the spring-embedder algorithms to avoid minimal energy states in which the system is collapsed in on itself in some manner. As a simple example of this consider a chain of 3 vertices $\{u, v, w\}$ connected by two edges $(u, v)$ and $(v, w)$ and a spring model of this graph where both springs have a natural length $k$. Perhaps the most intuitive zero energy layout for this system would have $u$ & $w$ placed a distance $2k$ apart with $v$ in the middle. However, with no global repulsive forces there is nothing to stop $u$ & $w$ from being placed in the *same* position and if this is a distance $k$ away from $v$ then once again the energy is zero. On a larger scale, repulsion is necessary to push whole regions, which are not immediately connected, away from each other.

The particular variant of force-directed placement that we use is based on an algorithm by Fruchterman & Reingold (FR), [7], itself a variation of Eades' original algorithm. From the point of view of the multilevel approach it is attractive as it is an incremental scheme which iterates to convergence and which can reuse a previously calculated initial layout. We have made a number of parameter modifications based on our experience with it and, in particular, because of the additional problems associated with drawing very large graphs. In principle however, it should be possible to use any iterative incremental algorithm for this part of the multilevel graph-drawing, although in practice different algorithms can be somewhat sensitive and require a certain amount of tuning.

Figure 1 shows the basic outline of our algorithm and is written in a similar fashion to the original FR algorithm, [7]. Thus $\Delta$ is shorthand notation for the difference vector between the positions of two vertices and $\Theta$ is short for the vector of displacements calculated for the current vertex $v$. There are two main differences (apart from the choice of parameters); the order of updating and the weighting of the repulsive forces (discussed in more detail below). One other fairly minor difference is that we do not impose any boundaries around the drawing (referred to as the frame in [7]); the layout can thus expand (or contract) as required by the forces within the system. The positions may be subsequently scaled to fit onto a computer screen or a hardcopy or indeed into any region required by the user, but this forms no part of the algorithm.

### 2.3.1 Updating

An important difference from the original FR algorithm is the order of updating of the vertex positions. The original algorithm used two vectors (of length $|V|$), one containing the position of the vertices and the second containing their displacement as calculated during the current iteration of the outer loop. The outer loop then contained three main inner loops, the first looping over the vertices to calculate displacement caused by (global) repulsive forces and the

C. Walshaw, *Multilevel Force-Directed Drawing*, JGAA, 7(3) 253–285 (2003)259

```
{ initialisation }
function f_r(x, w) := begin return −Cwk²/x end
function f_a(x) := begin return x²/k end
t := t_0;
Posn := NewPosn;

while (converged ≠ 1) begin
      converged := 1;

      for v ∈ V begin
            OldPosn[v] = NewPosn[v]
      end

      for v ∈ V begin
            { initialise Θ, the vector of displacements of v }
            Θ := 0;

            { calculate (global) repulsive forces }
            for u ∈ V, u ≠ v begin
                  Δ := Posn[u] − Posn[v];
                  Θ := Θ + (Δ/||Δ||) · f_r(||Δ||, |u|);
            end

            { calculate (local) attractive/spring forces }
            for u ∈ Γ_v begin
                  Δ := Posn[u] − Posn[v];
                  Θ := Θ + (Δ/||Δ||) · f_a(||Δ||);
            end

            { reposition v }
            NewPosn[v] = NewPosn[v] + (Θ/||Θ||) · min(t, ||Θ||);
            Δ := NewPosn[v] − OldPosn[v];
            if (||Δ|| > k·tol) converged := 0;
      end

      { reduce the temperature to reduce the maximum movement }
      t := cool(t);
end
```

Figure 1: Force-directed placement algorithm

C. Walshaw, *Multilevel Force-Directed Drawing*, JGAA, 7(3) 253–285 (2003)260

second looping over edges and calculating the displacement (on the vertices at either end of the edge) due to the local attractive forces. The final inner loop over the vertices updated the positions.

In our version however, we only calculate displacements for one vertex at a time, updating each at the end of the inner loop. At first it might seem as if this is less efficient, since the attractive forces are calculated twice for each edge. However, *Posn* is a pointer which points to *NewPosn*, the newly calculated position of each vertex which may have already been updated during the **current** iteration of the outer loop. In our experience this dramatically improves the performance of the algorithm (see §3.2). It is also very easy to recover the behaviour of the original FR algorithm (for comparison) by setting the pointer *Posn := OldPosn* in the initialisation section.

### 2.3.2 Vertex weighting.

We use a weighted version of the original FR repulsive function, computed by multiplying the repulsive force by the weight, $|u|$, of the vertex, $u$, which generates it, to give $f_r(x, |u|) = -C \cdot |u| \cdot k^2/x$. Although we are typically (but not exclusively) interested in drawing unweighted graphs, any of the coarsened graphs will have weights attached to both vertices & edges and in particular the vertex weight of a coarsened vertex $u$ will represent the sum of weights of vertices from the original graph contained in the cluster. If we then consider the repulsive forces in the original graph, all of the vertices in the cluster $u$ would act on any vertex from the cluster $v$ so it makes sense to multiply the repulsive force of $u$ on $v$ by $|u|$. This was also confirmed by experimentation and made a considerable improvement as compared with neglecting this factor. For unweighted graphs, and in particular the force-directed algorithm used in its standard single-level format then $|u| = 1$ for all $u \in V$ and this function reverts back to the original FR version from [7].

Finally the constant $C$ was determined by experimentation as suggested by Fruchterman & Reingold. We found that the smaller the value of $C$, the better the algorithms (both multilevel and the original single-level version) seemed to work, but the longer they took to run. This is presumably because, with the grid-variant in use (see below, §2.4), the smaller the value of $C$, the smaller the effect of the repulsive forces and hence the more vertices are used to calculate them. Thus the quality improves but the runtime increases. After extensive testing we settled on $C = 0.2$, although $C = 0.5$ & $C = 0.1$ could equally be used to give similar results.

### 2.3.3 Edge weighting.

Note that there is no simple equivalent edge weight analogue for the local attractive forces. To see this consider the three graphs shown in Figure 2(a)-(c) and suppose that in each case the ringed vertices are matched and merged to give the graph shown in Figure 2(d). The weight of the edge in Figure 2(d) would then be 3 if derived from Figure 2(a), 2 if derived from 2(b) and 1 if

C. Walshaw, *Multilevel Force-Directed Drawing*, JGAA, 7(3) 253–285 (2003)261

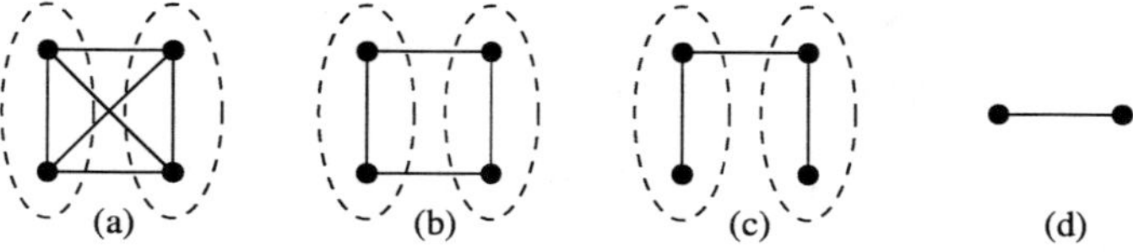

Figure 2: Examples of coarsening

from 2(c). Now consider the attractive force on each vertex of the cluster $v$ by looking at the attractive forces in the three original graphs and assuming that the matched vertices are placed at the same position. In the case of the graph in Figure 2(a) there are 2 attractive forces from each vertex in the cluster $w$ on each vertex in the cluster $v$ (this corresponds to the edge weight 3). Meanwhile, for Figure 2(b) there is 1 attractive force per vertex (corresponding to edge weight 2). However for the graph in Figure 2(c) it is not even clear what the attractive force from cluster $w$ on cluster $v$ should be, although arguably it should be less than 1 in some averaged sense (and this corresponds to edge weight 1). Hence there is no linear relation between edge weight and attractive forces and indeed for more complex cases (i.e., after multiple coarsenings) the relationship becomes even harder to evaluate.

The simplest way of dealing with this problem, and the one that we use for all the experiments in this paper, is just to ignore edge weights. However, we have tested two alternative schemes (using the same testing regime as that described in Section 3). The first scheme we tried was to multiply each attractive force by the weight of the edge along which it acts. In fact this produced very similar results to ignoring the edge weights altogether, except that the drawings were somewhat less extended and took around 10-20% longer to compute (essentially both of these effects arise because the attractive forces are stronger relative to the repulsive ones and similar results can be seen simply by reducing the size of the parameter $C$).

The second alternative was to average the attractive forces by again multiplying each force by the corresponding edge weight but also dividing by the weight of the cluster on which it acts (giving multipliers of $\frac{3}{2}$, 1 & $\frac{1}{2}$ respectively for the graphs in Figures 2(a), (b) & (c)). In fact this produced worse layouts than ignoring the edge weights, especially when using a fast cooling schedule (see §2.3.5), although for slow cooling schedules it made little difference. In the end, however, we decided to ignore edge weights.

### 2.3.4 Natural spring length, $k$

A crucial part of the algorithm is the choice of the natural spring length, $k$, (the length at which a spring or edge is neither extended nor compressed). At the start of the execution of the placement algorithm for graph $G_l$ the vertices will all be in positions determined by the layout calculated for graph $G_{l+1}$ (except for $G_L$, the coarsest graph). We must therefore somehow set the spring length

C. Walshaw, *Multilevel Force-Directed Drawing*, JGAA, 7(3) 253–285 (2003)262

relative to this existing layout in order not to destroy it. If, for example, we set $k$ too large, then the entire graph will have to expand from its current layout and potentially ruin any advantage gained from having calculated an initial layout via the multilevel process.

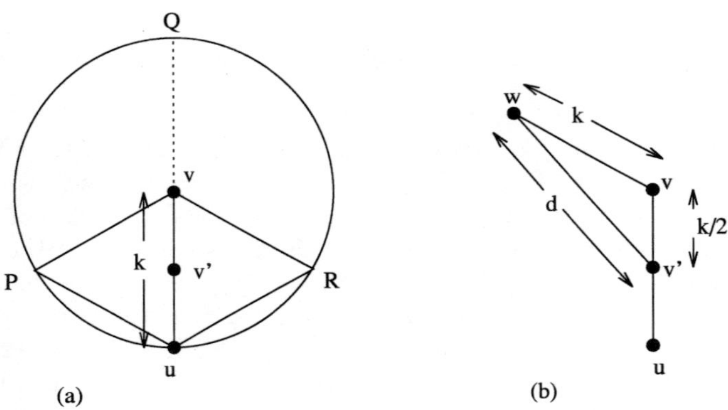


Figure 3: Calculation of natural spring length

In fact we derive the new value for $k$ by considering what happens when we coarsen a graph, $G_l$, with well placed vertices (i.e., all vertices are approximately at a distance $k$ from each other). Consider Figure 3(a) and suppose that $v$ and $u$ (at distance $k$ from each other) are going to be clustered to form a vertex $v'$ at the mid-point between them. Any vertex $w$ adjacent to $v$ should, if ideally spaced, lie somewhere on the arc $PQR$ of the circle of radius $k$ centred on $v$ (it should not be on the arc $PuR$ as that would place it too close to $u$). The distance between $w$ and $v'$ will then be $3k/2$ if $w$ lies at $Q$ or $\sqrt{3}k/2$ if $w$ lies at $P$ or $R$. If we take an average position for $w$ midway along the arc $PQ$ then from Figure 3(b) and the cosine rule, the length $d$ is given by

$$d^2 = k^2 + \left(\frac{k}{2}\right)^2 - 2 \cdot k \cdot \frac{k}{2} \cos(2\pi/3) = k^2 + \frac{1}{4}k^2 + \frac{1}{2}k^2 = \frac{7}{4}k^2$$

If we take $d$ as an estimate for the new natural spring length $k'$ then $k' = \sqrt{7/4} \cdot k$.

Reversing this process, given a graph $G_{l+1}$ with natural spring length $k_{l+1}$, we can estimate the natural spring length for the parent graph $G_l$ at the start of the placement algorithm to be

$$k_l = \sqrt{\frac{4}{7}} \cdot k_{l+1}.$$

Remarkably this simple formula works very robustly over all the examples that we have tested, certainly better than other functions we have tried (for example based on average edge length of the initial layout). Very occasionally on one or

C. Walshaw, *Multilevel Force-Directed Drawing*, JGAA, 7(3) 253–285 (2003)263

two of the examples the value for $k$ that it gives is too small for the existing layout and the graph placement expands rapidly for the first few iterations. However, this usually occurs on one of the coarser graphs and the multilevel procedure is still able to find a good layout. Nonetheless we feel that the choice of this parameter could do with further investigation.

For the initial coarsest graph, $G_L$, we simply set

$$k_L = \frac{1}{|E_L|} \sum_{(u,v)\in E_L} ||(u,v)||,$$

the average edge length. Typically we coarsen down to 2 vertices and 1 edge and so $k$ is set to the length of that edge.

### 2.3.5 Convergence

We retain the 'cooling schedule' used in the original FR algorithm. Notice from Figure 1 that when the positions are updated, the maximum movement is limited by the value $t$ (or temperature) and that $t$ is reduced at the end of each iteration of the outer loop. This idea, drawn from a graph-drawing algorithm due to Davidson & Harel, [1] and based on simulated annealing, allows large movements (high temperature $t$) at the beginning of the iterations but progressively reduces the maximum movement as the algorithm proceeds (and the temperature falls). Fruchterman & Reingold do not give the exact cooling schedule that they use, although they do recommend a two phase scheme, first cooling rapidly and steadily (possibly linearly) and the second phase at a constant low temperature. Here for simplicity we use the scheme $t_i = \lambda t_{i-i}$, or in pseudo-code

**function** $cool(t) :=$ **begin return** $\lambda t$ **end**

which operates similarly (i.e., initial rapid decay and a slow tail-off) but only involves one parameter, $\lambda$. After experimentation we then set $t_0 = k_l$ at each level $l$ and the algorithm is then deemed to have converged when the movement of every vertex is less than some tolerance, $tol$, times $k_l$. Again after extensive experimentation we set $tol = 0.01$. This also allows us to avoid explicitly setting any maximum number of iterations since eventually the temperature will drop below $tol \cdot k_l$ and so there is an implicit limit.

By varying the cooling rate, $\lambda$, and measuring performance against runtime for a range of values of $\lambda$, we are able to compare different algorithms in a more meaningful way (see §3.2). However in the examples following we then recommend a value of $\lambda = 0.9$ and this means that all movement ceases at iteration $i$ where $0.9^i < 0.01$ or in other words after 44 iterations. This is close to the 50 iterations recommended in the original FR algorithm, [7].

### 2.3.6 Coincident vertices

The algorithm needs minor exception handling if two vertices are found to be in exactly the same position. This can occasionally occur during the execution of

C. Walshaw, *Multilevel Force-Directed Drawing*, JGAA, 7(3) 253–285 (2003)264

the algorithm but it also always happens when the code commences on a graph $G_l$, having calculated the layout on $G_{l+1}$, since we initially place the vertices in a cluster at the same position as the cluster. In these cases the vertices are simply treated as if they were a small distance apart (the actual direction generated randomly with the distance no more than $0.001 \cdot k$) and the forces calculated accordingly. This allows us to extend the layout of one graph to its parent without any additional sophisticated mechanism.

## 2.4  Reducing the complexity

It is fairly clear from the description of the algorithm that the placement complexity for each iteration on graph $G_l(V_l, E_l)$ is $O(|V_l|^2 + |E_l|)$. For the types of sparse graphs in which we are interested, the $|V_l|^2$ heavily dominates this expression and we therefore use the FR grid variant for reducing the run-times, [7]. Their motivation was that over long distances the repulsive forces are sufficiently small to be neglected. If we set $R$ to be the maximum distance over which repulsive forces will act we can then modify the algorithm by changing the global force calculation to:

**function** $f_r(x, w) :=$
**begin**
        **if** $(x \leq R)$ **return** $-Cwk^2/x;$
        **else return** $0.0;$
**end**

In itself this modification will do little or nothing to speed up the calculation as the complexity is still $O(|V_l|^2)$. However Fruchterman & Reingold, [7], showed that if the domain is divided up into regular square cells (or cube shaped cells in 3D) of size $R^2$ (or $R^3$ in 3D) then each vertex will only be affected by repulsive forces from vertices in its own and adjacent cells (including those diagonally adjacent). To implement this efficiently we simply visit every vertex at the start of each outer loop and add each to a linked list of vertices for the cell to which it belongs. Repulsive forces can then be calculated for each vertex by using the linked lists of their own and adjacent cells. In practice this seems to work very well although we note that the number of grid cells can greatly exceed the number of vertices, particularly in 3D. However the implied memory limitations are not difficult to deal with by storing only the non-empty cells in a tree structure rather than storing all of them in an array.

We also note that, since we update vertex positions continuously throughout the outer loop, vertices are quite likely to move from one cell to another and thus not appear in the appropriate linked list. However we ignore the possible inaccuracies and do not transfer them during the course of an iteration and in practice it does not seem to matter.

Finally we must decide what value to give to $R$. In the original FR algorithm the value $R = 2k$ was used, but for the larger graphs in which we are interested,

C. Walshaw, *Multilevel Force-Directed Drawing*, JGAA, 7(3) 253–285 (2003)265

this did not prove sufficient to 'untangle' them in a global sense. Unfortunately the larger the value given to $R$ the longer the algorithm takes to run and so although assigning $R = 20k$ gave better results, it did so with a huge time penalty. Fortunately, however, the power of the multilevel paradigm comes to our aid once again and we can make $R$ a function of the level $l$. Thus for the initial coarse graphs we can set $R_l$ to be relatively large and achieve some impressive untangling without too much cost (since $|V_l|$ is very small for these graphs). Meanwhile, for the final large graphs, when most of the global untangling has already been achieved we can make $R_l$ relatively small without penalising the placement. In fact, provided this parameter is not too small it should be very robust (since it just determines a cut off point for tiny repulsive forces) and because the first such schedule that we tried worked very well, we have not experimented further.

The value that we use, therefore, is $R_l = 2(l + 1)k_l$ for each graph $G_l$. In this expression $l$ is just the graph number where $G_0$ is the original graph and $G_l$ the graph after $l$ coarsenings. Conveniently this also replicates the choice of $R = 2k$ for $G_0$ in the original single-level FR algorithm.

## 2.5  Complexity analysis

It is not easy to derive complexity results for the algorithm but we can state some bounds. Firstly the number of graph levels, $L$, is dependent on the rate of coarsening. At best the number of vertices will be reduced by a factor of 2 at every level (if the code succeeds in matching every vertex with another one) and in the worst case, the code may only succeed in matching 1 vertex at every level (e.g. if the graph is a star graph, a 'hub' vertex connected to every other vertex each of which is only connected to the hub). Thus we have $\log_2 |V| \leq L < |V|$. This probably indicates that the algorithm is not well suited to graphs with a small diameter relative to their size (such as star graphs) and in fact for the examples given in Section 3 the coarsening rate is close to 2.

The matching & coarsening parts of the algorithm are $O(|V_l| + |E_l|)$ for each level $l$ but in fact the total runtime is heavily dominated by the FDP algorithm. Using the above simplification (§2.4) of neglecting long range repulsive forces we can see that each iteration of the FDP algorithm is bounded below by $O(|V_l| + |E_l|)$ although with a large coefficient. In fact if the graph is dense, or in the worst case a complete graph, it may be that this is still $O(|V_l|^2 + |E_l|)$, dependent on the relative balance of attractive & repulsive forces. However, we suspect that no FDP algorithm is appropriate for dense graphs because the minimal energy state corresponds to a tightly packed 'hair-ball' and so no structure is discernible in the drawing.

In summary the total complexity at each level is close to $O(|V_l| + |E_l|)$ for sparse graphs and the runtime is heavily dominated by the FDP iterations.

Finally consider the FDP algorithm used, without coarsening, on a given sparse graph of size $N$ (i.e., standard single-level placement) and compare it with multilevel placement (MLFDP) used on the same graph. Let $T_p$ be the time for the FDP algorithm to run on the graph and for MLFDP let $T_c$ be

C. Walshaw, *Multilevel Force-Directed Drawing*, JGAA, 7(3) 253–285 (2003)266

the time to coarsen and contract it. If we suppose that the coarsening rate is close to 2 (which is true for most of the examples below) then for MLFDP this gives us a series of problems of size $N, N/2, \ldots, N/N$ whilst the (almost) linear complexity for the placement scheme at each level gives the total runtime for MLFDP as $T_c + T_p/N + \ldots + T_p/2 + T_p$. In all the examples we have tested $T_c \ll T_p$ and so we can neglect it giving a total runtime of approximately $T_p/N + \ldots + T_p/2 + T_p \approx 2T_p$. In other words MLFDP should take **only** twice as long as FDP to run (and yet in the examples below achieves far better results). In fact the final level of the MLFDP algorithm is likely to already have a very good initial layout which means that it should run even faster than FDP although this is neutralised somewhat by the fact that the coarsening rate is normally somewhat less than 2. Nonetheless this factor of 2 is a good 'rule of thumb' and note that if the chosen FDP algorithm were $O(N^2)$ or even $O(N^3)$ then a similar analysis suggests that the MLFDP runtime would be substantially **less** than twice that of FDP.

## 3   Experimental Results

We have implemented the algorithms described here within the framework of JOSTLE, a mesh partitioning software tool developed at the University of Greenwich[1]. We illustrate and test these schemes in a variety of ways and on a large number of problem instances including a suite of small random planar graphs together with some much bigger graphs from genuine applications. Firstly in §3.1 we demonstrate the multilevel scheme with an extended example of the technique in action. Next in §3.2 we present the results from extensive tests which show algorithmic performance against runtime and compare the behaviour of single-level and multilevel versions. In §3.3 we then present a test of runtime complexity and finally in §3.4 highlight the multilevel algorithm with some detailed individual layouts.

The experiments were all carried out using a 1 GHz Pentium III with 256 Mbytes of memory running Linux. (Although this is three times faster than the processor used for our original testing in [19], differences in floating point performance mean that it only runs about twice as fast on this sort of application. Other differences in runtime result from changes in the algorithms that we have made since the previous paper.)

### 3.1   An extended example

In this section we demonstrate in more detail how the multilevel scheme works. Figure 4 shows the original layout of a small mesh-based graph, 516 (with 516 vertices), drawn from a computational mechanics problem, and (lightly shaded) the underlying triangular mesh. Typically in such graphs the vertices can either represent mesh nodes (the nodal graph), mesh elements (the dual graph),

---

[1]freely available for academic and research purposes under a licensing agreement from `http://www.gre.ac.uk/jostle`

C. Walshaw, *Multilevel Force-Directed Drawing*, JGAA, 7(3) 253–285 (2003)267

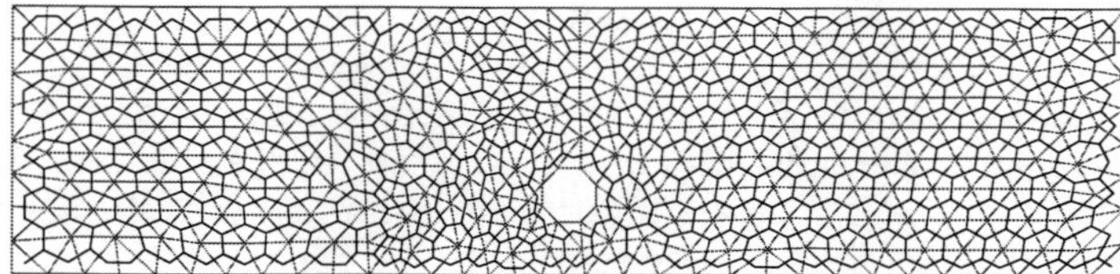

Figure 4: The original layout of 516 also showing the underlying triangular mesh elements

a combination of both (the full or combined graph) or some other special purpose representation. In this case the graph is a dual graph where each vertex represents a triangular element.

Table 1: The sizes of the coarsened graphs of 516

| $l$ | 0 | 1 | 2 | 3 | 4 | 5 | 6 | 7 | 8 | 9 |
|---|---|---|---|---|---|---|---|---|---|---|
| $|V_l|$ | 516 | 288 | 156 | 86 | 46 | 24 | 13 | 7 | 4 | 2 |
| $|E_l|$ | 729 | 501 | 319 | 190 | 97 | 48 | 23 | 9 | 3 | 1 |

The MLFDP algorithm was applied to this problem (ignoring the existing layout) and Table 1 lists the sizes of the graphs, $G_1(V_1, E_1)$ to $G_9(V_9, E_9)$, constructed by the coarsening. Notice that $|V_l| \geq |V_{l-1}|/2$ since no more than two vertices are clustered together and so the graph cannot shrink by more than a factor of two. The initial layout is computed by placing the two vertices of $G_9$ at random and setting the natural spring length, $k$, to be the distance between them. Starting from $G_l = G_8$ the layout is extended from $G_{l+1}$, by simply placing vertices at the same position as the cluster representing them in the coarser graph, and then refined.

Figure 5(a) shows the final layout on $G_4$ and it can been clearly seen that, although over 10 times smaller than the original, the layout is already beginning to take shape. Figures 5(b)-(d) meanwhile illustrate the placement algorithm on $G_2$. Figure 5(b) shows the initial layout as extended from $G_3$ and with many of the vertices coincident whilst Figure 5(c) then shows the layout after the first iteration and where the coincident vertices have been pushed apart. Figure 5(d) finally shows the layout after the placement algorithm has converged for $G_2$. Notice an important feature of the multilevel process (common with the partitioning counterpart) that the final layout (partition), does not differ greatly from the initial one and hence the placement scheme at each level need not be very powerful in a global sense, since the multilevel framework seems to impart this property. Figure 5(e) shows the final layout on the original graph, $G_0$. The small kink arises from the hole in the graph which distorts the layout slightly, but in general the final drawing is excellent. Finally note that the MLFDP algorithm took less than half a second to compute this layout (this is

C. Walshaw, *Multilevel Force-Directed Drawing*, JGAA, 7(3) 253–285 (2003)268

the time for the entire algorithm including reading the problem, coarsening and placement at each level).

For comparison, Figure 5(f) shows the placement algorithm used on a random initial layout of the same graph (in other words as the standard single-level placement algorithm, FDP). Possibly the algorithm is not well tuned for this problem, but what can be seen is that although the micro structure of the graph has been reconstructed reasonably well (at least this can be seen by examining the layout in more detail than Figure 5(f) allows), the single-level placement has not been able to 'untangle' the graph in a global sense. In fact, by adjusting the cooling schedule so that the algorithm runs for at least 2,150 iterations, the single-level scheme can achieve a similar layout to that shown in Figure 5(e); however this takes 9.76 seconds to run, about twenty times longer than the 0.48 seconds required by the multilevel scheme. We believe that this at least hints at the power of the multilevel framework.

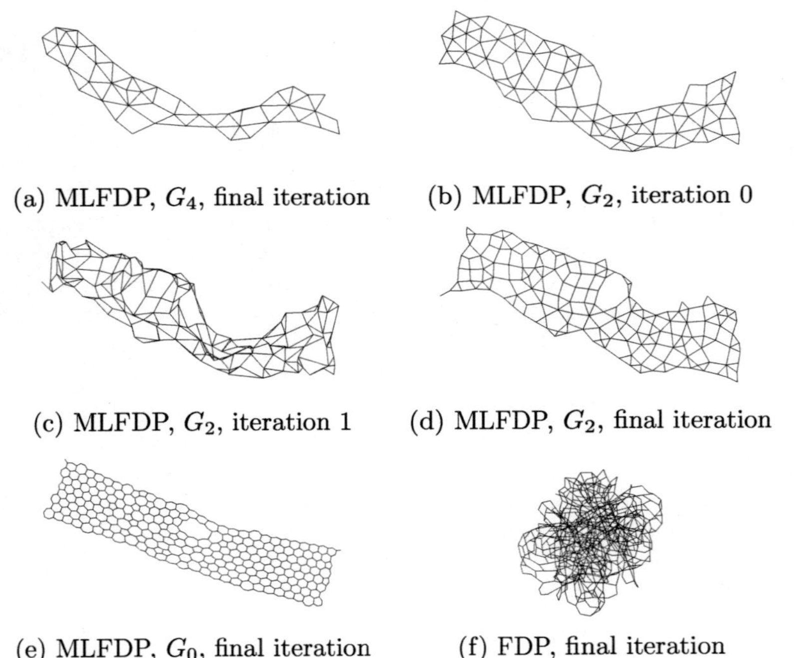

(a) MLFDP, $G_4$, final iteration     (b) MLFDP, $G_2$, iteration 0

(c) MLFDP, $G_2$, iteration 1     (d) MLFDP, $G_2$, final iteration

(e) MLFDP, $G_0$, final iteration     (f) FDP, final iteration

Figure 5: The multilevel algorithm illustrated for the graph 516

## 3.2 Comparison of single-level and multilevel algorithms

Although anecdotal evidence (above and in [19]) suggests that the multilevel framework can significantly enhance force-directed placement, we test this conclusion more thoroughly by comparing algorithmic performance on two test

C. Walshaw, *Multilevel Force-Directed Drawing*, JGAA, 7(3) 253–285 (2003)269

suites. The first suite[2] consists of 200 randomly generated planar graphs originally constructed to benchmark the GDT[3] software. There are 20 graphs each of size $|V| = 10, 20, \ldots, 100$ and we have subdivided the suite into two subclasses: 100 tiny graphs with $10 \leq |V| \leq 50$ and 100 small graphs with $60 \leq |V| \leq 100$.

Table 2: The test suite of mesh-based graphs

| | size | | degree | | | MLFDP placement ($\lambda = 0.9$) | |
|---|---|---|---|---|---|---|---|
| graph | $|V|$ | $|E|$ | max | min | avg | crossings | time (secs.) |
| 140 | 140 | 199 | 3 | 2 | 2.84 | 0 | 0.12 |
| grid1 | 252 | 476 | 4 | 2 | 3.78 | 121 | 0.29 |
| 399 | 399 | 573 | 3 | 1 | 2.87 | 0 | 0.38 |
| rj20 | 400 | 760 | 4 | 2 | 3.80 | 0 | 0.42 |
| 516 | 516 | 729 | 3 | 1 | 2.83 | 0 | 0.48 |
| 771 | 771 | 1133 | 3 | 1 | 2.94 | 0 | 0.86 |
| 788 | 788 | 1123 | 3 | 2 | 2.85 | 90 | 0.78 |
| mesh1024 | 1024 | 1504 | 3 | 2 | 2.94 | 0 | 1.18 |
| dime01 | 1095 | 1570 | 3 | 2 | 2.87 | 34 | 1.03 |
| sierpinski06 | 1095 | 2187 | 4 | 2 | 3.99 | 0 | 0.81 |
| grid2 | 3296 | 6432 | 5 | 2 | 3.90 | 0 | 3.82 |
| 3elt | 4720 | 13722 | 9 | 3 | 5.81 | 3391 | 7.57 |
| uk | 4824 | 6837 | 3 | 1 | 2.83 | 116 | 5.02 |
| 4970 | 4970 | 7400 | 3 | 2 | 2.98 | 0 | 6.86 |
| dime06 | 5343 | 7836 | 3 | 2 | 2.93 | 295 | 5.53 |
| ukerbe1 | 5981 | 7852 | 8 | 2 | 2.63 | 374 | 9.48 |
| whitaker3 | 9800 | 28989 | 8 | 3 | 5.92 | 0 | 11.48 |
| sierpinski08 | 9843 | 19683 | 4 | 2 | 4.00 | 158 | 7.71 |
| t10k | 10027 | 14806 | 3 | 2 | 2.95 | 0 | 10.65 |
| crack | 10240 | 30380 | 9 | 3 | 5.93 | 0 | 13.31 |

The second suite comprises 20 much larger planar graphs, listed in Table 2, and mostly drawn from genuine examples of computational mechanics meshes. The Table gives their sizes ($|V|$ & $|E|$) and the maximum, minimum & average degree of the vertices. It also shows the number of edge crossings and runtime required for a layout produced by the multilevel algorithm (using the cooling rate, $\lambda = 0.9$, suggested below, §3.3). Once again we have split the suite into two subclasses: 10 small mesh-based graphs of between 100 and around 1,000 vertices and 10 medium sized with between 1,000 to around 10,000 vertices. Note that although these graphs are all planar, some of them (in particular grid1, dime01, 3elt & dime06) are exceptionally difficult to draw with a planar layout because of extreme variations in mesh density. Meanwhile sierpinski06 & sierpinski08, despite their regular local structure, contain large holes which add to the drawing complexity. Both of these difficulties are explained further and illustrated in §3.4.

We use the test suites to compare three different algorithms: $FDP_0$, the original FR algorithm (or as close as we can get to it); FDP, our version of that algorithm; and MLFDP, the multilevel version of FDP. We have also tested

---

[2]available from `http://www.dia.uniroma3.it/~gdt/testsuite/GDT-testsuite-BUP.tgz`
[3]graph-drawing toolkit, see `http://www.dia.uniroma3.it/~gdt/`

C. Walshaw, *Multilevel Force-Directed Drawing*, JGAA, 7(3) 253–285 (2003)270

MLFDP$_0$, the multilevel version of FDP$_0$, but since it always seems to perform worse than MLFDP we do not present any results for it.

The variants are differentiated by parameter settings defined as follows (in approximate order of importance):

| FDP$_0$ | FDP | MLFDP |
|---|---|---|
| $Posn := OldPosn$ | $Posn := NewPosn$ | $Posn := NewPosn$ |
| $t_0 := 0.1\sqrt[d]{A}$ | $t_0 := k$ | $t_0 := k_l$ |
| $k := \sqrt[d]{A/|V|}$ | $k := \frac{1}{|E|}\sum_{e\in E}||e||$ | $k_L := \frac{1}{|E|}\sum_{e\in E}||e||$ |
| | | $k_l := \sqrt{\frac{4}{7}}\cdot k_{l+1} \qquad l = 0,\ldots,L-1$ |

Thus for FDP$_0$ (the original FR algorithm) the updating is based on vertex positions at the start of the loop rather than their current positions (see §2.3.1). The initial temperature is then given by $t_0 = 0.1\sqrt[d]{A}$ where $A$ is the area of the initial layout and $d$ $(= 2, 3)$ the dimension of the layout we wish to compute. Fruchterman & Reingold actually suggest $t_0$ as one tenth of the width of the drawing but since we generate the initial positions by random placement in the region $[0,1]^d$, for 2D drawings this amounts to the same thing. Similarly the original scheme uses $k = \sqrt{A/|V|}$ but for 3D drawings this should naturally be $k = \sqrt[3]{A/|V|}$ (where $A$ is then the volume of the region) to be dimensionally correct.

We compare the three algorithms by looking at how close each can come to some notionally 'optimal' layout in a given time. Although it is impossible to define an optimal drawing of any graph (because this is very much a subjective choice), nonetheless studies with real users have indicated that 'reducing the number of edge crossing is by far the most important aesthetic', [16], and we use this measure. Furthermore in order to compare different drawing algorithms it is necessary to bear in mind that for optimisation schemes such as these, typically the longer an algorithm is allowed for refinement, the better the layout it is likely to achieve. It is therefore insufficient to choose fixed parameter settings and compare results since the runtimes of the different algorithms are likely to be very different. We thus compare the algorithms over a range of different runtimes and fortunately, as described in §2.3.5, the cooling schedule, and in particular the cooling rate, $\lambda$, allows us an easy method to do this.

To assess a given algorithm then, we measure the runtime and solution quality (number of edge crossings) for a chosen group of problem instances and for a variety of values of $\lambda$. For problem instance $p$, at cooling rate $\lambda$, this gives a pair, $Q_{\lambda,p}$, the solution quality found, and $T_{\lambda,p}$, the runtime. We then normalise the runtime values and average over all problem instances to give a single data point of averaged solution quality, $Q_\lambda$, and runtime, $T_\lambda$, for a given cooling rate $\lambda$. By using several cooling rates, $\lambda$, we can then plot $Q_\lambda$ against $T_\lambda$ to give an indication of algorithmic performance over those instances.

Typically one might think of normalising solution quality by dividing the results by the quality of the optimal (or best known) solution, e.g. as in [20]. However all of the test graphs in this section are known to be planar and so an optimal layout (at least in terms of the performance measure) contains no edge crossings (i.e., quality 0). An alternative normalisation could be the number of

C. Walshaw, *Multilevel Force-Directed Drawing*, JGAA, 7(3) 253–285 (2003)271

edges in each graph, $|E|$, or even the number of edges squared since presumably $|E|^2$ is the maximum possible number of edge crossings. However we have not used either of these and, since the graphs in each subclass do not exhibit too much variation in size, we do not normalise the solution quality. The time normalisation is more simple and is calculated by $T_{\lambda,p}/T_p^A$ where $T_p^A$ is the runtime on an instance $p$ for some well known reference algorithm, A. In this case we use A = FDP$_0$ with $\lambda = 0.9$.

To summarise then, for a set of problem instances $P$, we plot averaged solution quality $Q_\lambda$ against averaged normalised runtime $T_\lambda$ for a variety of cooling rates, $\lambda$, and where:

$$Q_\lambda = \frac{1}{|P|} \sum_{p \in P} Q_{\lambda,p} \quad , \quad T_\lambda = \frac{1}{|P|} \sum_{p \in P} \frac{T_{\lambda,p}}{T_p^A}.$$

The particular cooling rates that we used for the tests shown here were

$$\lambda = 0.5, 0.8, 0.9, 0.95, 0.98, 0.99, 0.995, 0.998, 0.999$$

for MLFDP and additionally (because of the factor of two runtime overhead for MLFDP, §2.5) $\lambda = 0.9995$ for FDP$_0$ and FDP. In each case the runtime measurement includes reading in the problem, output of the solution and any initialisation required including an initial solution construction algorithm for the single-level local search schemes. It does not, however, include the time to count the edge crossings which forms no part of the algorithm and is only used here as a post-processed performance measure.

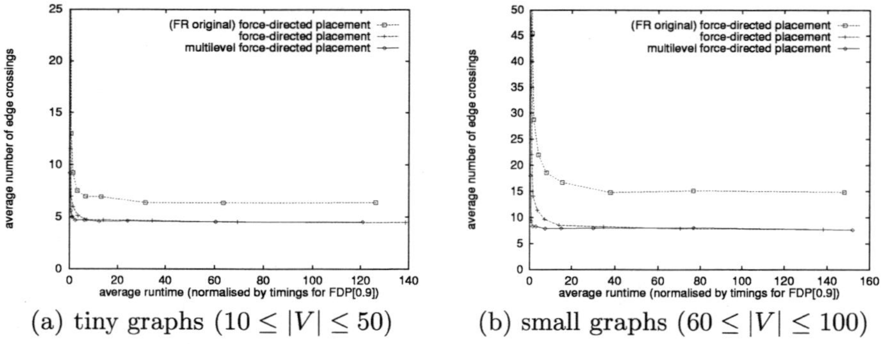

(a) tiny graphs ($10 \leq |V| \leq 50$)     (b) small graphs ($60 \leq |V| \leq 100$)

Figure 6: Plots of algorithmic behaviour on the random graphs

Figures 6 & 7 show the results on the two test suites and illustrate graphically the behaviour of the three algorithms. First of all we can see that the solution quality for FDP, our version of the FR algorithm, is far better than that for the original, FDP$_0$. The parameter settings certainly contribute to this improvement and hence this is a slightly unfair test for the original FR algorithm; firstly because Fruchterman & Reingold did not give complete parameter

C. Walshaw, *Multilevel Force-Directed Drawing*, JGAA, 7(3) 253–285 (2003)272

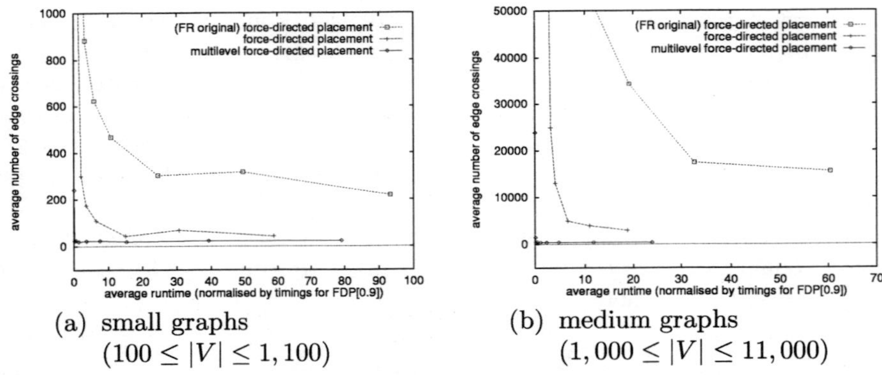

(a)  small graphs
$(100 \leq |V| \leq 1,100)$

(b)  medium graphs
$(1,000 \leq |V| \leq 11,000)$

Figure 7: Plots of algorithmic behaviour on the mesh-based graphs

settings and secondly because, even for those that they did give, our parameter settings were partially tuned on this set of problem instances and hence are likely to be more appropriate. On the other hand extensive testing of different parameter combinations revealed that by far the most significant contribution to the difference between FDP and $FDP_0$ was the fact that the positions are updated continuously rather than at the end of every outer iteration (see §2.3.1 for more details).

Next, comparing the single-level FDP algorithm with the multilevel version, we see that MLFDP can also significantly enhance the scheme and that the benefits increase with graph size. This is perhaps not a surprise since there is a much greater potential for global tangling in large graphs (rather than those with less than 50 vertices) and hence the multilevel scheme is more likely to be of assistance. Furthermore the larger the graph, the more coarsening, and hence the more refinement at different levels, takes place.

Looking at the curves in more detail in fact we see that for the random graphs, Figure 6, the MLFDP & FDP algorithms appear to have approximately the same asymptotic limit in solution quality. However the MLFDP curves bottom out more quickly.

The mesh-based graphs demonstrate this difference even more graphically in Figure 7 (where note that we have offset the $x$-axis away from 0 to avoid confusion with the MLFDP curve). MLFDP reaches its asymptotic limit almost immediately (for $\lambda \approx 0.9$) and it is not clear whether FDP can ever reach the same limit even after excessive runtimes (e.g. the 9,000 or so iterations represented by the final point on the FDP curves). Furthermore, even though the asymptotic limits for MLFDP & FDP do not seem that different, a look at individual layouts reveals that the single-level algorithm has not really untangled the graph properly. Of course the mesh-based test graphs are well structured locally and so we do not claim that the results would necessarily carry over to other graph types. On the other hand with such structure one might imagine that this sort of graph should be the *easiest* for a single-level force-directed

C. Walshaw, *Multilevel Force-Directed Drawing*, JGAA, 7(3) 253–285 (2003)273

approach to handle and that $FDP_0$ & FDP should be at their best for this suite.

Figure 7(a) demonstrates one further point; since the schemes are not directly trying to minimise the number of edge crossings, it is possible for the quality to depreciate rather than improve monotonically.

Finally the plots also suggest that the approximate runtime factor of two, suggested in §2.5, is fairly good. The final point on the FDP curve corresponds to $\lambda = 0.9995$ or 9,210 iterations whilst the final point on the MLFDP corresponds to $\lambda = 0.999$ or 4,604 iterations per level and, as can be seen, in all four plots these two points are fairly close together. Note that this analysis does not apply to $FDP_0$ because the initial temperature, $t_0$, is different.

## 3.3  Runtime complexity testing

From here on the tests are carried out at fixed cooling rate $\lambda = 0.9$ which we have found to be a good compromise between solution quality and runtime since it is close to the turning point in the MLFDP curves of Figures 6 & 7.

With this parameter fixed we then tested algorithmic complexity by comparing runtime against graph size using a set of 20 dual graphs, dime01, ..., dime20, ranging in size from $|V| = 1,095$ to $|V| = 224,843$ (note that some of these are used under different names in [19] and Laplace.0 $\equiv$ dime11, Laplace.2 $\equiv$ dime13, ..., Laplace.9 $\equiv$ dime20). They are somewhat unrepresentative, but of interest for complexity testing because each mesh is formed from the previous one using mesh refinement and so the underlying geometry is unchanged. Also, because they are duals of triangular meshes (as is 516 in §3.1), the average vertex degree approximately 3 which means that the number of edges for each graph scales linearly with the number of vertices. They are planar, but once again difficult for a spring-based placement method to draw because of the high variation in graph density (see §3.4 for further details). Two of them, dime01 & dime06, are also used in the mesh test suite (listed in Table 2 in §3.2), whilst another, dime20, is shown in §3.4.

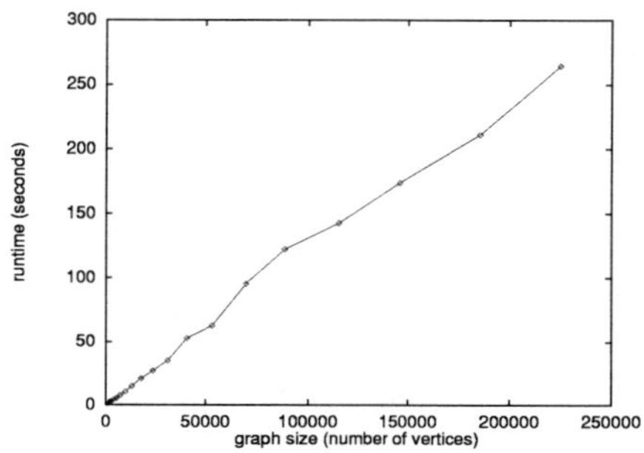


Figure 8: A plot of runtime against graph size

C. Walshaw, *Multilevel Force-Directed Drawing*, JGAA, 7(3) 253–285 (2003)274

Figure 8 shows a plot of runtime against graph size and, as can clearly be seen, the complexity is almost linear confirming one of the conclusions in §2.5. Furthermore, the runtime for the largest graph, dime20, of nearly a quarter of a million vertices is only 264 seconds.

## 3.4 Further examples

In this final results section we highlight the multilevel scheme with some interesting specific layouts that it has produced (including some for graphs which have a known layout). We discuss each example graph in a little more detail in the following sections but Table 3 gives a summary in the form of a list of the graphs, their sizes ($|V|$ & $|E|$), the maximum, minimum & average degree of the vertices, the time that the multilevel algorithm required to produce a layout (and whether it was computed in 2D or 3D) and a short description. Although all of the layouts were produced automatically, for the 3D examples, included to illustrate further points, the viewpoint has been selected manually by rotating the final drawing (the choice of optimal viewpoints is itself a subject for research, e.g. [22]).

Table 3: A summary of the illustrated graphs

| graph | size | | degree | | | placement | graph type |
| | $|V|$ | $|E|$ | max | min | avg | time (secs.) | |
|---|---|---|---|---|---|---|---|
| c-fat500-10 | 500 | 46627 | 188 | 185 | 186.5 | 5.6 (2D) | random clique test |
| 4970 | 4970 | 7400 | 3 | 2 | 3.0 | 6.4 (2D) | 2D dual |
| 4elt | 15606 | 45878 | 10 | 3 | 5.9 | 24.3 (2D) | 2D nodal |
| finan512 | 74752 | 261120 | 54 | 2 | 7.0 | 363.8 (2D) | linear programming |
| dime20 | 224843 | 336024 | 3 | 2 | 3.0 | 264.3 (2D) | 2D nodal |
| data | 2851 | 15093 | 17 | 3 | 10.6 | 6.6 (3D) | 3D nodal |
| add32 | 4960 | 9462 | 31 | 1 | 3.8 | 12.5 (3D) | 32-bit adder |
| sierpinski10 | 88575 | 177147 | 4 | 2 | 4.0 | 136.7 (3D) | 2D 'fractal' |
| mesh100 | 103081 | 200976 | 4 | 2 | 3.9 | 431.1 (3D) | 3D dual |

**c-fat500-10:** Despite the suggestion that FDP algorithms are best suited to sparse graphs, §2.5, Figure 9(a) shows a dense regular graph (originally generated to test algorithms for the maximum clique problem). The 2D layout took around 6 seconds to compute and demonstrates that here the MLFDP algorithm has captured the symmetries nicely.

**4970:** The next example, Figure 10(a), is a planar dual graph derived from a mesh (also used in the mesh-based test suite, §3.2), originally constructed to highlight a problem in the mesh generator which created it. In fact by most definitions this would be considered a very poor mesh as the triangles become extremely long and thin towards the bottom left hand corner. Figure 10(b) shows the (2D) layout calculated by the MLFDP algorithm and is useful in that, by trying to equalise the edge lengths, the

C. Walshaw, *Multilevel Force-Directed Drawing*, JGAA, 7(3) 253–285 (2003)275

drawing has actually revealed far more of the graph than was originally shown. This layout took around 6 seconds to compute.

**4elt:** A far more challenging task for any graph-drawing algorithm which seeks to equalise edge lengths is shown in Figure 11 (also showing the detail at the centre of the mesh). This is a planar nodal graph (a larger and more complex version of the 3elt graph used in the mesh-based test suite, §3.2) which represents the fluid around a 4 element airfoil. However, because the mesh has been created to study fluid behaviour close to the airfoil, the mesh exhibits extreme variations in nodal density and a far-field (the outer border of the mesh) containing very few edges. Figure 12(a) shows the (2D) layout generated by the MLFDP algorithm and illustrates some of the difficulties. Firstly the original outer border has become the smallest of the holes whilst the outer border of the new layout is actually the perimeter of one of the original holes. Furthermore, the perimeters of the other holes exhibit buckling and folding as too many vertices have to be crammed into a space constrained in size by the rest of the graph. Possibly we could eliminate some of this folding if we increased the strength of the repulsive forces, but the layout is nonetheless fairly good and arguably shows the whole of the graph at a single resolution better than the original layout. Figure 12(b) shows some of the folding in more detail and demonstrates that the micro structure is well captured. The runtime of the MLFDP algorithm for 4elt was around 24 seconds.

**finan512:** Perhaps the centre-piece of these examples is taken from a linear programming matrix with around 75,000 vertices and for which no existing layout is known. Figure 13(a) shows the highly illuminating layout found by the MLFDP algorithm; the graph is revealed to have a fairly regular structure and consists of a ring with 32 'handles' each of which has a number of fronds protruding. Figure 13(b) then shows a detailed view of one of the handles. It was an extremely useful picture from the point of view of partitioning the graph because it explained why there are good natural partitions of the graph (provided that the ring is cut between the handles). This 2D layout took about 6 minutes to compute.

Note that for this particular drawing we used a 2D layout (although a 3D layout looks identical from the right viewpoint) and this example illustrates well the memory problems that can arise with the grid based simplification of repulsive forces. As explained in §2.4 this modification divides the region into square or cube shaped cells with dimensions equal to some multiple of $k$. If a 3D layout is chosen and the ring happens by chance to more or less align itself with one of the $x$, $y$ or $z$ axes, then a box containing the graph is relatively flat and so the number of grid cells is not unreasonable. In the worst case however, if the ring happens to lie diagonally across all three axes then the box containing the graph will be cube shaped and the number of grid cells (most of which are empty) enormous relative to $|V|$. This reinforces our suggestion of using sparse

C. Walshaw, *Multilevel Force-Directed Drawing*, JGAA, 7(3) 253–285 (2003)276

data technology to only allocate memory for non-empty grid cells.

**dime20:** This is the largest graph which we have tested (also included as part of the complexity testing, §3.3, and, in smaller versions, as part of the mesh-based test suite, §3.2). Once again the original mesh, Figure 14(a), exhibits extreme variations in mesh density although it is perhaps easier to draw than 4elt. The runtime to calculate the layout, shown in Figure 14(b), for this huge graph was only 264 seconds (i.e., less than $4\frac{1}{2}$ minutes).

**data:** This is the first 3D layout we have shown and illustrates some interesting points. Figure 15(a) shows the original nodal graph which despite the appearance of being 2D is actually a segment of the thin shell of some aeronautical body. Figure 15(b) shows the 3D layout computed by the MLFDP algorithm which, despite looking nothing like the original, demonstrates some very interesting features not least of which are the three 'panels' only weakly connected to the rest of the mesh. Until seeing this layout we had no idea of the existence of such 'panels' – the original layout certainly gives no hint of them – although they could have considerable impact on any graph-based algorithm. This layout took around 6 seconds to compute.

**add32:** The next graph is a representation of an electronic circuit, a 32-bit adder, for which we do not know of an existing layout. Figure 16(a) shows the results of the 3D MLFDP algorithm whilst Figure 16(b) shows a detail of the micro structure. Although the graph is not a tree (because of the existence of loops) the placement has clearly demonstrated its tree like nature with many outlying branches or fronds. The 3D layout took about 12 seconds to compute.

**sierpinski10:** Figure 17(a) shows the original layout of sierpinski10, a self-similar 'fractal' type structure, constructed by splitting equilateral triangles of the previous graph in the series into four (two smaller examples, sierpinski06 & sierpinski08, are used in §3.2). This is a challenging problem for the drawing algorithms because of the large holes (so that repulsive forces do not act as uniformly as in a mesh-derived graph such as 4970). Here we have chosen to draw a 3D layout, shown in Figure 17(b), despite the fact that the original graph is planar. This helps to prevent vertices from becoming trapped in local optima (as discussed by Fruchterman & Reingold, [7]) since repulsive forces need not push vertices through groups of other vertices. The 3D layout adds approximately an additional 50% time penalty (as might be expected) but the runtime is still less than $2\frac{1}{2}$ minutes. Finally note that the bottom left-hand corner is not compacted in on itself, it is merely bent backwards along the line of vision; rotating the picture reveals this corner but hides other details. It is unfortunate that the drawing procedure has not managed to map the graph to a flat plane, but with an interactive visualisation tool this does not matter too much.

C. Walshaw, *Multilevel Force-Directed Drawing*, JGAA, 7(3) 253–285 (2003)277

**mesh100:** The final graph in this section is one of the largest that we have experimented with, over 100,000 vertices, and illustrates one of the problems that any graph-drawing algorithm faces. The graph is the dual of a 3D tetrahedral solid mesh and as such, with none of the face information that exists in the mesh, it is very difficult to draw meaningfully. Even in the original layout, Figure 18(a), 3D solid objects are seen to be very difficult to draw with a graph. Figure 18(b) shows the 3D MLFDP layout which took just over 7 minutes to compute. It suffers from the same problems as the original although it is splayed out because of the repulsive forces; however the symmetry is captured nicely.

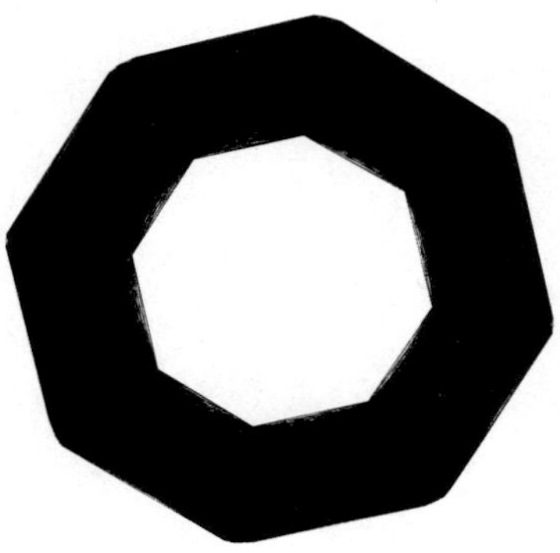

Figure 9: The graph c-fat500-10

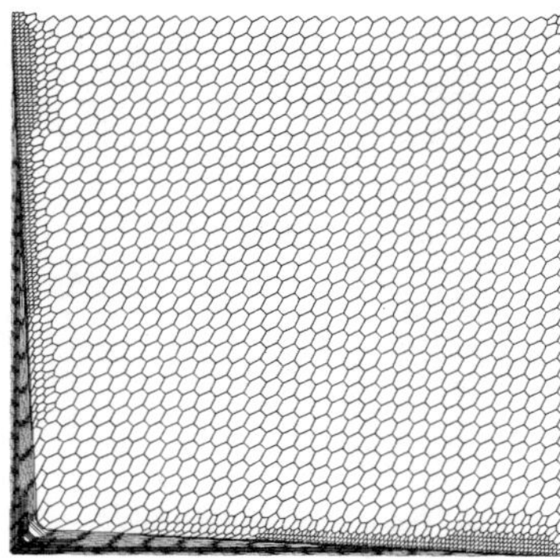

(a) original layout as derived from the mesh

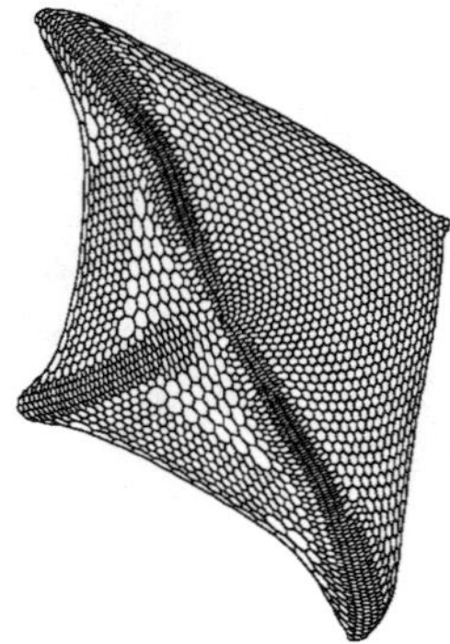

(b) the layout computed by multilevel placement

Figure 10: The graph 4970

C. Walshaw, *Multilevel Force-Directed Drawing*, JGAA, 7(3) 253–285 (2003)278

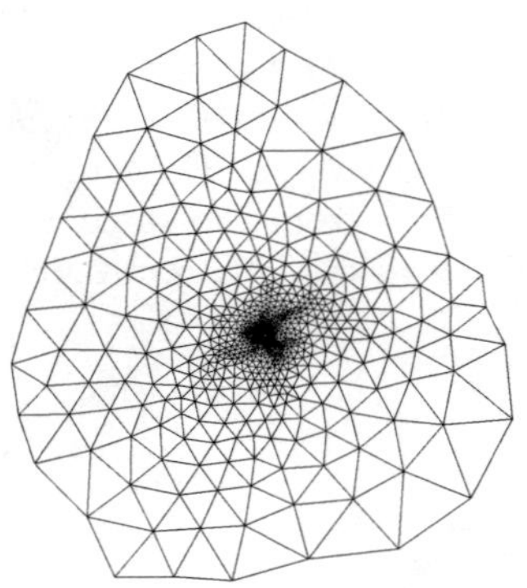

(a) original layout as derived from the mesh

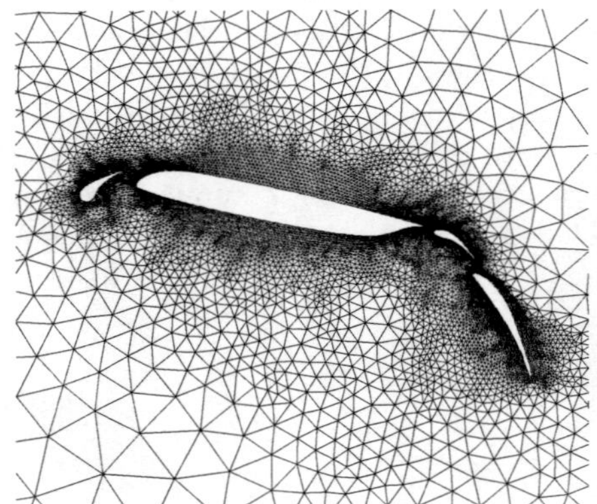

(b) detail of central region

Figure 11: The graph 4elt

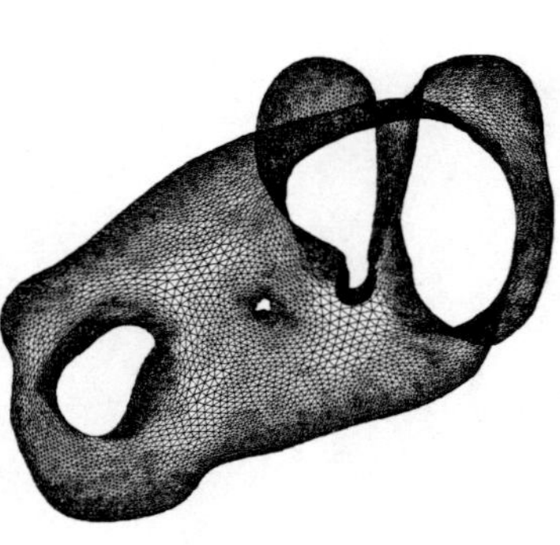

(a) the layout computed by multilevel placement

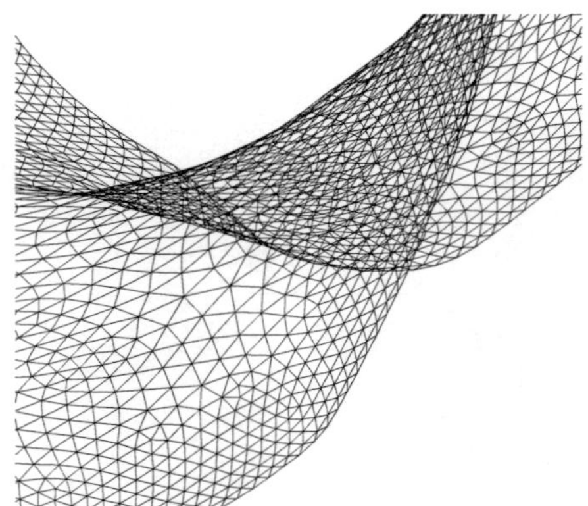

(b) detail of the folding

Figure 12: The graph 4elt

C. Walshaw, *Multilevel Force-Directed Drawing*, JGAA, 7(3) 253–285 (2003)279

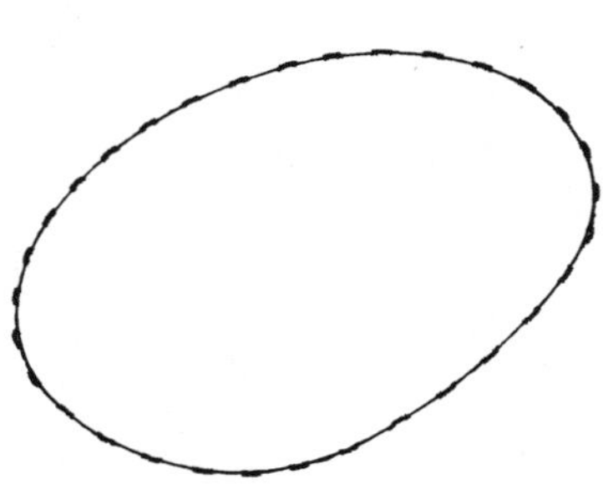

(a) the layout computed by multilevel placement

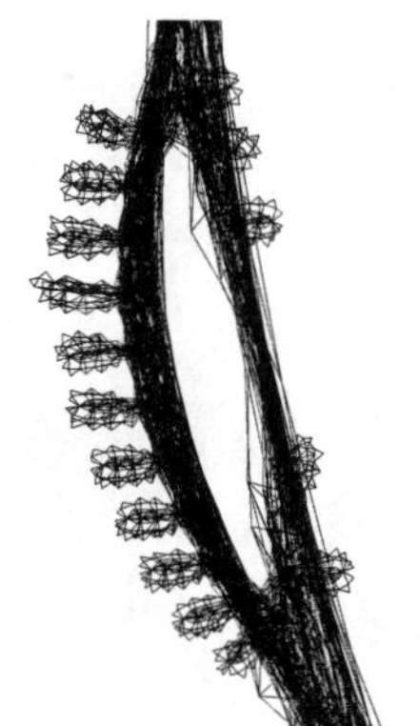

(b) detail of a "handle"

Figure 13: The graph finan512

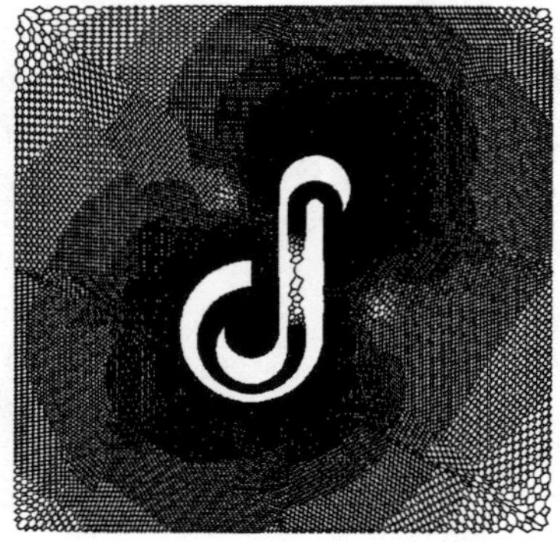

(a) original layout as derived from the mesh

(b) the layout computed by multilevel placement

Figure 14: The graph dime20

C. Walshaw, *Multilevel Force-Directed Drawing*, JGAA, 7(3) 253–285 (2003)280

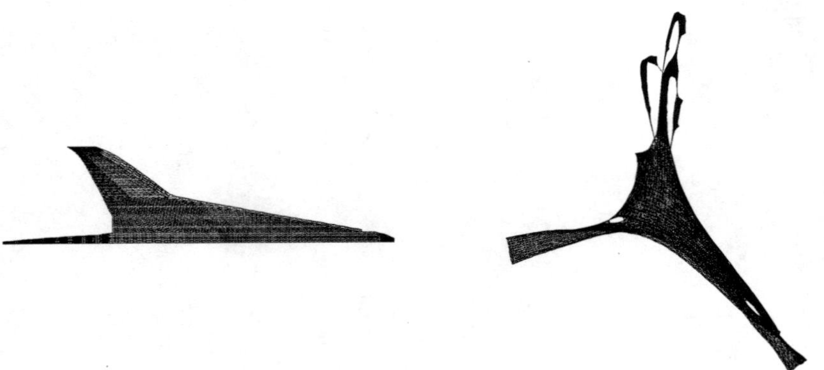

(a) original layout as derived from the mesh

(b) the layout computed by multilevel placement

Figure 15: The graph data

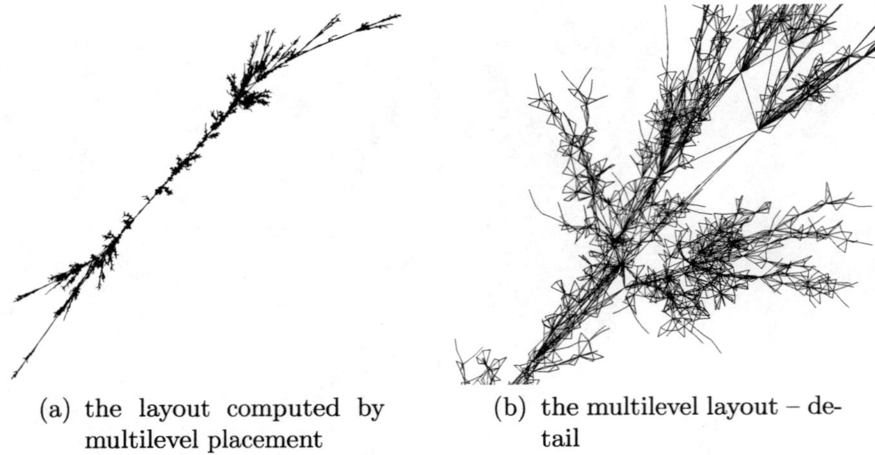

(a) the layout computed by multilevel placement

(b) the multilevel layout – detail

Figure 16: The graph add32

C. Walshaw, *Multilevel Force-Directed Drawing*, JGAA, 7(3) 253–285 (2003)281

(a) original layout

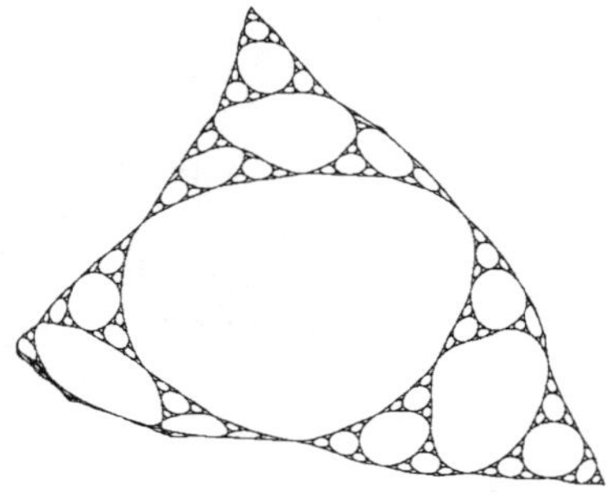

(b) the layout computed by multilevel placement

Figure 17: The graph sierpinski10

(a) original layout as derived from the mesh

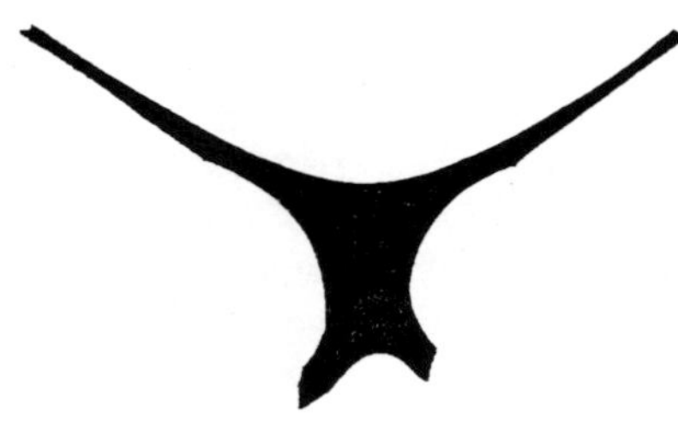

(b) the layout computed by multilevel placement

Figure 18: The graph mesh100

C. Walshaw, *Multilevel Force-Directed Drawing*, JGAA, 7(3) 253–285 (2003)282

# 4   Summary and further research

We have described a multilevel algorithm for force-directed graph-drawing. The algorithm does not actually operate simultaneously on multiple levels of the graph (as, for example, a multigrid algorithm might) but instead, inspired by the multilevel partitioning paradigm, refines the layout at each level and then extends the result onto the next level down. The algorithm is fast, e.g. for sparse graphs the runtime is about 12 seconds for a 2D layout of up to around 10,000 vertices and about 5-7 minutes for 75-100,000 vertices. At the time that the algorithm was originally devised (2000) this was an order of magnitude faster than existing single-level implementations of force-directed placement (e.g. 135 seconds for a 1,000 vertex sparse planar mesh-based graph in [18, pp. 421]; even taking into account the fact that the machine used for this calculation was notionally 8 times slower this is still 16 times slower than similar examples in Table 2 which took around 1 second). It also broadens the scope of physically based graph-drawing algorithms by imparting a more global element to the layout and seems to work robustly on a range of different graphs. Furthermore, although the number of coarse graphs is typically $O(\log_2 N)$, it only adds an approximate factor of two runtime overhead to the force-directed algorithm despite considerably enhancing the results. Finally it has sometimes been suggested that it is unnecessary to draw large graphs as the human eye can not distinguish more than about 500 vertices. However examples such as finan512 in §3.4 contradict this and indicate that graphs need to be drawn at the level of the structure contained within them, although this may suggest that it is fruitless to test drawing algorithms on very large random graphs (i.e., with no structure).

So far we have tested the algorithm on a number of different graphs including several derived from unstructured meshes which tend to be relatively homogeneous in both vertex degree and local adjacency patterns. An obvious source of further research is to test the technique on graphs arising from different areas (e.g. models of social or communications networks or the internet). Our algorithm also allows vertex weights and although we have only tested this in the context of the multilevel procedure, its use with weighted graphs might provide further interesting insights. In addition we believe, partly because of our experience in dynamic repartitioning algorithms, that the multilevel process is well suited to handling dynamically changing graphs and this looks to be a fruitful topic for future research. We have not addressed disconnected graphs but feel that this requires only minor modifications. Finally we suspect that further work on some of the parameters of the algorithm would enhance its robustness and efficiency. In particular the calculation of the natural spring length $k$ seems almost too simple to be effective.

We have not particularly tried to address graphs for which the technique might not work. It is likely that very dense graphs, or even those such as mesh100 which have a dense substructure, are never going to be good candidates for any graph-drawing algorithm, and ours is no exception. It is also likely that graphs of small diameter may not particularly suit the coarsening process (see

C. Walshaw, *Multilevel Force-Directed Drawing*, JGAA, 7(3) 253–285 (2003)283

§2.5) although it might be possible to develop modifications to the algorithm which could deal with hubs or star graphs (e.g. by contracting the whole star in one step). In summary, however, we believe that the multilevel process can accelerate and enhance FDP algorithms for a range of useful graphs and further testing on different types of graph is an important subject for further research.

## Acknowledgements

The author would like to acknowledge Yehuda Koren for interesting discussions relating to this work and thank the anonymous referees for their helpful comments.

C. Walshaw, *Multilevel Force-Directed Drawing*, JGAA, 7(3) 253–285 (2003)284

# References


[1] R. Davidson and D. Harel. Drawing Graphs Nicely using Simulated Annealing. *ACM Trans. Graphics*, 15(4):301–331, 1996.

[2] G. Di Battista, P. Eades, R. Tamassia, and I. G. Tollis. *Graph Drawing: Algorithms for the Visualization of Graphs*. Prentice-Hall, Englewood Cliffs, NJ, 1998.

[3] C. A. Duncan, M. T. Goodrich, and S. G. Kobourov. Planarity-Preserving Clustering and Embedding for Large Planar Graphs. In J. Kratochvíl, editor, *Proc. 7th Intl. Symp. Graph Drawing*, volume 1731 of *LNCS*. Springer, Berlin, 1999.

[4] P. Eades. A Heuristic for Graph Drawing. *Congr. Numer.*, 42:149–160, 1984.

[5] P. Eades and Q. Feng. Multilevel Visualization of Clustered Graphs. In *Proc. 4th Intl. Symp. Graph Drawing*, volume 1190 of *LNCS*, pages 101–112. Springer, Berlin, 1996.

[6] P. Eades, Q. Feng, X. Lin, and H. Nagamochi. Straight-Line Drawing Algorithms for Hierarchical Graphs and Clustered Graphs. Tech. rep. 98-03, Dept. Comp. Sci., Univ. Newcastle, Callaghan 2308, Australia, 1998.

[7] T. M. J. Fruchterman and E. M. Reingold. Graph Drawing by Force-Directed Placement. *Software — Practice & Experience*, 21(11):1129–1164, 1991.

[8] P. Gajer, M. T. Goodrich, and S. G. Kobourov. A Multi-dimensional Approach to Force-Directed Layouts of Large Graphs. In J. Marks, editor, *Graph Drawing, 8th Intl. Symp. GD 2000*, volume 1984 of *LNCS*, pages 211–221. Springer, Berlin, 2001.

[9] R. Hadany and D. Harel. A Multi-Scale Algorithm for Drawing Graphs Nicely. *Discrete Applied Mathematics*, 113(1):3–21, 2001.

[10] D. Harel and Y. Koren. A Fast Multi-Scale Algorithm for Drawing Large Graphs. *J. Graph Algorithms Appl.*, 6(3):179–202, 2002.

[11] D. Harel and Y. Koren. Graph Drawing by High-Dimensional Embedding. In S. G. Kobourov and M. T. Goodrich, editors, *Proc. Graph Drawing, 10th Intl. Symp. GD 2002*, volume 2528 of *LNCS*, pages 207–219. Springer, Berlin, 2002.

[12] B. Hendrickson and R. Leland. A Multilevel Algorithm for Partitioning Graphs. In S. Karin, editor, *Proc. Supercomputing '95, San Diego*. ACM Press, New York, 1995.

[13] T. Kamada and S. Kawai. An Algorithm for Drawing General Undirected Graphs. *Information Process. Lett.*, 31(1):7–15, 1989.

C. Walshaw, *Multilevel Force-Directed Drawing*, JGAA, 7(3) 253–285 (2003)285

[14] Y. Koren, L. Carmel, and D. Harel. ACE: A Fast Multiscale Eigenvector Computation for Drawing Huge Graphs. In *Proc. IEEE Symp. Information Visualization (InfoVis'02)*, pages 137–144. IEEE, Piscataway, NJ, 2002.

[15] C. H. Papadimitriou and K. Steiglitz. *Combinatorial Optimization: Algorithms and Complexity*. Prentice-Hall, Englewood Cliffs, NJ, 1982.

[16] H. Purchase. Which Aesthetic has the Greatest Effect on Human Understanding? In G. Di Battista, editor, *Proc. 5th Intl. Symp. Graph Drawing*, volume 1353 of *LNCS*, pages 248–261. Springer, Berlin, 1997.

[17] R. Sablowski and A. Frick. Automatic Graph Clustering. In *Proc. 4th Intl. Symp. Graph Drawing*, volume 1190 of *LNCS*, pages 395–400. Springer, Berlin, 1996.

[18] D. Tunkelang. JIGGLE: Java Interactive General Graph Layout Environment. In S. H. Whitesides, editor, *Proc. 6th Intl. Symp. Graph Drawing*, volume 1547 of *LNCS*, pages 413–422. Springer, Berlin, 1998.

[19] C. Walshaw. A Multilevel Algorithm for Force-Directed Graph Drawing. In J. Marks, editor, *Graph Drawing, 8th Intl. Symp. GD 2000*, volume 1984 of *LNCS*, pages 171–182. Springer, Berlin, 2001.

[20] C. Walshaw. Multilevel Refinement for Combinatorial Optimisation Problems. (To appear in Annals Oper. Res.; originally published as Univ. Greenwich Tech. Rep. 01/IM/73), 2001.

[21] C. Walshaw and M. Cross. Mesh Partitioning: a Multilevel Balancing and Refinement Algorithm. *SIAM J. Sci. Comput.*, 22(1):63–80, 2000. (originally published as Univ. Greenwich Tech. Rep. 98/IM/35).

[22] R. J. Webber. *Finding the Best Viewpoints for Three-Dimensional Graph Drawings*. PhD thesis, Dept. Comp. Sci., Univ. Newcastle, Callaghan 2308, Australia, 1998.

# Journal of Graph Algorithms and Applications

http://jgaa.info/

vol. 7, no. 3, pp. 287–303 (2003)

# Finding Shortest Paths With Computational Geometry

*Po-Shen Loh*

California Institute of Technology
http://www.caltech.edu/
po@caltech.edu

## Abstract

We present a heuristic search algorithm for the $\mathbb{R}^d$ Manhattan shortest path problem that achieves front-to-front bidirectionality in subquadratic time. In the study of bidirectional search algorithms, front-to-front heuristic computations were thought to be prohibitively expensive (at least quadratic time complexity); our algorithm runs in $O(n \log^d n)$ time and $O(n \log^{d-1} n)$ space, where $n$ is the number of visited vertices. We achieve this result by embedding the problem in $\mathbb{R}^{d+1}$ and identifying heuristic calculations as instances of a dynamic closest-point problem, to which we then apply methods from computational geometry.

Communicated by Joseph S.B. Mitchell: submitted October 2002; revised June 2003.

Research supported by Axline and Larson fellowships from the California Institute of Technology.

Po-Shen Loh, *Finding Shortest Paths*, JGAA, 7(3) 287–303 (2003)     288

# 1   Introduction

Let $\mathcal{E}$ be a finite set of axis-parallel line segments in $\mathbb{R}^d$ intersecting only at endpoints, let $\mathcal{V}$ be the set of all endpoints in $\mathcal{E}$, and let $\mathcal{G}$ be the graph defined by $\mathcal{V}$ and $\mathcal{E}$. The Manhattan shortest path problem is to find the shortest path in $\mathcal{G}$ between two distinct terminals $T_1, T_2 \in \mathcal{V}$. We will assume that the graph has already been processed and stored in memory; our algorithm is meant for applications in which multiple runs will be conducted on the same graph.

## 1.1   Notation

In this paper, we will use the following conventions.

- The symbol $XY$ denotes the Manhattan distance between two points $X$ and $Y$ in space.

- The symbol $\overline{XY}$ denotes the edge connecting two adjacent vertices $X, Y \in \mathcal{G}$.

- The symbol $\mathbf{XY}$ denotes the length of a shortest path in $\mathcal{G}$ between two vertices $X, Y \in \mathcal{V}$.

- The symbol $\mathcal{S}(\mathbf{XY})$ denotes the set of all shortest paths in $\mathcal{G}$ between two vertices $X, Y \in \mathcal{V}$.

- The function $l(\mathcal{P})$ denotes the length of a path $\mathcal{P}$ in $\mathcal{G}$. A path is a set of vertices and edges that trace a route between two vertices in a graph.

- The direct sum $\mathcal{S}(\mathbf{XY}) \oplus \mathcal{S}(\mathbf{YZ})$ forms all pairwise concatenations between paths in $\mathcal{S}(\mathbf{XY})$ and $\mathcal{S}(\mathbf{YZ})$. That is,

$$\mathcal{S}(\mathbf{XY}) \oplus \mathcal{S}(\mathbf{YZ}) = \{\mathcal{P} \cup \mathcal{Q} : \mathcal{P} \in \mathcal{S}(\mathbf{XY}), \mathcal{Q} \in \mathcal{S}(\mathbf{YZ})\}.$$

## 1.2   Previous Work

Dijkstra's algorithm solves this problem in $O(n \log n + dn)$ time, where $n$ is the number of vertices visited by an algorithm[1], which is a less expensive function of $n$ than that of our algorithm. However, the purpose of our algorithm's search heuristic is to reduce $n$ itself, thereby recapturing the additional log-power complexity factor.

Our algorithm is an extension of *A* heuristic search* [4, 5], which is a priority-first search in $\mathcal{G}$ from $T_1$ to $T_2$. A* maintains two dynamic point sets, a *wavefront* $\mathcal{W} \subseteq \mathcal{V}$ and a set $\Omega \subseteq \mathcal{V}$ of *visited* points, which begin as $\mathcal{W} = \{T_1\}$ and $\Omega = \varnothing$. The priority of a vertex $V$ is a conservative estimate of the length of a shortest path connecting the terminals via $V$; specifically:

$$\mathrm{priority}(V) = \text{``}\mathbf{T_1V}\text{''} + VT_2,$$

---

[1]Note: we do not define $n$ to be the number of edges on the shortest path.

Po-Shen Loh, *Finding Shortest Paths*, JGAA, 7(3) 287–303 (2003)        289

where "$\mathbf{T_1 T_1}$" $= 0$,

$$\text{``}\mathbf{T_1 V}\text{''} = \min_{U \in \mathcal{A}} \left\{ \mathbf{T_1 U} + UV \right\}, \tag{1}$$

and $\mathcal{A}$ is the set of all visited vertices that are adjacent to $V$. At each iteration, A* *visits* a point $P \in \mathcal{W}$ of minimal priority: it transfers $P$ from $\mathcal{W}$ to $\Omega$ and inserts $P$'s unvisited neighbors into $\mathcal{W}$. This technique is called *wave-propagation* because the wavefront $\mathcal{W}$ grows like a shock wave emitted from $T_1$.

The proofs of our algorithm's validity in Section 3 will show that "$\mathbf{T_1 V}$" approximates $\mathbf{T_1 V}$ and is in fact equal to $\mathbf{T_1 V}$ by the time $V$ is visited, hence the quotation marks. Once $T_2$ is visited, the length of the shortest path is then priority($T_2$). To recover the shortest path, *predecessor* information is stored: the predecessor of a point $V$ is defined to be the point $V^* \in \mathcal{A}$ that yields the minimum in equation (1). If there are multiple minimal points, then one of them is arbitrarily designated as the predecessor. Since $\mathcal{A}$ is a dynamic set, the predecessor of a point may vary as A* progresses, but it turns out that once a point $P$ has been visited, its predecessor is always adjacent to it along some path in $\mathcal{S}(\mathbf{T_1 P})$. Thus, after $T_2$ is visited, a shortest path from $T_1$ to $T_2$ can be obtained by tracing back through the predecessors from $T_2$ to $T_1$.

The purpose of a search heuristic is to help the wavefront grow in the direction of $T_2$. When no shortest paths from $T_1$ to $T_2$ approach $T_2$ directly, however, unidirectional search heuristics become misleading. This is not a rare occurrence; in computer chip design, one may wish to route wires between terminals that are accessible only through certain indirect channels. In order to develop a more intelligent heuristic, one must therefore explore the search space from both terminals. This motivates bidirectional search.

We concentrate here on bidirectional searches that are based on wave propagation, which expand one wavefront from each terminal. Such searches come in two types: *front-to-end* and *front-to-front*. In front-to-end searches, points are assigned heuristics without regard to information about points on the other wavefront. In contrast, the wavefronts cooperate in front-to-front searches; heuristic calculations for points on a given wavefront incorporate information from points on the opposing wavefront. This is illustrated in Figure 1.

The paper [6] by Kaindl and Kainz surveys many approaches to bidirectional search, and advocates front-to-end heuristics because of the apparently prohibitive computational complexity required for front-to-front calculations. In particular, the fastest such heuristic ran in time proportional to wavefront size, which would yield a worst-case running time that was at least quadratic. In this paper, we present a Manhattan shortest path algorithm that attains front-to-front bidirectionality for only a nominal log-power complexity cost over front-to-end search.

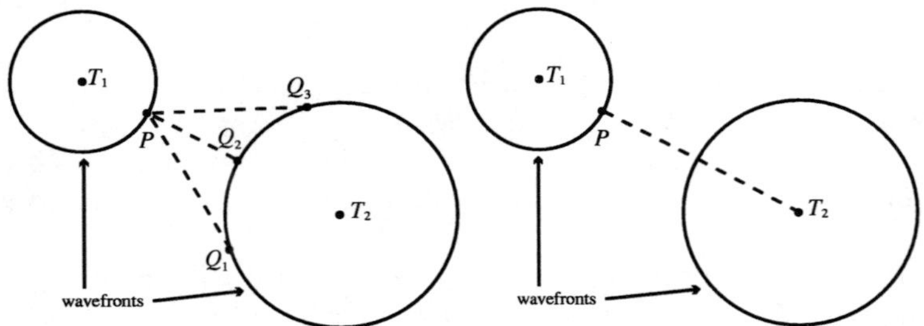


Figure 1: Two varieties of bidirectional heuristics.

## 1.3   Front-to-Front Bidirectionality

The natural front-to-front generalization of A* uses a two-parameter search heuristic that we shall call the *estimated total length*, or *ETL* for short. For any two points $P, Q \in \mathcal{G}$, the ETL function $\lambda$ is defined as follows:

$$\lambda(P, Q) = \text{``}\mathbf{T_1 P}\text{''} + PQ + \text{``}\mathbf{QT_2}\text{''},$$

where "$\mathbf{T_1 P}$" and "$\mathbf{QT_2}$" are defined along the lines of equation (1). This search algorithm, which we shall call *FF* (Front-to-Front), maintains two wavefronts, $\mathcal{W}_1$ and $\mathcal{W}_2$, and two visited sets, $\Omega_1$ and $\Omega_2$. Initially, $\mathcal{W}_1 = \{T_1\}$, $\mathcal{W}_2 = \{T_2\}$, and $\Omega_1 = \Omega_2 = \varnothing$, and at each iteration, a pair of vertices $P_1 \in \mathcal{W}_1$ and $P_2 \in \mathcal{W}_2$ with minimal $\lambda(P_1, P_2)$ is visited just as in A*. If several pairs tie with minimal ETL, then FF visits a pair with minimal $P_1 P_2$ among those tied. This helps to steer the wavefronts toward each other.

FF's termination condition is quite delicate, as it is possible for the wavefronts to intersect before a shortest path is found. Since our algorithm has a similar termination condition, we postpone the details until we present our algorithm in Figure 2.

## 2   Proposed Algorithm

Since FF finds the minimal $\lambda(P_1, P_2)$ at each iteration, it computes an accurate but costly heuristic. Our algorithm, which we name *FFF* (Fast Front-to-Front), accelerates FF by relaxing the minimality constraint on $\lambda(P_1, P_2)$. To motivate our approximation, we embed the wavefronts into $\mathbb{R}^{d+1}$ and apply computational geometry.

Po-Shen Loh, *Finding Shortest Paths*, JGAA, 7(3) 287–303 (2003)       291

## 2.1   Dimensional Augmentation

The first step is to identify our ETL minimization as a *dynamic bichromatic closest-pair* problem. To accomplish this, we enter $\mathbb{R}^{d+1}$ space, embedding our points $P \in \mathcal{W}_1 \cup \mathcal{W}_2$ as follows:

$$P(x_1, x_2, \ldots, x_d) \mapsto \begin{cases} P'(x_1, x_2, \ldots, x_d, +\sigma_1), & \text{if } P \in \mathcal{W}_1, \\ P'(x_1, x_2, \ldots, x_d, 0), & \text{if } P \text{ is } T_1 \text{ or } T_2, \\ P'(x_1, x_2, \ldots, x_d, -\sigma_2), & \text{if } P \in \mathcal{W}_2, \end{cases}$$

where

$$\sigma_k = \min_{V \in \mathcal{A}} \{ \mathbf{T_k V} + VP \},$$

and $\mathcal{A}$ is the set of vertices in $\Omega_k$ that are adjacent to $P$.

In this paper, we will use the convention that all primed points are in $\mathbb{R}^{d+1}$, all unprimed points are in $\mathbb{R}^d$, and if two points or sets in the same context have the same letter, a prime indicates passage between $\mathbb{R}^d$ and $\mathbb{R}^{d+1}$. If some point $P' \in \mathbb{R}^{d+1}$ is not in either embedded wavefront, then its unprimed form $P$ will still represent the projection of $P'$ onto its first $d$ coordinates. Finally, we will refer to the last coordinate of a point $P' \in \mathbb{R}^{d+1}$ by the shorthand $\sigma(P')$.

## 2.2   Identification and Approximation

Since we are working in a Manhattan metric, $\lambda(P_1, P_2)$ is equal to the distance between $P'_1$ and $P'_2$ in $\mathbb{R}^{d+1}$. Therefore, finding a minimal pair is equivalent to finding a closest pair between $\mathcal{W}'_1$ and $\mathcal{W}'_2$ in $\mathbb{R}^{d+1}$. At each iteration of FF, the point sets only change slightly, so each minimization is an instance of a dynamic bichromatic closest-pair problem.

Using the method of Eppstein in [3], one can solve the $\mathbb{R}^{d+1}$ Manhattan dynamic bichromatic closest-pair problem with $O(\log^{d+2} n)$ amortized time per insertion and $O(\log^{d+3} n)$ amortized time per deletion. In each of our iterations, we will need to perform at least one insertion and one deletion, so that method would give us an algorithm that ran in $O(n \log^{d+3} n)$ amortized time.

We are only looking for a search heuristic, though, so we can content ourselves with *almost-closest pairs* without compromising the validity of our algorithm. This in fact speeds up our algorithm by a factor of $\log^3 n$ and changes the complexity bounds from amortized to worst-case; in low dimensions ($d \leq 3$), it provides a significant performance boost. To accomplish this, we take a cue from the method of successive approximations: starting from a well-chosen seed point $P'_0 \in \mathbb{R}^{d+1}$, not necessarily in $\mathcal{W}'_1$ or $\mathcal{W}'_2$, we find a closest point $X'_1 \in \mathcal{W}'_1$, and then find an $X'_2 \in \mathcal{W}'_2$ that is closest to $X'_1$. We then use $X_1$ and $X_2$ instead of the $P_1$ and $P_2$ of FF. This pair may not have minimal $\lambda$, but it turns out to be close enough (cf. Theorems 4 and 5). Therefore, we only need *dynamic closest-point search*.

Po-Shen Loh, *Finding Shortest Paths*, JGAA, 7(3) 287–303 (2003) 

## 2.3  Dynamic Closest-Point Search

This problem can be reduced to that of *dynamic range search for minimum* (discussed in [7, 8, 10]) by adapting the technique [1] of Gabow, Bentley, and Tarjan in [9]. Let $\mathcal{S}'$ be a set of points in $\mathbb{R}^{d+1}$, let $m$ and $M$ be the respective minimum and maximum $(d+1)$-st coordinates in $\mathcal{S}'$, and let $P'(p_1, p_2, \ldots, p_{d+1}) \in \mathbb{R}^{d+1}$ be a *query point* with $p_{d+1} \leq m$ or $p_{d+1} \geq M$. We wish to find the point in $\mathcal{S}'$ that is closest to $P'$. The constraint on the last coordinate of $P'$ is introduced because FFF does not require the general case and this speeds up the search by a factor of $\log n$. In this section, we will only discuss the case when $p_{d+1} \leq m$; the other case can be treated similarly.

Let $\mathcal{I}$ be the set of all ordered $(d+1)$-tuples of the form $(\pm 1, \ldots, \pm 1)$, and for any $\epsilon(\epsilon_1, \ldots, \epsilon_{d+1}) \in \mathcal{I}$, let $\Omega'_\epsilon(P)$ refer to the following octant in $\mathbb{R}^{d+1}$:

$$X'(x_1, \ldots, x_{d+1}) \in \Omega'_\epsilon(P) \iff \forall i \in \{1, 2, \ldots, d+1\}, \begin{cases} x_i < p_i, & \text{if } \epsilon_i = -1, \\ x_i \geq p_i, & \text{if } \epsilon_i = +1. \end{cases}$$

Also, define the family of functions $f_\epsilon : \mathbb{R}^{d+1} \to \mathbb{R}$ for all $\epsilon \in \mathcal{I}$:

$$f_\epsilon : X'(x_1, \ldots, x_{d+1}) \mapsto \sum_{i=1}^{d+1} \epsilon_i x_i,$$

where the parameter $\epsilon$ is defined in the same way as used in $\Omega'_\epsilon(P)$. Then when $X' \in \mathcal{S}' \cap \Omega'_\epsilon(P)$, its distance from $P'$ is equal to

$$\sum_{i=1}^{d+1} |x_i - p_i| = \sum_{i=1}^{d+1} \epsilon_i (x_i - p_i) = f_\epsilon(X') - f_\epsilon(P'), \tag{2}$$

but since $p_{d+1} \leq m$, we must have $\mathcal{S}' \cap \Omega'_\epsilon(P) = \varnothing$ for all $\epsilon(\epsilon_1, \ldots, \epsilon_{d+1})$ that have $\epsilon_{d+1} = -1$. All of our computations on $X' \in \mathcal{S}' \cap \Omega'_\epsilon(P)$ in equation (2) must then turn out as

$$\sum_{i=1}^{d+1} |x_i - p_i| = \left( \sum_{i=1}^{d} \epsilon_i x_i \right) + x_{d+1} - \left( \sum_{i=1}^{d} \epsilon_i p_i \right) - p_{d+1},$$

so if we define the family of functions $g_\epsilon : \mathbb{R}^{d+1} \to \mathbb{R}$ for all $\epsilon \in \mathcal{I}$:

$$g_\epsilon : X'(x_1, \ldots, x_{d+1}) \mapsto \left( \sum_{i=1}^{d} \epsilon_i x_i \right) + x_{d+1},$$

then the distance from $P'$ to any $X' \in \mathcal{S}' \cap \Omega'_\epsilon(P)$ is $g_\epsilon(X') - g_\epsilon(P')$.

Now since the only $\Omega'_\epsilon(P)$ that contain points in $\mathcal{S}'$ are those with $\epsilon_{d+1} = +1$, we can repartition $\mathbb{R}^{d+1}$ into the $2^d$ distinct $\Theta'_\epsilon(P)$, defined as follows:

$$X'(x_1, \ldots, x_{d+1}) \in \Theta'_\epsilon(P) \iff \forall i \in \{1, 2, \ldots, d\}, \begin{cases} x_i < p_i, & \text{if } \epsilon_i = -1, \\ x_i \geq p_i, & \text{if } \epsilon_i = +1, \end{cases}$$

and the point in $\mathcal{S}'$ closest to $P'$ can be found by looking for the minimum $g_\epsilon(X') - g_\epsilon(P')$ in each of the $2^d$ distinct $\Theta'_\epsilon(P)$. Within each region, $g_\epsilon(P')$ is constant, and we can compute and store all $2^d$ distinct $g_\epsilon(X')$ whenever we insert a point $X'$ into $\mathcal{S}'$; thus, since the partition of $\mathbb{R}^{d+1}$ into $\bigcup \Theta'_\epsilon(P)$ ignores all $(d+1)$-st coordinates, we will have reduced our $\mathbb{R}^{d+1}$ dynamic closest-point problem to a $\mathbb{R}^d$ dynamic range search for minimum. According to [2, 3], range trees can solve this in $O(n \log^{d-1} n)$ space with $O(\log^d n)$ query/update time.

## 2.4   Implementation

FFF stores $\mathcal{W}'_1$ and $\mathcal{W}'_2$ in a pair of *Dynamic Closest-Point Structures*, or *DCPS's*, each of which holds a set of points in $\mathbb{R}^{d+1}$ and supports the following operations:

1. `insert(`$P'$`)`. Inserts $P'$ into the DCPS if the structure does not yet contain any point $Q'$ with $Q = P$. If it already has such a point $Q'$, then $Q'$ is replaced by $P'$ if $|\sigma(P')| < |\sigma(Q')|$; otherwise, $P'$ is ignored.

2. `delete(`$P'$`)`. Deletes $P'$ from the DCPS.

3. `query(`$P'$`)`. Let $m$ and $M$ be the minimum and maximum $(d+1)$-st coordinates in the DCPS, respectively. Then, for a point $P' \in \mathbb{R}^{d+1}$ with $\sigma(P') \leq m$ or $\sigma(P') \geq M$, this function performs a closest-point query; it returns a point $X'$ in the DCPS that is rectilinearly closest to $P'$. If several points $\{X'_i\}$ yield that minimal distance, then one with minimal $PX_i$ is returned.

The special features of our DCPS insertion and query can be added through simple modifications of implementation details, and do not affect its overall complexity.

We use a priority queue to select seed points for the successive approximations of Section 2.2. Our priority queue stores elements of the form $P'(\lambda, \delta)$, where $P' \in \mathbb{R}^{d+1}$ is the data and the ordered pair $(\lambda, \delta)$ is its associated priority. The $\lambda$ estimates the length of a shortest path through $P$ and the $\delta$ estimates how close $P$ is to $\mathcal{W}_2$. We compare priorities by interpreting the $(\lambda, \delta)$ as sorting keys; that is, $(\lambda_1, \delta_1) > (\lambda_2, \delta_2)$ if and only if $\lambda_1 > \lambda_2$ or $\lambda_1 = \lambda_2$ and $\delta_1 > \delta_2$. When we pop an element off the queue, we retrieve the one with minimal priority.

## 2.5   Pseudocode

The algorithm is presented in Figure 2.

# 3   Justification

Throughout this section, $k$ will be an index from the set $\{1, 2\}$. Also, the asterisk will be a shorthand for the predecessor of a point; for instance, $P^*$ will denote the point in $\mathbb{R}^d$ that prompted the insertion (line 43) of $P'$ into its $\mathcal{W}'_k$.

Po-Shen Loh, *Finding Shortest Paths*, JGAA, 7(3) 287–303 (2003)  294

```
1:  algorithm FFF {
2:      insert T₁′ and T₂′ into W₁′ and W₂′, respectively;
3:      MIN := ∞;
4:      push T₁′(T₁′T₂′, T₁T₂) onto priority queue;
5:      while (priority queue, W₁′, and W₂′ are all nonempty) {
6:          pop P′(λ, δ) off priority queue;
7:          if (λ ≥ MIN) { end algorithm; }
8:          if (P ∉ W₁) { return to line 5; }
9:          P₀′ := P′ with last coordinate set to zero;
10:         query:  X₁′ := point in W₁′ that is closest to P₀′;
11:         if (P ≠ X₁) { PriorityQueueInsert(P′); }
12:         if (WavefrontsCrossed(X₁, Ω₂)) { return to line 5; }
13:         Visit(X₁′, 1);
14:         query:  X₂′ := point in W₂′ that is closest to X₁′;
15:         if (WavefrontsCrossed(X₂, Ω₁)) { return to line 5; }
16:         Visit(X₂′, 2);
17:     }
18:     Assertion 1: MIN ≠ ∞ ⇔ S(T₁T₂) ≠ ∅ ⇔ S(T₁S) ⊕ S(ST₂) ⊆ S(T₁T₂);
19: }
20:
21: procedure PriorityQueueInsert(P′) {
22:     if (P ∈ Ω₂) {
23:         Assertion 2: PT₂ is known;
24:         Y′ := P′ with last coordinate set to −PT₂;
25:     }
26:     else { query:  Y′ := point in W₂′ that is closest to P′; }
27:     push P′(P′Y′, PY) into priority queue;
28: }
29:
30: function WavefrontsCrossed(X, Ω) {
31:     if (X ∈ Ω) {
32:         Assertion 3: T₁X + XT₂ is known;
33:         if (T₁X + XT₂ < MIN) { S := X;  MIN := T₁X + XT₂; }
34:         return TRUE;
35:     }
36:     else { return FALSE; }
37: }
38:
39: procedure Visit(X′, k) {
40:     Assertion 4: T_kX = |σ(X′)|;
41:     move X from W_k to Ω_k;
42:     for (all {V ∉ Ω_k : X̄V ∈ E}) {
43:         insert V′ into W_k′, where we use T_kX + XV for |σ(V′)|;
44:         if (k = 1) { PriorityQueueInsert(V′); }
45:     }
46: }
```

Figure 2: FFF pseudocode

Po-Shen Loh, *Finding Shortest Paths*, JGAA, 7(3) 287–303 (2003)          295

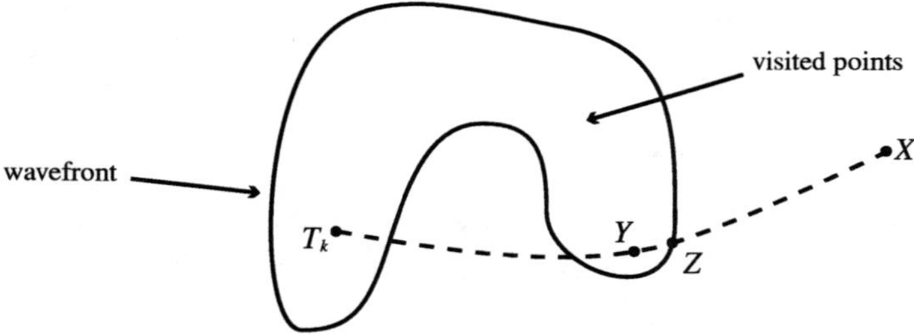


Figure 3: Lemma 1

**Lemma 1** *At the beginning of some iteration after the first, let $X$ be a point not in $\Omega_k$ and let $\mathcal{P}$ be a path in $\mathcal{G}$ connecting $T_k$ and $X$. Then there exist two points $Y, Z \in \mathcal{P}$ such that $\overline{YZ} \in \mathcal{P}$, $Y \in \Omega_k$, and $Z \in \mathcal{W}_k$.*

**Proof:** The first iteration moves $T_k$ into $\Omega_k$, so $\mathcal{P} \cap \Omega_k \neq \varnothing$ in every iteration thereafter. Let $Y$ be the point in that intersection whose distance from $X$ along edges in $\mathcal{P}$ is minimal, and let $Z \in \mathcal{P}$ be the point adjacent to $Y$ that is closer to $X$ along $\mathcal{P}$ than $Y$ was (fig. 3). From the minimality of $Y$, we know that $Z \notin \Omega_k$, so when $Y$ was inserted into $\Omega_k$ (line 41), FFF must have inserted $Z$ into $\mathcal{W}_k$ (lines 42–43). Again by the minimality of $Y$, $Z$ cannot currently be in $\Omega_k$, so this $Y$ and this $Z$ satisfy the statement of the lemma. $\qquad\square$

**Theorem 1** *In every iteration after the first, when FFF visits (lines 13, 16) a point $X \in \mathcal{W}_k$,*

$$|\sigma(X')| = \mathbf{T_k X^*} + X^*X = \min_{M \in \mathcal{W}_k} \{\mathbf{T_k M^*} + M^*M + MX\} = \mathbf{T_k X}, \qquad (3)$$

*and a path in $\mathcal{S}(\mathbf{T_k X})$ can be obtained by tracing back through predecessors from $X$.*

**Proof:** We proceed by induction.

*Base Case.* FFF begins with $\Omega_k = \varnothing$, $\mathcal{W}_k = \{T_k\}$, and $|\sigma(T'_k)| = 0$, so in the second iteration, $X^* = T_k = M^*$, $\mathbf{T_k X} = T_k X$, and $\mathbf{T_k M^*} = 0$. Now $|\sigma(X')| = \mathbf{T_k T_k} + T_k X = \mathbf{T_k X}$ and by the triangle inequality $\mathbf{T_k M^*} + M^*M + MX \geq T_k X = \mathbf{T_k X}$; our base case is true.

*Induction Step.* By line 43 and the induction hypothesis, $|\sigma(X')| = \mathbf{T_k X^*} + X^*X$, so the left equality is true. To prove the center equality, we use indirect proof; suppose, for the sake of contradiction, that

$$\mathbf{T_k X^*} + X^*X \neq \min_{M \in \mathcal{W}_k} \{\mathbf{T_k M^*} + M^*M + MX\}.$$

Then there must currently exist some point $M' \in \mathcal{W}'_k$ for which $\mathbf{T_k M^*} + M^*M + MX < |\sigma(X')|$, and $\mathbf{T_k M^*} + M^*M = |\sigma(M')|$ by line 43. Let $Q'$ be the query point (the $P'_0$ or the $X'_1$) that led to $X'$ in line 10 or 14. Since $\sigma(Q')$ and $\sigma(M')$ are not of the same sign, $Q'M' = |\sigma(Q')| + QM + |\sigma(M')|$. Combining all of our information:

$$\begin{aligned}
Q'M' &= |\sigma(Q')| + QM + |\sigma(M')| \\
&< |\sigma(Q')| + QM - MX + |\sigma(X')| \\
&\leq |\sigma(Q')| + QX + |\sigma(X')| \\
&= Q'X'.
\end{aligned}$$

Thus $X'$ is not the point in $\mathcal{W}'_k$ that is closest to $Q'$; contradiction.

Next, we prove the rightmost equality in equation (3). Let $\mathcal{P}$ be a path in $\mathcal{S}(\mathbf{T_k X})$. Since $X \in \mathcal{W}_k$ and $\mathcal{W}_k \cap \Omega_k = \varnothing$, Lemma 1 implies that there exists some point $K$ in $\mathcal{P} \cap \mathcal{W}_k$ with adjacent point $K_0 \in \mathcal{P} \cap \Omega_k$. Since both $K_0$ and $K^*$ are in $\Omega_k$ and are adjacent to $K$, they each must have attempted to insert an embedding of $K$ into $\mathcal{W}'_k$ by lines 42–43. Our DCPS insertion operation has the special property that it retains only the best embedding, so $\mathbf{T_k K^*} + K^*K \leq \mathbf{T_k K_0} + K_0 K$. Yet $K_0$ is adjacent to $K$ along $\mathcal{P} \in \mathcal{S}(\mathbf{T_k K})$, so $\mathbf{T_k K_0} + K_0 K = \mathbf{T_k K}$. Furthermore, $\overline{K^*K} \in \mathcal{E}$, so $K^*K = \mathbf{K^*K}$ and thus $\mathbf{T_k K^*} + K^*K \geq \mathbf{T_k K}$. Hence $\mathbf{T_k K^*} + K^*K = \mathbf{T_k K}$ so

$$\begin{aligned}
\min_{M \in \mathcal{W}_k} \left\{ \mathbf{T_k M^*} + M^*M + MX \right\} &\leq \mathbf{T_k K^*} + K^*K + KX \\
&= \mathbf{T_k K} + KX \\
&\leq \mathbf{T_k K} + \mathbf{KX} \\
&= l(\mathcal{P}) \\
&= \mathbf{T_k X},
\end{aligned}$$

and

$$\mathbf{T_k X} \leq \mathbf{T_k X^*} + X^*X = \min_{M \in \mathcal{W}_k} \left\{ \mathbf{T_k M^*} + M^*M + MX \right\}$$

by the center equality. Therefore, the rightmost equality is true.

Only the last claim of our theorem remains to be proven. By equation (3), $\mathbf{T_k X^*} + X^*X = \mathbf{T_k X}$, so since $\overline{X^*X} \in \mathcal{E}$, we know that $\mathcal{S}(\mathbf{T_k X^*}) \oplus \mathcal{S}(\mathbf{X^*X}) \subseteq \mathcal{S}(\mathbf{T_k X})$. Now $X^* \in \Omega_k$, so it must have been visited at some earlier time. From the induction hypothesis, a path in $\mathcal{S}(\mathbf{T_k X^*})$ can be obtained by following predecessors back from $X^*$; if we append the edge $\overline{X^*X}$ to this path, then we will have a path in $\mathcal{S}(\mathbf{T_k X})$, so we are done. $\qquad\square$

**Corollary 1** *At the start of every iteration, for every $O \in \Omega_k$, a path in $\mathcal{S}(\mathbf{T_k O})$ can be obtained by tracing back through predecessors from $O$.*

**Proof:** This follows immediately from Theorem 1. $\qquad\square$

**Corollary 2** *Assertions 2, 3, and 4 are true.*

Po-Shen Loh, *Finding Shortest Paths*, JGAA, 7(3) 287–303 (2003)      297

**Proof:** These follow from Theorem 1 and Corollary 1.      □

**Corollary 3** *Let $\mathcal{W}'_k$ be one of the embedded wavefronts at the beginning of some iteration, and let $P'$ be a point that is not on the same side of the $\sigma = 0$ hyperplane. That is, if $k = 1$, $\sigma(P')$ should be nonpositive, and if $k = 2$, $\sigma(P')$ should be nonnegative. Furthermore, suppose that $P \notin \Omega_k$ and let $Q'$ be the point in $\mathcal{W}'_k$ that is closest to $P'$. Then $QQ^* + \mathbf{Q}^*\mathbf{T_k} = |\sigma(Q')| = \mathbf{QT_k}$ and $\mathbf{PT_k} \geq PQ + \mathbf{QT_k}$.*

**Proof:** The first claim follows from the induction step of Theorem 1, since that proof only used the fact that $X$ resulted from a query by a point that was not on the same side of the $\sigma = 0$ hyperplane. To prove the second claim, observe that if we replace $X$ with $P$ in the proof of the rightmost equality in Theorem 1, we find a point $K' \in \mathcal{W}'_k$ for which $\mathbf{T_k}\mathbf{K}^* + K^*K + KP \leq \mathbf{T_k}\mathbf{P}$. Yet $|\sigma(K')| = \mathbf{T_k}\mathbf{K}^* + K^*K$ by line 43, and $|\sigma(Q')| = \mathbf{QT_k}$ by the first claim, so since $Q' \in \mathcal{W}'_k$ is closer to $P'$ than $K' \in \mathcal{W}'_k$,

$$
\begin{aligned}
Q'P' &\leq K'P', \\
|\sigma(Q')| + QP + |\sigma(P')| &\leq |\sigma(K')| + KP + |\sigma(P')|, \\
|\sigma(Q')| + QP &\leq |\sigma(K')| + KP, \\
\mathbf{QT_k} + QP &\leq \mathbf{T_k}\mathbf{K}^* + K^*K + KP \leq \mathbf{T_k}\mathbf{P}
\end{aligned}
$$

as desired.      □

**Lemma 2** *Let $X'$ be a point in $\mathcal{W}'_1$ at the start of some iteration after the first. If $\mathbf{T_1}\mathbf{X}^* + X^*X + \mathbf{XT_2} < \text{MIN}$, then the priority queue contains an element $X'(\lambda, \delta)$ with $\lambda \leq \mathbf{T_1}\mathbf{X}^* + X^*X + \mathbf{XT_2}$.*

**Proof:** We first show that when a point $X$ is visited and a neighbor $V'$ is inserted into $\mathcal{W}'_1$, the priority queue element $V'(\lambda, \delta)$ that FFF inserts (line 44) has $\lambda \leq \mathbf{T_1}\mathbf{X} + XV + \mathbf{VT_2}$.

In the call to `priorityQueueInsert`, if line 24 is used, then by Corollary 2, $\lambda = |\sigma(V')| + \mathbf{VT_2} = \mathbf{T_1}\mathbf{X} + XV + \mathbf{VT_2}$. If line 26 is used, we can apply Corollary 3 to see that $\mathbf{VT_2} \geq VY + \mathbf{YT_2} = VY + |\sigma(Y')|$, so

$$
\begin{aligned}
\mathbf{T_1}\mathbf{X} + XV + \mathbf{VT_2} &\geq \mathbf{T_1}\mathbf{X} + XV + VY + |\sigma(Y')| \\
&= |\sigma(V')| + VY + |\sigma(Y')| \\
&= V'Y' = \lambda.
\end{aligned}
$$

Thus the lemma's condition holds at the moment of insertion, and we are left to deal with the case where an element $P'(\lambda, \delta)$ is popped off the priority queue but $P'$ is not removed from $\mathcal{W}'_1$.

Note that we may pop off an element that is not in $\mathcal{W}'_1$. This happens when a point $V$ is inserted into $\mathcal{W}_1$ more than once, with different $V'$ embeddings. In

Po-Shen Loh, *Finding Shortest Paths*, JGAA, 7(3) 287–303 (2003)          298

such cases, our DCPS only retains the one with lesser $|\sigma(V')|$, although both remain in the priority queue. We wish to ignore all other $V'$ since they represent longer paths from $T_k$ to $V$; such is the purpose of line 8.

If our iteration passes line 8 but $P$ is not visited, then we may need to reinsert $P'$ into the priority queue to satisfy the lemma. If $P \neq X_1$, this job is successfully accomplished by line 11, since the above analysis of `priorityQueueInsert` applies here. The only case left to consider is when the iteration terminates via line 12 with $P = X_1 \in \Omega_2$. Now, $\mathbf{T_1 X_1^*} + X_1^* X_1 + \mathbf{X_1 T_2} = \mathbf{T_1 X_1} + \mathbf{X_1 T_2}$ by Corollary 3 since $X_1'$ resulted from a DCPS query by $P_0'$. Yet line 33 makes this at least MIN and the lemma makes no claim when $\mathbf{T_1 X_1^*} + X_1^* X_1 + \mathbf{X_1 T_2} \geq$ MIN, so we are done. $\qquad\square$

**Lemma 3** *After each iteration except the last, $\Omega_1$ has grown or the number of elements in the priority queue with $\lambda <$ MIN has decreased by at least one.*

**Proof:** If an iteration makes it to line 13, it inserts a point into $\Omega_1$, and if the iteration does not satisfy line 11, then the size of the priority queue decreases by one. Therefore, the only nontrivial case is when the iteration satisfies line 11 but fails to reach line 13; this is when $P = X_1 \in \Omega_2$. The iteration will pass lines 22 and 33, setting $\lambda = \mathbf{T_1 X_1} + \mathbf{X_1 T_2} \geq$ MIN, but by line 7, $\lambda$ was originally less than MIN, so the lemma is true. $\qquad\square$

**Lemma 4** *FFF terminates; it eventually reaches assertion 1.*

**Proof:** We have a finite graph, so Lemma 3 implies that each iteration that does not decrease the number of elements with $\lambda <$ MIN must grow $\Omega_1$ and insert finitely many elements into the priority queue. The set $\Omega_1$ cannot grow indefinitely, however, because $|\mathcal{V}| < \infty$; hence only a finite number of insertions occur. Once the priority queue runs out of elements with $\lambda <$ MIN, FFF will terminate via line 7. Therefore, we cannot loop forever. $\qquad\square$

**Theorem 2** *FFF finds a shortest path when one exists; assertion 1 is true.*

**Proof:** By Lemma 4, FFF terminates, so it suffices to show that assertion 1 is true. We begin with the first equivalence. The reverse implication is trivial, so we move on to prove the forward direction. Suppose for the sake of contradiction that $\mathcal{S}(\mathbf{T_1 T_2}) \neq \varnothing$ but FFF terminates with MIN $= \infty$. Let $\mathcal{P}$ be a path in $\mathcal{S}(\mathbf{T_1 T_2})$; lines 12 and 15 ensure that $\Omega_1 \cap \Omega_2 = \varnothing$, so since $T_1 \in \Omega_1$ and $T_2 \in \Omega_2$, Lemma 1 applies to $\mathcal{P}$. Thus both wavefronts must be nonempty. By Lemma 2, the priority queue contains an element with finite $\lambda$, so FFF could not have terminated via line 5. Yet MIN $= \infty$, so FFF could not have terminated via line 7 and we have a contradiction.

Next we prove the second equivalence. Note that $S$ is undefined until a path is found, so if $\mathcal{S}(\mathbf{T_1 S}) \oplus \mathcal{S}(\mathbf{S T_2}) \subseteq \mathcal{S}(\mathbf{T_1 T_2})$, then a shortest path exists. It remains to prove the forward implication. If $\mathcal{S}(\mathbf{T_1 T_2}) \neq \varnothing$, then from the first equivalence, MIN must be eventually be set to some finite value, at which point $S$ is defined. We must show that $\mathcal{S}(\mathbf{T_1 S}) \oplus \mathcal{S}(\mathbf{S T_2}) \subseteq \mathcal{S}(\mathbf{T_1 T_2})$. Once we reach

Po-Shen Loh, *Finding Shortest Paths*, JGAA, 7(3) 287–303 (2003)      299

this stage, we will indeed know a shortest path in $\mathcal{S}(\mathbf{T_1T_2})$ because Corollaries 1 and 3 apply to $S$.

We proceed by contradiction; suppose that FFF does not find a minimal path, terminating with MIN $> \mathbf{T_1T_2}$. Consider the situation at the beginning of the last iteration. Let $\mathcal{P}$ be a path in $\mathcal{S}(\mathbf{T_1T_2})$; by Lemma 1, we can find $X_1 \in \mathcal{W}_1 \cap \mathcal{P}$ and $Y_1 \in \Omega_1 \cap \mathcal{P}$ with $X_1Y_1 \in \mathcal{P}$. This is illustrated in Figure 4.

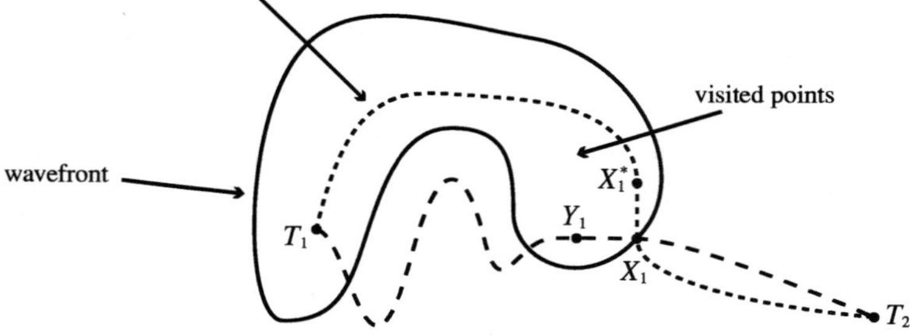


Figure 4: Theorem 2: the dashed curves represent two different shortest paths between $T_1$ and $T_2$, and $\mathcal{P}$ is the curve with longer dashes. Wavefront $\mathcal{W}_2$ is not shown.

Since $Y_1 \in \Omega_1$ and $Y_1X_1 \in \mathcal{P} \in \mathcal{S}(\mathbf{T_1T_2})$, the portion of $\mathcal{P}$ between $T_1$ and $X_1$ is a shortest path, and hence the embedding $X_1'$ with $|\sigma(X_1')| = \mathbf{T_1X_1}$ is in $\mathcal{W}_1'$. Then $\mathbf{T_1X_1^*} + X_1^*X_1 + \mathbf{X_1T_2} = \mathbf{T_1X_1} + \mathbf{X_1T_2} = \mathbf{T_1T_2}$ and by Lemma 2, there must be an element in the priority queue with $\lambda \le \mathbf{T_1T_2} <$ MIN. We have a priority queue that looks for minimal $\lambda$, so the loop could not have terminated via line 7. Furthermore, the wavefronts are nonempty by the reasoning in the proof of the first equivalence; thus the loop could not have terminated via line 5. Hence we have a contradiction, and FFF is indeed valid.                    $\square$

## 4   Performance

**Lemma 5** *The main loop runs at most $(2d + 1)n + 1$ times, where $d$ is the dimension of the search space and $n$ is the number of visited vertices.*

**Proof:** By definition of $\mathcal{G}$, the degree of each vertex is at most $2d$, because each vertex only has $2d$ axis-parallel directions of approach and no two edges overlap in more than one point. We can apply this fact to the argument used in the proof of Lemma 4 to classify iterations into three types: either an iteration adds a point to $\Omega_1$ and inserts up to $2d$ priority queue elements with $\lambda <$ MIN, or it reduces the number of such priority queue elements by at least one, or it is the last iteration.

Po-Shen Loh, *Finding Shortest Paths*, JGAA, 7(3) 287–303 (2003)     300

Since $n = |\Omega_1| + |\Omega_2|$, $\Omega_1$ can grow no more than $n$ times, so we can have no more than $n$ iterations of the first type. The number of priority queue insertions with $\lambda < \text{MIN}$ must then be bounded by $2dn$, so we can have no more than $2dn$ iterations of the second type. We can only have one iteration of the third type, and thus the number of iterations is bounded by $(2d+1)n + 1$.     $\square$

**Theorem 3** *FFF's worst-case space and time complexities are $O(n \log^{d-1} n)$ and $O(n \log^d n)$, respectively, where $n$ is the number of vertices visited.*

**Proof:** Using the method in Section 2.3, our DCPS can achieve $O(n \log^{d-1} n)$ space and $O(\log^d n)$ time for updates and queries, where $n$ is the number of points in the structure. In our algorithm, the DCPS's determine the space complexity, so FFF takes $O(n \log^{d-1} n)$ storage. As for speed, the dominant operations in the main loop are the DCPS update/query, both of which take $O(\log^d n)$. Lemma 5 bounds the number of iterations by $(2d+1)n + 1$, so the overall time complexity is $O(n \log^d n)$.     $\square$

**Theorem 4** *Suppose that in some graph, $\mathcal{S}(\mathbf{T_1 T_2}) = \varnothing$ and the $A^*$ search described in Section 1.2 visits $N$ vertices. Then FFF will visit at most $2N$ vertices.*

**Proof:** $A^*$ terminates after it visits all $N$ vertices that are connected to $T_1$ through paths in $\mathcal{G}$. Similarly, FFF will have exhausted $\mathcal{W}_1$ by the time it has visited all of these vertices, so it will have terminated via line 5. In each iteration, $X_1$ is visited before $X_2$, so the the number of vertices visited by FFF must be no more than $2N$.     $\square$

**Theorem 5** *Suppose that in some graph, $\mathcal{S}(\mathbf{T_1 T_2}) \neq \varnothing$. Let $\mathcal{N}_0$ be the set of all vertices $X \in \mathcal{G}$ for which $\mathbf{T_1}X + XT_2 < \mathbf{T_1 T_2}$, and let $\mathcal{N}_1$ be the set of all vertices $X \in \mathcal{G}$ for which $\mathbf{T_1}X + XT_2 \leq \mathbf{T_1 T_2}$. Then $A^*$ will visit between $|\mathcal{N}_0|$ and $|\mathcal{N}_1|$ vertices, while FFF will visit no more than $2|\mathcal{N}_1|$ vertices.*

**Proof:** The set $\mathcal{N}_0$ consists of vertices with $A^*$ priority less than $\mathbf{T_1 T_2}$, so since the priority of $T_2$ is $\mathbf{T_1 T_2}$, the priority-first search must visit all vertices in $\mathcal{N}_0$. There is no estimate, however, of how many vertices with priority $\mathbf{T_1 T_2}$ are visited by $A^*$; the method could visit as few as one or as many as all, depending on the search space.

We now prove the second claim. By the reasoning in the proof of the previous theorem, it suffices to show that $\Omega_1 \subseteq \mathcal{N}_1$. Suppose that $X_1 \in \Omega_1$; by lines 10 and 9, $X_1$ resulted from a query by some point $P_0'$, which was derived from some $P'$. When $P'$ was inserted into $\mathcal{W}_1'$, it had a priority $\lambda = P'Y' = |\sigma(P')| + PY + |\sigma(Y')| \geq \mathbf{T_1}P + PY + |\sigma(Y')|$ for some $Y$. If $Y$ came from line 24, then $\mathbf{T_1}P + PY + |\sigma(Y')| = \mathbf{T_1}P + 0 + \mathbf{P T_2} \geq \mathbf{T_1}P + PT_2$. If $Y$ came from line 26, Corollary 3 implies:

$$
\begin{aligned}
\mathbf{T_1}P + PY + |\sigma(Y')| &= \mathbf{T_1}P + PY + \mathbf{Y T_2} \\
&\geq \mathbf{T_1}P + PY + YT_2 \\
&\geq \mathbf{T_1}P + PT_2.
\end{aligned}
$$

Po-Shen Loh, *Finding Shortest Paths*, JGAA, 7(3) 287–303 (2003)        301

Therefore, in both cases we get $\lambda \geq \mathbf{T_1 P} + PT_2$.

Corollary 3 applied to line 10 tells us that $\mathbf{T_1 X_1} + X_1 P \leq \mathbf{T_1 P}$, so

$$\begin{aligned}
\mathbf{T_1 X_1} + X_1 T_2 &\leq \mathbf{T_1 X_1} + X_1 P + PT_2 \\
&\leq \mathbf{T_1 P} + PT_2 \leq \lambda.
\end{aligned}$$

By the argument in the last paragraph of the proof of Theorem 2, until MIN is set to $\mathbf{T_1 T_2}$, there always exists a priority queue element with $\lambda \leq \mathbf{T_1 T_2}$. At the beginning of every iteration thereafter (except for the last), line 7 ensures that such a priority queue element exists. Since we popped off $P'$ and will visit $X_1$, $\lambda \leq \mathbf{T_1 T_2} \Rightarrow \mathbf{T_1 X_1} + X_1 T_2 \leq \mathbf{T_1 T_2}$. Hence $X_1 \in \mathcal{N}_1$, as desired, and we are done. $\qquad\square$

## 5   Future Work

The theoretical bounds of the previous section are all upper limits; however, the purpose of front-to-front bidirectionality was to provide a more accurate heuristic that would reduce $n$, the number of visited nodes. Since the complexity is $O(n \log^d n)$, a significant reduction in $n$ would justify the additional log-power complexity factor.

This amounts to classifying the spaces on which front-to-front searches out-perform other search algorithms. Unfortunately, that is beyond the scope of this paper; instead, we just provide a simple thought-experiment that illustrates the existence of such spaces.

Figures 5 and 6 are ideal spaces for front-to-front algorithms because the terminals are only accessible through indirect channels that would confuse other heuristics. In fact, they yield relative reductions in $n$ that follow $O(\sqrt{n})$. In light of these simple examples, we can identify one class of spaces for which our front-to-front search is favorable: if source and destination terminals are located in intricate, separated networks that are linked by a large, simple network, then the $d$-th root reduction in $n$ can be attained.

In this paper, we have established the feasibility of subquadratic front-to-front bidirectional heuristic search. We close by posing a few questions that are opened by this result. Our algorithm is specific to lattice graphs; do there exist analogous tricks that produce subquadratic front-to-front searches in other situations? What types of graphs favor front-to-front heuristics? Do these types of graphs arise in "real-world" situations?

## Acknowledgements

Special thanks to Alain Martin and Mika Nyström for introducing this problem to the author, and to Charles Leiserson for providing pointers toward related literature. Thanks also to Po-Ru Loh for providing many valuable suggestions that significantly improved the clarity of this paper.

Po-Shen Loh, *Finding Shortest Paths*, JGAA, 7(3) 287–303 (2003)      302

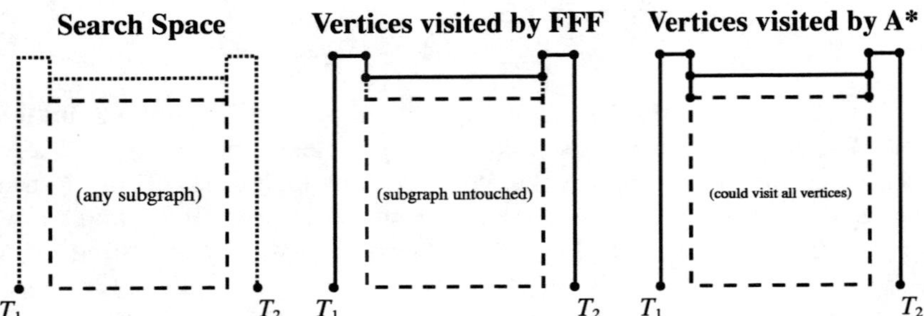

Figure 5: The subgraph can be arbitrarily complex, but FFF will always visit 8 vertices while A* must visit all subgraph vertices with priority less than $T_1T_2$. Those vertices include all points that can be reached from the top-left vertex by moving down and to the right along edges in the graph.

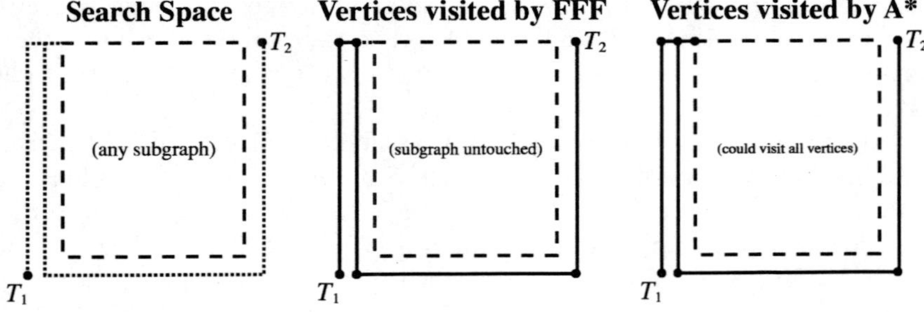

Figure 6: This is another family of search spaces in which FFF always visits a fixed number of vertices while A* can visit arbitrarily many.

Po-Shen Loh, *Finding Shortest Paths*, JGAA, 7(3) 287–303 (2003) 303

# References


[1] H. GABOW, J. BENTLEY, AND R. TARJAN, *Scaling and Related Techniques for Geometry Problems*, Proc. 16th Annual ACM Sympos. Theory of Computing (1984), pp. 135–143.

[2] Y.-J. CHIANG AND R. TAMASSIA, *Dynamic Algorithms in Computational Geometry*, Proc. IEEE, 80 (1992), pp. 1412–1434.

[3] D. EPPSTEIN, *Dynamic Euclidean Minimum Spanning Trees and Extrema of Binary Functions*, Discrete Comput. Geom., 13 (1995), pp. 111–122.

[4] F. O. HADLOCK, *A Shortest Path Algorithm for Grid Graphs*, Networks, 7 (1977), pp. 323–334.

[5] P. HART, N. NILSSON, AND B. RAPHAEL, *A Formal Basis for the Heuristic Determination of Minimum Cost Paths*, IEEE Trans. Systems Science and Cybernetics, 4 (1968), pp. 100–107.

[6] H. KAINDL AND G. KAINZ, *Bidirectional Heuristic Search Reconsidered*, J. Artificial Intelligence Res., 7 (1997), pp. 283–317.

[7] G. LUEKER, *A Data Structure for Orthogonal Range Queries*, Proc. 19th IEEE Sympos. Foundations of Computer Science (1978), pp. 28–34.

[8] G. LUEKER AND D. WILLARD, *A Data Structure for Dynamic Range Queries*, Inform. Process. Lett., 15 (1982), pp. 209–213.

[9] M. SMID, *Closest-Point Problems in Computational Geometry*, in Handbook of Computational Geometry, J. Sack and J. Urrutia, eds., Elsevier, 2000, pp. 877–935.

[10] D. WILLARD AND G. LUEKER, *Adding Range Restriction Capability to Dynamic Data Structures*, J. ACM, 32 (1985), pp. 597–617.

# Volume 7:4 (2003)

Journal of Graph Algorithms and Applications

http://jgaa.info/

vol. 7, no. 4, pp. 307–309 (2003)

# Advances in Graph Drawing
## Special Issue on Selected Papers from the Ninth International Symposium on Graph Drawing, GD 2001
### Guest Editors' Foreword

*Petra Mutzel*

Institut für Computergraphik und Algorithmen
Technische Universität Wien
http://www.ads.tuwien.ac.at/
mutzel@ads.tuwien.ac.at

*Michael Jünger*

Institut für Informatik
Universität zu Köln
http://www.informatik.uni-koeln.de/ls_juenger/
mjuenger@informatik.uni-koeln.de

P. Mutzel and M. Jünger, *Editors' Foreword*, JGAA, 7(4) 307–309 (2003)   308

This Special Issue brings together selected papers from the Ninth Annual Symposium on Graph Drawing, held in Vienna, Austria, on September 23–26, 2001. We have invited the strongest papers in the ranking generated by the GD program committee and we are glad that the following five papers could be included into this special issue after a strong refereeing process.

All papers in the issue deal with planar graphs and trees, respectively, and their crossing free representation. Planar graph drawing is getting increasing attention with the availability of software libraries such as AGD, PIGALE, and GDToolkit that have implemented the planarization approach and a variety of planar graph drawing algorithms.

Classical algorithms for angular resolution are all based on unit vertex and unit bend separation in the Cartesian coordinate representation. The paper by C. A. Duncan and S. G. Kobourov suggests to use a polar coordinate representation for drawing planar graphs, thus allowing independent control over the vertex resolution, bend-point resolution, and edge separation. The authors also provide a family of algorithms demonstrating the strength of the polar coordinate representation in comparison to the standard (Cartesian) representation.

Bend minimization is an important topic in planar graph drawing and even more in planar orthogonal graph drawing. So far there is no characterization for those planar graphs with maximum degree four that can be drawn orthogonally without any bends at all. The paper by Md. S. Rahman, T. Nishizeki, and M. Naznin provides a necessary and sufficient condition for a plane graph, i.e., a planar graph with given planar embedding, of degree at most three to have an orthogonal drawing without bends. The authors also provide a linear time algorithm for constructing such a drawing if it exists.

Drawings without bend points in an alternative setting are considered in the paper by S. Felsner, G. Liotta, and S. Wismath. They investigate the question which graphs can be drawn straight-line and crossing free on a strip, i.e., a grid of size $n \times k$. They give a characterization for trees satisfying this condition and prove lower bounds for the height $k$ for arbitrary planar graphs. Moreover, they show that every outerplanar graph can be drawn crossing-free with straight lines in linear volume on a 3-dimensional restricted grid called *prism*. This is not true for general planar graphs, not even if the prism is extended to a so-called *box*, i.e., an integer grid of size $n \times 2 \times 2$.

The paper by R. Babilon, J. Matoušek, J. Maxová, and P. Valtr deals with the problem of low-distortion embeddings of trees. They show that every tree on $n$ vertices with edges of unit length can be embedded in the plane with distortion $O(\sqrt{n})$, i.e., the distance between each pair $u, v \in V$ of vertices in the embedding correlates with the length of the path from $u$ to $v$ in the tree distorted by a factor up to $O(\sqrt{n})$. This embedding can be found by a simple formula. This result is best possible in the sense that it is asymptotically optimal in the worst case.

Finally, the paper by H. de Fraysseix and P. Ossona de Mendez investigates minimal non-planar structures of non-planar graphs in a depth-first-search setting. Kuratowski characterized the minimal forbidden substructures in a planar graph, namely the subdivisions of $K_5$ and $K_{3,3}$. In quite some applications in

P. Mutzel and M. Jünger, *Editors' Foreword*, JGAA, 7(4) 307–309 (2003)   309

graph drawing it is essential to find either one or many of these Kuratowski subdivisions. The authors provide a characterization of so-called *DFS cotree-critical graphs* and give a simple algorithm for finding one ore more Kuratowski subdivisions which is useful for the planarity testing and planarization algorithms based on depth-first-search.

We would like to thank the JGAA editors for inviting us to compile this special issue, the referees for their invaluable help, and all authors for the considerable extra effort they put into making their GD 2001 contributions into the journal articles contained in this issue.

## Journal of Graph Algorithms and Applications

http://jgaa.info/

vol. 7, no. 4, pp. 311–333 (2003)

# Polar Coordinate Drawing of Planar Graphs with Good Angular Resolution

*Christian A. Duncan*

Department of Computer Science
University of Miami
Coral Gables, FL 33124
http://www.cs.miami.edu/
duncan@cs.miami.edu

*Stephen G. Kobourov*

Department of Computer Science
University of Arizona
Tucson, AZ 85721
http://www.cs.arizona.edu/
kobourov@cs.arizona.edu

### Abstract

We present a novel way to draw planar graphs with good angular resolution. We introduce the polar coordinate representation and describe a family of algorithms for constructing it. The main advantage of the polar representation is that it allows independent control over grid size and bend positions. We first describe a standard (Cartesian) representation algorithm, CRA, which we then modify to obtain a polar representation algorithm, PRA. In both algorithms we are concerned with the following drawing criteria: angular resolution, bends per edge, vertex resolution, bend-point resolution, edge separation, and drawing area.

The CRA algorithm achieves 1 bend per edge, unit vertex and bend resolution, $\sqrt{2}/2$ edge separation, $5n \times \frac{5n}{2}$ drawing area and $\frac{1}{2d(v)}$ angular resolution, where $d(v)$ is the degree of vertex $v$. The PRA algorithm has an improved angular resolution of $\frac{\pi}{4d(v)}$, 1 bend per edge, and unit vertex resolution. For the PRA algorithm, the bend-point resolution and edge separation are parameters that can be modified to achieve different types of drawings and drawing areas. In particular, for the same parameters as the CRA algorithm (unit bend-point resolution and $\sqrt{2}/2$ edge separation), the PRA algorithm creates a drawing of size $9n \times \frac{9n}{2}$.

Communicated by: P. Mutzel and M. Jünger;
submitted June 2002; revised December 2002.

The work by S. G. Kobourov was partially supported by NSF grant ACR-0222920.

Duncan & Kobourov, *Polar Coordinate Drawing*, JGAA, 7(4) 311–333 (2003)312

# 1   Introduction

In the area of planar graph drawing there has been considerable interest in algorithms that produce readable drawings [3]. Among the many properties which contribute to the readability of planar graphs, edge smoothness, vertex resolution, bend-point resolution, angular resolution, and edge separation are of great importance. Edges are often drawn as straight-line segments connecting two vertices. An edge can also be drawn as a sequence of straight-line segments, in which case the smallest number of bends is desirable. An edge may also be drawn as a smooth curve. These three types of edges generally provide aesthetically pleasing drawings.

## 1.1   Definitions

A graph drawing has good *vertex resolution* if vertices cannot get arbitrarily close to one another, that is, if vertices are well distributed in the drawing. As a result, a great deal of research has been concentrated on graph drawing algorithms which place vertices on the integer grid such that the *drawing area* is proportional to the number of vertices $n$ of the graph, typically $O(n) \times O(n)$. If there are bends in the edges, then the bend-points are also placed on the integer grid. The *bend-point resolution* of a graph refers to the minimum distance between two bends. The *edge separation* of a graph refers to the minimum distance between two edges that are sufficiently away from their endpoints (since incident edges can get arbitrarily close to each other near their common endpoint).

A graph drawing has good *angular resolution* if adjacent edges cannot form arbitrarily small angles. This is achieved by ensuring that the edges emanating from a given vertex "fan out" evenly around the vertex. Note, however, that good angular resolution cannot always be achieved while simultaneously guaranteeing straight-line edges and small sub-exponential drawing area [10]. By introducing bends in the edges, however, we can guarantee both good resolution and small drawing area.

## 1.2   Previous Work

Garg and Tamassia [6] consider the problem of drawing with good angular resolution, and Kant [9] shows how to create drawings with angular resolution of $\Theta(1/d(v))$ in an $O(n) \times O(n)$ area grid, using edges with at most three bends each. Gutwenger and Mutzel [8] describe an improved algorithm with better constant factors which produces very aesthetically pleasing drawings in a $(2n - 5) \times (3n/2 - 7/2)$ grid with at least $2/d(v)$ angular resolution using at most three bends per edge. The algorithm of Goodrich and Wagner [7] requires one less bend per edge and guarantees angular resolution of $\Theta(1/d(v))$ for each vertex $v$, but at the expense of larger area, $(20n - 48) \times (10n - 24)$. Cheng, Duncan, Goodrich, and Kobourov [1] improve the above algorithm so that every edge has at most one bend while the angular resolution is $\Theta(1/d(v))$ for each vertex $v$ and maximum area is $30n \times 15n$.

Duncan & Kobourov, *Polar Coordinate Drawing*, JGAA, 7(4) 311–333 (2003)313

## 1.3  Our Results

We first present a new Cartesian representation algorithm (CRA) which improves the bounds of previous algorithms. In particular, CRA guarantees 1 bend per edge, unit vertex resolution, unit bend-point resolution, $\sqrt{2}/2$ edge separation, $5n \times \frac{5n}{2}$ drawing area, and $\frac{1}{2d(v)}$ angular resolution, where $d(v)$ is the degree of vertex $v$.

We then present a novel polar representation algorithm (PRA). The PRA algorithm also guarantees $\frac{\pi}{4d(v)}$ angular resolution, 1 bend per edge, and unit vertex resolution. The bend-point resolution and edge separation are parameters that can be modified to achieve different types of drawings and drawing areas. In particular, for the same parameters as the CRA algorithm (unit bend-point resolution and $\sqrt{2}/2$ edge separation), the PRA algorithm creates a drawing of size $9n \times \frac{9n}{2}$. Note that in some situations the vertex resolution is more important than the bend-point resolution or the edge separation. In such situations, all of the previous algorithms perform poorly since they are designed to maintain constant resolution particularly between vertices and bend-points. Using the PRA algorithm, we can relax the bend-point resolution constraints and get significant improvements.

The PRA algorithm relies on a novel approach for representing bends and vertices. Traditionally, vertices and bend-points are restricted to lie on integer grid coordinates. One reason for this is that the points are defined by a pair of integers. In this way, all operations on the points (for example, shifting) are performed with integer arithmetic. At the drawing stage, the integer coordinates are mapped to pixels on the screen.

Another reason for placing vertices and bend-points on integer grid coordinates is that this approach guarantees good vertex resolution, good bend-point resolution, and good edge separation [1, 7, 8, 9]. Rather than insisting that bend-points lie on integer grid coordinates, we propose an alternative approach which allows bend-points to be located on a grid represented by polar coordinates. We call this a *polar representation* approach because both the vertices and the bend-points are represented using polar coordinates.

At the exact moment of drawing the graph onto the screen, an algorithm using polar representation requires a rounding calculation to determine the exact pixel location for the bend-points. Note, however, that the traditional approach also uses a rounding calculation for scaling from the integer grid space to the pixel space.

The main advantage of using a polar representation is that it allows us to independently control grid size and bend positions. Polar coordinates allow us to specify different vertex resolution, bend-point resolution, and edge separation. We achieve this added flexibility at the expense of slightly increased storage for the graph representation. A Cartesian representation requires exactly two integers for each point while the polar representation requires up to five integers per point.

Both of our algorithms assume that the graph is a fully triangulated, undirected, planar graph. If the graph were not fully triangulated, one can still

Duncan & Kobourov, *Polar Coordinate Drawing*, JGAA, 7(4) 311–333 (2003)314

solve the problem by fully triangulating the graph, embedding this new graph, and then removing the inserted edges. Approached properly, this scheme incurs at most a constant factor decrease in angular resolution as the modified degree of a vertex can triple in size, e.g. fully-triangulating a path. As a result, for the remainder of this paper when we say "graph" we mean a fully triangulated, undirected, planar graph. We leave it as an open exercise to modify the algorithm to work more effectively for general undirected planar graphs.

In Section 2 and Section 3 we present the Cartesian Representation Algorithm (CRA) and argue its correctness. CRA is an improved version of the algorithm from Cheng *et al* [1] for drawing with good angular resolution. In Section 4 we introduce the concept of embedding graphs using a polar coordinate system and then present the Polar Representation Algorithm (PRA) which is a modification of the CRA.

## 2  The CRA Algorithm

The Cartesian Representation Algorithm is a natural extension of some previous algorithms that guarantee good angular resolution [9, 8, 7, 1]. In our algorithm the vertices of the graph are inserted sequentially by their canonical ordering, generating subgraphs $G_1, G_2, \ldots, G_n$. The canonical ordering [5] for a planar graph $G$ orders the vertices of $G$ so that they can be inserted one at a time without creating any crossings. We define $G_k$ at step $i$ to be the graph induced by vertices $1, 2, \ldots, k$. From our ordering, we shall see that $G_1$, $G_2$, and $G_3$ are basic graphs, a vertex, a line, and a triangle respectively. Graph $G_{k+1}$ is created from $G_k$ by inserting the next vertex $v_{k+1}$ in the canonical order. Before we show the details of our algorithm we need several definitions. Following the notation of [5], let $w_1 = v_1, w_2, \ldots, w_m = v_2$ be the vertices of the exterior face $C_k$ of graph $G_k$ in order. For a particular subgraph $G_k$ with $k > 2$ and vertex $v_{k+1}$, we refer to $w_l$ and $w_r$ as the leftmost and rightmost neighbors of $v_{k+1}$ on $C_k$, see Fig. 1. We also say that $v_{k+1}$ *dominates* $w_i$ for $l < i < r$. That is these vertices on $C_k$ are no longer on $C_{k+1}$.

When referring to vertices and points, we often need to use the (current) coordinates of the vertices and points on the grid. Let $v(x)$ and $v(y)$ represent the $x$ and $y$ coordinates of some vertex $v$.

### 2.1  Vertex Regions

In the immediate vicinity of every vertex there are two types of regions: *free regions* and *port regions*. The free and port regions alternate around the vertex, see Fig. 2(a). For each free region there is at most one edge passing through it to $v$. Each port region is bounded by a line segment with a number of ports and every edge inside the port region passes through a unique port. The number of ports in a port region is as small as possible. We define the six regions around $v$ based on rays extending at certain angles or slopes from $v$. For convenience, we

Duncan & Kobourov, *Polar Coordinate Drawing*, JGAA, 7(4) 311–333 (2003)315

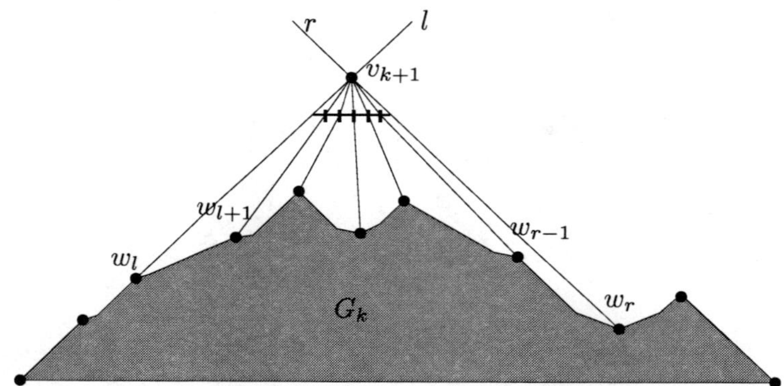

Figure 1: Graph $G_{k+1}$ after inserting $v_{k+1}$. The shaded part is $G_k$. Vertices $w_l$ and $w_r$ are the leftmost and rightmost neighbors of $v_{k+1}$. The horizontal line segment below $v_{k+1}$ is the middle port region through which all the edges $(v_{k+1}, w_i)$, $l < i < r$, are routed.

assume that $0°$ is pointing in the vertical direction. As illustrated in Figure 2(a), the six regions around $v$ are defined as follows:

- Free region $M^f$: between $-45°$ and $45°$

- Free region $R^f$: between $90°$ and $135°$

- Free region $L^f$: between $-135°$ and $-90°$

- Port region $M^p$: between $L^f$ and $R^f$

- Port region $L^p$: between $L^f$ and $M^f$

- Port region $R^p$: between $R^f$ and $M^f$

The algorithm draws each edge in $E$, except the initial edge $(v_1, v_2)$, by "routing" it through a port of one of the two vertices in a fashion similar to Cheng *et al* [1]. Each edge consists of two connected edge segments. One edge segment, the *port edge segment*, connects a vertex with one of its ports. The other segment, the *free edge segment*, connects a vertex to one of its neighbor's ports. For example, for an edge $e = (u, v)$, if we route $e$ through the leftmost port in $u$'s middle port region $M^p$, we would draw two line segments, see Fig. 2(b): the *port edge segment* would pass from $u$ to the port, and the *free edge segment* would pass from the port to $v$. This method of construction guarantees that the free edge segments always pass through free regions and that each port transmits at most one port edge segment.

We perform our construction in incremental stages, where each stage corresponds to the insertion of a new vertex. Observe that at each stage, for every vertex $v$ except those on the external face, $w_1 = v_1, w_2, \ldots, w_m = v_2$, there

Duncan & Kobourov, *Polar Coordinate Drawing*, JGAA, 7(4) 311–333 (2003)316

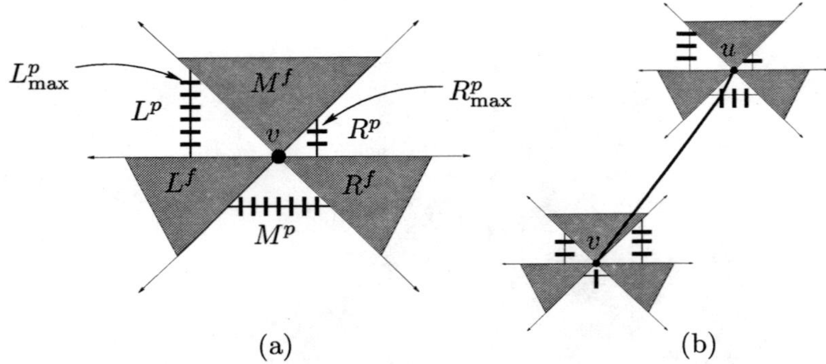

Figure 2: Vertex regions and edge routing: the number of ports along each port region is determined by the number of edges that need to be routed through that port region. (a) The area around a vertex $v$ is divided into 6 regions. The free regions are shaded and at most one free edge segment goes through each one of them. All the port segments use ports in the port regions of $v$. (b) Routing an edge $e = (u, v)$, where the port edge segment connects $u$ to one of its ports and the free edge segment connects the port to $v$, going through one of $v$'s free regions.

are exactly three free edge segments. The remaining edges are connected to $v$ via port segments. These remaining edges can be grouped into three classes based on which port region they are routed through, $L^p$, $R^p$, or $M^p$. Count the number of edges in each of these groups and let $d_l(v)$ be the number of port edge segments using port region $L^p$. Similarly, define $d_r(v)$ and $d_m(v)$ to be the number of port segments using port regions $R^p$ and $M^p$. Observe that in the final stage, there are exactly three vertices on the exterior face, $v_1, v_2, v_n$, and then $\sum_{v \in V}(d_l(v) + d_r(v) + d_m(v)) = |E| - 1$. That is, for every edge, there is a corresponding port and free edge segment, except for the edge $(v_1, v_2)$. This initial edge is only a single free (horizontal) edge segment. We could also, of course, remove the port edge segments for the final external face as well and thus the summation would be $|E| - 3$.

For a vertex $v$ we define the *maximal right port* $R^p_{\max}$ as follows. Let $v$ have coordinates $(v_x, v_y)$. Then the $R^p_{\max}$ of $v$ has coordinates $(v_x + d_r(v) + 1, v_y + d_r(v))$ if $d_r(v) > 0$ and $(v_x, v_y)$ otherwise. We define the *maximal left port* $L^p_{\max}$ of $v$ in a similar fashion, see Fig. 2(a).

## 2.2 Invariants of the CRA Algorithm

By design, our algorithm is incremental with $n$ stages, where each stage corresponds to the insertion of the next vertex in the canonical order. Thus it is natural to define several key invariants to be maintained at every stage. The four invariants below are similar in flavor to those of Cheng *et al* [1] except that here we do not need to maintain any joint boxes.

Duncan & Kobourov, *Polar Coordinate Drawing*, JGAA, 7(4) 311–333 (2003)317

1. All vertices and ports have integer coordinates.

2. Let $w_1 = v_1, w_2, \ldots, w_m = v_2$ be the vertices of the exterior face $C_k$ of $G_k$ in order. Then $w_1(x) < w_2(x) < \ldots < w_m(x)$.

3. The free edge segment of edge $e = (w_i, w_{i+1})$, $0 < i < m$, has slope $\pm 1$ and $e$'s port edge segment goes through a maximal port.

4. For every vertex $v$ there is at most one (free) edge segment crossing each of its free regions. All other edge segments are port edge segments.

## 2.3  Vertex Shifting

In the algorithms that maintain good angular resolution with the aid of vertex joint boxes [1, 7], every time a new vertex is inserted, already placed vertices need to be shifted a great deal so that the joint box can fit amongst them. The amount of shifting required is typically of the order of the degree of the vertex. Invariably this leads to large constants behind the $O(n) \times O(n)$ area, e.g. $(20n - 48) \times (10n - 24)$ in [7] and $30n \times 15n$ in [1]. In our algorithm we never need to shift any vertex by more than five grid units allowing us to draw $G$ in a $5n \times \frac{5n}{2}$ grid. When a new vertex $v$ is inserted, we must create enough space so that the leftmost $w_l$ and rightmost $w_r$ neighbors of $v$ can "see" $v$ through their respective maximal port regions. Note that the previous $R^p_{\max}$ port of $w_l$ and $L^p_{\max}$ of $w_r$ were used at an earlier stage. Thus, we must create an additional port along the $R^p$ region of $w_l$. Similarly, additional space is necessary along the $L^p$ region of $w_r$.

In order to create more space we need to move $w_l$ and $w_r$. We also have to ensure that the four invariants and the planarity of the graph are maintained. This is achieved by shifting the "shifting set" of the vertex as well as the vertex itself. Using the definition of de Fraysseix *et al* [5], define the *shifting set* $M_k(w_i)$ for a vertex $w_i$ on the external face of $G_k$ to be a subset of the vertices of $G$ such that:

1. $w_j \in M_k(w_i)$ iff $j \geq i$

2. $M_k(w_1) \supset M_k(w_2) \supset \ldots \supset M_k(w_m)$

3. Let $\delta_1, \delta_2, \ldots, \delta_m > 0$; if we sequentially translate all vertices in $M_k(w_i)$ by distance $\delta_i$ to the right ($i = 1, 2, \ldots, m$), then the embedding of $G_k$ remains planar.

These shifting sets can be defined recursively. Let $w_l$ and $w_r$ be the leftmost and rightmost neighbors of $v$ on $C_k$. Then construct $M_{k+1}(w_i)$ recursively as follows:

$$M_{k+1}(w_i) = M_k(w_i) \cup v_{k+1}, \text{ for } i \leq l,$$

$$M_{k+1}(v_{k+1}) = M_k(w_{l+1}) \cup v_{k+1},$$

$$M_{k+1}(w_j) = M_k(w_j), \text{ for } j \geq r.$$

Duncan & Kobourov, *Polar Coordinate Drawing*, JGAA, 7(4) 311–333 (2003)318

For convenience, define a *right-shift of $m$ units* for a vertex $w_i$ as shifting $M_k(w_i)$ by $m$ units to the right so that all ports for every vertex in $M_k(w_i)$ also shift except the ports in the $L^p$ region of $w_i$. Define a *left-shift of $m$ units* for vertex $w_i$ as shifting $M_k(w_{i+1})$ by $m$ units to the right so that all ports for every vertex in $M_k(w_{i+1})$ also shift including the ports in the $R^p$ region of $w_i$.

## 2.4  CRA Overview

The CRA algorithm constructs the graph one vertex at a time, by creating the graphs $G_1, G_2, \ldots, G_n$. Constructing $G_i$, $1 \leq i \leq 3$ is straightforward (see Figure 4(a)), so assume that $G_k$, for $k \geq 3$, has been constructed with exterior face $C_k = (v_1 = w_1, w_2, \ldots, w_m = v_2)$. Suppose we have embedded $G_k$ with exterior face $C_k$. To construct $G_{k+1}$, let $v_{k+1}$ be the next vertex in the canonical ordering and recall that $w_l$ and $w_r$ are, respectively, the leftmost and rightmost neighbors of $v_{k+1}$ on the exterior face $C_k$.

Recall that $d_r(w_l)$ is the current number of port edge segments using $R^p$ of $w_l$, and that $d_l(w_r)$ is the current number of port edge segments using $L^p$ of $w_r$. There are two cases to consider:

- **case (a)** $d_r(w_l) = 0$, see Fig. 3(a).

- **case (b)** $d_r(w_l) > 0$, see Fig. 3(b).

In case (a) perform a left-shift of 2 units on $w_l$ in order to free space for a port in the $R^p$ region of $w_l$. In case (b) perform a left-shift of 1 unit on $w_l$. Similarly, if $d_l(w_r) = 0$ then perform a right-shift of 2 units on $w_r$. Otherwise perform a right-shift of 1 unit on $w_r$.

Insert $v_{k+1}$ at the intersection of lines $l$ and $r$, where $l$ is the line with slope $+1$ through $w_l$'s maximal right port and $r$ is the line with slope $-1$ through $w_r$'s maximal left port, see Fig. 1. In the case where lines $l$ and $r$ do not intersect in a grid point it suffices to shift all the elements in $M_k(w_r)$ one additional unit to the right.

The edges from $v_{k+1}$ to $w_l$ and $w_r$ are routed through $w_l$'s maximal right port and $w_r$'s maximal left port, respectively. The remaining edges go from $v_{k+1}$ to vertices $w_i$, $l < i < r$.

Before placing the $M^p$ region of $v_{k+1}$ it is necessary to ensure that there are enough ports on it that can be used to connect $v_{k+1}$ to $w_{l+1}, w_{l+2}, \ldots, w_{r-1}$. The $M^p$ region is a horizontal line segment with $1, 3, \ldots, 2m - 1$ ports when the line segment is respectively $1, 2, \ldots, m$ grid units below $v_{k+1}$. To allocate enough space then we simply locate the horizontal line segment $\lceil (r-l)/2 \rceil$ units below $v_{k+1}$.

As shown in the next section the $M^p$ region can be placed correctly, that is, placed so that it does not lie below any of the vertices $w_i$, $l < i < r$. We now need to route the edges from $v_{k+1}$ to $w_i$. In the case where $r - l$ is an even amount there are exactly enough ports for each of the vertices, so the routing is simple, the first (leftmost) port goes to $w_{l+1}$ and the last (rightmost) port to $w_{r-1}$. If it is odd, there is one extra port. Ideally, we would simply skip

Duncan & Kobourov, *Polar Coordinate Drawing*, JGAA, 7(4) 311–333 (2003)319

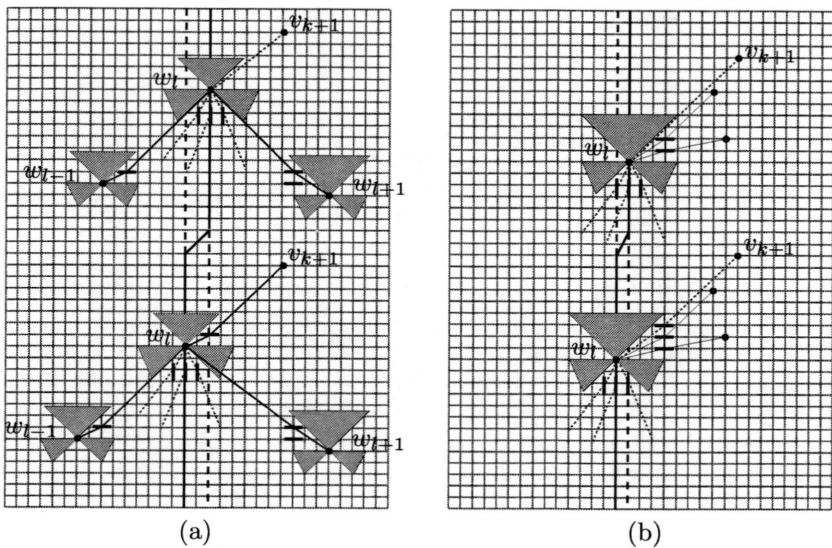

(a)        (b)

Figure 3: Adding the current vertex $v_{k+1}$. Here $w_l$ is the leftmost neighbor of $v_{k+1}$ on the exterior face of $G_k$. (a) If $d_r(w_l) = 0$, then we need to shift $w_l$ two grid units to the left. (b) If $d_r(w_l) > 0$, then it suffices to shift $w_l$ only one unit to the left. Note that the shifting set $M_k(w_l)$ also shifts with $w_l$.

the rightmost (or leftmost) port. However, it is possible that this would force the last edge (among others) to have the free edge segment be outside the valid region. Therefore, we proceed as follows, assign leftmost port to $w_{l+1}$, then $w_{l+2}$, and so on until either all are assigned or one vertex, $w_a$, has a free edge segment that is outside of the free region. We then assign ports from rightmost port to $w_{r-1}$, then $w_{r-2}$, until $w_a$ is assigned. Note this is identical to simply skipping one port and continuing left. In the next section, we shall show that this correctly routes the edges. That is, all edges go through ports and the free edge segments lie in free regions.

It is important to point out that in the interest of saving space, being as compact as possible, we allow free edge segments to initially have length zero. That is a vertex $w_i$ can actually lie on a port of another vertex $v$. This is not a problem so long as the port is used only to route an edge between $v$ and $w_i$. During shifting, the vertex and the port are treated separately. That is, they are not necessarily confined to be in the same location. See Figure 4.

## 3 Correctness of the Algorithm

The algorithm works correctly if all four invariants are maintained. We show that free edge segments always remain in free edge regions and that there is at most one free edge segment per free region. We then need to bound the

Duncan & Kobourov, *Polar Coordinate Drawing*, JGAA, 7(4) 311–333 (2003)320

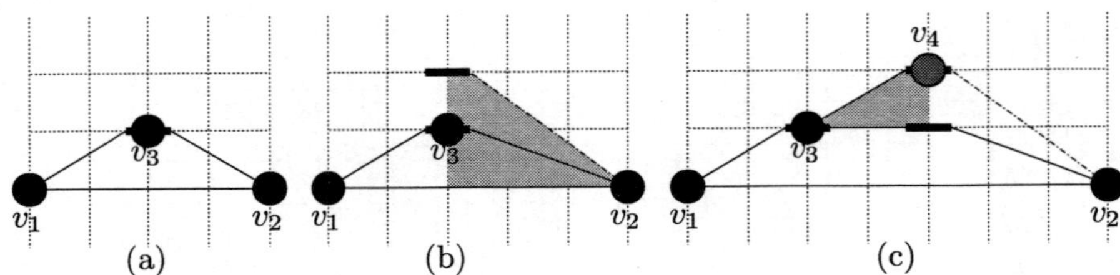

Figure 4: After initial settings of $v_1$, $v_2$, and $v_3$, inserting $v_4$ to share edges with $v_2$ and $v_3$. (a) Initial configuration. Notice that $v_3$ lies on both $R^p_{\max}$ of $v_1$ and $L^p_{\max}$ of $v_2$. (b) Shifting vertex $v_2$ one unit to the right. The shaded region indicates the $L^p$ region of $v_2$. Notice that $v_2$'s port did not move but a new port was inserted above it, the new $L^p_{\max}$ of $v_2$. (c) Shifting vertex $v_3$ two units to the left. The shaded region indicates the $R^p$ region of $v_3$. Notice that $v_2$'s port did not move and so now $v_3$ does not overlap that port though it still overlaps $v_1$'s port. The new vertex $v_4$ is placed at the intersection of the diagonals from $v_3$'s $R^p_{\max}$ port and $v_2$'s $L^p_{\max}$ port. In this case, they are the same point and once again $v_4$ overlaps both maximal ports.

drawing area required by the algorithm and show that good angular resolution is maintained. Finally, we have to bound the number of bends created and analyze the running time.

**Lemma 1** *If a free edge segment lies in a free region in $G_k$, then it remains in the free region in $G_{k+1}$.*

*Proof:* The initial edge $(v_1, v_2)$ is treated as a special edge. It is a free edge segment that is not connected to any ports. As the vertices shift this edge remains horizontal and thus remains inside its free regions. It can also be seen that the lemma holds from $G_2$ to $G_3$.

For $k \geq 3$, we must consider how a free edge segment could "move". When inserting vertex $v_{k+1}$, the graph $G_k$ changes by performing shifts (right or left shifts). These shifts move vertices and ports and possibly cause edge segments to change slope. Let $e$ be a free edge segment in $G_k$ which connects two vertices $w$ and $v$ via one of $w$'s port regions. The slope of $e$ determines whether $e$ lies inside a free region of $v$ or not. Therefore, we need to prove that upon shifting either the slope of $e$ does not change or that the change does not allow $e$ to leave the free region.

There are a few important points to remember about shifting before we proceed with the proof. First, shifting is only done with the shifting sets of vertices on $C_k$ and for any such operation, all vertices and ports which move are shifted the same amount. In addition, all ports are shifted along with their respective vertices except for certain ports belonging to vertices on $C_k$. Second, if a vertex $w \in G_k$ is not in $C_k$ then it must have been previously dominated by another vertex $v_{k'}, k' \leq k$. At that time, $w$ is added to the shifting set of

Duncan & Kobourov, *Polar Coordinate Drawing*, JGAA, 7(4) 311–333 (2003)321

$M_{k'}(v_{k'})$. From the recursive construction of the shifting sets, for all $k \geq k'$ and any vertex $w_i \in C_k$, $v_{k'} \in M_k(w_i)$ if and only if $w \in M_k(w_i)$. That is, $v_{k'}$ shifts if and only if $w$ shifts.

Recall that there are three types of free regions: $M^f, L^f, R^f$. Let us first assume that $e$ lies in the $M^f$ region of $v$. Edge segments in the $M^f$ regions are created by a vertex $w$ dominating another vertex $v$. But from the arguments above, $v$ must then belong to the shifting set of $w$ and so both $v$ and $w$ are always shifted together. Since $e$ connects to $w$'s $M^p$ region, the port always shifts with $v$ and $w$. Therefore, the slope of $e$ cannot change and it must remain in $M^f$.

As the two remaining cases are symmetric to each other, without loss of generality, let us now assume that $e$ lies in the $L^f$ region of $v$. This implies that the slope of $e$ is between 0 and $+1$. Edges lying in the $L^f$ region are created by neighboring vertices on some prior external face. That is, at some previous stage $k' \leq k$, we connected $v_{k'} = v$ and $w = w_l \in C_{k'}$. In this situation, $e$ is routed from a port in $R^p$ of $w_l$ to $v_{k'}$. If we define $v_{k'}, w_l, w_r$ in the usual manner, for all $k \geq k'$ and all $w \in C_k$, if $w_l \in M_k(w)$ then $v_{k'} \in M_k(w)$. That is, if $w_l$ shifts then so must $v_{k'}$. Note that $v_{k'}$ can shift without $w_l$ shifting. As for $w_l$'s $R^p$ region, it is possible that the port region shifts without $w_l$ but only on a left shift of $w_l$. In this case, $v_{k'}$ again still shifts with the $R^p$ region of $w_l$. Therefore, shifting affects the slope of $e$ only if $v_{k'}$ shifts and $w_l$'s $R^p$ region does not. Since $v_{k'}$ moves farther away from $w_l$'s port region, the slope of $e$ becomes more horizontal (approaches 0). Consequently, $e$ still remains within the $L^f$ region of $v$. ∎

In order to prove our next main lemma (Lemma 3) we need to present a few smaller issues describing the relationship between the vertices and the lines of slope $\pm 1$. Each of these lemmas relies on the fact that $G_k$ maintains the key invariants as described in Section 2.2.

**Definition 1** *Let $v$ be a vertex in $G_k$. Define $v^+$ (respectively) $v^-$) to be the line of slope $+1$ (resp. -1) passing through $v$.*

**Property 1** *Suppose we are given a graph $G_k$ maintaining the key invariants. For any $w_i \in C_k$, the region above $w_i$ and between $w^+$ and $w^-$ is empty of any vertices in $G_k$.*

Note, this property comes directly from the fact that the external face has free edge segments of slope $\pm 1$ and that $x(w_1) < x(w_2) < \ldots < x(w_m)$. See Figure 1.

**Lemma 2** *Suppose we are given a graph $G_k$ maintaining the key invariants. For any $w_i, w_j \in C_k$ and $i \leq j$, let $p_i$ and $p_j$ be any two points on $w_i^+$ and $w_j^+$ such that $p_i(y) = p_j(y)$. Then $p_j(x) - p_i(x) \geq j - i$. By symmetry, if we use $w_i^-$ and $w_j^-$, then we still have $p_j(x) - p_i(x) \geq j - i$.*

*Proof:* See Figure 5 for a simple example. We shall prove this lemma inductively. It is certainly true for the case when $i = j$. So, let us assume the lemma

Duncan & Kobourov, *Polar Coordinate Drawing*, JGAA, 7(4) 311–333 (2003)322

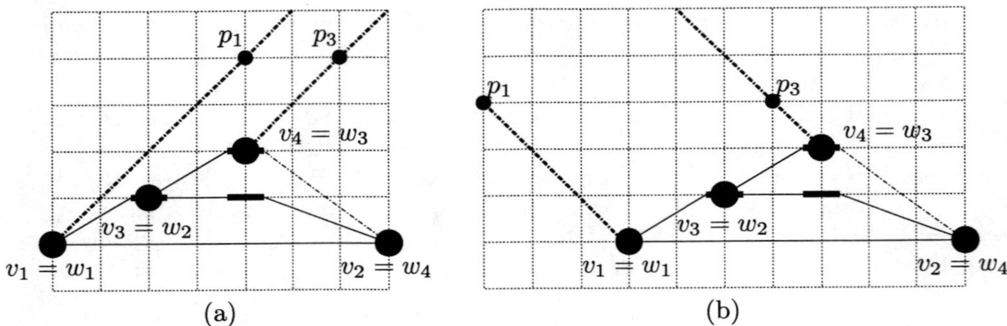

Figure 5: A simple example between two vertices on $C_k$, $w_1$ and $w_3$. (a) Points $p_1$ and $p_3$ are on $w_1^+$ and $w_3^+$ respectively. Notice that $p_3(x) - p_1(x) = 2 = 3 - 1$. (b) Same scenario except now points are on $w_1^-$ and $w_3^-$.

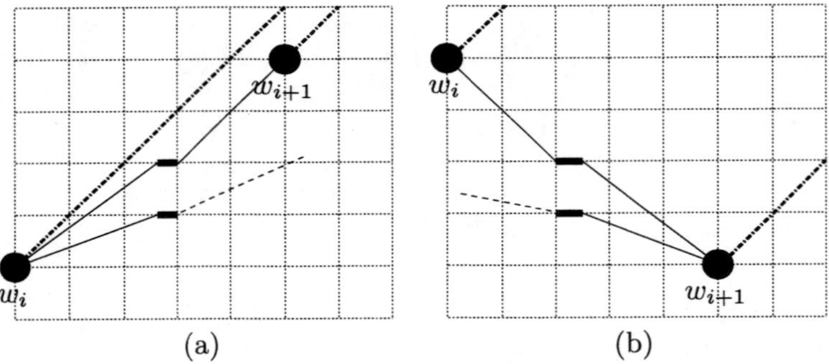

Figure 6: Notice the relationship between $w_i$, $w_{i+1}$, and their respective lines of slope $+1$ when $w_{i+1}$ is (a) increasing and (b) decreasing.

holds for all $j' < j$. There are two possibilities for $w_j$, either $w_j(y) > w_{j-1}(y)$, increasing along the external face, or $w_j(y) < w_{j-1}(y)$, decreasing along the external face. In the first case, recall that the port edge segment connecting $w_j$ to $w_{j-1}$ must go through the $R_{\max}^p$ of $w_{j-1}$. Therefore, $w_j^+$ is shifted over one unit and hence $p_j(x) - p_{j-1}(x) = 1$. In the second case, the connection is through the $L_{\max}^p$ of $w_j$ and therefore $p_j(x) - p_{j-1}(x) \geq 3$. See Figure 6. From our assumption then, we have

$$p_j(x) - p_i(x) = (p_j(x) - p_{j-1}(x)) + (p_{j-1}(x) - p_i(x)) \geq 1 + (j - 1 - i) = j - i.$$

To see the symmetric argument, notice that if we flip the graph about the $y$-axis, we have the same problem. ∎

**Corollary 1** *Suppose we are given a graph $G_k$ maintaining the key invariants.*

Duncan & Kobourov, *Polar Coordinate Drawing*, JGAA, 7(4) 311–333 (2003)323

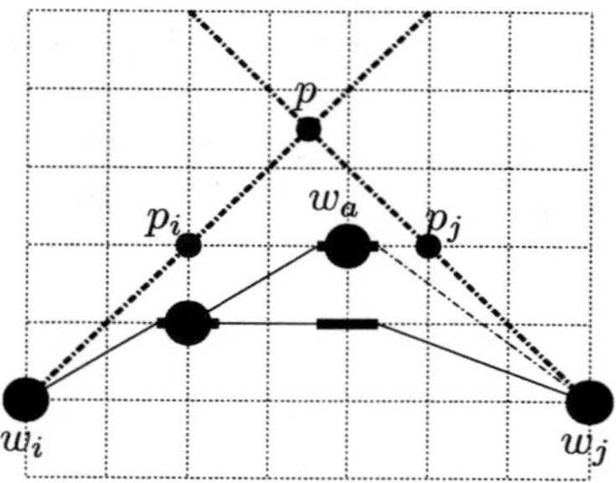

Figure 7: The intersection, $p$, of $w_i^+$ and $w_j^-$ and its relationship with some $w_a$ between $w_i$ and $w_j$. Notice the distances between $p_i(x)$, $p_j(x)$, and $w_a(x)$.

*For any $w_i, w_j \in C_k$ and $i \leq j$, let $p$ be the intersection of $w_i^+$ and $w_j^-$. Then $p(y) \geq \max_{i \leq a \leq j}(w_a(y)) + (j - i)/2$.*

*Proof:*   Let $w_a$ be any vertex with $i \leq a \leq j$. Let $p_i$ and $p_j$ be any two points on $w_i^+$ and $w_j^-$ such that $p_i(y) = p_j(y) = w_a(y)$. That is, we are looking at points on the horizontal line passing through $w_a$. For notation, let $p_a = w_a$ which by definition is on both $w_a^+$ and $w_a^-$.

Since $p_i$ and $p_j$ satisfy the assumptions of Lemma 2, we can see that $p_j(x) - p_i(x) \geq j - i$. Let us now look at $p$, the intersection of $w_i^+$ and $w_j^-$. See Figure 7. From the above inequality and the fact that $p_i(y) = p_j(y) = w_a(y)$ we have

$$p(y) = p_i(y) + (p_j(x) - p_i(x))/2 \geq w_a(y) + (j - i)/2.$$

Therefore, the corollary holds for the maximum of all $w_a(y)$. ∎

**Lemma 3** *Every free edge segment passes through a free region which contains no other edges.*

*Proof:*   From Lemma 1, we know that once a free edge segment lies within a free region it remains inside. Therefore, we only need to be concerned about ensuring that free edge segments are initially routed through a free region. This, of course, happens only with edges extending from a new vertex $v = v_{k+1}$.

For $k \geq 2$, when $v$ is inserted there are two types of new edges added: the *outside edges* between $v$ and the outside neighbors, $w_l$ and $w_r$, and the *inside edges* between $v$ and the inside neighbors $w_i$ where $l < i < r$. In both cases the new edge is routed through a port creating one free edge segment and one port edge segment. A free edge segment of an outside edge has slope either $+1$ or $-1$ by construction; therefore it lies inside the free regions $L^f$ and $R^f$ of vertex $v$. Since $v$ is a new vertex, there are no other segments inside these two free regions.

Dealing with the inside edges is more complex. We first need to show that there is sufficient space between the vertices on the exterior face of $G_k$ and the

Duncan & Kobourov, *Polar Coordinate Drawing*, JGAA, 7(4) 311–333 (2003)324

new vertex $v$. Second, we need to show that $v$ has enough ports in its middle port region $M^p$ for each of the vertices on $G_k$ that it is connected to. Third, we need to show that the free edge segments of the inside edges initially lie inside their respective free regions. We shall show that for every inside neighbor $w_i$, $l < i < r$,

- vertex $w_i$ lies on or below $M^p$, the middle port region for $v$, and

- we can assign a unique port along the $M^p$ port region of $v$, such that the edge segment connecting $w_i$ to that port fits inside $w_i$'s middle free region $M^f$

The first part is fairly easy, we chose the middle port region $M^p$ to be $\lceil (r-l)/2 \rceil$ units below $v$. That is, the $y$-coordinate of $M^p$ is $v(y) - \lceil (r-l)/2 \rceil$. Recall that when inserting $v$, it is placed at the intersection $p$ of $w_l^+$ and $w_r^-$, unless such an intersection is not on a grid point, in which case $w_l$ is shifted left one unit to place $p$ on a grid point. Note that in actuality, $w_l$ and $w_r$ are also shifted one or two units to make the connection fall on a port but the end result is that $v$ is located at the intersection of $w_l^+$ and $w_r^-$ prior to shifting. From Corollary 1, then we know that $v(y) = p(y) \geq \max_{l \leq a \leq r}(w_a(y)) + (r-l)/2$ and it follows that all inside neighbors $w_i$, $l < i < r$, lie completely below (or on) the $M^p$ port region. Note that it is only possible for one inside neighbor $w_a$ (the maximum vertex) to actually lie directly on the port region.

We now show that our assignment strategy from Section 2.4 properly routes edges through free regions. First note that if $r - l$ is even, then there are exactly $r - l - 1$ ports on $M^p$ and if it is odd there are exactly $r - l$ ports. As there are two cases, let us look at the odd case, which has one "extra" port and is a bit trickier to prove. The other case follows a nearly identical (though simpler) argument. The assignment is done in two phases, a left to right assignment, $w_{l+1}, w_{l+2}, \ldots, w_{a-1}$, for some vertex $w_a$ followed by a right to left assignment, $w_{r-1}, w_{r-2}, \ldots w_a$. The vertex $w_a$ is defined to be the first time in the left to right assignment where the free edge segment in the routing would lie outside the free region. We call this the *skip* vertex because it essentially skips one port. Since there are exactly $r - l$ ports for $r - l - 1$ vertices, there can only be one possible "skip".

Let $w_i$ be one of the vertices routed. If $i < a$, the edge $e$ connecting $w_i$ to $v$ is routed through the $(i - l)^{\text{th}}$ port, $p_i$. Otherwise, $e$ is routed through the $(i - l + 1)^{\text{th}}$ port. Let $p_l$ be the intersection of $w_l^+$ with the port region. And, let $p_r$ be the intersection of $w_r^-$ with the port region. Then we know that if $i < a$,

$$\begin{aligned} p_i(x) - p_l(x) &= i - l \\ p_r(x) - p_i(x) &= r - i + 1 \end{aligned} \qquad (1)$$

where the $+1$ term comes because $r - l$ is odd. If $i \geq a$,

$$\begin{aligned} p_i(x) - p_l(x) &= i - l + 1 \\ p_r(x) - p_i(x) &= r - i. \end{aligned} \qquad (2)$$

Duncan & Kobourov, *Polar Coordinate Drawing*, JGAA, 7(4) 311–333 (2003)325

The free edge segment of $e$ lies in a free region only if its slope is between $-1$ and $+1$. If we let $p_i^+$ and $p_i^-$ be the intersection of $w_i^+$ and $w_i^-$ with the port region, then $e$'s free edge segment is in a free region if and only if $p_i(x)$ lies on or between $p_i^+(x)$ and $p_i^-(x)$. Applying Lemma 2, we know

$$p_i^+(x) - p_l(x) \;\geq\; i - l, \text{ and} \tag{3}$$
$$p_r(x) - p_i^-(x) \;\geq\; r - i. \tag{4}$$

Because $w_i$ lies on or below the port region, we know that

$$p_i^+(x) \geq p_i^-(x) \tag{5}$$

Given that $w_a$ is the first vertex which lies outside of the free region in the first phase, we know that, if $i < a$, $w_i$'s edge segment must lie in a free region. Let us then look at the case where $i = a$. Combining Equations (2), (3), and ( 4), we see that

$$\begin{aligned}
p_a^+(x) - p_l(x) &\;\geq\; a - l \\
&\;=\; p_a(x) - p_l(x) - 1 \Rightarrow \\
p_a^+(x) + 1 &\;\geq\; p_a(x) \\
p_r(x) - p_a^-(x) &\;\geq\; r - a \\
&\;=\; p_r(x) - p_a(x) \Rightarrow \\
p_a(x) &\;\geq\; p_a^-(x). \\
p_a^-(x) &\;\leq\; p_a(x) \leq p_a^+(x) + 1 \tag{6}
\end{aligned}$$

Notice that $p_a(x)$ is (on or) between $p_a^-(x)$ and $p_a^+(x)$ except for the case when $p_a^+(x) + 1 = p_a(x)$, i.e. $p_a$ lies one unit to the left of $p_a^+$.

So, let us assume that $p_a$ does not lie (on or) between the two slopes. Therefore, $p_a^+(x) + 1 = p_a(x)$. Now, since $w_a$ is the skip vertex we know that the port $q$ lying just to the left of $p_a$ is free. Since $p_a(x) = q(x) + 1$, we substitute in to Equation (6) yielding

$$p_a^-(x) - 1 \leq q(x) \leq p_a^+(x). \tag{7}$$

But, $q$ was not a valid port so it must not lie (on or) between $w^-$ and $w^+$. The only possibility is that $q(x) = p_a^-(x) - 1$ which implies that $p_a^+(x) + 1 = p_a(x) = q(x) + 1 = p_a^-(x)$. This in turn immediately implies that $p_a^+(x) < p_a^-(x)$ which contradicts Equation (5). Hence, $p_a$ must lie between $w_a^-$ and $w_a^+$, more precisely,

$$p_a^-(x) \leq p_a(x) \leq p_a^+(x). \tag{8}$$

Let us now look at the case where $i > a$. Observe that since the ports are assigned consecutively $p_i(x) - p_a(x) = i - a$. Applying Lemma 2 and Equation (8), for $w_a$ and $w_i$, we see that

$$\begin{aligned}
p_i^+(x) - p_a^+(x) &\;\geq\; i - a \\
&\;=\; p_i(x) - p_a(x) \\
&\;\geq\; p_i(x) - p_a^+(x) \Rightarrow \\
p_i^+(x) &\;\geq\; p_i(x). \tag{9}
\end{aligned}$$

Duncan & Kobourov, *Polar Coordinate Drawing*, JGAA, 7(4) 311–333 (2003)326

Applying Equations (2) and (4), we see that

$$
\begin{aligned}
p_r(x) - p_i^-(x) &\geq r - i \\
&= p_r(x) - p_i(x) \Rightarrow \\
p_i(x) &\geq p_i^-(x).
\end{aligned}
$$

Therefore $p_i$ lies between $w_i^-$ and $w_i^+$ and the edge to $w_i$ is properly routed.

The argument for the case when $r - l$ is even is identical except one does not have to deal with the issue of a skip vertex. Therefore, we know that all free edge segments are properly routed through free regions. ∎

**Lemma 4** *If $G_k$ maintains invariants one through four, then $G_{k+1}$ maintains invariants one through four.*

*Proof:* By definition of the shifting set, invariants one and two hold, see [7]. By construction of the algorithm, invariant three holds as well. Also by construction every edge, except $(v_1, v_2)$, inserted has a port edge segment and a free edge segment. By lemmas 1 and 3 invariant four also holds. ∎

**Lemma 5** *The angular resolution for vertex $v \in G$ as produced by the algorithm is $1/2d(v)$, where $d(v)$ is the degree of vertex $v$.*

*Proof:* The worst angle is achieved between a free edge segment for some edge $f$ and a port edge segment for some edge $e$, where $f$ is located at the boundary of its free region and $e$ is the neighboring port edge segment. There are six possible cases but the argument is the same for all of them, so without loss of generality consider the case in Fig. 8. Let $v$ be the vertex and $d(v) = d$ be its degree. Also let $s$ and $t$ be the lengths as shown in Fig. 8. Let $\theta$ be the angle between $f$ and $e$, and $x$ the number of ports as shown in the figure. Note that all vertices have at least one edge connected to them via free edge segments.[1] So, the number of ports, $x$, in any port region is at most $d - 1$. From the figure, observe that $\tan(\theta) = t/(s - t)$ and hence $\arctan(t/(s - t)) = \theta$. But

$$
\frac{t}{s - t} = \frac{\sqrt{2}/2}{\sqrt{2}(x + 1) - \sqrt{2}/2} = \frac{1}{2x + 1}
$$

Using the Maclaurin expansion for $\arctan(y)$, where $y < 1$ we have

$$
\arctan(y) = y - y^3/3 + y^5/5 - \ldots > y - y^3/3 > y - y^2/(y+1) = y/(y+1) = 1/(1+1/y)
$$

Here, the last inequality comes from the fact that for $0 < y < 1$, $y^3/3 < y^3/(y+1) < y^2/(y+1)$. Since $0 < x \leq d - 1$ and $0 < 1/(2x + 1) < 1$, this yields

$$
\theta = \arctan(1/(2x + 1)) > 1/(1 + 2x + 1) = 1/(2x + 2) \geq 1/2d.
$$

Therefore, the angular resolution is strictly greater than $1/2d$. ∎

---

[1] In fact, all but the three external vertices have three free edge segments connected to them and it is a simple matter to make the external vertices have two free edge segments connected to them.

Duncan & Kobourov, *Polar Coordinate Drawing*, JGAA, 7(4) 311–333 (2003)327

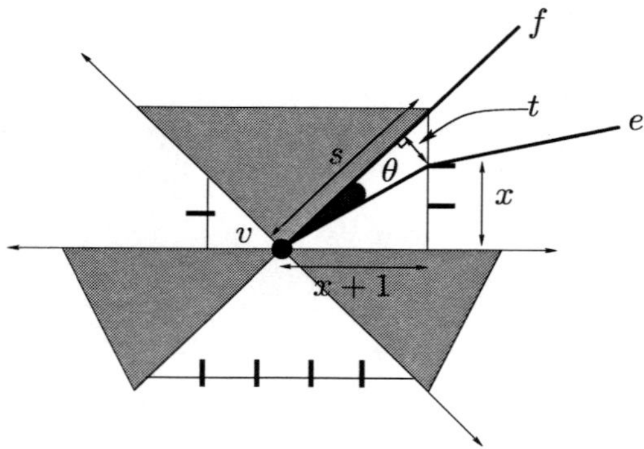

Figure 8: The minimum angle between two edges adjacent to vertex $v$ is proportional to the degree $d$ of the vertex. Using our algorithm the angle cannot be smaller than $1/2d$.

**Theorem 1** *For a given planar graph $G$, the algorithm produces in $O(n)$ time a planar embedding with grid size $5n \times 5n/2$, using at most one bend. The angular resolution for every vertex $v$ of $G$ is $1/2d(v)$.*

*Proof:*  Since every edge has only two segments, there can be at most one bend per edge. Chrobak and Payne [2] show how to implement the algorithm of De Fraysseix, *et al.* [5] in linear time. Their approach can be easily extended to our algorithm. By invariants three and four and by lemma 5 the angular resolution is at most $1/2d(v)$.

It remains to show that the drawings produced by the algorithm fit on the $5n \times 5n/2$ grid. Every time we insert a vertex $v_k$, we increase the grid size by at most 5 units, which implies that the width of the drawing is at most $5n$. The final drawing fits inside an isosceles triangle with sides of slope $0, +1, -1$. The width of the base is $5n$ and so the height is less than $5n/2$. ∎

## 4 The PRA Algorithm

In this section, we introduce a novel approach to represent bends and vertices. Rather than insisting that bends lie on integer grid coordinates, we propose an alternative approach which allows bends to be located on a grid represented by polar coordinates. Using a polar representation allows us to independently control the grid size and edge bend positions. We begin by considering the polar representation in general and then present the PRA algorithm that uses the new approach.

A point $p$ in the polar grid system is represented by a set of integers. For the vertices we only need two integers $(p_x, p_y)$. For the bend-points we may need up to five integers. We shall see in the PRA algorithm that these five integers

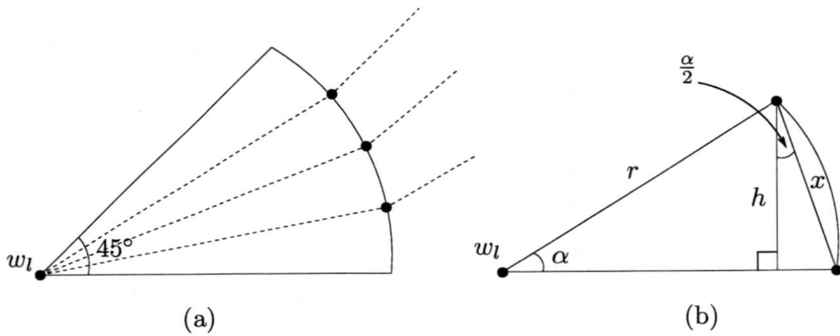



Figure 9: Vertex $w_l$ is the left-most neighbor of the next vertex $v_{k+1}$ along the exterior face of $G_k$. The $d_r(w_l)$ ports of $w_l$ are evenly spaced on the arc of a circle of radius $2d_r(w_l)$ bounded by the middle free region $M^f$ and the right free region $R^f$. (a) An example of the layout for the $R^p$ region with $d_r(w_l) = 3$. (b) The distance $x$ between two adjacent ports or a port and an adjacent free region can be computed given the radius of the circle and the angle between the edges connecting the ports to $w_l$: $x = 2r \sin \frac{\alpha}{2}$.

need not be explicitly stored for every bend-point. In general, a bend-point is given by:

- $(p_x, p_y)$, the origin of the polar system

- $p_r$, the radius of the circle around the origin $(p_x, p_y)$

- $p_d$ and $p_n$, the angle ($p_\theta$) of the circle where the point is located, i.e., $p_\theta = 2\pi p_n/p_d$. For convenience, we consider $p_\theta = 0$ to be the vertical direction.

The PRA algorithm places vertices at integer grid coordinates, thus guaranteeing unit vertex resolution. As it is based on the CRA algorithm it also uses only 1 bend per edge. The main difference in the two algorithms is in the placement of the bend-points. In the PRA algorithm, bend-points will be placed on a circle around the vertex (rather than on a straight-line segment). Therefore, the origin, $(p_x, p_y)$ for each bend-point need not be explicitly stored – it suffices to store the origin of the vertex that the bend-point is associated with. Similarly, groups of bend-points around a given vertex will have the same radius and hence each of the bend-points need not explicitly store $p_r$. Since the points will be evenly spaced in a port region, the values for $p_\theta$ need also not be explicitly stored for each bend-point.

Consider the leftmost neighbor, $w_l$, of the next vertex in the canonical order, $v_{k+1}$. The ports are evenly spaced in the $R^p$ region for $w_l$, Fig. 9(a). The length of the straight-line segment separating two bend-points or a bend-point and an adjacent free region can be computed as follows. Consider the example in Fig. 9(b). We would like to compute the length $x$ in terms of the radius of

Duncan & Kobourov, *Polar Coordinate Drawing*, JGAA, 7(4) 311–333 (2003)329

the circle and the angle between the two line segments connecting consecutive ports to $w_l$. From basic trigonometry, the angle between $h$ and $x$ is $\alpha/2$. We can express $h$ in terms of $r$ and $\alpha$: $h/r = \sin\alpha$ and we can express $x$ in terms of $h$ and $\alpha$: $h/x = \cos\alpha/2$. Combining the two expressions we obtain

$$x = \frac{h}{\cos\frac{\alpha}{2}} = \frac{r\sin\alpha}{\cos\frac{\alpha}{2}} = \frac{2r\sin\frac{\alpha}{2}\cos\frac{\alpha}{2}}{\cos\frac{\alpha}{2}} = 2r\sin\frac{\alpha}{2}.$$

Assume we have inserted $v_1, v_2, \ldots, v_k$ and have a drawing of $G_k$ with exterior face $C_k$. Consider inserting the next vertex $v_{k+1}$ in the canonical order. Let $w_l$ and $w_r$ be the leftmost and rightmost neighbors of $v_{k+1}$ on the exterior face $C_k$. Define $f_b$ and $f_e$ to be the *bend-point resolution* and *edge separation* respectively. Observe that in the standard Cartesian representation algorithms $f_b = 1$ and $f_e = \sqrt{2}/2$. Let $d_r(w_l)$, respectively $d_l(w_r)$, be the number of port edge segments using $R^p$ of $w_l$, respectively $L^p$ of $w_r$. When inserting $v_{k+1}$, the degrees for $w_l$ and $w_r$ affect the amount of shifting necessary to ensure proper resolution. As the cases for $d_r(w_l)$ and $d_l(w_r)$ are symmetrical, we shall concentrate on $d_r(w_l)$. There are two cases to consider:

- **case (a)** $d_r(w_l) = 0$ prior to insertion

- **case (b)** $d_r(w_l) \geq 1$ prior to insertion

In **case (a)** we insert the first edge in the port region $R^p$ between the two free regions $R^f$ and $M^f$ of $w_l$. We place the port in the middle of the arc of a circle connecting $R^f$ and $M^f$. Since there are no other bends yet in $R^p$ we are only concerned with maintaining the edge separation. We need to place the port sufficiently away from the vertex $w_l$. Consider the relationship between the radius of the circle and the edge separation, see Fig. 9.

The edge separation $f_e = x = 2r\sin\frac{\alpha}{2}$. But since there is only one port and it is in the middle of the arc, $\alpha = \pi/8$. We are interested in the radius necessary to achieve the edge separation $f_e$ which is given by

$$r = \frac{f_e}{2\sin\frac{\alpha}{2}} = \frac{f_e}{2\sin\frac{\pi}{16}} < \frac{4f_e}{\sqrt{2}} = 2\sqrt{2}f_e.$$

Since we maintain that the vertices are at integer coordinates and the radii are also integers, then the minimum radius required in case (a) is

$$r < \lceil 2\sqrt{2}f_e\rceil.$$

In **case (b)** we insert an additional port in the port region $R^p$ which already has at least one port. In this case, we must ensure that both the edge separation $f_e$ and bend-point resolution $f_b$ are preserved. In this case the radius required is given by:

$$\max\left\{\left\lceil\frac{f_e}{2\sin\frac{\pi}{8(d_r(w_l)+1)}}\right\rceil, \left\lceil\frac{f_b}{2\sin\frac{\pi}{8(d_r(w_l)+1)}}\right\rceil\right\}.$$

Duncan & Kobourov, *Polar Coordinate Drawing*, JGAA, 7(4) 311–333 (2003)330

| Algorithm | $f_v$ | $f_b$ | $f_e$ | drawing area | resolution |
|:---------:|:-----:|:-----:|:-----:|:------------:|:----------:|
| CRA | 1 | 1 | $\sqrt{2}/2$ | $5n \times 5n/2$ | $1/2d(v)$ |
| PRA1 | 1 | 1 | $\sqrt{2}/2$ | $9n \times 9n/2$ | $\pi/4d(v)$ |
| PRA2 | 1 | 1/2 | 1/2 | $7n \times 7n/2$ | $\pi/4d(v)$ |

Table 1: Fixing specific values for the vertex resolution $f_v$, bend-point resolution $f_b$, and edge separation $f_e$ allows us to compare the PRA and CRA algorithms.

Typically, $f_b \geq f_e$, so we can assume that the bend-point resolution determines the radius in case (b). Using this together with the fact that $\sin \alpha > 0.97\alpha$ for $\alpha < \pi/8$, the minimum radius required is

$$r < \lceil \frac{f_b}{2\sin\frac{\pi}{8(d_r(w_l)+1)}} \rceil < \lceil \sqrt{2}f_b(d_r(w_l) + 1) \rceil$$

Summing over all vertices in the graph, the sum of the radii used for the right port regions, $R$, yields:

$$R = \sum_{v_i \in V : d_r(v_i)=1} \lceil 2\sqrt{2}f_e \rceil + \sum_{v_i \in V : d_r(v_i)>1} \lceil \sqrt{2}f_b(d_r(v_i) + 1) \rceil. \qquad (10)$$

With $R$ we bounded the number of shifts required because of "right" neighbors. Similarly, we can define $L$, the shifts necessary due to "left" neighbors:

$$L = \sum_{v_i \in V : d_l(v_i)=1} \lceil 2\sqrt{2}f_e \rceil + \sum_{v_i \in V : d_l(v_i)>1} \lceil \sqrt{2}f_b(d_l(v_i) + 1) \rceil. \qquad (11)$$

$L$ and $R$ bound the number of shifts required due to left and right neighbor visibility. Note, however, that if we shift by the minimum amount required by the $f_e$ and $f_b$ parameters, the location of the next vertex $v_{k+1}$ may not be at integer coordinates. We can guarantee that $v_{k+1}$ is placed on the integer grid by performing some additional shifts. By shifting at most 3 more units, we are guaranteed to find an integer location for $v_{k+1}$. Then the total shifting required is at most $L + R + 3n$. Since the final drawing fits inside an isosceles right-angle triangle, the total area required for the drawing is $(L + R + 3n) \times (\frac{L+R+3n}{2})$.

In order to compare the PRA algorithm to the CRA algorithm, we evaluate equations 11 and 10 using two sets of parameters, Table 1. In all three cases the algorithms guarantee at most one bend per edge. The PRA algorithms place all the vertices on grid points and each bend-point is determined by at most five integer polar coordinates.

## 5  Conclusion and Open Problems

In this paper we present two algorithms for drawing planar graphs with good angular resolution while maintaining small drawing area. Other drawing criteria optimized by the algorithms include number of bends, vertex resolution,

Duncan & Kobourov, *Polar Coordinate Drawing*, JGAA, 7(4) 311–333 (2003)331

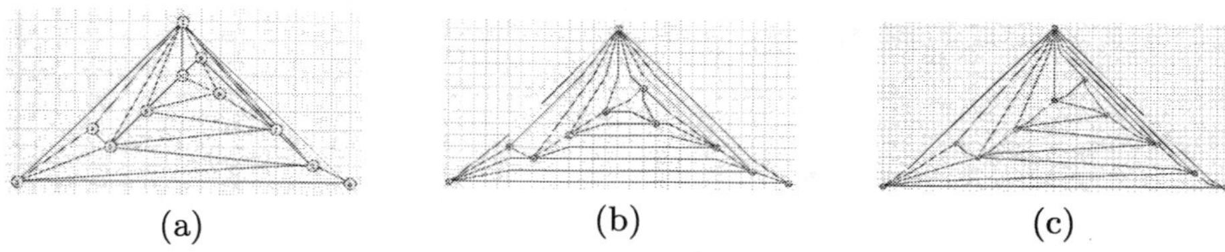

(a)       (b)       (c)

Figure 10: A graph with 11 vertices drawn using (a) the canonical ordering on the $10 \times 19$ grid; (b) the CRA algorithm on the $14 \times 29$ grid; (c) the PRA algorithm on the $23 \times 45$ grid.

bend-point resolution, and edge separation. The first algorithm, CRA, is a traditional algorithm in which vertices and bend-points are represented using Cartesian coordinates. It improves on the best known simultaneous bounds for the six drawing criteria. In the PRA algorithm vertices and bend-points are represented using polar coordinates. It is based on the CRA algorithm but allows for independent control over the grid size and bend positions.

Using a polar coordinate representation yields slightly worse area bounds compared to the CRA algorithm, see Fig 10 and Fig. 11. We believe, however, that the PRA approach is more promising. The angular resolution of the PRA algorithm is better and it provides greater control over the drawing process.

The PRA bounds presented in this paper can be further improved. Using two integers to represent the radius (similar to the way the angles are currently represented) will most likely result in smaller drawing area. Our current estimates indicate that certain (small) values of edge separation and bend-point resolution yield grids of size $4n \times 2n$. It is likely that when using only one bend per edge, the best angular resolution will be achieved for vertex regions in which each of the port and free regions have angles $\pi/3$ rather than a combination of $\pi/4$ and $\pi/2$. The biggest challenge, however, to the success of the PRA algorithm deals with the three potential shifts needed to align a new vertex onto an integer grid. If we can reduce this bottleneck, we feel that the PRA algorithm can significantly surpass the bounds of the CRA algorithm.

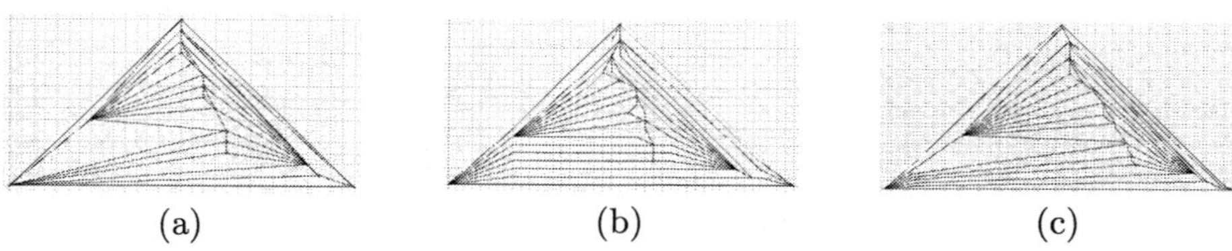

(a)       (b)       (c)

Figure 11: A graph with 17 vertices drawn using (a) the canonical ordering on the $16 \times 31$ grid; (b) the CRA algorithm on the $21 \times 41$ grid; (c) the PRA algorithm on the $43 \times 85$ grid.

Duncan & Kobourov, *Polar Coordinate Drawing*, JGAA, 7(4) 311–333 (2003)332

# Acknowledgments

A preliminary version of this paper appeared in the Proceedings of the 9th Symposium on Graph Drawing [4].

Duncan & Kobourov, *Polar Coordinate Drawing*, JGAA, 7(4) 311–333 (2003)333

# References


[1] C. C. Cheng, C. A. Duncan, M. T. Goodrich, and S. G. Kobourov. Drawing planar graphs with circular arcs. *Discrete and Computational Geometry*, 25:405–418, 2001.

[2] M. Chrobak and T. Payne. A linear-time algorithm for drawing planar graphs. *Inform. Process. Lett.*, 54:241–246, 1995.

[3] G. Di Battista, P. Eades, R. Tamassia, and I. G. Tollis. *Graph Drawing: Algorithms for the Visualization of Graphs*. Prentice Hall, Englewood Cliffs, NJ, 1999.

[4] C. A. Duncan and S. G. Kobourov. Polar coordinate drawing of planar graphs with good angular resolution. In *Proc. 9th Symposium on Graph Drawing*, volume 2265 of *Lecture Notes in Computer Science*, pages 407–421, 2002.

[5] H. Fraysseix, J. Pach, and R. Pollack. How to draw a planar graph on a grid. *Combinatorica*, 10(1):41–51, 1990.

[6] A. Garg and R. Tamassia. Planar drawings and angular resolution: Algorithms and bounds. In *Proc. 2nd European Symposium on Algorithms*, pages 12–23, 1994.

[7] M. T. Goodrich and C. G. Wagner. A framework for drawing planar graphs with curves and polylines. In *Proc. 6th Symposium on Graph Drawing*, volume 1547 of *Lecture Notes in Computer Science*, pages 153–166, 1998.

[8] C. Gutwenger and P. Mutzel. Planar polyline drawings with good angular resolution. In *Proc. 6th Symposium on Graph Drawing*, volume 1547 of *Lecture Notes in Computer Science*, pages 167–182, 1998.

[9] G. Kant. Drawing planar graphs using the canonical ordering. *Algorithmica*, 16:4–32, 1996. (special issue on Graph Drawing, edited by G. Di Battista and R. Tamassia).

[10] S. Malitz and A. Papakostas. On the angular resolution of planar graphs. *SIAM J. Discrete Math.*, 7:172–183, 1994.

# Journal of Graph Algorithms and Applications

http://jgaa.info/

vol. 7, no. 4, pp. 335–362 (2003)

# Orthogonal Drawings of Plane Graphs Without Bends

*Md. Saidur Rahman*    *Takao Nishizeki*

Graduate School of Information Sciences
Tohoku University
Aoba-yama 05, Sendai 980-8579, Japan
http://www.nishizeki.ecei.tohoku.ac.jp/
saidur@nishizeki.ecei.tohoku.ac.jp, nishi@ecei.tohoku.ac.jp

*Mahmuda Naznin*

Department of Computer Science
North Dakota State University
Fargo, ND 58105-5164, USA
Mahmuda.Naznin@ndsu.nodak.edu

### Abstract

In an orthogonal drawing of a plane graph each vertex is drawn as a point and each edge is drawn as a sequence of vertical and horizontal line segments. A bend is a point at which the drawing of an edge changes its direction. Every plane graph of the maximum degree at most four has an orthogonal drawing, but may need bends. A simple necessary and sufficient condition has not been known for a plane graph to have an orthogonal drawing without bends. In this paper we obtain a necessary and sufficient condition for a plane graph $G$ of the maximum degree three to have an orthogonal drawing without bends. We also give a linear-time algorithm to find such a drawing of $G$ if it exists.

Communicated by: P. Mutzel and M. Jünger;
submitted May 2002; revised November 2002.

---

Part of this work was done while the first and the third authors were in Bangladesh University of Engineering and Technology (BUET). This work is supported by the grants of Japan Society for the Promotion of Science (JSPS).

M. S. Rahman et al., *Orthogonal Drawings*, JGAA, 7(4) 335–362 (2003)    336

# 1   Introduction

Automatic graph drawings have numerous applications in VLSI circuit layout, networks, computer architecture, circuit schematics etc. For the last few years many researchers have concentrated their attention on graph drawings and introduced a number of drawing styles. Among the styles, "orthogonal drawings" have attracted much attention due to their various applications, specially in circuit schematics, entity relationship diagrams, data flow diagrams etc. [1]. An *orthogonal drawing* of a plane graph $G$ is a drawing of $G$ with the given embedding in which each vertex is mapped to a point, each edge is drawn as a sequence of alternate horizontal and vertical line segments, and any two edges do not cross except at their common end. A *bend* is a point where an edge changes its direction in a drawing. Every plane graph of the maximum degree four has an orthogonal drawing, but may need bends. For the cubic plane graph in Fig. 1(a) each vertex of which has degree 3, two orthogonal drawings are shown in Figs. 1(b) and (c) with 6 and 5 bends respectively. Minimization of the number of bends in an orthogonal drawing is a challenging problem. Several works have been done on this issue [2, 3, 8, 13]. In particular, Garg and Tamassia [3] presented an algorithm to find an orthogonal drawing of a given plane graph $G$ with the minimum number of bends in time $O(n^{7/4}\sqrt{\log n})$, where $n$ is the number of vertices in $G$. Rahman *et al.* gave an algorithm to find an orthogonal drawing of a given triconnected cubic plane graph with the minimum number of bends in linear time [8].

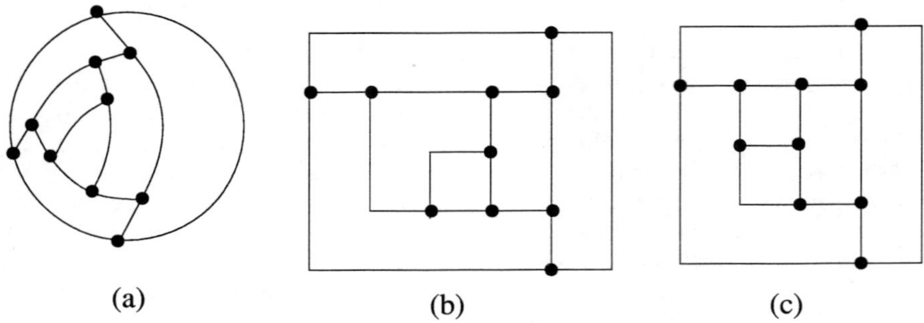

(a)          (b)          (c)

Figure 1: (a) A plane graph $G$, (b) an orthogonal drawing of $G$ with 6 bends, and (c) an orthogonal drawing of $G$ with 5 bends.

In a VLSI floorplanning problem, an input is often a plane graph of the maximum degree 3 [4, 9, 10]. Such a plane graph $G$ may have an orthogonal drawing without bends. The graph in Fig. 2(a) has an orthogonal drawing without bends as shown in Fig. 2(b). However, not every plane graph of the maximum degree 3 has an orthogonal drawing without bends. For example, the cubic plane graph in Fig. 1(a) has no orthogonal drawing without bends, since any orthogonal drawing of an outer cycle have at least four convex corners which must be bends in a cubic graph. One may thus assume that there are

M. S. Rahman et al., *Orthogonal Drawings*, JGAA, 7(4) 335–362 (2003)    337

four or more vertices of degree two on the outer cycle of $G$. It is interesting to know which classes of such plane graphs have orthogonal drawings without bends. However, no simple necessary and sufficient condition has been known for a plane graph to have an orthogonal drawing without bends, although one can know in time $O(n^{7/4}\sqrt{\log n})$ by the algorithm [3] whether a given plane graph has an orthogonal drawing without bends.

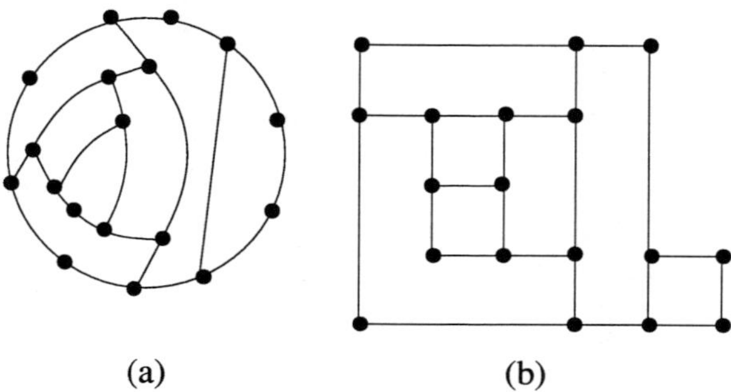

(a)                              (b)

Figure 2: (a) A plane graph $G$ and (b) an orthogonal drawing of $G$ without bends.

In this paper we obtain a simple necessary and sufficient condition for a plane graph $G$ of the maximum degree 3 to have an orthogonal drawing without bends. The condition is a generalization of Thomassen's condition for the existence of "rectangular drawings" [12]. Our condition leads to a linear-time algorithm to find an orthogonal drawing of $G$ without bends if it exists.

The rest of paper is organized as follows. Section 2 describes some definitions and presents known results. Section 3 presents our results on orthogonal drawings of biconnected plane graphs without bends. Section 4 deals with orthogonal drawings of arbitrary (not always biconnected) plane graphs without bends. Finally Section 5 gives the conclusion. A preliminary version of this paper is presented in [11].

## 2  Preliminaries

In this section we give some definitions and preliminary known results.

Let $G$ be a connected simple graph with $n$ vertices and $m$ edges. The *degree* of a vertex $v$ is the number of neighbors of $v$ in $G$. A vertex of degree 2 in $G$ is called a *2-vertex* of $G$. We denote the maximum degree of graph $G$ by $\Delta(G)$ or simply by $\Delta$. The *connectivity* $\kappa(G)$ of a graph $G$ is the minimum number of vertices whose removal results in a disconnected graph or a single vertex graph. We say that $G$ is *k-connected* if $\kappa(G) \geq k$. We call a vertex of $G$ a *cut vertex* if its removal results in a disconnected graph.

A graph is *planar* if it can be embedded in the plane so that no two edges intersect geometrically except at a vertex to which they are both incident. A

M. S. Rahman et al., *Orthogonal Drawings*, JGAA, 7(4) 335–362 (2003)     338

*plane graph* $G$ is a planar graph with a fixed planar embedding. A plane graph $G$ divides the plane into connected regions called *faces*. We refer the *contour* of a face as a cycle formed by the edges on the boundary of the face. We denote the contour of the outer face of $G$ by $C_o(G)$.

An edge of a plane graph $G$ is called a *leg* of a cycle $C$ if it is incident to exactly one vertex of $C$ and located outside $C$. The vertex of $C$ to which a leg is incident is called a *leg-vertex* of $C$. A cycle in $G$ is called a *k-legged cycle* of $G$ if $C$ has exactly $k$ legs in $G$ and there is no edge which joins two vertices on $C$ and is located outside $C$.

An *orthogonal drawing* of a plane graph $G$ is a drawing of $G$ with the given embedding in which each vertex is mapped to a point, each edge is drawn as a sequence of alternate horizontal and vertical line segments, and any two edges do not cross except at their common end. A *bend* is a point where an edge changes its direction in a drawing. Any cycle $C$ in $G$ is drawn as a rectilinear polygon in an orthogonal drawing $D(G)$ of $G$. The polygon is denoted by $D(C)$. A (polygonal) vertex of the rectilinear polygon is called a *corner of the drawing* $D(C)$. A corner has an interior angle 90° or 270°. A corner of an interior angle 90° is called a *convex corner* of $D(C)$, while a corner of an interior angle 270° is called a *concave corner*. A vertex $v$ on $C$ is called a *non-corner* of $D(C)$ if $v$ is not a corner of $D(C)$. Thus any vertex on $C$ is a convex corner, a concave corner, or a non-corner of $D(C)$.

A *rectangular drawing* of a plane biconnected graph $G$ is a drawing of $G$ such that each edge is drawn as a horizontal or a vertical line segment, and each face is drawn as a rectangle. (See Fig. 9.) Thus a rectangular drawing is an orthogonal drawing in which there is no bends and each face is drawn as a rectangle. The rectangular drawing of $C_o(G)$ is called the *outer rectangle*. The following result is known on rectangular drawings.

**Lemma 1** *Assume that $G$ is a plane biconnected graph with $\Delta \leq 3$, and that four 2-vertices on $C_o(G)$ are designated as the four (convex) corners of the outer rectangle. Then $G$ has a rectangular drawing if and only if $G$ satisfies the following two conditions [12]:*

*(r1) every 2-legged cycle contains at least two designated vertices, and*

*(r2) every 3-legged cycle contains at least one designated vertex.*

*Furthermore one can examine in linear time whether $G$ satisfies the condition above, and if $G$ does then one can find a rectangular drawing in linear time [7].*

Consider two examples in Fig. 3, where the four designated corner vertices are drawn by white circles in each graph. Cycles $C_1$, $C_2$ and $C_3$ are 2-legged, and $C_4$, $C_5$ and $C_6$ are 3-legged. $C_3$, $C_5$ and $C_6$ do not violate the conditions in Lemma 1. On the other hand, cycles $C_1$, $C_2$ and $C_4$ violate the conditions.

A cycle in $G$ violating (r1) or (r2) is called a *bad cycle*: a 2-legged cycle is *bad* if it contains at most one designated vertex; a 3-legged cycle is *bad* if it contains no designated vertex.

M. S. Rahman et al., *Orthogonal Drawings*, JGAA, 7(4) 335–362 (2003)     339

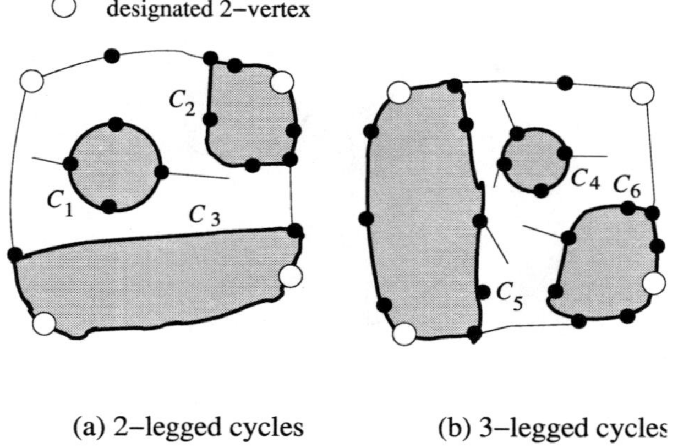



(a) 2–legged cycles          (b) 3–legged cycles

Figure 3: (a) 2-legged cycles $C_1$, $C_2$ and $C_3$, and (b) 3-legged cycles $C_4$, $C_5$ and $C_6$.

Rahman *et al.* [7] have obtained a linear-time algorithm to find a rectangular drawing of a plane graph $G$ if $G$ satisfies the conditions in Lemma 1 for four designated corner vertices on $C_o(G)$. We call it Algorithm **Rectangular-Draw** and use it in our orthogonal drawing algorithm of this paper.

For a cycle $C$ in a plane graph $G$, we denote by $G(C)$ the plane subgraph of $G$ inside $C$ (including $C$). A bad cycle $C$ in $G$ is called a *maximal bad cycle* if $G(C)$ is not contained in $G(C')$ for any other bad cycle $C'$ of $G$. In Fig. 4 $C_1, C_3, C_4, C_5$ and $C_6$ are bad cycles, but $C_2$ is not a bad cycle, where $C_2$ and $C_4$ are drawn by thick lines. $C_1, C_4, C_5$ and $C_6$ are the maximal bad cycles. $C_3$ is not a maximal bad cycle because $G(C_3)$ is contained in $G(C_4)$ for a bad cycle $C_4$. We say that cycles $C$ and $C'$ in a plane graph $G$ are *independent* of each other if $G(C)$ and $G(C')$ have no common vertex. We now have the following lemma.

**Lemma 2** *If $G$ is a biconnected plane graph of $\Delta \le 3$ and four 2-vertices on $C_o(G)$ are designated as corners, then the maximal bad cycles in $G$ are independent of each other.*

**Proof:** Assume for a contradiction that a pair of maximal bad cycles $C_1$ and $C_2$ in $G$ are not independent. Then the subgraphs $G(C_1)$ and $G(C_2)$ have a common vertex. In particular, the cycles $C_1$ and $C_2$ have a common vertex, because $C_1$ and $C_2$ are maximal bad cycles. Since $\Delta \le 3$, $C_1$ and $C_2$ share a common edge; $C_1$ contains two legs of $C_2$, and $C_2$ contains two legs of $C_1$. There are two cases to consider.

**Case 1:** $C_1$ and $C_2$ have a common vertex not on $C_o(G)$.

There are three cases; (i) both $C_1$ and $C_2$ are 2-legged cycles, (ii) one of $C_1$ and $C_2$ is a 2-legged cycle and the other is a 3-legged cycle, and (iii) both

M. S. Rahman et al., *Orthogonal Drawings*, JGAA, 7(4) 335–362 (2003)    340

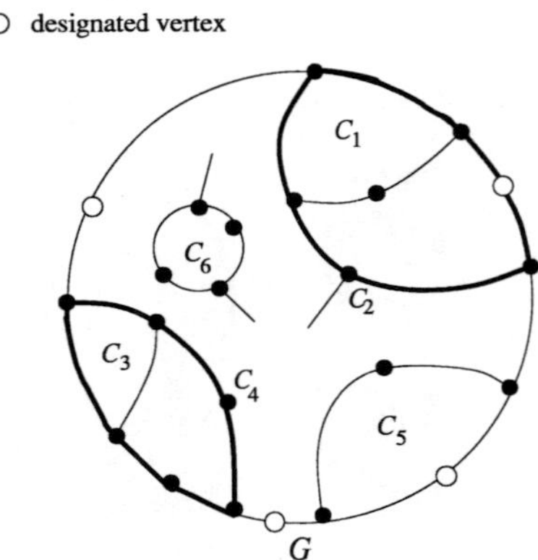


Figure 4: Maximal bad cycles $C_1, C_4, C_5$ and $C_6$.

$C_1$ and $C_2$ are 3-legged cycles. If both $C_1$ and $C_2$ are 2-legged cycles, then $G$ would be a disconnected graph as illustrated in Fig. 5(a), a contradiction to the assumption that $G$ is biconnected. If one of $C_1$ and $C_2$ is a 2-legged cycle and the other is a 3-legged cycle, then $G$ would have a cut-vertex $v$ as illustrated in Fig. 5(b), a contradiction to the assumption that $G$ is biconnected. If both $C_1$ and $C_2$ are 3-legged cycles, then there would exist a 2-legged bad cycle $C^*$ in $G$ such that $G(C^*)$ contains both $C_1$ and $C_2$, a contradiction to the assumption that $C_1$ and $C_2$ are maximal bad cycles. In Fig. 5(c) $C^*$ is drawn by thick lines.

**Case 2:** $C_1$ and $C_2$ have a common vertex on $C_o(G)$.

If both $C_1$ and $C_2$ are 2-legged cycles, then one of $G(C_1)$ and $G(C_2)$ would be contained in the other as illustrated in Fig. 5(d), a contradiction to the assumption that both $C_1$ and $C_2$ are maximal bad cycles. If one of $C_1$ and $C_2$ is a 2-legged cycle and the other is a 3-legged cycle, then one of $G(C_1)$ and $G(C_2)$ would be contained in the other, as illustrated in Fig. 5(e) and Fig. 5(f), contrary to the assumption. If both $C_1$ and $C_2$ are 3-legged cycles, then they have no designated vertex and there would exist a bad 2-legged cycle $C^*$ such that $G(C^*)$ contains both of $C_1$ and $C_2$, a contradiction to the assumption. In Fig. 5(g) $C^*$ is drawn by thick lines. □

# 3  Orthogonal Drawings of Biconnected Plane Graphs

In this section we present our results on biconnected plane graphs. From now on we assume that $G$ is a biconnected plane graph with $\Delta \leq 3$ and there are four or more 2-vertices on $C_o(G)$. The following theorem is the main result of this section.

**Theorem 1** *Assume that $G$ is a plane biconnected graph with $\Delta \leq 3$ and there are four or more 2-vertices of $G$ on $C_o(G)$. Then $G$ has an orthogonal drawing*

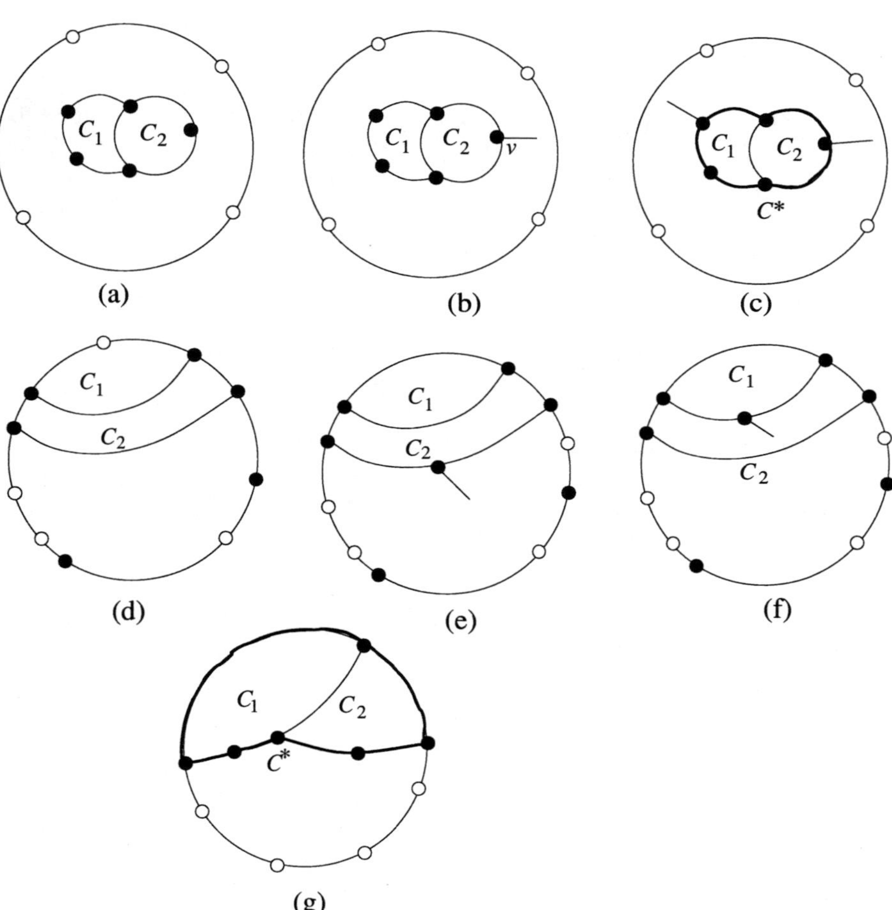

Figure 5: Illustration for the proof of Lemma 2.

M. S. Rahman et al., *Orthogonal Drawings*, JGAA, 7(4) 335–362 (2003)    342

*without bends if and only if every 2-legged cycle contains at least two 2-vertices of $G$ and every 3-legged cycle contains at least one 2-vertex of $G$.*

Note that Theorem 1 is a generalization of Thomassen's condition for rectangular drawings in Lemma 1; applying Theorem 1 to a plane biconnected graph $G$ in which all vertices have degree 3 except the four 2-vertices on $C_o(G)$, one can derive the condition.

It is easy to prove the necessity of Theorem 1, as follows.

**Necessity of Theorem 1** Assume that a plane biconnected graph $G$ has an orthogonal drawing $D$ without bends.

Let $C$ be any 2-legged cycle. Then the rectilinear polygon $D(C)$ in $D$ has at least four convex corners. These convex corners must be vertices since $D$ has no bends. The two leg-vertices of $C$ may serve as two of the convex corners. However, each of the other convex corners must be a 2-vertex of $G$. Thus $C$ must contain at least two 2-vertices of $G$.

One can similarly show that any 3-legged cycle $C$ in $G$ contains at least one 2-vertex of $G$. $\qquad\qquad\square$

In the rest of this section we give a constructive proof for the sufficiency of Theorem 1 and show that the proof leads to a linear-time algorithm to find an orthogonal drawing without bends if it exists.

Assume that $G$ satisfies the condition in Theorem 1. We now need some definitions. Let $C$ be a 2-legged cycle in $G$, and let $x$ and $y$ be the two leg-vertices of $C$. We say that an orthogonal drawing $D(G(C))$ of the subgraph $G(C)$ is *feasible* if $D(G(C))$ has no bend and satisfies the following condition (f1) or (f2).

**(f1)** The drawing $D(G(C))$ intersects neither the first quadrant with the origin at $x$ nor the third quadrant with the origin at $y$ after rotating the drawing and renaming the leg-vertices if necessary, as illustrated in Fig. 6. Note that $C$ is not always drawn by a rectangle.

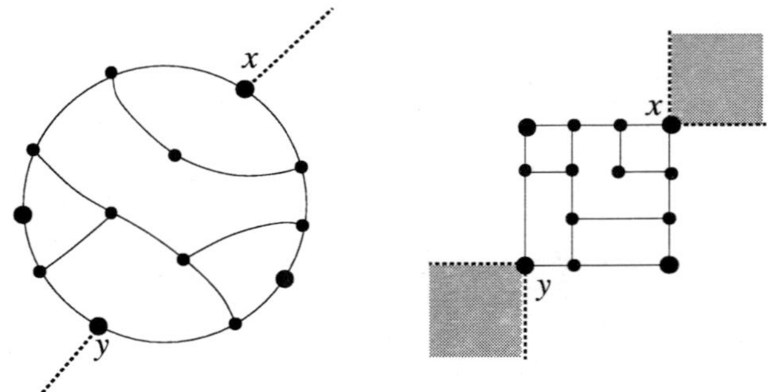


Figure 6: Illustration of (f1) for a 2-legged cycle.

M. S. Rahman et al., *Orthogonal Drawings*, JGAA, 7(4) 335–362 (2003)    343

**(f2)** The drawing $D(G(C))$ intersects neither the first quadrant with the origin at $x$ nor the fourth quadrant with the origin at $y$ after rotating the drawing and renaming the leg-vertices if necessary, as illustrated in Fig. 7.

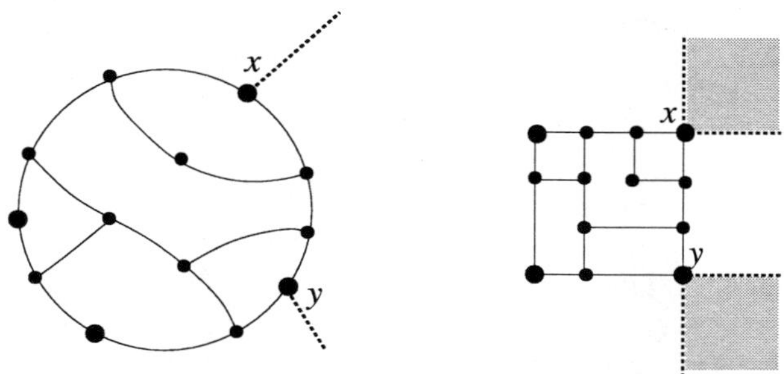

Figure 7: Illustration of (f2) for a 2-legged cycle.

Let $C$ be a 3-legged cycle in $G$, and let $x$, $y$ and $z$ be the three leg-vertices. One may assume that $x$, $y$ and $z$ appear clockwise on $C$ in this order. We say that an orthogonal drawing $D(G(C))$ is *feasible* if $D(G(C))$ has no bend and $D(G(C))$ satisfies the following condition (f3).

**(f3)** The drawing $D(G(C))$ intersects none of the following three quadrants: the first quadrant with origin at $x$, the fourth quadrant with origin at $y$, and the third quadrant with origin at $z$ after rotating the drawing and renaming the leg-vertices if necessary, as illustrated in Fig. 8.

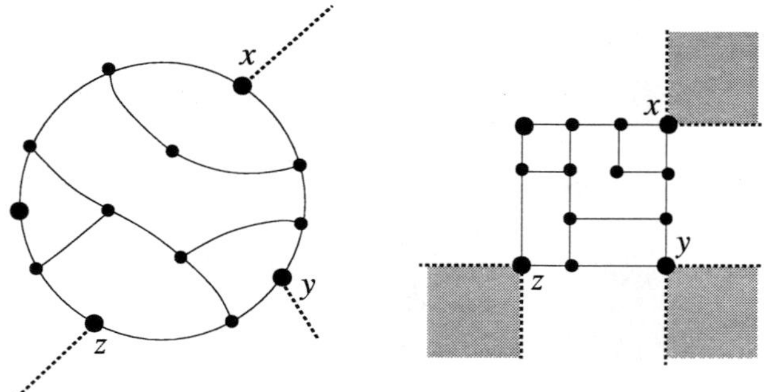

Figure 8: Illustration of (f3) for a 3-legged cycle.

Each of Conditions (f1), (f2) and (f3) implies that, in the drawing of $G(C)$, any vertex of $G(C)$ except the leg-vertices is located in none of the shaded quadrants in Figs. 6, 7 and 8, and hence a leg incident to $x$, $y$ or $z$ can be drawn by a horizontal or vertical line segment without edge-crossing as indicated by dotted lines in Figs. 6, 7 and 8.

We now have the following lemma.

M. S. Rahman et al., *Orthogonal Drawings*, JGAA, 7(4) 335–362 (2003)     344

**Lemma 3** *If $G$ satisfies the condition in Theorem 1, that is, every 2-legged cycle in $G$ contains at least two 2-vertices of $G$ and every 3-legged cycle in $G$ contains at least one 2-vertex of $G$, then $G(C)$ has a feasible orthogonal drawing for any 2- or 3-legged cycle $C$ in $G$.*

**Proof:** We give a recursive algorithm to find a feasible orthogonal drawing of $G(C)$. There are two cases to be considered.

**Case 1:** $C$ is a 2-legged cycle.

Let $x$ and $y$ be the two leg-vertices of $C$, and let $e_x$ and $e_y$ be the legs incident to $x$ and $y$, respectively. Since $C$ satisfies the condition in Theorem 1, $C$ has at least two 2-vertices of $G$. Let $a$ and $b$ be any two 2-vertices of $G$ on $C$. We now regard the four vertices $x$, $y$, $a$ and $b$ as the four designated corner vertices of $C$.

We first consider the case where $G(C)$ has no bad cycle with respect to the four designated vertices. In this case, by Lemma 1 $G(C)$ has a rectangular drawing $D$ with the four designated corner vertices, as illustrated in Fig. 9. Such a rectangular drawing $D$ of $G(C)$ can be found by the algorithm **Rectangular-Draw** in [7]. The outer cycle $C$ of $G(C)$ is drawn as a rectangle in $D$, and $x$, $y$, $a$ and $b$ are the convex corners of the rectangle. Hence $D$ satisfies Condition (f1) or (f2). Since $D$ is a rectangular drawing, $D$ has no bend. Thus $D$ is a feasible orthogonal drawing of $G(C)$.

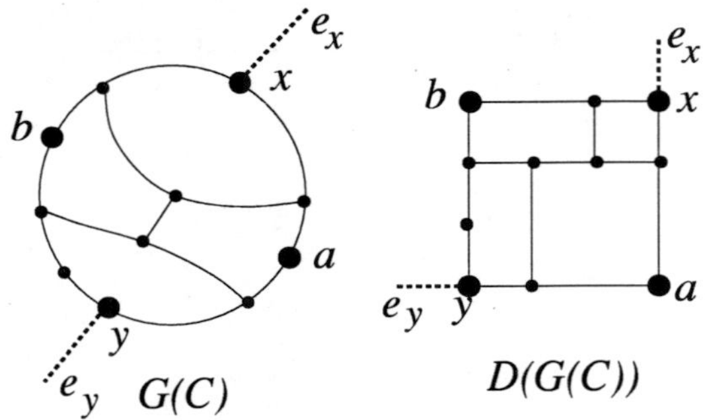


Figure 9: Subgraph $G(C)$ and its rectangular drawing $D(G(C))$.

We then consider the case where $G(C)$ has a bad cycle. Let $C_1, C_2, \cdots, C_l$ be the maximal bad cycles of $G(C)$. By Lemma 2 $C_1, C_2, \cdots, C_l$ are independent of each other. Construct a plane graph $H$ from $G(C)$ by contracting each subgraph $G(C_i), 1 \le i \le l$, to a single vertex $v_i$, as illustrated in Figs. 10(a) and (b). Clearly $H$ is a plane biconnected graph with $\Delta \le 3$. Every bad cycle $C_i$ in $G(C)$ contains at most one designated vertex. If $C_i$ contains a designated vertex, then we newly designate $v_i$ as a corner vertex of $H$ in place of the designated vertex. Thus $H$ has exactly four designated vertices. (In Fig. 10 $H$ has four designated vertices $a$, $b$, $x$, and $v_2$ since the bad cycle $C_2$ contains $y$.) Since all maximal bad cycles are contracted to single vertices in $H$, $H$ has no bad cycle with respect to the four designated vertices, and hence by Lemma 1 $H$ has a

M. S. Rahman et al., *Orthogonal Drawings*, JGAA, 7(4) 335–362 (2003)     345

rectangular drawing $D(H)$, as illustrated in Fig. 10(c). Such a drawing $D(H)$ can be found by Algorithm **Rectangular-Draw**. Clearly there is no bend on $D(H)$. The shrunken outer cycle of $G(C)$ is drawn as a rectangle in $D(H)$, and hence $D(H)$ satisfies Conditions (f1) or (f2). If $C_i$ is a 2-legged cycle, then the two legs $e_{x_i}$, $e_{y_i}$ and vertex $v_i$ are embedded in $D(H)$ as illustrated in Figs. 11(b) and 12(b) or as in their rotated ones, and the two legs $e_{x_i}$, $e_{y_i}$ and $C_i$ can be drawn as illustrated in Figs. 11(c) and 12(c) or as in their rotated ones for a feasible orthogonal drawing $D(G(C_i))$ of $G(C_i)$. If $C_i$ is a 3-legged cycle, then $v_i$ and the three legs $e_{x_i}$, $e_{y_i}$ and $e_{z_i}$ are embedded in $D(H)$ as illustrated in Fig. 13(b) or as in their rotated ones, and $C_i$ and three legs $e_{x_i}$, $e_{y_i}$ and $e_{z_i}$ can be drawn as illustrated in Fig. 13(c) or as in their rotated ones for a feasible orthogonal drawing $D(G(C_i))$ of $G(C_i)$. One can obtain a drawing $D(G(C))$ of $G(C)$ from $D(H)$ and $D(G(C_i))$ $1 \leq i \leq l$, as follows. Replace each $v_i$, $1 \leq i \leq l$, in $D(H)$ with one of the feasible drawings of $G(C_i)$ in Fig. 11(c), Fig. 12(c) and Fig. 13(c) and their rotated ones that corresponds to the embedding of $v_i$ and the legs of $C_i$ in $D(H)$, and draw each leg of $C_i$ in $D(G(C))$ by a straight line segment having the same direction as the leg in $D(H)$, as illustrated in Fig. 10(d). We call this operation a *patching operation.*

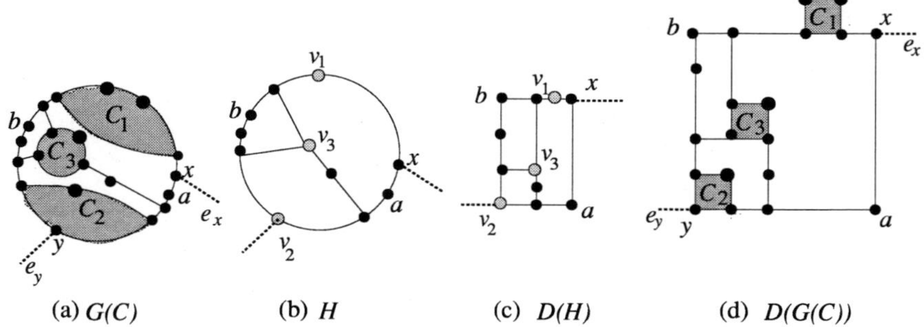

(a) *G(C)*          (b) *H*          (c) *D(H)*          (d) *D(G(C))*

Figure 10: Illustration for Case 1 where $C$ has the maximal bad cycles $C_1$, $C_2$ and $C_3$.

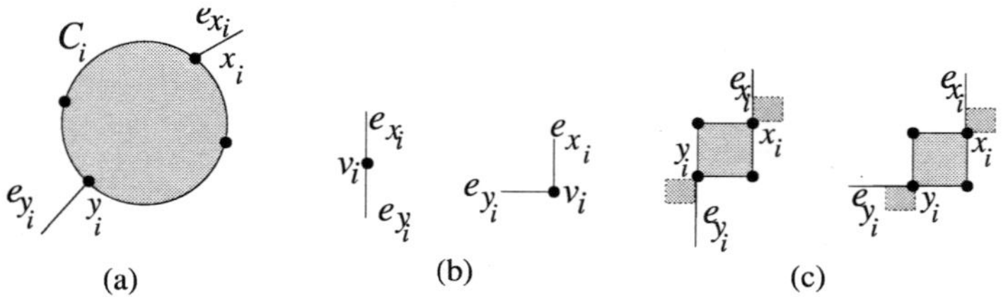

(a)          (b)          (c)

Figure 11: (a) A 2-legged cycle $C_i$ having a feasible orthogonal drawing satisfying (f1), (b) embeddings of a vertex $v_i$ and two legs $e_{x_i}$ and $e_{y_i}$ incident to $v_i$, and (c) feasible orthogonal drawings of $G(C_i)$ with two legs.

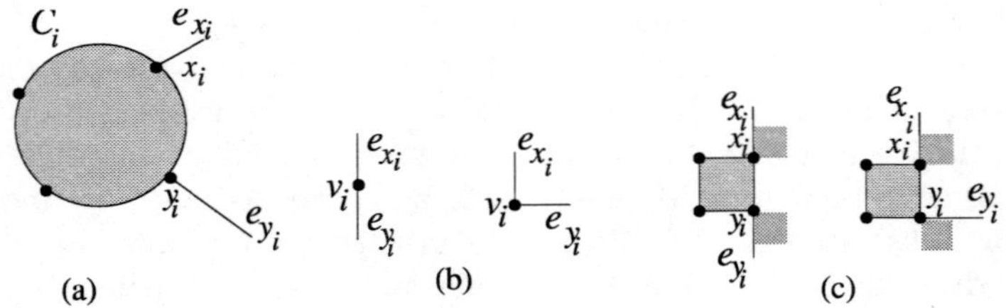

Figure 12: (a) A 2-legged cycle $C_i$ having a feasible orthogonal drawing satisfying (f2), (b) embeddings of a vertex $v_i$ and two legs $e_{x_i}$ and $e_{y_i}$ incident to $v_i$, and (c) feasible orthogonal drawings of $G(C_i)$ with two legs.

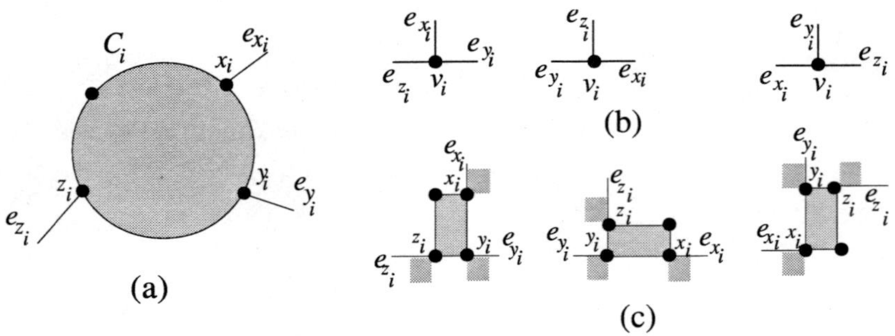

Figure 13: (a) A 3-legged cycle $C_i$ having feasible orthogonal drawings satisfying (f3), (b) embeddings of a vertex $v_i$ and three legs $e_{x_i}$, $e_{y_i}$ and $e_{z_i}$ incident to $v_i$, and (c) feasible orthogonal drawings of $G(C_i)$ with three legs.

M. S. Rahman et al., *Orthogonal Drawings*, JGAA, 7(4) 335–362 (2003)     347

We find a feasible orthogonal drawing $D(G(C_i))$ of $G(C_i), 1 \leq i \leq l$, in a recursive manner. We then patch the drawings $D(G(C_1)), D(G(C_2)), \cdots, D(G(C_l))$ into $D(H)$ by patching operation. Since there is no bend in any of $D(G(C_1))$, $D(G(C_2)), \cdots, D(G(C_l))$, there is no bend in the resulting drawing $D(G(C))$. Since the outer cycle of $D(H)$ is a rectangle and the resulting drawing $D(G(C))$ always expands outwards, $D(C)$ is not always a rectangle but $D(G(C))$ satisfies (f1) or (f2). Hence $D(G(C))$ is a feasible orthogonal drawing.

**Case 2:** $C$ is a 3-legged cycle.

Let $x$, $y$ and $z$ be the three leg-vertices of $C$, and let $e_x$, $e_y$ and $e_z$ be the legs incident to $x$, $y$ and $z$, respectively. Since $C$ satisfies the condition in Theorem 1, $C$ has at least one 2-vertex of $G$. Let $a$ be any 2-vertex of $G$ on $C$. We now regard the four vertices $x$, $y$, $z$ and $a$ as designated corner vertices.

We first consider the case where $G(C)$ has no bad cycle with respect to the four designated vertices. In this case by Lemma 1 $G(C)$ has a rectangular drawing $D$ with the four designated vertices as illustrated in Fig. 14. Since the outer cycle $C$ of $G(C)$ is drawn as a rectangle in $D$, $D$ satisfies Condition (f3). Since $D$ is a rectangular drawing, $D$ has no bend. Thus $D$ is a feasible orthogonal drawing of $G(C)$.

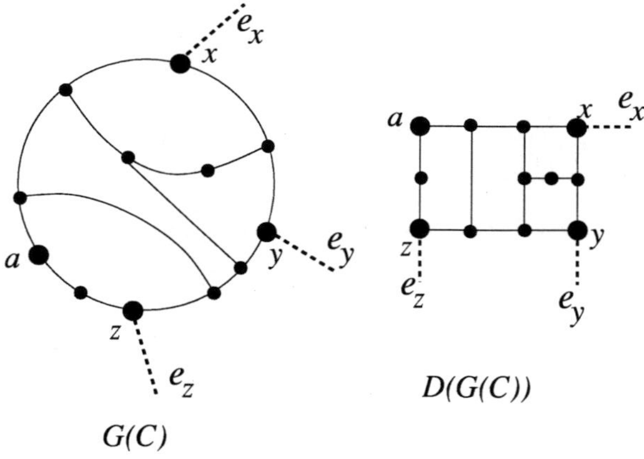


Figure 14: Illustration for Case 2 where $C$ has no bad cycle.

We then consider the case where $G(C)$ has a bad cycle. Let $C_1, C_2, \cdots, C_l$ be the maximal bad cycles of $G(C)$. By Lemma 2 $C_1, C_2, \cdots, C_l$ are independent of each other. Construct a plane graph $H$ from $G(C)$ by contracting each subgraph $G(C_i), 1 \leq i \leq l$, to a single vertex $v_i$, as illustrated in Figs. 15(a) and (b). Clearly $H$ is a plane biconnected graph with $\Delta \leq 3$, $H$ has no bad cycle with respect to the four designated vertices, and hence $H$ has a rectangular drawing $D(H)$ as illustrated in Fig. 15(c). Clearly there is no bend on $D(H)$. Since the outer cycle of $H$ is drawn as a rectangle in $D(H)$, $D(H)$ satisfies Condition (f3).

We then find a feasible orthogonal drawing $D(G(C_i))$ of $G(C_i), 1 \leq i \leq l$, in a recursive manner, and patch the drawings $D(G(C_1)), D(G(C_2)), \cdots, D(G(C_l))$

M. S. Rahman et al., *Orthogonal Drawings*, JGAA, 7(4) 335–362 (2003)    348

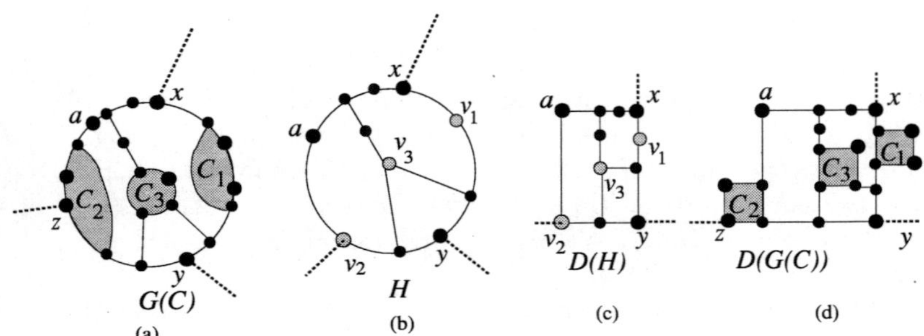

Figure 15: Illustration for Case 2 where $C$ has bad cycles $C_1$, $C_2$ and $C_3$.

into $D(H)$ as illustrated in Fig. 15(d). Since there is no bend in any of $D(G(C_1))$, $D(G(C_2)), \cdots, D(G(C_l))$, there is no bend in the resulting drawing $D(G(C))$. Since the outer boundary of $D(H)$ is a rectangle and $D(G(C))$ expands outwards, $D(G(C))$ satisfies (f3). Thus $D(G(C))$ is a feasible orthogonal drawing of $G(C)$. □

We call the algorithm for obtaining a feasible orthogonal drawing of $G(C)$ as described in the proof of Lemma 3 Algorithm **Feasible-Draw.** We now have the following lemma.

**Lemma 4** *Algorithm* **Feasible-Draw** *finds a feasible orthogonal drawing of* $G(C)$ *in time* $O(n(G(C))$, *where* $n(G(C))$ *is the number of vertices in* $G(C)$.

**Proof:** Let $T_R(G)$ be the computation time of **Rectangular-Draw** for graph $G$. Then $T_R(G) = O(n)$ by Lemma 1, and hence there is a positive constant $c$ such that

$$T_R(G) \ \leq \ c \cdot m(G) \tag{1}$$

for any plane graph $G$, where $m(G)$ is the number of edges in $G$.

We first consider the computation time needed for contraction and patching operations in Algorithm **Feasible-Draw.** During the traversal of all inner faces of $G(C)$ we can find the leg-vertices for each bad cycle [8]. Given the leg-vertices of a bad cycle, we can contract the bad cycle to a single vertex in constant time. Therefore the contraction operations in **Feasible-Draw** take $O(n(G(C)))$ time in total. Similarly the patching operations in **Feasible-Draw** take $O(n(G(C)))$ time in total.

We then consider the time needed for operations in **Feasible-Draw** other than the contractions and patchings. Let $T(G(C))$ be the computation time of **Feasible-Draw** for finding a feasible orthogonal drawing of $G(C)$ excluding the time for the contractions and patchings. We claim that $T(G(C)) = O(n(G(C)))$. Since $G$ is a plane graph, $m(G(C)) \leq 3n(G(C))$, where $m(G(C))$ denotes the number of edges in $G(C)$. Therefore it is sufficient to show that

$$T(G(C)) \ \leq \ c \cdot m(G(C)). \tag{2}$$

M. S. Rahman et al., *Orthogonal Drawings*, JGAA, 7(4) 335–362 (2003)    349

We prove Eq. (2) by induction.

We first consider the case where $G(C)$ has no bad cycle. In this case Algorithm **Feasible-Draw** finds a rectangular drawing of $G(C)$ by **Rectangular-Draw**. Hence, by Eq. (1) we have

$$T(G(C)) = T_R(G(C)) \leq c \cdot m(G(C)).$$

We next consider the case where $G(C)$ has the maximal bad cycles $C_1, C_2, \cdots, C_l$ where $l \geq 1$. Suppose inductively that Eq. (2) holds for each $C_i, 1 \leq i \leq l$, that is,

$$T(G(C_i)) \leq c \cdot m(G(C_i)) \tag{3}$$

for $1 \leq i \leq l$. Algorithm **Feasible-Draw** constructs a plane graph $H$ from $G(C)$ by contracting $G(C_i), 1 \leq i \leq l$, to a single vertex. $H$ has no bad cycles, and the rectangular drawing $D(H)$ can be found by **Rectangular-Draw**. Therefore, by Eq. (1)

$$T_R(H) \leq c \cdot m(H). \tag{4}$$

Algorithm **Feasible-Draw** recursively finds drawings of $G(C_i), 1 \leq i \leq l$, and patches them into the rectangular drawing $D(H)$. Therefore,

$$T(G(C)) = T_R(H) + \sum_{i=1}^{l} T(G(C_i)). \tag{5}$$

One can observe that

$$m(H) + \sum_{i=1}^{l} m(G(C_i)) \;=\; m(G(C)). \tag{6}$$

Using Eqs. (3), (4), (5), and (6), we have

$$\begin{aligned}
T(G(C)) \;&\leq\; c \cdot m(H) + \sum_{i=1}^{l} c \cdot m(G(C_i)) \\
&=\; c \cdot m(G(C)).
\end{aligned}$$

$\square$

We are now ready to prove the sufficiency of Theorem 1; we actually prove the following lemma.

**Lemma 5** *If $G$ satisfies the condition in Theorem 1, then $G$ has an orthogonal drawing without bends.*

M. S. Rahman et al., *Orthogonal Drawings*, JGAA, 7(4) 335–362 (2003)    350

**Proof:** Since there are four or more 2-vertices on $C_o(G)$, we designate any four of them as (convex) corners.

Consider first the case where $G$ does not have any bad cycle with respect to the four designated (convex) corners. Then by Lemma 1 there is a rectangular drawing of $G$. Since the rectangular drawing of $G$ has no bends, it is an orthogonal drawing $D(G)$ of $G$ without bends.

Consider next the case where $G$ has bad cycles. Let $C_1, C_2, \cdots, C_l$ be the maximal bad cycles in $G$. By Lemma 2 $C_1, C_2, \cdots, C_l$ are independent of each other. We contract each $G(C_i)$, $1 \leq i \leq l$, to a single vertex $v_i$. Let $G^*$ be the resulting graph. Clearly, $G^*$ has no bad cycle with respect to the four designated vertices, some of which may be vertices resulted from the contraction of bad cycles. By Lemma 1 $G^*$ has a rectangular drawing $D(G^*)$, which can be found by the algorithm **Rectangular-Draw.** We recursively find a feasible orthogonal drawing of each $G(C_i)$, $1 \leq i \leq l$, by **Feasible-Draw.** Patch the feasible orthogonal drawings of $G(C_1), G(C_2), \cdots, G(C_l)$ into $D(G^*)$ by patching operations. The resulting drawing is an orthogonal drawing $D$ of $G$. Note that $D(G^*)$ and $D(G(C_i))$, $1 \leq i \leq l$, have no bend. Furthermore, patching operation introduces no new bend. Thus $D$ has no bend.    □

We now formally describe our algorithm as follows.

> **Algorithm Bi-Orthogonal-Draw$(G)$**
> **begin**
> 1  Select any four 2-vertices on $C_o(G)$ as designated corners;
> 2  Find the maximal bad cycles $C_1, C_2, \cdots, C_l$ in $G$;
> 3  For each $i$, $1 \leq i \leq l$, contract cycle $C_i$ to a single vertex $v_i$;
> 4  Let $G^*$ be the resulting graph;
> 5  Find a rectangular drawing of $G^*$ by **Rectangular-Draw**;
> 6  Find a feasible orthogonal drawing of each $C_1, C_2, \cdots, C_l$ by
>    **Feasible-Draw**;
> 7  Patch the drawings $D(G(C_1)), D(G(C_2)), \cdots, D(G(C_l))$ into $D(G^*)$;
> 8  The resulting drawing $D(G)$ is an orthogonal drawing of $G$
>    without bends.
> **end.**

We then have the following theorem.

**Theorem 2** *If $G$ satisfies the condition in Theorem 1, then Algorithm* **Bi-Orthogonal-Draw** *finds an orthogonal drawing of $G$ without bends in linear time.*

**Proof:** Using a method similar to one in [7, 8, 9], one can find all the maximal bad cycles in $G$ in linear-time. Algorithms **Rectangular-Draw** and **Feasible-Draw** take linear-time. Patching operations take linear time. Therefore the overall time complexity of the algorithm **Bi-Orthogonal-Draw** is linear.    □

Using an algorithm similar to Algorithm **Bi-Orthogonal-Draw**, one can find a particular orthogonal drawing without bends, which we call a "four-corner

M. S. Rahman et al., *Orthogonal Drawings*, JGAA, 7(4) 335–362 (2003)     351

drawing" and define as follows. Let $C_o(G)$ contain four or more 2-vertices of $G$, and let $x, y, z$ and $w$ be any four of them. Then an orthogonal drawing $D(G)$ of $G$ is called a *four-corner orthogonal drawing* for $x$, $y$, $z$ and $w$ if the drawing intersects none of the four quadrants, the first quadrant with the origin at $x$, the fourth quadrant with the origin at $y$, the third quadrant with the origin at $z$, and the second quadrant with the origin at $w$, after rotating the drawing and renaming vertices $x, y, z$ and $w$ if necessary. In Fig. 16 the four quadrants are shaded. Vertices $x$, $y$, $z$, and $w$ must be convex corners of the drawing $D(C_o(G))$ of the outer cycle $C_o(G)$, and $D(G)$ should intersect neither the horizontal open halfline with left end at $x$ nor the vertical halfline with the lower end at $x$, and so on for $y$, $z$, and $w$. Clearly, a rectangular drawing with four designated corners is a four-corner drawing for the designated corners. A four-corner drawing has applications in finding an orthogonal drawing with the minimum number of bends [5].

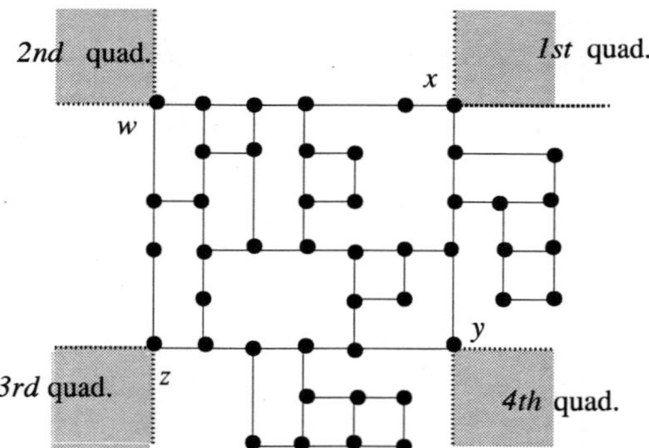


Figure 16: Illustration for a four-corner orthogonal drawing.

We now have the following corollary.

**Corollary 6** *Assume that $G$ is a plane biconnected graph with $\Delta \leq 3$ and there are four or more 2-vertices on $C_o(G)$. If every 2-legged cycle in $G$ contains at least two 2-vertices of $G$ and every 3-legged cycle in $G$ contains at least one 2-vertex of $G$, then one can find a four-corner orthogonal drawing $D(G)$ without bends for any four 2-vertices $x$, $y$, $z$ and $w$ on $C_o(G)$ in linear time.*

**Proof:** One can find a four-corner orthogonal drawing $D(G)$ without bends for any four 2-vertices $x$, $y$, $z$ and $w$ on $C_o(G)$ by using an algorithm similar to Algorithm **Bi-Orthogonal-Draw**. The algorithm selects $x$, $y$, $z$ and $w$ as designated corners in Step 1 and expands the drawing outwards, if necessary, after each patching operation in Step 7. Other steps of Algorithm **Bi-Orthogonal-Draw** remain unchanged in the algorithm. Clearly the algorithm takes linear time, since Algorithm **Bi-Orthogonal-Draw** takes linear time.     $\square$

M. S. Rahman et al., *Orthogonal Drawings*, JGAA, 7(4) 335–362 (2003)     352

# 4  Orthogonal Drawings of Arbitrary Plane Graphs

In this section we extend our result on biconnected plane graphs in Theorem 1 to arbitrary (not always biconnected) plane graphs with $\Delta \leq 3$ as in the following theorem.

**Theorem 3** *Let $G$ be a plane graph with $\Delta \leq 3$. Then $G$ has an orthogonal drawing without bends if and only if every $k$-legged cycle $C$ in $G$ contains at least $4 - k$ 2-vertices of $G$ for any $k$, $0 \leq k \leq 3$.*

Theorem 3 is a generalization of both Theorem 1 and Thomassen's condition [12].

The proof for the necessity of Theorem 3 is similar to the proof for the necessity of Theorem 1. In the rest of this section we give a constructive proof for the sufficiency of Theorem 3. We need some definitions.

We may assume that $G$ is connected. We call a subgraph $B$ of $G$ a *biconnected component* of $G$ if $B$ is a maximal biconnected subgraph of $G$. We call an edge $(u, v)$ a *bridge* of $G$ if the deletion of $(u, v)$ results in a disconnected graph. Any graph can be decomposed to biconnected components and bridges. The graph $G$ in Fig. 17(a) has three biconnected components $B_1$, $B_2$ and $B_3$ depicted in Fig. 17(b) and six bridges $(v_3, v_{24})$, $(v_{24}, v_{25})$, $(v_{24}, v_{26})$, $(v_4, v_{16})$, $(v_{10}, v_{11})$ and $(v_7, v_{27})$.

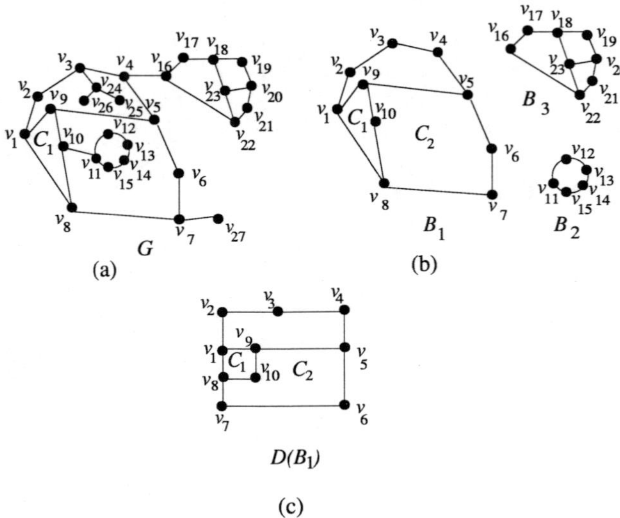

Figure 17: (a) A connected plane graph $G$, (b) three biconnected components $B_1$, $B_2$ and $B_3$ of $G$, and (c) a feasible orthogonal drawing of $B_1$.

Let $C$ be a cycle in $G$, and let $v$ be a cut vertex of $G$ on $C$. We call $v$ an *outcut vertex* for $C$ if $v$ is a leg-vertex of $C$ in $G$, otherwise we call $v$ an *incut*

M. S. Rahman et al., *Orthogonal Drawings*, JGAA, 7(4) 335–362 (2003)     353

*vertex* for $C$. (See Fig. 18.) Any outcut vertex for $C$ is either a convex corner or a non-corner of $D(C)$ in any orthogonal drawing $D(G)$ of $G$, because if it were a concave corner then the leg of $C$ could not be drawn as a horizontal or vertical line segment without edge-crossing. Similarly, any incut vertex for $C$ is either a concave corner or a non-corner of $D(C)$. Thus any orthogonal drawing of $G$ must satisfy the following condition (f4).

**(f4)** If $v$ is an outcut vertex for a cycle $C$ in $G$ then $v$ is either a convex corner or a non-corner of $D(C)$, and if $v$ is an incut vertex for a cycle $C$ then $v$ is either a concave corner or a non-corner of $D(C)$.

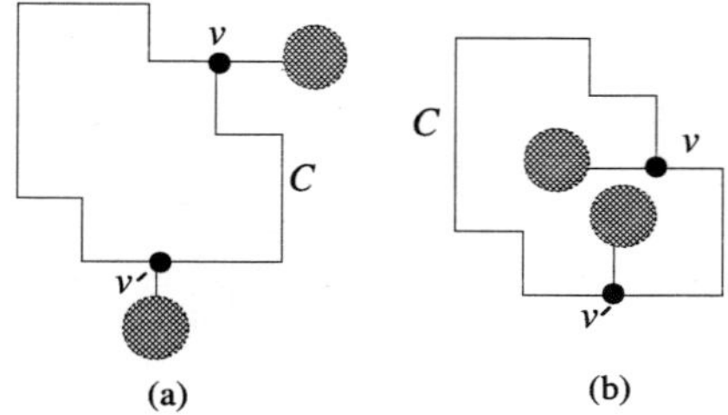

(a)      (b)

Figure 18: (a) Outcut vertices $v$ at a convex corner and $v'$ at a non-corner, and (b) incut vertices $v$ at a concave corner and $v'$ at a non-corner.

Of course, for any subgraph $B$ of $G$, the drawing $D(B)$ in an orthogonal drawing $D(G)$ of $G$ satisfies Condition (f4). In the plane graph $G$ in Fig. 17(a), vertices $v_4$ and $v_7$ are outcut vertices for the cycle $C_o(B_1)$, and $v_3$ is an incut vertex for the cycle $C_o(B_1)$. Vertex $v_{10}$ is an outcut vertex for the cycle $C_1 = v_1, v_9, v_{10}, v_8$, but is an incut vertex for the cycle $C_2 = v_5, v_6, v_7, v_8, v_{10}, v_9$. The orthogonal drawing $D(B_1)$ of the biconnected component $B_1$ in Fig. 17(c) satisfies (f4), because the outcut vertices $v_4$, $v_7$ for $C_o(B_1)$ and $v_{10}$ for $C_1$ are convex corners while the incut vertices $v_3$ for $C_o(B_1)$ is a non-corner and $v_{10}$ for $C_2$ is a concave corner. We call an orthogonal drawing $D(B)$ of a subgraph $B$ of $G$ a *mergeable* orthogonal drawing if $D(B)$ satisfies (f4) and has no bends.

Let $B$ be a biconnected subgraph of $G$, let $C$ be a cycle in $B$, and let $v$ be a 2-vertex of $B$ on $C$. We say that vertex $v$ is *good* for $C$ if $v$ is not an incut vertex for $C$ in $G$. For $B_1$ in Fig. 17(b), $v_2$, $v_4$, $v_6$ and $v_7$ are the good vertices for $C_o(B_1)$ while $v_3$ is not a good vertex. Only a good vertex for $C$ can be drawn as a convex corner of the rectilinear polygon $D(C)$ in a mergeable orthogonal drawing $D(B)$.

We now have the following lemmas.

**Lemma 7** *If $G$ satisfies the condition in Theorem 3 and $B$ is a biconnected component of $G$, then*

*(a) there are at least four good vertices for $C_o(B)$,*

*(b) there are at least two good vertices for every 2-legged cycle $C$ in $B$, and*

M. S. Rahman et al., *Orthogonal Drawings*, JGAA, 7(4) 335–362 (2003)     354

*(c) there are at least one good vertex for every 3-legged cycle $C$ in $B$.*

**Proof:** (a) Assume that $C_o(B)$ is a $k$-legged cycle in $G$ for some $k \geq 0$. The $k$ leg-vertices are outcut vertices, are not incut vertices, have degree 2 in $B$, and hence are good vertices for $C_o(B)$. Thus, if $k \geq 4$, then there are at least four good vertices for $C_o(B)$.

If $k = 1$, then the condition in Theorem 3 implies that $C_o(B)$ contains at least three 2-vertices of $G$. Since these vertices have degree 2 in the biconnected component $B$, they are not incut vertices for $C_o(B)$ and hence are good vertices for $C_o(B)$. The leg-vertex of $C_o(B)$ is a good vertex, too. Thus there are at least four good vertices for $C_o(B)$.

Similarly we can prove the claim when $k = 0$, 2 or 3.

(b) Let $C$ be a 2-legged cycle in $B$. The two leg-vertices have degree 3 in $B$, and hence they are not good for $C$. Let $C$ be a $k$-legged cycle in $G$ for some $k \geq 2$. If $k = 2$, then the condition in Theorem 3 implies that $C$ contains at least two 2-vertices of $G$, which are good for $C$. If $k \geq 4$, then $C$ contains at least two outcut vertices which are 2-vertices of $B$ and hence are good for $C$. Thus one may assume that $k = 3$. Then $C$ contains an outcut vertex which is a 2-vertex of $B$ and hence is good for $C$. Furthermore, the condition in Theorem 3 implies that $C$ contains at least one 2-vertex of $G$, which is good for $C$. Thus $C$ contains at least two good vertices.

(c) Similar to (b).     $\square$

We now have the following lemmas.

**Lemma 8** *Let $G$ be a connected plane graph of $\Delta \leq 3$ satisfying the condition in Theorem 3, let $B$ be a biconnected component of $G$, and let $C$ be a 2- or 3-legged cycle in $B$. Then the plane subgraph $B(C)$ of $B$ inside $C$ has a mergeable feasible orthogonal drawing $D(B(C))$.*

**Proof:** We can recursively find a mergeable feasible orthogonal drawing $D(B(C))$ by an algorithm similar to Algorithm **Feasible-Draw** for finding a feasible orthogonal drawing. However, in each recursive step, we have to choose the four designated corner vertices carefully in a way that none of the incut vertices for an outer cycle is chosen as a designated (convex) corner. This can be done, because by Lemma 7(b) every 2-legged cycle in $B$ contains at least two good vertices, by Lemma 7(c) every 3-legged cycle in $B$ contains at least one good vertex, and hence one can choose leg-vertices and good vertices as the four designated corner vertices. These vertices are convex corners in the drawing of the cycle, while any cut vertex which is not chosen as a designated corner is drawn as a non-corner. Hence $D(B(C))$ satisfies (f4), and $D(B(C))$ is a mergeable feasible orthogonal drawing.     $\square$

**Lemma 9** *If $G$ is a connected plane graph of $\Delta \leq 3$ and satisfies the condition in Theorem 3, then every biconnected component $B$ of $G$ has a mergeable orthogonal drawing.*

M. S. Rahman et al., *Orthogonal Drawings*, JGAA, 7(4) 335–362 (2003)    355

**Proof:** One can find a mergeable orthogonal drawing of $B$ as follows.

By Lemma 7(a) one can select four good vertices on $C_o(B)$ as designated corners.

Consider first the case where $B$ has no bad cycle with respect to the four designated corners. Then by Lemma 1 there is a rectangular drawing $D(B)$ of $B$. Of course, $D(B)$ has no bend. Any vertex of $B$ which is an outcut or incut vertex of $G$ has degree 2 in $B$. In $D(B)$, the four designated vertices on $C_o(B)$ are convex corners, and every 2-vertex of $B$ except the four designated vertices is a non-corner. Hence the rectangular drawing $D(B)$ satisfies Condition (f4). Therefore $D(B)$ is a mergeable orthogonal drawing.

Consider next the case where $B$ has bad cycles. Let $C_1, C_2, \cdots, C_l$ be the maximal bad cycles in $B$. Then by Lemma 2 $C_1, C_2, \cdots, C_l$ are independent of each other. We contract each $B(C_i)$, $1 \leq i \leq l$, to a single vertex $v_i$. Let $B^*$ be the resulting graph. Clearly, $B^*$ has no bad cycle with respect to the four designated vertices, some of which may be vertices resulted from the contraction of bad cycles. By Lemma 1 $B^*$ has a rectangular drawing $D(B^*)$. We recursively find a mergeable feasible orthogonal drawing $D(B(C_i))$ of each $B(C_i)$, $1 \leq i \leq l$, by the method described in the proof of Lemma 8. Patch the mergeable feasible orthogonal drawings of $D(B(C_i))$, $1 \leq i \leq l$, into $D(B^*)$ by patching operations. Clearly the resulting drawing is an orthogonal drawing $D(B)$ of $B$. $D(B^*)$ is a mergeable drawing and and $D(B(C_i))$, $1 \leq i \leq l$, are mergeable feasible drawings. Furthermore, the patching operation introduces neither a convex corner at any incut vertex nor a concave corner at any outcut vertex in the drawing of a cycle in $B$. Hence $D(B)$ is a mergeable orthogonal drawing. $\square$

The algorithm described in the proof of Lemma 9 takes linear time similarly as Algorithm **Bi-Orthogonal-Draw** in Section 3.

Based on the algorithm described in the proof of Lemma 9, we now present a result on four-corner orthogonal drawing in Lemma 10. The result described in Lemma 10 is used to obtain a bend-minimum orthogonal drawing of a plane graph with $\Delta \leq 3$ in [5].

**Lemma 10** *Let $G$ be a connected plane graph with $\Delta \leq 3$, let $B$ be a biconnected subgraph of $G$, and let $x, y, z$ and $w$ be any four 2-vertices of $B$ on $C_o(B)$. Then $B$ has a mergeable four-corner orthogonal drawing $D(B)$ for $x$, $y$, $z$ and $w$ if and only if the following (a), (b) and (c) hold:*

*(a) all the vertices $x$, $y$, $z$ and $w$ are good for the cycle $C_o(B)$ in $G$;*

*(b) there are at least two good vertices for every 2-legged cycle $C$ in $B$; and*

*(c) there are at least one good vertex for every 3-legged cycle $C$ in $B$.*

*Furthermore the drawing above can be found in linear time.*

**Proof: Necessity:** Suppose that $B$ has a mergeable four-corner orthogonal drawing $D(B)$. Then $D(B)$ has no bends and satisfies Condition (f4).

M. S. Rahman et al., *Orthogonal Drawings*, JGAA, 7(4) 335–362 (2003)     356

(a) All the vertices $x$, $y$, $z$ and $w$ are convex corners of $D(C_o(B))$, and hence by Condition (f4) none of $x$, $y$, $z$ and $w$ is an incut vertex for $C_o(B)$. Therefore $x$, $y$, $z$ and $w$ are good vertices for $C_o(B)$.

(b) The rectilinear polygonal drawing $D(C)$ of a 2-legged cycle $C$ in $D(B)$ has at least four convex corners. Since $D(B)$ has no bend, every convex corner of $D(C)$ is either a 2-vertex of $B$ or a leg-vertex of $C$ in $B$. The two leg-vertices of $C$ may serve as two convex corners. Any other convex corner of $D(C)$ is a 2-vertex of $B$ and is not an incut vertex for $C$ in $G$ by Condition (f4). Hence there are at least two good vertices for $C$.

(c) Similar to (b).

**Sufficiency:** Assume that $B$ satisfies Conditions (a)–(c) in Lemma 10. Then we can obtain a mergeable orthogonal drawing $D(B)$ of $B$ in linear time by the algorithm described in the proof of Lemma 9. To ensure that $D(B)$ is a four-corner orthogonal drawing for $x$, $y$, $z$ and $w$, the algorithm must select $x$, $y$, $z$ and $w$ as the four designated corners, and the drawing may be needed to expand outwards after each patching operation.  $\square$

A *block* of a connected graph $G$ is either a biconnected component or a bridge of $G$. The graph in Fig. 19(a) has the blocks $B_1, B_2, \cdots, B_9$ depicted in Fig. 19(b). The blocks and cut vertices in $G$ can be represented by a tree $T$,

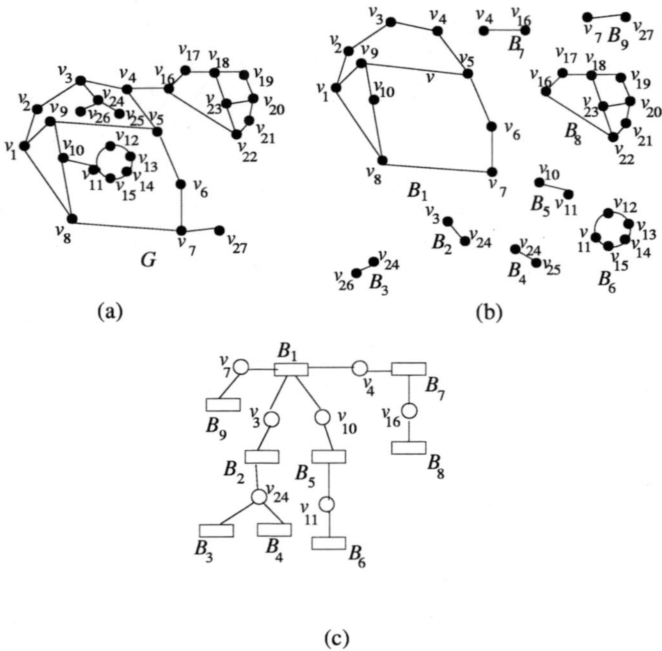

Figure 19: (a) $G$, (b) blocks, and (c) $BC$-tree $T$.

called the *BC-tree* of $G$. In $T$ each block is represented by a $B$-node and each

M. S. Rahman et al., *Orthogonal Drawings*, JGAA, 7(4) 335–362 (2003)     357

cut vertex of $G$ is represented by a $C$-node. The $BC$-tree $T$ of the plane graph $G$ in Fig. 19(a) is depicted in Fig. 19(c), where each $B$-node is represented by a rectangle and each $C$-node is represented by a circle.

We call a cycle $C$ in a plane graph $G$ a *maximal cycle* of $G$ if $G(C)$ is not contained in $G(C')$ for any other cycle $C'$ in $G$. Thus a maximal cycle is an outer cycle of a biconnected component of $G$. The graph $G$ in Fig. 20(a) has two maximal cycles $C_1$ and $C_2$ drawn by thick lines. $G(C)$ is called a *maximal closed subgraph* of $G$ if $C$ is a maximal cycle of $G$. We now have the following lemma.

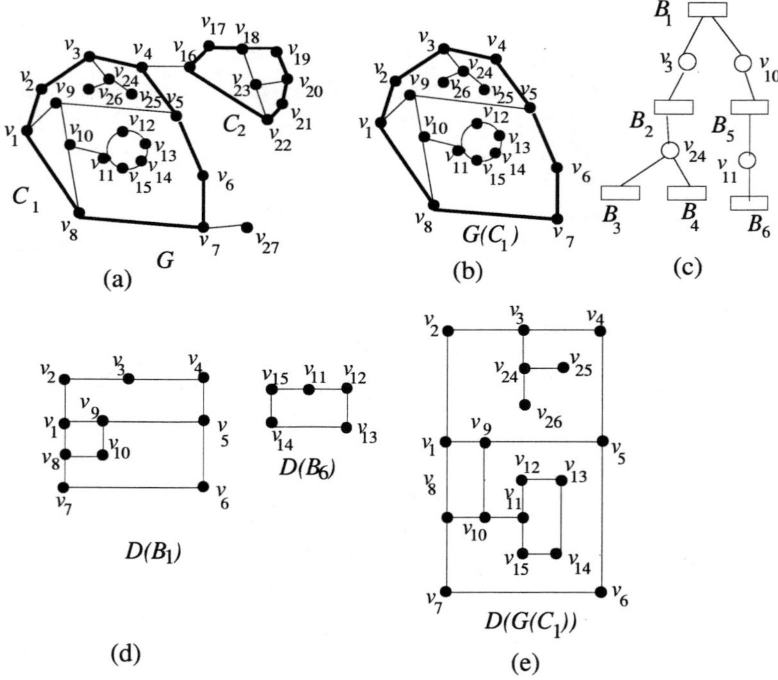

Figure 20: (a) A plane graph $G$ with two maximal cycles $C_1$ and $C_2$, (b) $G(C_1)$, (c) BC-tree of $G(C_1)$, (d) drawings of the two biconnected components $B_1$ and $B_6$ of $G(C_1)$, and (e) the final drawing of $G(C_1)$.

**Lemma 11** *If $G$ is a connected plane graph of $\Delta \leq 3$ and satisfies the condition in Theorem 3, then $G(C)$ has a mergeable orthogonal drawing for any maximal cycle $C$ in $G$.*

**Proof:** We give an algorithm for finding a mergeable orthogonal drawing of $G(C)$, that is, an orthogonal drawing of $G(C)$ which has no bends and satisfies (f4).

If $G(C)$ is a biconnected component of $G$, then by Lemma 9 $G(C)$ has a mergeable orthogonal drawing. One may thus assume that $G(C)$ is not a

M. S. Rahman et al., *Orthogonal Drawings*, JGAA, 7(4) 335–362 (2003)    358

biconnected component of $G$. Then $G(C)$ has some biconnected components and bridges. Clearly the biconnected components of $G(C)$ are vertex-disjoint with each other. For each biconnected component we can find a mergeable orthogonal drawing by the algorithm described in the proof of Lemma 9, while we draw each bridge by a horizontal or vertical line segment. We then merge the drawings of biconnected components and bridges without introducing new bends and edge crossings as follows.

We construct a *BC*-tree $T$ of $G(C)$. Let $B_1$ be the node in the BC-tree corresponding to the biconnected component of $G(C)$ whose outer cycle is $C$. We consider $T$ as a rooted tree with root $B_1$. Starting from the root $B_1$ we visit the tree by depth-first search and merge the orthogonal drawings of the blocks in the depth first-search order.

Let $B_1, B_2, \cdots, B_b$ be the ordering of the blocks following a depth-first search order starting from $B_1$. $G(C_1)$ for the graph $G$ in Fig. 20(a) is depicted in Fig. 20(b) and the BC-tree of $G(C_1)$ is depicted in Fig. 20(c), where $B_1$ is the root of the tree and the other $B$-nodes are numbered according to a depth-first search order starting from $B_1$.

We assume that we have obtained a mergeable orthogonal drawing $D_i$ by merging the orthogonal drawings of the blocks $B_1, B_2, \cdots, B_i$, and that we are now going to obtain a mergeable orthogonal drawing $D_{i+1}$ by merging $D_i$ with an orthogonal drawing of the block $B_{i+1}$. Let $v_t$ be the cut vertex corresponding to the C-node which is the parent of $B_{i+1}$ in $T$. Let $B_x$ be the parent of $v_t$ in $T$. Then both $B_x$ and $B_{i+1}$ contain $v_t$, and $D_i$ contains the drawing of $B_x$. We have the following three cases to consider.

**Case 1:** $B_x$ is a biconnected component and $B_{i+1}$ is a bridge.

In this case $B_{i+1}$ is an edge and will be drawn inside an inner face of the drawing $D_i$. Let $C_f$ be the facial cycle of $B_x$ corresponding to the inner face. Then $v_t$ is an incut vertex for $C_f$. Since we have obtained a mergeable orthogonal drawing $D(B_x)$ of $B_x$, $v_t$ is a concave corner or a non-corner of the drawing of $C_f$ in $D(B_x)$, and hence the two edges incident to $v_t$ are drawn in $D_i$ as in Fig. 21 or as a rotated one. We can draw the bridge $B_{i+1}$ as a horizontal or a vertical line segment started from $v_t$ as illustrated by dotted lines in Fig. 21. We thus obtain a drawing $D_{i+1}$. Clearly no new bend is introduced in $D_{i+1}$ and $D_i$ may be expanded outwards to avoid edge crossings. In Fig. 20(e) the bridge $B_2 = (v_3, v_{24})$ is merged with a biconnected component $B_1$ at vertex $v_3$.

**Case 2:** Both $B_x$ and $B_{i+1}$ are bridges.

In this case $v_t$ is drawn in an inner face of $D_i$ and has degree 1 or 2 in $D_i$. (See Fig. 22.)

We first consider the case where $v_t$ has degree 1. We then draw $B_{i+1}$ as indicated by the dotted line in Fig. 22(a).

We next consider the case where $v_t$ has degree 2 in $D_i$. Then $v_t$ has degree 3 in $G(C)$, and let $x$, $y$, and $z$ be the three neighbors of $v_t$ in $G$. We may assume without loss of generality that edges $(v_t, x)$ and $(v_t, y)$ are bridges and are already drawn in $D_i$ and that $B_x$ is either $(v_t, x)$ or $(v_t, y)$. We now merge the drawing of bridge $B_{i+1} = (v_t, z)$ to $D_i$. It is evident from the drawing described above that bridges $(v_t, x)$ and $(v_t, y)$ are drawn on a (horizontal or

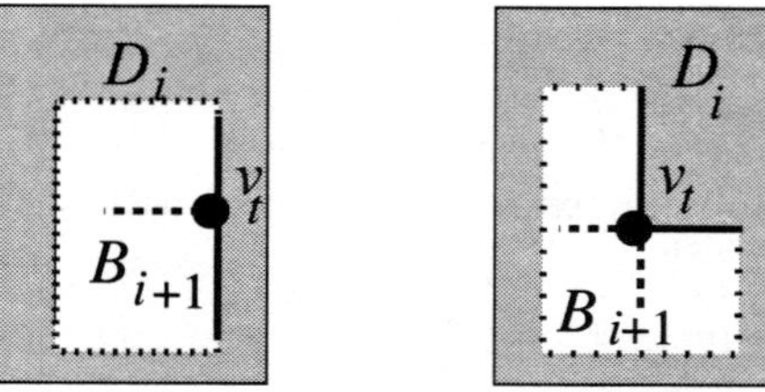

Figure 21: Drawing of edges incident to $v_t$ in $D_i$ when $B_x$ is a biconnected component and $B_{i+1}$ is a bridge.

vertical) straight line segment. We draw $B_{i+1}$ as indicated by a dotted line in Fig. 22(b).

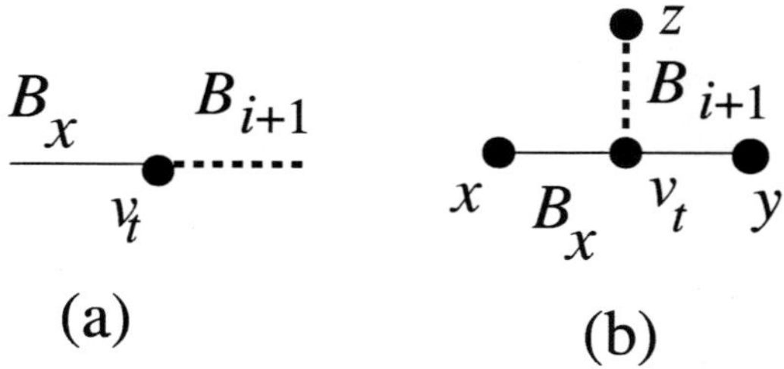

Figure 22: Drawings of $B_i$ when both $B_x$ and $B_{i+1}$ are bridges

**Case 3:** $B_x$ is a bridge and $B_{i+1}$ is a biconnected component.

In this case $v_t$ is drawn in $D_i$ as an end of a horizontal or vertical line segment inside an inner face of $D_i$. Vertex $v_t$ has degree 2 in $B_{i+1}$ and is an outcut vertex for $C_o(B_{i+1})$. By Lemma 9 $D(B_{i+1})$ is a mergeable orthogonal drawing, and hence $v_t$ is a convex corner or a non-corner of the drawing of $C_o(B_{i+1})$ in $D(B_{i+1})$. Therefore $D(B_{i+1})$ can be easily merged with $D_i$ by rotating $D(G(B_{i+1}))$ by 90° or 180° or 270° and expanding the drawing $D_i$ if necessary. In Fig. 20(e) the orthogonal drawing of $B_6$ is merged to $D_5$ at vertex $v_{11}$ where $D(B_6)$ in Fig. 20(d) has been rotated by 90° and the drawing $D_5$ is expanded outwards. $\square$

We call the algorithm described in the proof of Lemma 11 Algorithm **Maximal-Orthogonal-Draw**. Clearly Algorithm **Maximal-Orthogonal-Draw** takes linear time.

We are now ready to give a proof for sufficiency of Theorem 3.

**Proof for sufficiency of Theorem 3**

We decompose $G$ into maximal closed subgraphs and bridges. We find an orthogonal drawing of each maximal closed subgraph by Algorithm **Maximal-**

**Orthogonal-Draw.** Each of the bridges can be drawn by a horizontal or a vertical line segment. Using a technique similar to one in the proof of Lemma 11, we merge the drawings of the maximal closed subgraphs and bridges. The resulting drawing is an orthogonal drawing of $G$ without bends.     □

We call the algorithm described in the proof for sufficiency of Theorem 3 Algorithm **No-bend-Orthogonal-Draw**. An execution of the algorithm **No-bend-Orthogonal-Draw** is illustrated in Fig. 23. We now have the following theorem.

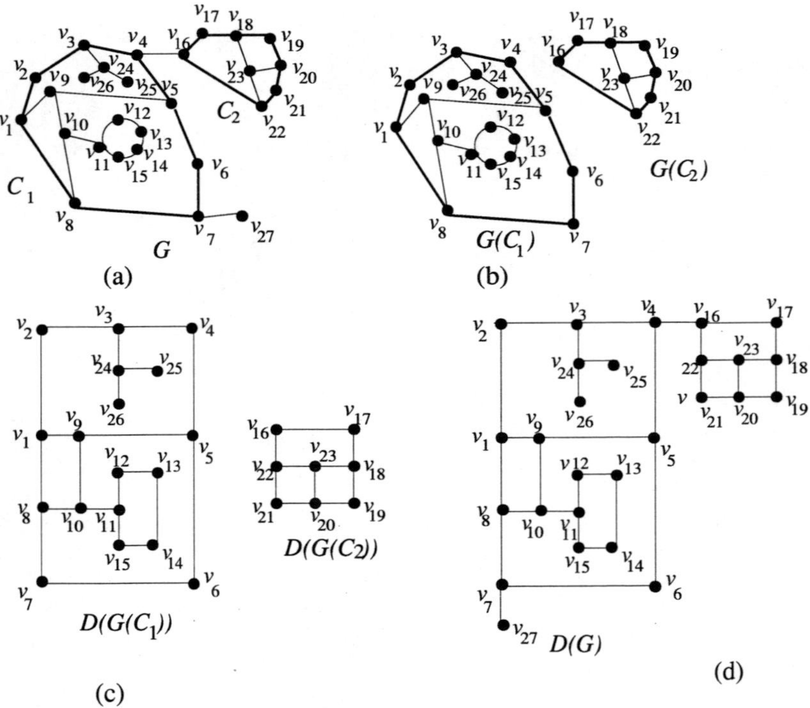

Figure 23: (a) A plane graph $G$, (b) two maximal closed subgraphs $G(C_1)$ and $G(C_2)$ of $G$, (c) orthogonal drawings of $G(C_1)$ and $G(C_2)$ without bends, and (d) orthogonal drawings of $G$ without bends.

**Theorem 4** *If $G$ is a plane connected graph of $\Delta \le 3$ and satisfies the condition in Theorem 3, then Algorithm* **No-bend-Orthogonal-Draw** *finds an orthogonal drawing of $G$ without bends in linear time.* □

# 5 Conclusions

In this paper we established a necessary and sufficient condition for a plane graph $G$ of $\Delta \le 3$ to have an orthogonal drawing without bends, and gave a linear-

M. S. Rahman et al., *Orthogonal Drawings*, JGAA, 7(4) 335–362 (2003)    361

time algorithm to examine whether $G$ has an orthogonal drawing without bends and find such a drawing of $G$ if it exists. The condition is a generalization of Thomassen's condition for rectangular drawings [12]. The algorithm presented in this paper has applications in finding an orthogonal drawing of a plane graph of $\Delta \leq 3$ with the minimum number of bends in linear time [5]. It is remained as a future work to establish a necessary and sufficient condition for a plane graph of $\Delta \leq 4$ to have an orthogonal drawing without bends.

An orthogonal drawing of a plane graph $G$ without bends is called a *rectangular drawing* of $G$ if each face of $G$ including the outer face is drawn as a rectangle. A planar graph is said to have a rectangular drawing if at least one of its plane embeddings has a rectangular drawing. Recently Rahman *et al.* [6] gave a necessary and sufficient condition for a planar graph of $\Delta \leq 3$ to have a rectangular drawing which leads to a linear time algorithm to find a rectangular drawing of a planar graph, if it exists. It is thus an interesting future work to generalize the condition of Rahman *et al.* [6] for orthogonal drawings of planar graphs of $\Delta \leq 3$ without bends.

M. S. Rahman et al., *Orthogonal Drawings*, JGAA, 7(4) 335–362 (2003)    362

# References


[1]  G. Di Battista, P. Eades, R. Tamassia, I. G. Tollis, *Graph Drawing: Algorithms for the Visualization of Graphs*, Prentice-Hall Inc., Upper Saddle River, New Jersey, 1999.

[2]  A. Garg and R. Tamassia, *On the computaional complexity of upward and rectilinear planarity testing*, SIAM J. Comput., 31(2), pp. 601-625, 2001.

[3]  A. Garg and R. Tamassia, *A new minimum cost flow algorithm with applications to graph drawing*, Proc. of Graph Drawing'96, Lect. Notes in Computer Science, 1190, pp. 201-206, 1997.

[4]  T. Lengauer, *Combinatorial Algorihms for Integrated Circuit Layout*, Wiley, Chichester, 1990.

[5]  M. S. Rahman and T. Nishizeki, *Bend-minimum orthogonal drawings of plane 3-graphs*, Proc. of WG'02, Lect. Notes in Computer Science, 2573, pp. 365-376, 2002.

[6]  M. S. Rahman, T. Nishizeki and S. Ghosh, *Rectangular drawings of planar graphs*, Journal of Algorithms, 50, pp. 62-78, 2004.

[7]  M. S. Rahman, S. Nakano and T. Nishizeki, *Rectangular grid drawings of plane graphs*, Comp. Geom. Theo. Appl., 10(3), pp. 203-220, 1998.

[8]  M. S. Rahman, S. Nakano and T. Nishizeki, *A linear algorithm for bend-optimal orthogonal drawings of triconnected cubic plane graphs*, Journal of Graph Alg. and Appl., http://jgaa.info, 3(4), pp. 31-62, 1999.

[9]  M. S. Rahman, S. Nakano and T. Nishizeki, *Box-rectangular drawings of plane graphs*, Journal of Algorithms, 37, pp. 363-398, 2000.

[10]  M. S. Rahman, S. Nakano and T. Nishizeki, *Rectangular drawings of plane graphs without designated corners*, Comp. Geom. Theo. Appl., 21(3), pp. 121-138, 2002.

[11]  M. S. Rahman, M. Naznin and T. Nishizeki, *Orthogonal drawings of plane graphs without bends*, Proc. of Graph Drawing'01, Lect. Notes in Computer Science, 2265, pp. 392-406, 2002.

[12]  C. Thomassen, *Plane representations of graphs*, (Eds.) J.A. Bondy and U.S.R. Murty, Progress in Graph Theory, Academic Press Canada, pp. 43-69, 1984.

[13]  R. Tamassia, *On embedding a graph in the grid with the minimum number of bends*, SIAM J. Comput., 16, pp. 421-444, 1987.

# Journal of Graph Algorithms and Applications

http://jgaa.info/

vol. 7, no. 4, pp. 363–398 (2003)

# Straight-Line Drawings on Restricted Integer Grids in Two and Three Dimensions

*Stefan Felsner*[1], *Giuseppe Liotta*[2], *Stephen Wismath*[3]

[1] Freie Universität Berlin, Fachbereich Mathematik und Informatik, Takustr. 9, 14195 Berlin, Germany.
felsner@inf.fu-berlin.de

[2] Dipartimento di Ingegneria Elettronica e dell'Informazione, Università di Perugia, via G. Duranti 93, 06125 Perugia, Italy.
liotta@diei.unipg.it

[3] Dept. of Mathematics and Computer Science, U. of Lethbridge, Alberta, T1K-3M4 Canada.
wismath@cs.uleth.ca

### Abstract

This paper investigates the following question: Given a grid $\phi$, where $\phi$ is a proper subset of the integer 2D or 3D grid, which graphs admit straight-line crossing-free drawings with vertices located at (integral) grid points of $\phi$? We characterize the trees that can be drawn on a strip, *i.e.*, on a two-dimensional $n \times 2$ grid. For arbitrary graphs we prove lower bounds for the height $k$ of an $n \times k$ grid required for a drawing of the graph. Motivated by the results on the plane we investigate restrictions of the integer grid in 3D and show that every outerplanar graph with $n$ vertices can be drawn crossing-free with straight lines in linear volume on a grid called a prism. This prism consists of $3n$ integer grid points and is *universal* – it supports all outerplanar graphs of $n$ vertices. We also show that there exist planar graphs that cannot be drawn on the prism and that extension to an $n \times 2 \times 2$ integer grid, called a box, does not admit the entire class of planar graphs.

Communicated by: P. Mutzel and M. Jünger;
submitted May 2002; revised October 2003.

---

Research supported in part by: the MIUR project "Progetto ALINWEB: Algoritmica per Internet e per il Web"; and by the Natural Sciences and Engineering Research Council of Canada.

Felsner *et al.*, *Restricted Integer Grids*, JGAA, 7(4) 363–398 (2003)    364

# 1  Introduction

This paper deals with crossing-free straight-line drawings of planar graphs in two and three dimensions. Given a graph $G$, we constrain the vertices in a drawing of $G$ to be located at integer grid points and aim at computing drawings whose area/volume is small. The interest in these two requirements is motivated in part by the fact that the screen of a computer is an integer grid of limited size and by the fact that a drawing algorithm should not be affected by round-off errors when representing the output coordinates. Also, the increasing demand of visualization algorithms to draw and browse very large networks makes it natural to study what families of graphs can be entirely visualized on a two dimensional screen and to investigate how much benefit can be obtained from the third dimension to represent the overall structure of a huge graph in a small portion of a virtual 3D environment. A rich body of literature has been published on computing straight-line drawings of graphs, such that the vertices are the intersection points of an integer 2D grid and the overall area of the drawing is kept small. Typically, papers that deal with this subject focus on lower bounds on the area required by straight-line drawings of specific classes of graphs and on the design of algorithms that possibly match these lower bounds. A very limited list of mile-stone papers in this field includes the works by de Fraysseix, Pach, and Pollack [11, 12] and by Schnyder [38] who independently showed that every $n$-vertex triangulated planar graph has a crossing-free straight-line drawing such that the vertices are at grid points, the size of the grid is $O(n) \times O(n)$, and that this is worst case optimal; the work by Kant [26, 27], Chrobak and Kant [6], Schnyder and Trotter [39], Felsner [21] and Chrobak, Goodrich, and Tamassia [7] who studied convex grid drawings of triconnected planar graphs in an integer grid of quadratic area; and the many papers proving that linear or almost-linear area bounds can be achieved for classes of trees, including the result by Garg, Goodrich and Tamassia [23] and the result by Chan [5]. Summarizing tables and more references can be found in the book by Di Battista, Eades, Tamassia, and Tollis [14].

While the problem of computing small-sized crossing-free straight-line drawings in the plane has a long tradition, its 3D counterpart has become the subject of much attention only in recent years. Chrobak, Goodrich, and Tamassia [7] gave an algorithm for constructing 3D convex drawings of triconnected planar graphs with $O(n)$ volume and non-integer coordinates. Cohen, Eades, Lin and Ruskey [9] showed that every graph admits a straight-line crossing-free 3D drawing on an integer grid of $O(n^3)$ volume, and proved that this is asymptotically optimum. Calamoneri and Sterbini [3] showed that all 2-, 3-, and 4-colorable graphs can be drawn in a 3D grid of $O(n^2)$ volume with $O(n)$ aspect ratio and proved a lower bound of $\Omega(n^{1.5})$ on the volume of such graphs. For $r$-colorable graphs, Pach, Thiele and Tóth [31] showed a bound of $\theta(n^2)$ on the volume. Garg, Tamassia, and Vocca [24] showed that all 4-colorable graphs (and hence all planar graphs) can be drawn in $O(n^{1.5})$ volume and with $O(1)$ aspect ratio but using a grid model where the coordinates of the vertices may not be integer. In this paper we study the problem of computing drawings of graphs on integer

Felsner *et al.*, *Restricted Integer Grids*, JGAA, 7(4) 363–398 (2003)     365

2D or 3D grids that have small area/volume. The area/volume of a drawing $\Gamma$ is measured as the number of grid points contained in or on a *bounding box* of $\Gamma$, *i.e.* the smallest axis-aligned box enclosing $\Gamma$. Note that along each side of the bounding box the number of grid points is one more than the actual length of the side. We approach the drawing problem with the following point of view: Instead of "squeezing" a drawing onto a small portion of a grid of unbounded dimensions, we assume that a grid of *specified* dimensions (involving a function of $n$) is given and we consider what the graphs are whose drawings fit that restricted grid. For example, it is well-known that there are families of graphs that require $\Omega(n^2)$ area to be drawn in the plane, the canonical example being a sequence of $n/3$ nested triangles (see [12, 8, 38]). Such graphs can be drawn on the surface of a three dimensional triangular prism of linear volume and using integer coordinates. Thus a natural question is whether there exist specific restrictions of the 3D integer grid of linear volume that can support straight-line crossing-free drawings of meaningful families of graphs. For planar graphs the best known results for three dimensional crossing-free straight-line drawings on an integer grid are by Calamoneri and Sterbini [3] who show $O(n^2)$ volume for general planar graphs and by Eades, Lin and Ruskey [9] who show $O(n \log n)$ volume for trees.

The main contributions of the present paper are investigations concerning the drawability of graphs on 2D and 3D restricted integer grids and new drawing algorithms for some classes of graphs. An overview of the results is as follows.

- We characterize those trees that can be drawn on a strip, *i.e.*, an integer 2D grid restricted to two consecutive horizontal grid lines. From the characterization we derive a linear time algorithm to generate such drawings, if possible. This result was independently obtained by Schank [36] in his Master's thesis.

- Generalizing the result for strips, we present a lower bound "the strictness of a tree" for the number $k$ of horizontal grid lines required for grid drawings of trees. A consequence of this bound is that for any given $k$ there always exist some trees that are not drawable on the $n \times k$ grid.

- We show that the strictness of a tree is closely related to the well-known parameter *path-width*. For general graphs the path-width is shown to be a lower bound for the height of grid drawings.

- Motivated by the results on restricted integer 2D grids we explore the capability of restricted 3D integer grids for supporting linear volume drawings of graphs. In particular, we focus on two types of 3D integer grids to be defined subsequently, both having linear volume, called the *prism* and the *box*. We show that all outerplanar graphs can be drawn in linear volume on a prism. Note that this is the first result on 3D straight-line drawings of a significant class of planar graphs that achieves linear volume with integer coordinates.

Felsner *et al.*, *Restricted Integer Grids*, JGAA, 7(4) 363–398 (2003)    366

- We further explore the class of graphs that can be drawn on a prism by asking whether the prism is a *universal* 3D integer grid for all planar graphs. We answer this question in the negative by exhibiting examples of planar graphs that cannot be drawn on a prism. We also investigate the relationship between prism-drawable and hamiltonian graphs.

- We extend our study to box-drawability and present a characterization of the box-drawable graphs. While the box would appear to be a much more powerful grid than the prism, we prove that not all planar graphs are box-drawable.

Several recent related results about 3D straight line drawings of limited volume have been published after the conference version of this paper was presented at the Symposium on Graph Drawing GD 2001 [22]. Dujmović, Morin, and Wood [20] present $O(n \log^2 n)$ volume drawings of graphs with bounded tree-width and $O(n)$ volume for graphs with bounded path-width. Wood [42] shows that also graphs with bounded queue number have 3D straight-line grid drawings of $O(n)$ volume. A very recent result by Dujmović and Wood [17] shows that linear volume can also be achieved for graphs with bounded tree-width; they show 3D straight-line grid drawings of volume $c \times n$ for these graphs, where $c$ is a constant whose value exponentially depends on the tree-width. Di Giacomo, Liotta, and Wismath [16] show $4 \times n$ volume for a subclass of series-parallel graphs. The problem of computing straight-line 3D drawings of planar graphs on an integer grid of $o(n^2)$ volume is still open. A recent lower bound on the volume of 3D straight line drawings as a function of the number of edges is obtained by Bose, Czyzowicz, Morin, and Wood [2].

The remainder of the paper is organized as follows. Preliminaries and basic definitions are in Section 2. The study of trees drawable on restricted integer 2D grids is the topic of Section 3. In this section we also investigate the connection of grid drawings and path-width. Section 4 presents the linear volume algorithm for outerplanar graphs. Combinatorial properties of the graphs that can be drawn on the surface of a prism and on a box are studied in Sections 5 and 6. Final remarks, directions for further research and open problems can be found in Section 7.

## 2    Preliminaries

We assume familiarity with basic graph drawing, and computational geometry terminology; see for example [33, 14]. Since in the remainder of the paper we shall only study crossing-free straight-line drawings of planar graphs, from now on we shall simply talk about "graphs" to mean "planar graphs" and about "drawings" to mean "crossing-free straight-line drawings". We use the terms "vertex" and "edge" for both the graph and its drawing. We will draw graphs such that vertices are located at integer grid points. The dimensions of a grid are specified as the number of different grid points along each side of a bounding box of the grid. In two dimensions, a $p \times q$ grid consists of all points $(i, j)$ with

Felsner *et al.*, *Restricted Integer Grids*, JGAA, 7(4) 363–398 (2003)      367

$1 \leq i \leq p$ and $1 \leq j \leq q$ that have integer coordinates. In three dimensions, a $p \times q \times r$ grid consists of all points $(i, j, k)$ with $1 \leq i \leq p$, $1 \leq j \leq q$ and $1 \leq k \leq r$ that have integer coordinates; $p$, $q$ and $r$ are referred to as the $x$-, $y$-, and $z$-*dimension* of the grid, respectively.

We shall deal with the following grids and drawings.

- A 2D 1-*track* (or simply a *track*) is an $\infty \times 1$ grid; a 1-*track* drawing of a graph $G$ is a drawing of $G$ where the vertices are at distinct grid points of the track.

- A 2D *strip* is an $\infty \times 2$ grid; note that a strip contains two tracks. A *strip drawing* of a graph $G$ is a drawing of $G$ with the vertices located at distinct grid points of the strip and the edges either connect vertices on the same track or connect vertices on different tracks.

- Next we extend the notion of a strip to multiple overlapping strips. Let $k$ be a given positive integer value. A 2D $k$-*track* grid is an $\infty \times k$ grid consisting of $k$ consecutive parallel tracks. A $k$-*track drawing* of a graph $G$ is a drawing of $G$ where the vertices are at distinct grid points of the $k$-track and edges are only permitted between vertices that are either on the same track or that are one unit apart in their $y$-coordinates. Note that the previous two grids are the specific cases of $k = 1$ and $k = 2$.

- Let $k$ be a given positive integer value. In an $n \times k$-*grid drawing* of a graph $G$, the vertices are located at distinct grid points and the edges may connect any pair of vertices on that grid. To avoid confusion with track drawings, we refer to the value $k$ as the number of *grid lines* in the grid drawing.

- We will also study two different types of $n \times 2 \times 2$ grids. A *box* is an $n \times 2 \times 2$ grid where each side of the bounding box is also a grid line. Therefore, a box has four tracks which lie on two parallel planes and are one grid unit apart from each other. A *prism* is a subset of an $n \times 2 \times 2$ grid obtained by removing a track from a box. Figure 1 shows an example of a box of size $6 \times 2 \times 2$ and an example of a prism.

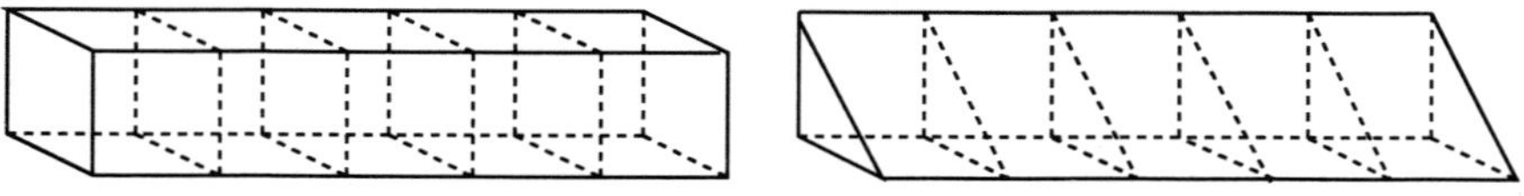

Figure 1: A box and a prism

Note that $k$-track drawings differ from the so-called $k$-level drawings (see, e.g. [25]) as in a $k$-track drawing (consecutive) vertices on the same track are permitted to be joined by an edge and the given graph is undirected. Let $\phi$ be one

Felsner *et al.*, *Restricted Integer Grids*, JGAA, 7(4) 363–398 (2003)        368

of the grids defined above. We say that a graph $G$ is $\phi$ *drawable* if $G$ admits a $\phi$ drawing $\Gamma$ where each vertex is mapped to a distinct grid point of $\phi$.

**Property 1** *A graph is 1-track drawable if and only if it is a forest whose vertices have degree at most two.*

While in a $k$-track drawing no edge can connect vertices that are on non-consecutive tracks, in an $n \times k$-grid this is allowed. As the following property shows, this difference has immediate consequences on the families of $k$-track drawable and $n \times k$-grid drawable graphs.

**Property 2** *Let $k \geq 3$ be a fixed positive integer. There exist graphs with $n$ vertices that are $n \times k$-grid drawable but are not $k$-track drawable.*

**Proof.** The graph $K_4$ has an $n \times 3$-grid drawing but it does not have a 3-track drawing – indeed $K_4$ is not drawable on tracks. Furthermore any graph containing $K_4$ as a subgraph is not track-drawable. Given a drawing of a graph on an $n \times k$-grid (for $k > 3$), we can attach a copy of $K_4$ which makes the resulting graph not track-drawable. $\square$

In the extended abstract for the graph drawing conference GD'01[22] we incorrectly argued that for trees $n \times k$-grid drawable is equivalent to $k$-track drawable. It was first observed by Matthew Suderman that this was false. In a recent manuscript Suderman [40] describes a family $S^k$ of trees, such that $S^k$ can be drawn on the $n \times (k+1)$ grid but requires $2k - 1$ tracks. Suderman's results are actually stated in terms of the pathwidth pw of a tree (a notion introduced in the next section). He shows that $\mathsf{pw}(S^k) = k$ and every tree $T$ with $\mathsf{pw}(T) \leq k$ admits a drawing on $2k - 1$ tracks.

## 3  Grids, Path-width and Trees

In this section we investigate the connections between drawability of a graph $G$ on grids and the path-width of $G$. The notion of the *path-width* of a graph $G$ was introduced by Robertson and Seymour[34] in the first paper of their series on graph minors. A *path decomposition* of a graph $G = (V, E)$ is a sequence $W_1, W_2, \ldots, W_t$ of subsets of $V$ such that

- $\forall (u, v) \in E, \quad \exists i$ such that $u, v \in W_i$

- $\forall v \in V$ the set $I(v) = \{i : v \in W_i\}$ is an interval of $\{1, \ldots, t\}$,
  *i.e.* if $a < b < c$ and $a, c \in I(v)$ then $b \in I(v)$.

The width of the path decomposition is $\max(|W_i| - 1 : i = 1, \ldots, t)$ and the path-width of $G$, denoted $\mathsf{pw}(G)$ is the minimum width of a path decomposition of $G$. The path-width of an independent set is defined as zero.

Several graph parameters have been shown to be equivalent to path-width. The *interval thickness* $\Theta(G)$ is the smallest max-clique over all interval supergraphs of $G$. Since interval graphs are perfect this can also be stated in terms of the chromatic number: $\Theta(G) = \min\{\chi(H) : H$ is an interval graph with $E(G) \subseteq E(H)\}$. Möhring [30] has shown that $\Theta(G) = \mathsf{pw}(G) + 1$. In *node searching*, an undirected graph is considered as a system of tunnels in which a fugitive is hidden. The *node search number* $\mathsf{ns}(G)$ is the least number of searchers required to capture the fugitive when the search is governed by the following rules: A search move consists of placing a searcher at a node or removing a searcher from a node. The fugitive is captured if both ends of the edge where he hides are simultaneously occupied by a searcher. The fugitive is allowed to move (at any speed) along edges subject to the condition that he never passes a node occupied by a searcher. Kiriousis and Papadimitriou [28] have shown that $\Theta(G) = \mathsf{ns}(G)$. Moreover there is an optimal search, *i.e.*, a search requiring only $\mathsf{ns}(G)$ searchers, such that after an edge has been cleared by two searchers simultaneously guarding its end-nodes it never recontaminates, *i.e.*, there never appears a path that carries no searcher connecting the cleared edge with a contaminated (uncleared) one.

We now show that the path-width is a lower bound on the number $k$ of grid lines needed for an $n \times k$ grid drawing for general planar graphs. We note that a similar result was obtained in [19] however in the context of $h$-layer graph drawings.

**Theorem 1** *Let $G$ be a planar graph. Then we have*
$$\mathsf{pw}(G) \leq \min_k \ (G \ \text{is drawable on an } n \times k \ \text{grid}).$$

**Proof.** We prove the inequality in the node searching context; recall $\mathsf{ns}(G) = \mathsf{pw}(G) + 1$. Given a planar graph $G$ which is drawn on an $n \times k$ grid, we show that the layout can be used to design a node search strategy using $k+1$ searchers for $G$. At the beginning a grid-line-searcher is placed on the leftmost node of each of the $k$ grid lines of the drawing. The invariant is that at any intermediate step of the search there is a searcher on each grid line and the $k$ nodes occupied by these searchers form a node separator. Left of the searchers there are cleared edges and nodes; edges and nodes to the right are not yet cleared and there is no edge connecting a cleared with an uncleared node. A *move* consists of identifying a searcher $s$ sitting on a node $v(s)$ on grid line $t(s)$, such that either there is no edge connecting $v(s)$ to an uncleared node or there is exactly one such edge and this edge connects $v(s)$ to the next node right of $v(s)$ on the same grid line. In both cases the $(k+1)$st searcher $s^*$ is placed on the next node $v'$ on the same grid line $t(s)$. Then $s$ moves to $v'$ thus setting $s^*$ free again. We claim that a move as described is possible as long as not all the $k$ grid-line-searchers are sitting on the rightmost node of their line. Since a move clearly keeps the invariant valid the claim implies that the graph can be decontaminated from left to right in a sequence of moves. It remains to prove the existence of a move. Label the grid lines $t_1, t_2, \ldots, t_k$ from lowermost to topmost. Denote the node occupied by the grid-line-searcher on $t_i$ by $v_i$. With $v_i$ associate the number $h_i$ of the highest grid line such that there is an uncleared edge from $v_i$ to a node on grid line $h_i$; if there is no uncleared edge leaving $v_i$ we set $h_i = 0$. Since $h_k \leq k$ there is a least index $i$ such that $h_i \leq i$. We claim that we can choose

Felsner *et al.*, *Restricted Integer Grids*, JGAA, 7(4) 363–398 (2003)      370

the searcher of grid line $t_i$ for the move. If $i = 1$ condition $h_1 \leq 1$ implies that we can advance the searcher on the first grid line. If $i > 1$ consider the edge emanating from node $v_{i-1}$ to a node on grid line $h_{i-1} \geq i$. By planarity and since $h_{i-1} \geq i$ this edge shields $v_i$ from all uncleared nodes on grid lines $t_j$ for $j < i$. Therefore, there can be at most one edge leaving $v_i$ to an uncleared node to the right and this node must sit on the same grid line. $\qquad\square$

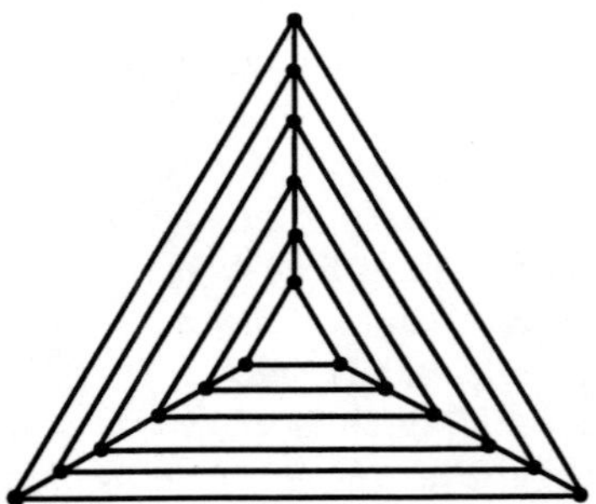

Figure 2: Six nested triangles.

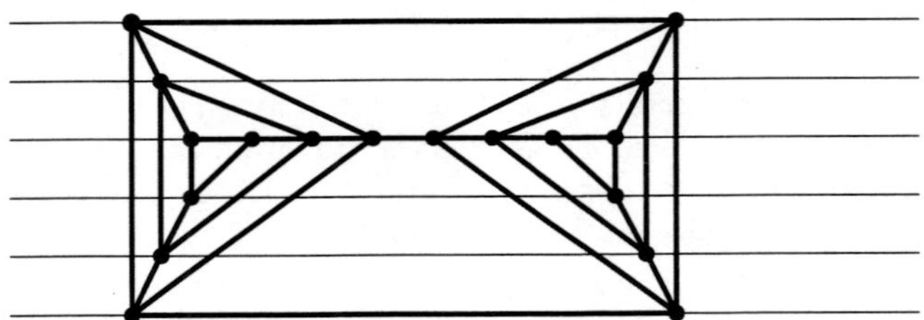

Figure 3: A drawing of $G_3$ on an $n \times 6$ grid.

For the case of trees Suderman [40] has given tight bounds for the gap between path-width and the required height of the grid. Depending on the drawing model chosen, this height is bounded by the $\beta$ fold of the path-width with $\beta \in \{\frac{3}{2}, 2, 3\}$.

In contrast to the situation with trees, the gap in the inequality of Theorem 1 can be arbitrarily large for other classes of planar graphs. Let $G_k$ be the graph consisting of $2k$ nested triangles (see Figure 2 for the case $k = 3$). It is not difficult to see that a grid drawing of $G_k$ requires at least $2k$ grid lines (*i.e.* an $n \times 2k$ grid). Namely, since $G_k$ is three-connected it is enough to study the drawings given by all choices of outer faces. If the outer face is a three cycle, by induction we have that a drawing of $G_k$ requires a number of grid lines that is at least twice the number of nested triangles, *i.e.* it requires at least $4k$ grid lines. If the outer face is a four cycle, we still have to draw a three cycle with at least $k$ nested triangles inside, and therefore at least $2k$ grid lines are required. However, the node-search number of $G_k$ is 4, independent of $k$. In an optimal

Felsner *et al.*, *Restricted Integer Grids*, JGAA, 7(4) 363–398 (2003)      371

search the searchers sweep the triangles starting at the innermost and moving out. An optimal drawing for $G_3$ is given in Figure 3.

## 3.1 Minor monotone issues

The contraction of an edge $e = (x, y)$ in a graph means replacing $x$ and $y$ by a single vertex $z$ which is made adjacent to all the remaining neighbours of $x$ and $y$.

Given a graph $G$ we can generate smaller graphs by repeatedly deleting and contracting edges and deleting isolated vertices. These smaller graphs are called *minors* of $G$. Define $G' \prec G$ if $G'$ is a minor of $G$. The relation $\prec$ is an order relation on the set of all graphs. A set $\mathcal{P}$ of graphs is *minor monotone* if $G \in \mathcal{P}$ and $G' \prec G$ implies $G' \in \mathcal{P}$.

Important examples of minor monotone sets of graphs are: forests, outer-planar graphs, planar graphs and for any fixed $k$, the set of graphs $G$ with $\mathsf{pw}(G) \leq k$. A fundamental result of Robertson and Seymour [35] asserts that every minor monotone set of graphs $M$ is characterized by a finite set of obstructions, *i.e.* there is an integer $t$ and a list of graphs $O_1, O_2, \ldots, O_t$ such that $G \in M$ iff $O_i \not\prec G$ for $i = 1, \ldots, t$. A classical instance is the theorem of Wagner [41]: $G$ is planar iff it has no minor isomorphic to $K_5$ or $K_{3,3}$.

From Theorem 1 it is conceivable to view the minimal height $k$ such that a planar graph admits a drawing on an $n \times k$ grid as a more discriminating, *i.e.* refined, version of path-width. The following theorem shows that this parameter 'grid-height' lacks one very important property.

**Theorem 2** *Being drawable on an $n \times k$ grid is not a minor monotone graph property for $k \geq 3$.*

**Proof.** The graph $G$ shown in Figure 4 is drawn on three grid lines. By contracting the dotted edge we obtain the graph $G'$ shown to the left in Figure 5. We show that $G'$ requires a grid of height at least four.

In Figure 5 the vertex resulting from the contraction is emphasized and labeled $c$, another vertex of degree two is labeled $t$. Let $G'_t$ be the graph obtained by deleting $t$.

The first step of the proof is to show that in every drawing of $G'_t$ the bold vertices are drawn on the middle line. This is based on an observation which is interesting in its own right.

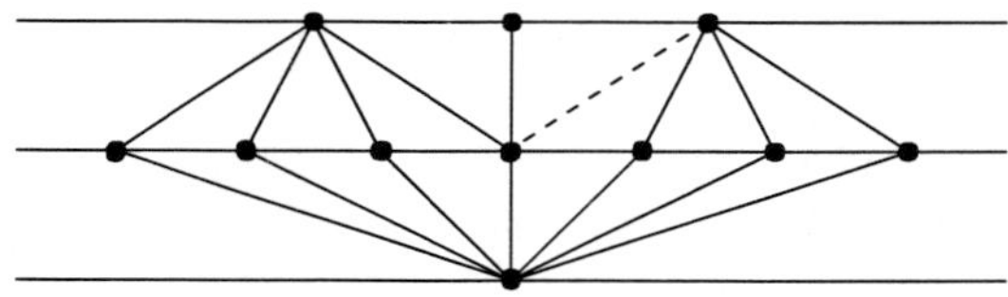

Figure 4: The graph $G$ drawn on a grid of height 3.

Felsner *et al.*, *Restricted Integer Grids*, JGAA, 7(4) 363–398 (2003)      372

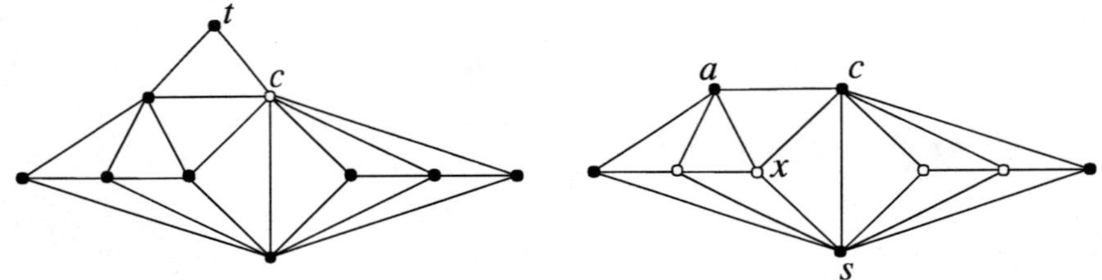

Figure 5: The contraction graph $G'$ and its subgraph $G'_t$

Let $G$ be embedded on a grid of height $k$ and let $C$ be the outer cycle of $G$. The interior vertices of $G$, *i.e.*, those not on $C$, are embedded on the grid lines from 2 to $k-1$.

For the case of three grid lines, $k = 3$, this implies that all vertices not on the outer cycle must be drawn on the middle grid line. As a consequence, the graph induced by the inner vertices must be a subgraph of a path.

In the case of $G'_t$ there are two candidates for the outer cycle namely, the one shown in the figure and the cycle of length four including the edge $(s, c)$. In both cases the four black vertices are interior and must be drawn on the middle grid line. Assume that $s$ is placed on grid line 1; then both vertices $a$ and $c$ have to go on grid line 3. A drawing of $G'$ contains a drawing of $G'_t$, hence, for a drawing of $G'$ on a grid of height three we also have: If $s$ is on line 1 then $a$ and $c$ are on line 3. However, since $x$ and $t$ are common neighbours of $a$ and $c$ this placement of $a$ and $c$ makes a crossing of edges unavoidable. In conclusion the minor $G'$ of $G$ has no drawing on a grid of height 3.  □

An interesting open problem suggested by one of the referees is whether for trees and forests the required height for grid drawings might be a minor monotone parameter.

## 3.2  Drawings of Trees

Next we consider drawings of trees. As a first step we characterize the family of strip-drawable trees and give linear time recognition and drawing algorithms for such trees.

The approach taken applies to strip-drawable trees and has a natural generalization which leads to the notion of a $k$-strict tree. We show that a $(k+1)$-strict tree cannot be drawn on an $n \times k$ grid. We then show that the strictness of a tree is closely related to the path-width. More precisely, there is a difference of at most one between the two parameters. In his recent paper Matthew Suderman [40] has obtained tight bounds on the height of a grid required for drawings of trees of path-width $k$ in several drawing models.

Felsner *et al.*, *Restricted Integer Grids*, JGAA, 7(4) 363–398 (2003)       373

### 3.2.1   Strip-Drawable Trees

Property 1 establishes that all paths are strip-drawable, since they are in fact 1-track drawable. We define a tree as *2-strict* if it contains a vertex of degree greater than or equal to three. An immediate consequence of Property 1 is the following.

**Property 3** *A 2-strict tree is not 1-track drawable.*

An edge is defined as a *core edge* if its removal results in two 2-strict components. For an edge $e = (u, v)$, we refer to the two subtrees resulting from its removal (but including vertices $u$, $v$) as $T_u$ and $T_v$.

**Lemma 1** *Core edges are connected.*

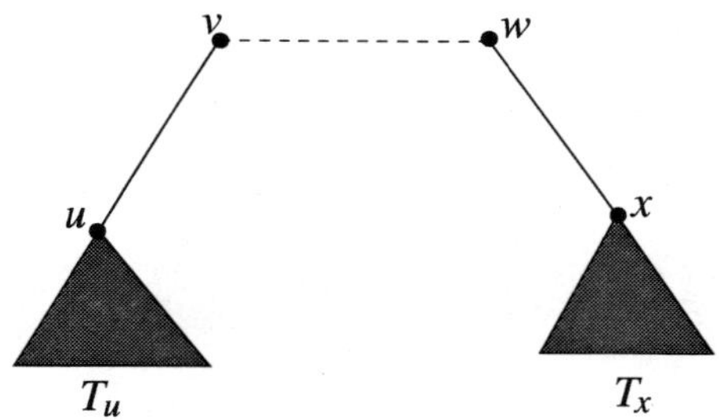


Figure 6: Core edges are connected

**Proof.** Let $e_1 = (u, v)$ and $e_2 = (w, x)$ be two core edges and consider any edge $e$ on the path connecting $e_1$ to $e_2$. Refer to Figure 6. Edge $e$ receives one 2-strict component from $T_u$ and one from $T_x$ and thus must be core.  □

**Lemma 2** *A tree is strip drawable if and only if its core edges form a path.*

**Proof.** ($\Rightarrow$) (by contradiction) By the previous lemma, if the core edges do not form a path, then there is a vertex $v$ with at least three incident core edges $(v, a), (v, b), (v, c)$ – see Figure 7. If the subtrees $T_a, T_b, T_c$ are drawable then by Property 3 their associated drawings $\Gamma_a, \Gamma_b, \Gamma_c$ each require two tracks. There is no location for $v$ that permits a crossing-free connection to all three subdrawings. ($\Leftarrow$) Refer to Figure 8. If the core edges form a non-degenerate path (*i.e.* a non-zero length path), then draw them consecutively on track $t_1$. Consider an arbitrary non-core edge $e = (u, v)$ with $u$ on track $t_1$. Since $e$ is non-core, $T_v$ must *not* be 2-strict and is thus 1-track drawable. Therefore $v$ can be placed on track $t_2$ with the drawing of $T_v$ also on the same track, as in Figure 8. There is one degenerate case to consider. If there are no core edges (*i.e.* a path of length 0), then either the tree has no vertex of degree three and is in fact drawable on a single track, or there exists at most one vertex $v$ with

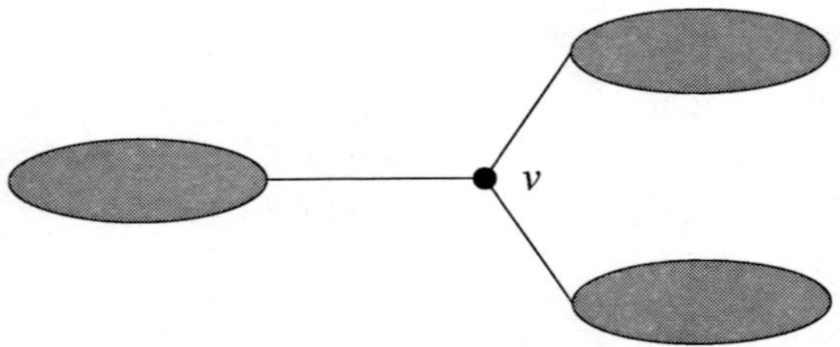

Figure 7: T is not strip-drawable if the core edges are not a path

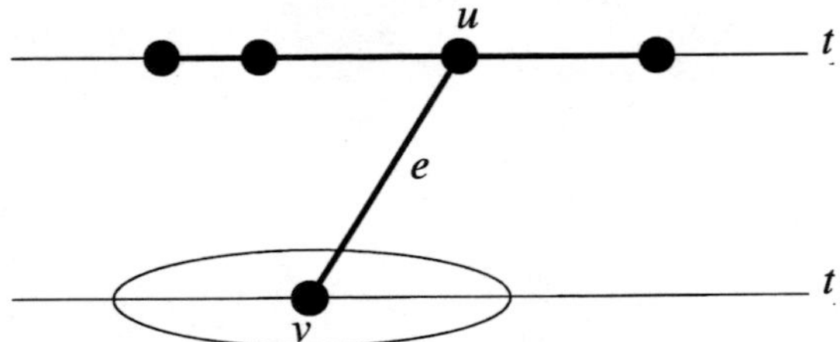

Figure 8: Drawing a tree on a strip

neighbours $w_1, w_2, ...w_k$ and each $T_{w_i}$ is *not* 2-strict. Each of the subtrees can thus be drawn on track $t_2$ and $v$ on track $t_1$ as in Figure 9.     □

Based on this characterization, we now consider the complexity of recognizing and drawing the trees that are strip-drawable.

**Lemma 3** *Let $T$ be a tree with $n$ vertices. There exists an $O(n)$-time algorithm that recognizes whether $T$ is strip-drawable and, if so, computes a strip drawing of $T$.*

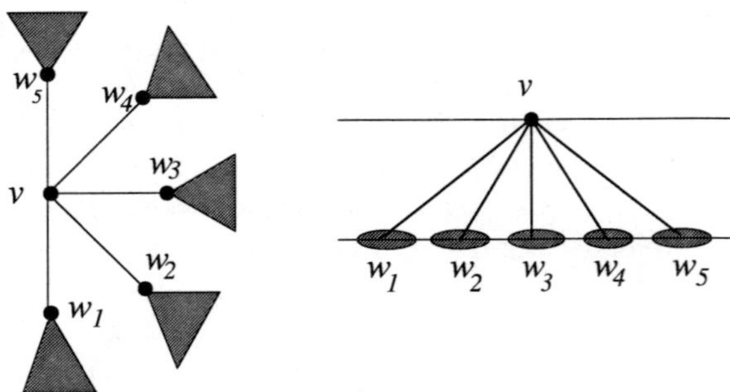

Figure 9: A degenerate core path

Felsner *et al.*, *Restricted Integer Grids*, JGAA, 7(4) 363–398 (2003)     375

**Proof.** Note that a tree is 2-strict if and only if it has more than two leaves; thus counting leaves is the crucial operation. First the core edges must be established and then the path condition on the core edges checked. With each edge $e = (u, v)$ we associate two counters: $l_u$ will be the number of leaves in $T_u$, and $l_v$ will be the number of leaves in $T_v$. Let $l$ be the number of leaves in the entire tree $T$. Since $l_u + l_v = l$ it follows that $e$ is a core edge if and only if both $l_u$ and $l_v$ are larger than 2. Choose an arbitrary non-leaf vertex $r$ as a root. Each vertex $v$ reports the number of leaves in the subtree below it to its parent $u$ – thus establishing $l_v$ for the edge $(u, v)$ and hence $l_u$. If $v$ has no children then it is a leaf and reports 1. A simple recursive function can be used to implement this counting step in linear time. Finally, checking that the core edges form a path is easily accomplished in linear time and the proof of the previous Lemma described the construction of the strip drawing. $\qquad\square$

We can summarize Lemmas 2 and 3 as follows.

**Theorem 3** *A tree $T$ with $n$ vertices is strip drawable if and only if its core edges form a path. Furthermore, there exists an $O(n)$-time algorithm that determines whether $T$ is strip drawable and, if so computes a strip drawing of $T$.*

### 3.2.2   $k$-Strict Trees

The results of Theorem 3 can be extended to give a necessary condition for trees to be drawable on an $n \times k$ grid by generalizing the concepts of the previous section. A tree is $k$-*strict* if it contains a vertex adjacent to at least three vertices whose subtrees are $(k - 1)$-strict. For example, the tree of Figure 10 is 3-strict since the vertex $u$ is adjacent to three 2-strict subtrees.

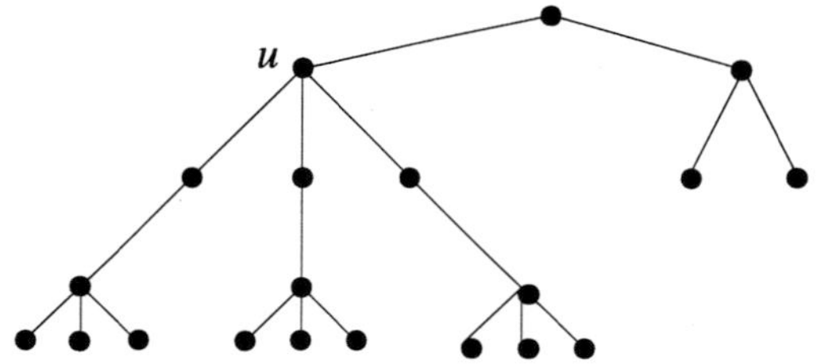

Figure 10: A 3-strict tree

**Lemma 4** *A $(k + 1)$-strict tree is not drawable on an $n \times k$ grid.*

**Proof.** The proof is by induction on $k$ and Property 3 provides the base case. If a tree $T$ is $(k + 1)$-strict then it contains a vertex $v$ adjacent to at least three vertices whose subtrees are $k$-strict and by the inductive hypothesis each subtree requires at least $k$ tracks to be drawn. In this case, there is no location for $v$ on the $k$ tracks that allows it to connect to the three subtrees without creating an intersection. $\qquad\square$

Felsner *et al.*, *Restricted Integer Grids*, JGAA, 7(4) 363–398 (2003)     376

**Corollary 1** *The complete ternary tree of height[1] $k+1$ is not drawable on an $n \times k$ grid.*

**Proof.** Such a tree is $(k+1)$-strict and hence not drawable on the $n \times k$ grid. $\square$

### 3.2.3 $k$-strict trees and Path-width

In this subsection we show that the strictness of a tree is tightly related to the well known parameter path-width.

**Theorem 4** *Let $T$ be a tree. Then*
$$\mathsf{pw}(T) \leq \max_k (T \text{ is } k\text{-strict}) \leq \mathsf{pw}(T) + 1.$$

**Proof.** For the proof we compare the strictness of a tree $T$ to its node search number $\mathsf{ns}(T) = \mathsf{pw}(T) + 1$ (refer to the beginning of this section for the definitions). In the proof we make use of a lemma attributed to Parsons [32] in [29]

> [Parsons' Lemma] *For any tree $T$ and integer $k \geq 1$, $\mathsf{ns}(T) \geq k+1$ if and only if $T$ has a vertex $v$ at which there are three or more branches that have search number $k$ or more.*

First we show that for a tree $T$ $k$-strictness implies $\mathsf{ns}(T) \geq k$, by induction on $k$. Assume $T$ is 2-strict. Then $T$ contains a vertex of degree 3; in particular therefore at least one edge and $\mathsf{ns}(T) \geq 2$. Now assume $T$ is $k$-strict, $k > 2$. Then $T$ contains a vertex $v$ at which three branches $T_1, T_2, T_3$ are $(k-1)$-strict. By induction, each $T_i$ satisfies $\mathsf{ns}(T_i) \geq k-1$, and by Parsons' lemma $\mathsf{ns}(T) \geq k$.

Next we show that $\mathsf{ns}(T) \geq k$ implies that $T$ is $(k-1)$-strict, again by induction on $k$. Assume that $\mathsf{ns}(T) = 2$. Then $T$ contains an edge, and therefore it is 1-strict. Now $\mathsf{ns}(T) = k > 2$. By Parson's lemma $T$ contains a vertex $v$ with three branches $T_1, T_2, T_3$ such that $\mathsf{ns}(T_i) \geq k-1$. By induction, $T_i$ is $(k-2)$-strict, and therefore $T$ is $(k-1)$-strict. $\square$

## 4    Three-Dimensional Drawings of Outerplanar Graphs

In Section 3, Corollary 1 showed that, for a fixed $k$, there is no $n \times k$ grid that supports all trees of $n$ vertices. This motivates us to investigate the existence of three-dimensional restricted grids that support all trees. As it turns out, the situation in three dimensions is distinctly different. Namely, we show that all outerplanar graphs are prism-drawable by providing a linear time algorithm

---

[1]The height is measured as the number of vertices on the path from the root to the deepest leaf.

Felsner *et al.*, *Restricted Integer Grids*, JGAA, 7(4) 363–398 (2003)     377

that computes this drawing. This is the first known three-dimensional straight-line drawing algorithm for the class of outerplanar graphs that achieves $O(n)$ volume on an integer grid.

A high level description of our drawing algorithm, called **Algorithm Prism Draw**, is as follows. Let $G$ be an outerplanar graph with a specified *outerplanar embedding*, *i.e.* a circular ordering of the edges incident around each vertex such that all vertices of $G$ belong to the external face. (Such an embedding can be computed in linear time). **Algorithm Prism Draw** computes a prism drawing of $G$ by executing two main steps. Firstly a 2D drawing of $G$ is computed on a grid that consists of $O(n)$ horizontal tracks and such that adjacent vertices are at grid points whose $y$-coordinates differ by at most one. This is done by visiting $G$ in a breadth-first fashion and setting the $x$-coordinate to be the breadth-first search (BFS) number and the $y$-coordinate to be the depth in the BFS tree. Secondly, the drawing is "wrapped" onto the faces of a prism by folding it along the tracks. Refer to Figure 11.

Figure 12 shows an example of the output of Step 1 of the algorithm; for consistency with track layout terminology, the $Y$ axis points downwards. The following results establish that **Algorithm Prism Draw** computes a prism drawing of any outerplanar graph $G$. First observe that *currx* is incremented each time a vertex is drawn, and therefore we have the following proposition.

**Proposition 1** *No two vertices of* $\Gamma$ *are assigned the same X-coordinate.*

Also, since the unmarked neighbours of a vertex $u$ are all drawn on the track consecutive to that of $u$ during Step 1, we have the following.

**Proposition 2** *A vertex is assigned to track* $t_{i+1}$ *if and only if it has not yet been marked/assigned and has a neighbour on track* $t_i$, *for* $i \geq 0$.

The following lemmas establish that the drawing between any two consecutive tracks forms a strip drawing, and therefore Step 1 of **Algorithm Prism Draw** computes a $k$-track drawing of the input graph.

**Lemma 5** *Let $G$ be an outerplanar graph with a given embedding and let $\Gamma$ be the drawing computed by Step 1 of* **Algorithm Prism Draw**. *Then $\Gamma$ is a $k$-track drawing of $G$ for some $k \leq n$.*

**Proof.** Step 1 of **Algorithm Prism Draw** draws $G$ on a 2D $k$-track, where $k \leq n$. Also, by Proposition 2 we have that an edge of $\Gamma$ can connect only vertices that are drawn either on the same track or on two consecutive tracks. In order to complete the proof we must show that $\Gamma$ satisfies the following properties.

1. No two edges connecting vertices on consecutive tracks intersect.

2. Let $u$ and $v$ be two vertices of $\Gamma$ which are drawn on the same track $t$. If $u$ and $v$ are adjacent in $G$, then they appear consecutively on $t$.

We start by proving that $\Gamma$ has the first property. Suppose there exist four vertices $a$, $b$, $c$, $d$ such that

Felsner *et al.*, *Restricted Integer Grids*, JGAA, 7(4) 363–398 (2003)     

**Algorithm Prism Draw**

*input:* An outerplanar graph $G$ with a given outerplanar embedding.
*output:* A prism drawing of $G$.

**Step 1.** **The 2D Drawing Phase:** A 2D track drawing $\Gamma$ of $G$ where vertices are assigned to different $x$-coordinates is computed as follows.

- Add a dummy vertex $d$ on the external face and an edge connecting $d$ to an arbitrary vertex $v$.
- mark $d$
- $i := 0$
- $currx := 0$
- draw $v$ on track $t_0$ by setting $X(v) := currx$; $Y(v) := i$
- $currx := currx + 1$
- mark $v$
- while there are unmarked vertices of $G$ do
    - visit the vertices on track $t_i$ from left to right and for each encountered vertex $u$ do
        * let $w$ be a marked neighbour of $u$ in $G$
        * visit the neighbours of $u$ in counterclockwise order starting from $w$, and for each encountered vertex $r$ such that $r$ is unmarked do order
            · draw $r$ on track $t_{i+1}$ by setting $X(r) := currx$; $Y(r) := i+1$
            · $currx := currx + 1$
            · mark $r$
    - $i := i+1$

**Step 2:** **The 3D Wrapping Phase:** A prism drawing $\Gamma'$ is obtained by folding $\Gamma$ along its tracks as follows.

- for each vertex $v$ of $\Gamma$ define its coordinates $X'(v)$, $Y'(v)$ and $Z'(v)$ in $\Gamma'$ by setting:
    - $X'(v) := X(v)$
    - if $Y(v) = 0, 1 \bmod 3$ then $Y'(v) := 0$, else $Y'(v) := 1$
    - if $Y(v) = 0, 2 \bmod 3$ then $Z'(v) := 0$, else $Z'(v) := 1$

Figure 11: **Algorithm Prism Draw.**

Felsner *et al.*, *Restricted Integer Grids*, JGAA, 7(4) 363–398 (2003)        379

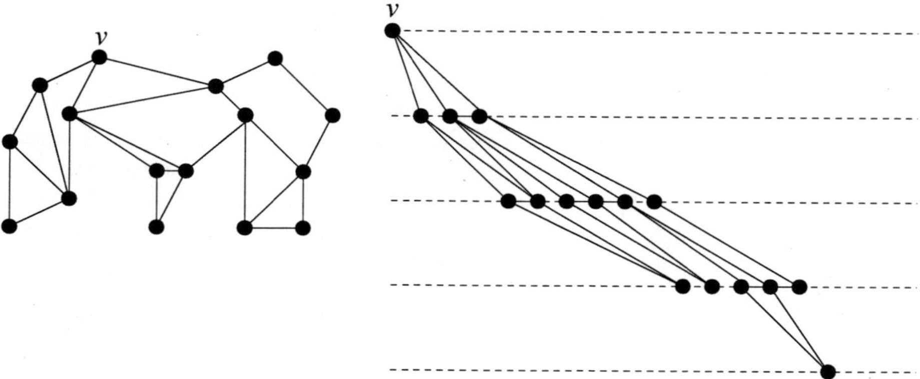

Figure 12: An outerplanar graph drawn by **Step 1** of **Algorithm Prism Draw**.

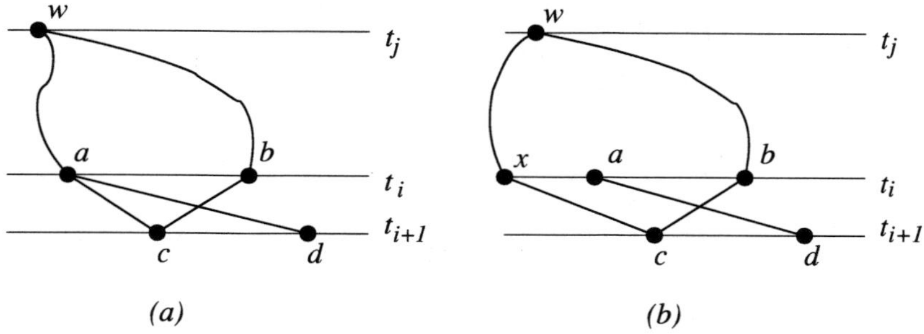

Figure 13: Two cases for the proof of Lemma 5.

- $a$ and $b$ are on track $t_i$ and $a$ is to the left of $b$.

- $c$ and $d$ are on track $t_{i+1}$.

- There is an edge $(a,d)$ intersecting an edge $(b,c)$.

Note that **Algorithm Prism Draw** draws all the unmarked neighbours of $a$ on track $t_{i+1}$ by following the counterclockwise order of the edges around $a$ given by the outerplanar embedding of $G$. Also note that all the neighbours of $a$ on track $t_{i+1}$ are assigned an $X$-coordinate that is strictly smaller than the $X$-coordinates assigned to the neighbours of any vertex drawn to the right of $a$ on track $t_i$. Therefore we may conclude that if $(a,d)$ and $(b,c)$ cross, then one of the two cases must hold: Case 1: Vertex $c$ is a neighbour of both $b$ and $a$ (see Figure 13a). Case 2: There is a vertex $x$, drawn on track $t_i$ to the left of $a$ and such that $c$ is a neighbour of both $x$ and $b$ (see Figure 13b). Consider Case 1. By Proposition 2 and by the fact that **Algorithm Prism Draw** places only vertex $v$ on track $t_0$, it follows that there exists a lowest common ancestor of both $a$

Felsner *et al.*, *Restricted Integer Grids*, JGAA, 7(4) 363–398 (2003)      380

and $b$, say $w$, drawn on some track $t_j$ with $0 \leq j < i$. Let $\Pi_{wa}$ and $\Pi_{wb}$ be two disjoint paths connecting $w$ to $a$ and $w$ to $b$ respectively. Observe that Step 1 of **Algorithm Prism Draw** computes a drawing that preserves the given outerplanar embedding of $G$ since it draws the vertices on consecutive tracks with increasing $x$ values and by following the circular ordering of the edges around the vertices. Therefore, if edge $(a, d)$ follows edge $(a, c)$ in the counterclockwise ordering of the edges around $a$, then in the outerplanar embedding of $G$ there must be a cycle (namely that formed by the path $\Pi_{wa}$, edge $(a, c)$, edge $(c, b)$ and path $\Pi_{wb}$) with vertex $d$ in its interior. But this is a contradiction. Now consider Case 2. By the same argument used for the previous case, there is a lowest common ancestor $w$ of $x$ and $b$ on some track $t_j$ with $0 \leq j < i$. Let $\Pi_{wx}$ and $\Pi_{wb}$ be two disjoint paths connecting $w$ to $x$ and $w$ to $b$ respectively. Now observe that vertex $d$ would necessarily lie in the interior of the cycle defined by $\Pi_{wx}$, edge $(x, c)$, edge $(c, b)$ and path $\Pi_{wb}$ thus contradicting the fact that $\Gamma$ preserves the outerplanar embedding of $G$. Finally, we prove that $\Gamma$ satisfies the

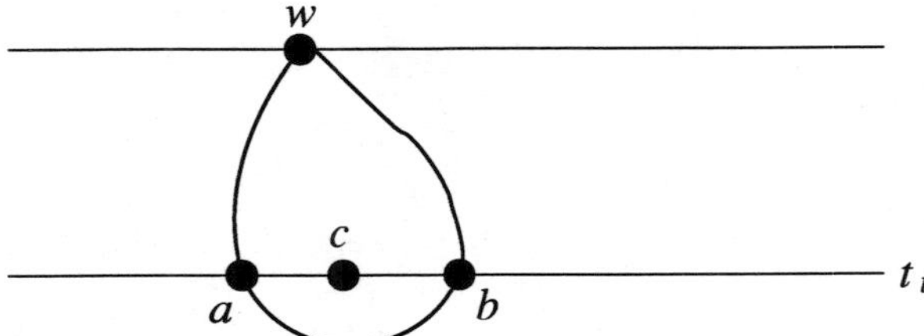


Figure 14: Edges on a given track only join consecutive vertices.

second property and again the proof is by contradiction. Suppose there exist two vertices $a$ and $b$ on track $t_i$ $(0 < i \leq k)$ such that:

- $a$ and $b$ are adjacent in $G$.

- There exists a vertex $c$ in $\Gamma$ such that $c$ is drawn on track $t_i$ between $a$ and $b$. See also Figure 14.

By the same reasoning as that used in the previous cases, let $w$ be the lowest common ancestor of $a$ and $b$ on some track $t_j$ with $0 \leq j < i$ and let $\Pi_{wa}$ and $\Pi_{wb}$ be two disjoint paths connecting $w$ to $a$ and $w$ to $b$ respectively. Consider the cycle formed by $\Pi_{wa}$, $\Pi_{wb}$ and edge $(a, b)$. This cycle has vertex $c$ in its interior contradicting the fact that $\Gamma$ preserves the outerplanar embedding of $G$. $\qquad\square$

**Theorem 5** *Every outerplanar graph $G$ with $n$ vertices admits a crossing-free straight-line grid drawing in three dimensions in optimal $O(n)$ volume. Furthermore, **Algorithm Prism Draw** computes such a drawing of $G$ in $O(n)$ time and with the vertices of $G$ drawn on the grid points of a prism.*

**Proof.** Lemma 5 ensures there are no crossings in the $k$-track drawing. Concerning **Step 2 (3D Wrapping Phase)**, observe that the coordinate assignment is such that the vertices of $\Gamma'$ are grid points of a $n \times 2 \times 2$ grid and they all belong to just three of the four possible tracks of the $n \times 2 \times 2$ grid. Therefore, the vertices of $\Gamma'$ are drawn on grid points of a prism. Also, no two edges of $\Gamma'$ intersect because, as shown above, each subdrawing of $\Gamma'$ induced by vertices on two different tracks is a strip drawing and because the vertices on each track have distinct X-coordinates. Finally, note that **Algorithm Prism Draw** runs in linear time since it is essentially a breadth-first traversal of the graph. $\qquad\square$

**Remark:** Note that Step 2 of the algorithm is applicable given any track-drawing of a graph after suitable shifting to ensure increasing $x$-coordinates. Thus, graph $G$ is track-drawable implies that $G$ is prism-drawable. The converse however does not hold; $K_4$ is an example of a graph that is prism-drawable but not track-drawable.

## 5   Prism-Drawable Graphs

Motivated by Theorem 5, we study in this section whether the prism is a universal grid for planar graphs. For example, Figure 15 shows a maximal planar graph, and its prism drawing. As another example, note that the family of maximal planar graphs consisting of a sequence of nested triangles (as in Fig-

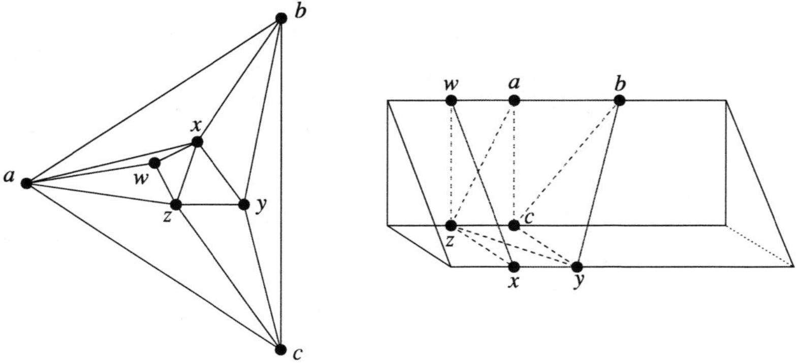

Figure 15: A prism-drawable graph $G$ and its drawing

ure 2) and that are known to require $\Omega(n^2)$ area in the plane [11], can easily be drawn on the prism in $O(n)$ volume. Unfortunately, it turns out that not all planar graphs are prism-drawable. In Section 5.1 we give a characterization of prism-drawable graphs and in Sections 5.2 and 5.3 we illustrate two different approaches for constructing planar graphs that violate the characterization. The first approach is based on the concept of *strictly-prism drawable* graphs and the second exploits the relationship between hamiltonicity and prism-drawability.

Felsner *et al.*, *Restricted Integer Grids*, JGAA, 7(4) 363–398 (2003)     382

## 5.1 Characterization of Prism-Drawable Graphs

An essential prerequisite of our characterization of prism-drawable graphs, is the study of the strip-drawable graphs since a prism effectively consists of three strips. Independently, Cornelsen, Schank and Wagner [10] developed a linear time algorithm for determining if a graph is strip-drawable. Clearly such graphs must be (a subset of the) outerplanar graphs since the two tracks, which contain all vertices, form the exterior face. Our characterization differs significantly from that contained in [10].

We define a *spine* in a graph $G$ as a sequence $v_0, v_1, \ldots v_m$ of vertices such that the subgraph induced by $v_0, v_1, \ldots v_m$ is a path. The definition of spine precludes any edge between non-consecutive vertices; we refer to such an edge as a *chordal* edge. The characterization of prism-drawable graphs is based on the observation that in a strip drawing, there must exist two sub-spines (each defined by the vertices on one of the two tracks of the strip) and that edges connecting vertices on these two sub-spines must not intersect. Since every graph with less than four vertices is clearly strip-drawable (a three cycle is strip-drawable and therefore every subgraph of a three cycle is strip drawable), the next theorem considers graphs with at least four vertices. Note that although this characterization is not efficiently implementable, it is a basis for characterizing the prism-drawable graphs, the box-drawable graphs and provides a means for showing that planar graphs are not necessarily prism-drawable.

**Theorem 6** *A graph $G$ with at least four vertices is strip-drawable if and only if it is possible to augment $G$ with edges to produce a graph $G'$ such that:*

- *$G'$ contains two edges $(r_0, b_0)$ and $(r_z, b_t)$.*

- *There are two vertex-disjoint spines $r_0, r_1, \ldots r_z$ and $b_0, b_1, \ldots b_t$ in $G'$ such that all vertices of $G$ are on the two spines.*

- *If there exists an edge $(r_i, b_j)$ with $0 \le i \le z$ and $0 \le j \le t$ then there are no edges of the form $(r_k, b_l)$ with $(0 \le k < i$ and $j < l \le t)$ or $(i < k \le z$ and $0 \le l < j)$.*

**Proof.** $(\Rightarrow)$We show how to construct a graph $G'$ that satisfies the statement. Let $\Gamma$ be a strip drawing of $G$ with $t_1$ and $t_2$ as the two tracks of the strip. Let $r_0, b_0$ be the leftmost pair of vertices and let $r_z, b_t$ be the rightmost pair of vertices of $\Gamma$ such that $b_0$ and $b_t$ are on track $t_1$. If $r_0, b_0$ are not adjacent in $\Gamma$, then edge $(r_0, b_0)$ is added; similarly, if $r_t, b_t$ are not adjacent in $\Gamma$, then edge $(r_t, b_t)$ is added. Also, for each pair of consecutive non-adjacent vertices encountered when walking along each track an edge is added so to form two paths $\Pi_1$ and $\Pi_2$. Let $\Gamma'$ be the new drawing and let $G'$ be the graph represented by $\Gamma'$. Note that $G'$ has two edges $(r_0, b_0)$ and $(r_t, b_t)$. Since there are no chordal edges between any two non consecutive vertices of $\Gamma$ that are on the same track, it follows from the construction that $\Pi_1$ and $\Pi_2$ are spines for $G'$. Each vertex of $G$ is drawn either on track $t_1$ or on track $t_2$ and by construction of $\Gamma'$ it belongs either to $\Pi_1$ or to $\Pi_2$; it follows that all vertices of $G$ are on the two spines of

Felsner *et al.*, *Restricted Integer Grids*, JGAA, 7(4) 363–398 (2003)    383

$G'$. Also, since in $\Gamma$ there are no crossings between any two edges connecting vertices on different tracks and since in $\Gamma'$ edges $(r_0, b_0)$ and $(r_t, b_t)$ do not cross any other edge, it follows that if there exists an edge $(r_i, b_j)$ in $G'$ with $0 \leq i \leq z$ and $0 \leq j \leq t$ then there cannot be edges of the form $(r_k, b_l)$ with $(0 \leq k < i$ and $j < l \leq t)$ or $(i < k \leq z$ and $0 \leq l < j)$.

($\Leftarrow$)Given an augmented graph $G'$, a strip drawing $\Gamma'$ of $G'$ is obtained as follows. Spine $r_0, r_1, \ldots r_z$ is drawn on one track such that $r_j$ is drawn to the right of $r_i$ for $0 \leq i < j \leq z$. Spine $b_0, b_1, \ldots b_t$ is drawn on the second track such that $b_j$ is drawn to the right of $b_i$ for $0 \leq i < j \leq t$. Edges connecting vertices along the same track and between the two tracks are drawn as straight-line segments. Since the subgraph induced by each spine is a path and since for each edge $(r_i, b_j)$ with $0 \leq i \leq z$ and $0 \leq j \leq t$ there are no edges of the form $(r_k, b_l)$ with $(0 \leq k < i$ and $j < l \leq t)$ or $(i < k \leq z$ and $0 \leq l < j)$, it follows that the drawing of $G'$ does not have crossings. Finally, a strip drawing of $G$ is obtained by deleting edges from $\Gamma'$. $\qquad\square$

The characterization of prism drawable graphs generalizes Theorem 6 to three dimensions. Intuitively, it must be possible to augment a given graph to obtain three spines with two "lids" (three cycles) and between each pair of spines the strip drawability condition must hold. Since every strip-drawable graph is also prism-drawable, the next theorem assumes that $G$ has at least four vertices.

**Theorem 7** *A graph $G$ with at least four vertices is prism-drawable if and only if it is possible to augment $G$ with edges to produce a graph $G'$ such that:*

- *$G'$ contains two three-cycles $r_0, b_0, g_0$ and $r_z, b_t, g_s$, where $z, t, s \geq 0$.*

- *There are three vertex-disjoint spines, denoted by $r_0, r_1, \ldots r_z$, $b_0, b_1, \ldots b_t$ and $g_0, g_1, \ldots g_s$ in $G'$ such that all vertices of $G$ are on the three spines.*

- *For each pair of spines $x_0, x_1, \ldots x_m$ and $y_0, y_1, \ldots y_p$ ($x, y \in \{r, b, g\}$, $x \neq y$, $m, p \in \{z, t, s\}$) , if $(x_i, y_j)$ is an edge, then there are no edges of the form $(x_k, y_l)$ with $(0 \leq k < i$ and $j < l \leq p)$ or $(i < k \leq m$ and $0 \leq l < j)$.*

**Proof.** ($\Rightarrow$) We show how to construct a graph $G'$ that satisfies the statement. Let $\Gamma$ be a prism drawing of $G$ and let $t_1$, $t_2$, and $t_3$ be the three tracks of the prism. Consider the subgraph $G_{ij}$ induced by the vertices that are on two different tracks $t_i$ and $t_j$ ($i, j = 1, 2, 3$, $i \neq j$); $G_{ij}$ is strip-drawable and therefore there exists an augmented graph $G'_{ij}$ with the properties stated by Theorem 6. A graph $G'$ that satisfies the statement is then defined as $G' = G'_{12} \cup G'_{13} \cup G'_{23}$.

($\Leftarrow$) Given an augmented graph $G'$, a prism drawing $\Gamma'$ of $G'$ is obtained as follows. Spine $r_0, r_1, \ldots r_z$ is drawn on track $t_1$ such that $r_j$ is drawn to the right of $r_i$ for $0 \leq i < j \leq z$. Spine $b_0, b_1, \ldots b_t$ is drawn on track $t_2$ such that $b_j$ is drawn to the right of $b_i$ for $0 \leq i < j \leq t$. Spine $g_0, g_1, \ldots g_s$ is drawn on track $t_3$ such that $g_j$ is drawn to the right of $g_i$ for $0 \leq i < j \leq s$. Edges connecting vertices along the same track and between two different tracks are drawn as

Felsner *et al.*, *Restricted Integer Grids*, JGAA, 7(4) 363–398 (2003)      384

straight-line segments. Since the subgraph induced by each spine is a path and since for each edge $(x_i, y_j)$ $(x, y \in \{r, b, g\};\ x \neq y;\ m, p \in \{z, t, s\})$ there are no edges of the form $(x_k, y_l)$ with $0 \leq k < i$ and $j < l \leq p$ or $i < k \leq m$ and $0 \leq l < j$, it follows that $\Gamma'$ does not have crossings. Finally, a prism drawing of $G$ is obtained by deleting edges from $\Gamma'$. $\qquad\square$

In the rest of the paper it will be convenient to imagine the three spines $r_0, r_1, \ldots r_z$, $b_0, b_1, \ldots b_t$, and $g_0, g_1, \ldots g_s$ of Theorem 7 as colored red, blue and green, respectively. Also, we shall refer to a graph $G'$ described in Theorem 7 as an *augmented graph* of $G$.

## 5.2   Prism-Drawability and Planarity

In this section we show that prism drawable graphs are a proper subset of planar graphs. In the next section we shall further restrict the set of prism-drawable graphs.

**Theorem 8** *Let $G$ be a prism-drawable graph. Then $G$ is planar.*

**Proof.** Any prism-drawing of $G$ can be augmented by edges to form a convex polytope and therefore by the theorem of Steinitz [43] only planar graphs are prism-drawable. $\qquad\square$

**Corollary 2** *If $G$ is a maximal planar graph and is prism drawable, then the augmented graph $G'$ coincides with $G$.*

One approach for constructing planar *prism-forbidden* graphs, *i.e.* planar graphs which do not admit a prism drawing, is based on the following definition and lemma. A graph $G$ is *strictly prism-drawable* if it is prism-drawable and all prism drawings of $G$ have at least three edges $(x, y)$, $(y, z)$ and $(z, x)$ such that $x$, $y$ and $z$ are on different tracks.

**Lemma 6** *Let $G$ be a 1-connected planar graph that has a cut vertex $v$ whose removal separates the graph into $h$ strictly prism-drawable components $G_1, \ldots G_h$ ($h \geq 3$). Then $G$ is prism-forbidden.*

**Proof.** Consider any prism drawing $\Gamma_i$ of $G_i$ ($0 \leq i \leq h$). $\Gamma_i$ has a three-cycle which defines a plane that intersects all three facets of the prism, because $G_i$ is strictly prism-drawable. Thus, $\Gamma_1, \ldots, \Gamma_h$ slice the prism into $h + 1$ slices (see Figure 16). Now there is no location for $v$ that permits it to be connected to all $\Gamma_i$ ($0 \leq i \leq h$) without crossing at least one three-cycle. $\qquad\square$

Lemma 6, provides the key to creating a prism-forbidden graph. Although $K_4$ can easily be shown to be strictly prism-drawable we choose to show that the graph in figure 17 is strictly prism-drawable for three reasons: the extension to the box-forbidden case is more natural, the series-parallel case follows as a consequence, and the proof portrays the importance of the spine characterization of Theorem 7.

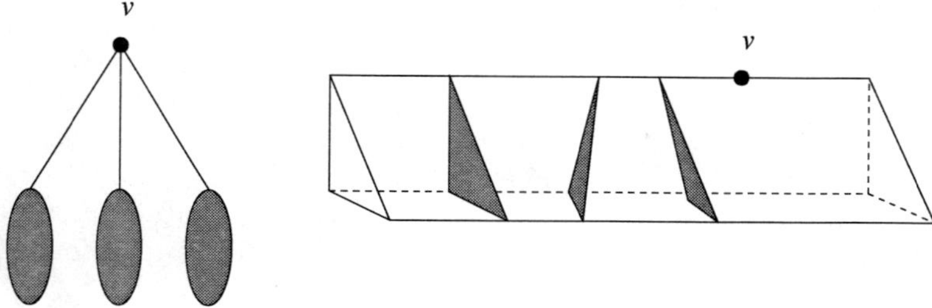

Figure 16: The "slicing" argument in the proof of Lemma 6.

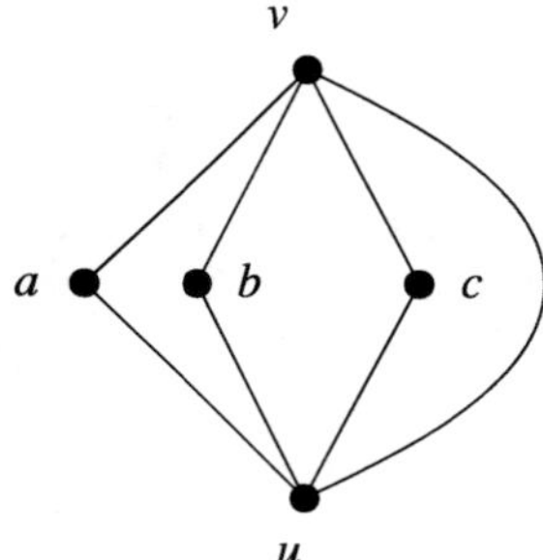

Figure 17: A strictly prism-drawable graph.

**Lemma 7** *The graph in Figure 17 is strictly prism-drawable.*

**Proof.** Let $G$ be the graph of Figure 17. We first show that $G$ satisfies Theorem 7. The augmented graph $G'$ is defined by adding to $G$ edges $(a, b)$ and $(b, c)$. We choose the two three-cycles of $G'$ as follows: One three-cycle consists of edges $(u, v)$, $(v, a)$, $(a, u)$; the second three-cycle consists of edges $(u, v)$, $(u, c)$, $(c, v)$. The three spines of $G'$ are as follows: the red spine is the path of vertices $a, b, c$, the blue spine consists of vertex $u$ and the green spine consists of vertex $v$. Since $G'$ is planar and two of the three spines consist of a single vertex, the non-crossing condition stated by Theorem 7 among edges connecting vertices on different spines is trivially verified. It follows that $G$ is prism-drawable. It remains to show that $G$ is *strictly* prism-drawable. This is done by proving the following two claims.

1. In any prism drawing of $G$ vertices $u$ and $v$ cannot be on the same track.

2. In any prism drawing of $G$ at least one of vertices $a$, $b$, $c$ is drawn on a track different from that of $u$ and different from that of $v$.

To prove Claim 1, suppose that there existed a prism drawing $\Gamma$ of $G$ with $u$ and $v$ on the same track $t_1$ (see Figure 18(a)). Since $u$ and $v$ are adjacent in $G$, there

Felsner *et al.*, *Restricted Integer Grids*, JGAA, 7(4) 363–398 (2003)          386

exists in $\Gamma$ a straight-line segment connecting $u$ to $v$. As a consequence neither $a$, nor $b$, nor $c$ can be drawn on $t_1$ or else there would be an edge overlapping edge $(u,v)$. It follows that vertices $a$, $b$, and $c$ are drawn as points of the other two tracks, and therefore at least two of them are on the same track. Suppose without loss of generality that $a$ and $b$ are both drawn on track $t_2$. Observe that there is no way of drawing edges $(a,u)$, $(a,v)$, $(b,u)$, $(b,v)$ avoiding a crossing. It follows that in any prism drawing of $G$ $u$ and $v$ must appear on different tracks.

To prove Claim 2 we assume that $u$ is drawn on track $t_1$ and that $v$ is drawn on track $t_2$. Let $t_3$ be the third track of the prism. Assume there existed a prism drawing with vertices $a$, $b$, and $c$ all on tracks $t_1$ and $t_2$. Assume without loss of generality that both $a$ and $b$ are on track $t_1$. In order to avoid crossings it must be that one vertex, say $a$, is on the right-hand side and the other is on the left-hand side of $u$ (see Figure 18(b)). Note however that $c$ cannot be drawn on track $t_1$ or else edge $(u,c)$ would intersect one of the edges $(a,u)$, $(b,u)$. But if $c$ were drawn on track $t_2$, then edge $(c,u)$ would intersect either $(a,v)$ or $(c,v)$. It follows that $c$ is drawn on track $t_3$ and therefore $G$ is strictly prism-drawable.

$\square$

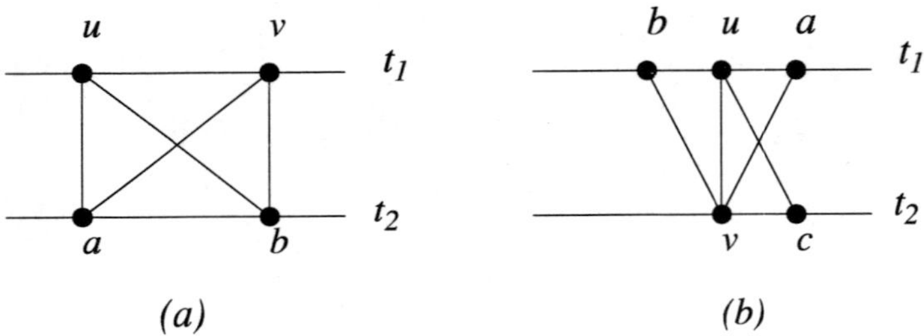

Figure 18: Illustration for the proof of Lemma 7.

**Theorem 9** *There exist prism-forbidden planar graphs.*

**Proof.** Let $G$ be a planar graph with a vertex $v$ adjacent to three copies of the graph of Figure 16. Let $G_0$ be the subgraph of $G$ induced by $v$ and by these three copies. By Lemma 6 $G_0$ is prism-forbidden. It follows that also $G$ is prism-forbidden. $\square$

A consequence of the previous lemmas is the following.

**Corollary 3** *There exist prism-forbidden series-parallel graphs.*

**Proof.** Let $G$ be the graph in Figure 19. By using the same argument in the proof of Theorem 9 the corollary follows. $\square$

Felsner *et al.*, *Restricted Integer Grids*, JGAA, 7(4) 363–398 (2003)    387

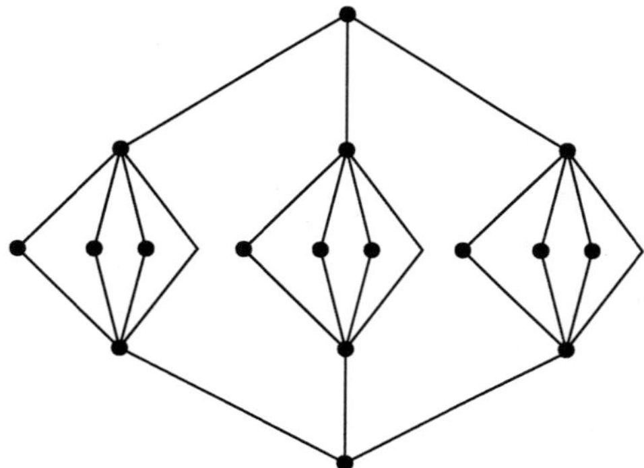

Figure 19: A prism-forbidden series-parallel graph.

## 5.3  Prism-Drawability and Hamiltonicity

Theorem 9 shows that not all planar graphs admit a prism drawing. In this section we further restrict the family of drawable graphs by exploiting the relation between prism-drawability and hamiltonicity.

A graph is *subhamiltonian* if it can be augmented with edges to produce a planar graph having a hamiltonian cycle.

**Lemma 8** *Let $G$ be a prism-drawable graph. Then $G$ is subhamiltonian.*

**Proof.** Let $\Gamma$ be a prism drawing of $G$ and let $G'$ be a maximally augmented graph obtained from $G$ by adding edges to satisfy the conditions of Theorem 7. Let the three tracks of the prism be labeled $t_1$, $t_2$ and $t_3$ (see for example Figure 20(a)). Label the vertices on track $t_2$, as $r_0, \ldots r_z$ and the vertices on track $t_3$ as $b_0, \ldots b_y$. The hamiltonian cycle begins at the start vertex of $t_1$, visits the start vertex of $t_2$ (denoted as $r_0$), the start vertex of $t_3$ (denoted as $b_0$), and then alternates between the $t_2$ and $t_3$ track visiting all vertices on those two tracks and ending at the end vertex of either tracks $t_2$ or $t_3$ (see for example Figure 20(b)). In either case, the cycle then visits the end vertex of $t_1$ and then the entire $t_1$ spine in reverse order.

We now give a more formal description of the cycle's traversal of the strip between tracks $t_2$ and $t_3$. Note that $r_0$ is adjacent to $b_0$ (and 0 or more consecutive vertices on track $t_3$). Since $G'$ is maximally augmented, each vertex $b_i$ ($0 < i < y$) is adjacent to $b_{i-1}$ and $b_{i+1}$ and to a non-empty set of consecutive vertices on the $r$-track $r_j, \ldots r_{j+k}$. Furthermore, $r_j$ is adjacent to $b_{i-1}$, and $r_{j+k}$ is adjacent to $b_{i+1}$. The hamiltonian cycle goes from $r_0$ to $b_0$ and then applies a greedy-like approach. In general, from $b_i$, the cycle goes to the first (*i.e.* lowest-indexed) neighbour $r_j$ that has not previously been visited. Only if all the neighbours of $b_i$ on the $r$-track have been visited, does the cycle go to $b_{i+1}$. The rule on the $r$ track is symmetric. Since at each vertex, there is always

Felsner *et al.*, *Restricted Integer Grids*, JGAA, 7(4) 363–398 (2003)      388

at least one neighbour that has not been visited, namely the next vertex on the same track, it is clear that the cycle can always proceed. To establish that all vertices are visited, a straightforward proof by contradiction can be applied. The (partial) hamiltonian cycle thus ends at the endpoint of either track $t_2$ or $t_3$ and can be completed as described above. $\qquad\square$

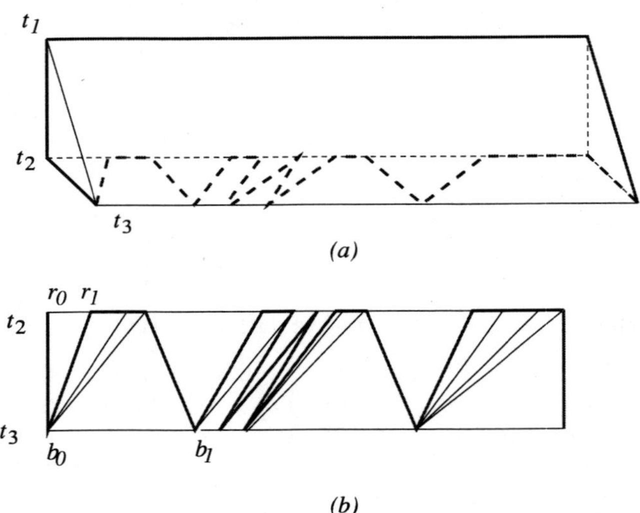



Figure 20: Illustration for the proof of Lemma 8

A natural question arising from Lemma 8 is whether all subhamiltonian planar graphs are prism drawable. This is not the case, for example the graph of Figure 19 is subhamiltonian but not prism drawable. The following lemma shows that even subhamiltonian graphs with only 9 vertices may not be prism drawable. In this instance it is the proof technique that is of primary interest.

**Lemma 9** *The hamiltonian maximal planar graph $G$ of Figure 21 is prism-forbidden.*

**Proof.** Suppose for a contradiction that $G$ were prism-drawable and let $G'$ be the augmented graph of $G$ described in Theorem 5.

Consider the vertex $a$ of $G$ displayed in Figure 21. We start by showing that vertex $a$ must be an endvertex in one of the three spines of $G'$.

Let $\Gamma$ be a prism drawing of $G$. Since $G$ is maximal planar, by Corollary 2 it follows that $G$ coincides with its augmented graph $G'$ and that $\Gamma$ is also a prism drawing of $G'$. If $a$ were not an endvertex of a spine, then the point representing $a$ in $\Gamma$ would be adjacent to two other points on the same track. But all neighbours of $a$ are mutually adjacent in $G$ and this would imply a chordal edge in $\Gamma$, which is impossible.

Felsner *et al.*, *Restricted Integer Grids*, JGAA, 7(4) 363–398 (2003)        389

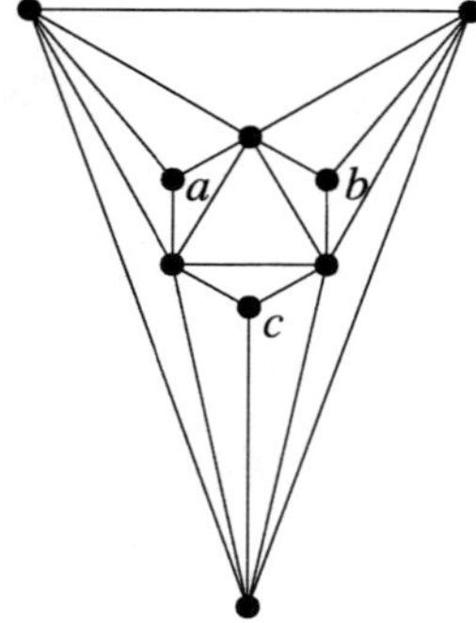

Figure 21: A hamiltonian planar graph $G$ that is not prism-drawable.

By the same argument, also vertices $b$ and $c$ of $G$ (see Figure 21) are end-vertices of some spine in $G'$.

By Theorem 5, the endvertices of the spines in $G'$ belong to two three-cycles. Therefore, at least two vertices among $a$, $b$, and $c$ must be connected in $G'$. But $G$ is identical to $G'$ by Corollary 2 and no pair of these vertices are adjacent in Figure 21. This provides the required contradiction and proves that $G$ is indeed prism-forbidden. □

Based on Lemmas 8 and 9 we can summarize the discussion of this section as follows.

**Theorem 10** *The family of prism-drawable graph is a proper subset of the family of subhamiltonian planar graphs.*

# 6 Box-Drawable Graphs

Motivated by Theorems 9 and 10, we consider a restricted integer 3D grid consisting of four tracks, namely the *box*, and ask whether this grid supports all planar graphs. Clearly, the class of box-drawable graphs is larger than the class of prism-graphs: Every prism-drawable graph can be drawn on the box and it is easy to draw some non-planar graphs on a box. For example, Figure 22 shows a box drawing of $K_5$ and a box drawing of $K_{3,3}$. However we will show that even the box is not a universal grid for planar graphs.

## 6.1 Characterization of Box-Drawable Graphs

We start our investigation by characterizing the family of box-drawable graphs with more than five vertices. Note that any graph with at most five vertices is box drawable and its box drawing can be obtained by deleting edges and vertices from the drawing of $K_5$ depicted in Figure 22. The next theorem follows the same pattern as Theorem 7.

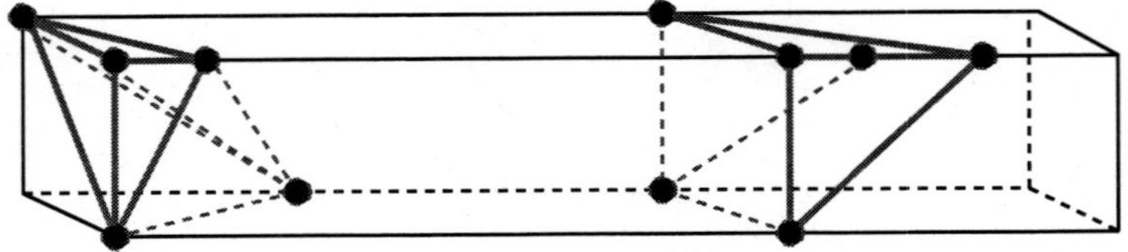

Figure 22: $K_5$ and $K_{3,3}$ drawn on a box

**Theorem 11** *A graph $G$ with at least six vertices is box-drawable if and only if it is possible to augment $G$ with edges to produce a graph $G'$ such that:*

- *$G'$ contains two four-cycles $r_0, b_0, g_0, w_0$ and $r_z, b_t, g_s, w_q$, where $z, t, s, q \geq 0$.*

- *The graph $G'$ contains four vertex-disjoint spines $r_0, r_1, \ldots r_z$, $b_0, b_1, \ldots b_t$, $g_0, g_1, \ldots g_s$, and $w_0, w_1, \ldots w_q$ such that all vertices of $G$ are on the four spines.*

- *For each pair of spines $x_0, x_1, \ldots x_m$ and $y_0, y_1, \ldots y_p$ $(x, y \in \{r, b, g, w\}$, $x \neq y$, $m, p \in \{z, t, s, q\})$ , if $(x_i, y_j)$ is an edge, then there are no edges of the form $(x_k, y_l)$ with $(0 \leq k < i$ and $j < l \leq p)$ or $(i < k \leq m$ and $0 \leq l < j)$.*

**Proof.** ($\Rightarrow$) We show how to construct a graph $G'$ that satisfies the statement. Let $\Gamma$ be a box drawing of $G$ and let $t_1$, $t_2$, $t_3$, and $t_4$ be the four tracks of the box. Consider the subgraph $G_{ij}$ induced by the vertices that are on two different tracks $t_i$ and $t_j$ $(i, j = 1, 2, 3, 4$, $i \neq j)$; $G_{ij}$ is strip-drawable and therefore there exists an augmented graph $G'_{ij}$ with the properties stated by Theorem 6. A graph $G'$ that satisfies the statement is then defined as $G' = G'_{12} \cup G'_{13} \cup G'_{14} \cup G'_{23} \cup G'_{24} \cup G'_{34}$ .

($\Leftarrow$) Among the four tracks $t_1$, $t_2$, $t_3$, and $t_4$ of the box, we assume that the pairs $t_1, t_3$ and $t_2, t_4$ are diagonally opposite and that the four tracks are horizontal lines. Given an augmented graph $G'$, a box drawing $\Gamma'$ of $G'$ is constructed by spiralling the vertices on the four tracks as follows.

- Spine $r_0, r_1, \ldots r_z$ is drawn on track $t_1$ such that $r_j$ is drawn to the right of $r_i$ for $0 \leq i < j \leq z$.

- Spine $b_0, b_1, \ldots b_t$ is drawn on the diagonally opposite track $t_3$ such that $b_0$ is given an $x$-coordinate larger than the $x$-coordinate of $r_z$ and $b_j$ is drawn to the right of $b_i$ for $0 \leq i < j \leq t$.

- Spine $g_0, g_1, \ldots g_s$ is drawn on track $t_2$ such that $g_0$ is given an $x$-coordinate larger than the $x$-coordinate of $b_z$ and $g_j$ is drawn to the right of $g_i$ for $0 \leq i < j \leq s$.

- Spine $w_0, w_1, \ldots w_q$ is drawn on track $t_4$ such that $w_0$ is given an $x$-coordinate larger than the $x$-coordinate of $g_s$ and $w_j$ is drawn to the right of $w_i$ for $0 \leq i < j \leq q$.

Then, edges connecting vertices along the same track and between two different tracks are drawn as straight-line segments. In order to see that the computed drawing does not have edge crossings observe the following.

- There are no edges between non consecutive vertices on the same track: For each spine, the subgraph of $G'$ induced by the spine is a path which is drawn sequentially on a track.

- For each pair of tracks $t_h, t_f$ $(h, f \in \{1, 2, 3, 4\}, h \neq f)$ the edges connecting vertices on $t_h$ with vertices on $t_f$ do not cross with each other. Namely, for an edge $(x_i, y_j)$ connecting a vertex on $t_h$ with a vertex on $t_l$, there are no edges of the form $(x_k, y_l)$ with $(k < i$ and $j < l)$ or $(i < k$ and $l < j)$.

- For each pair of diagonally opposite edges $e_1, e_2$ such that $e_1$ connects a vertex on track $t_1$ with a vertex on track $t_3$ and $e_2$ connects a vertex on track $t_2$ with a vertex on track $t_4$ there are no crossings. By construction, the coordinates of the endpoints of $e_2$ are both strictly larger than those of the endpoints of $e_1$.

Since $\Gamma'$ is a box drawing and $G'$ is a supergraph of $G$, a box drawing of $G$ can be obtained by deleting edges from $\Gamma'$. $\qquad\square$

## 6.2    Box-Drawability and Planarity

We extend the approach of Section 5.2 to construct planar graphs that cannot be drawn on a box. We call these graphs *box-forbidden*. In order to construct a box-forbidden graph, we need a preliminary lemma.

**Lemma 10** *In any box drawing of the graph of Figure 23 vertices $u$ and $v$ are on different tracks.*

**Proof.** Let $G$ be the graph of Figure 23. It is trivial to see that $G$ is prism-drawable and hence it is also box-drawable. Let $\Gamma$ be a box drawing of $G$. Suppose for a contradiction that vertices $u$ and $v$ in $\Gamma$ were represented as points of the same track $t_1$. Observe that no other vertex of $\Gamma$ can be a point of $t_1$ or else there would be a crossing. Since $G$ has six vertices, a box consists of four tracks, and track $t_1$ cannot contain more than two vertices, it follows that at least two other vertices of $\Gamma$ must be on another track, say $t_2$ of the box. But each vertex on $t_2$ is adjacent to both $u$ and $v$, which forces a crossing in $\Gamma$; contradiction. $\qquad\square$

A graph $G$ is *strictly box-drawable* if it is box-drawable and there are four mutually adjacent vertices $a$, $b$, $c$ and $d$ and in all box drawings of $G$, $a$, $b$, $c$ and $d$ appear on separate tracks.

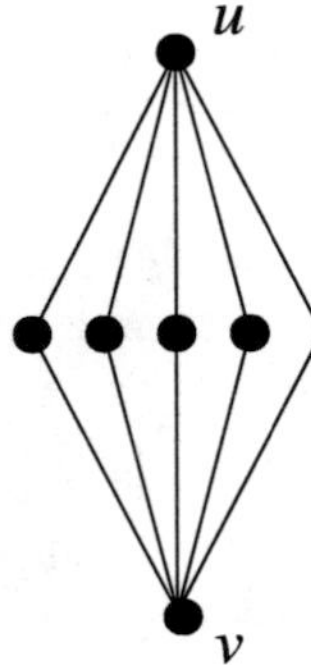

Figure 23: Graph for Lemma 10.

**Lemma 11** *The graph of Figure 24(a) is strictly box-drawable.*

**Proof.** Let $G$ be the graph of Figure 24(a). We first prove that $G$ is box-drawable and then that it is strictly box-drawable. We adopt the notation of Figure 24(a).

We apply Theorem 11 to $G$. The four spines of $G'$ are $r_0, .., r_8$, $b_0, .., b_4$, $g_0, .., g_8$, and $w_0, .., w_4$. The two four-cycles have the vertices $r_0, b_0, g_0, w_0$ and $r_8, b_4, g_8, w_4$. Also, as Figures 24(b), (c), (d), (e), (f), and (g) show, the subgraphs of $G'$ induced by vertices on two different spines are strip-drawable. It follows that $G$ is box-drawable.

We now prove that $G$ is strictly box-drawable. Graph $G$ consists of six copies of the graph in Figure 23. By Lemma 10 in any box drawing of $G$ vertices $r_0, g_4, b_4, w_4$ must be on four different tracks. Furthermore, those four vertices are mutually adjacent. It follows that $G$ is strictly box-drawable.  □

**Theorem 12** *There exist box-forbidden planar graphs.*

**Proof.** Let $G$ be a planar graph with a vertex $v$ adjacent to three copies of the graph of Figure 24. Let $G_0$ be the subgraph of $G$ induced by $v$ and by these three copies; see also Figure 25. Removing $v$ and all its incident edges from the graph $G_0$ splits it into three components that we name $G_1$, $G_2$, and $G_3$. In any box drawing $\Gamma_i$ of $G_i$ ($0 \leq i \leq 3$) there are four mutually adjacent vertices on four different tracks because by Lemma 11 $G_i$ is strictly box-drawable. Thus, $\Gamma_1$, $\Gamma_2$, and $\Gamma_3$ slice the box into 4 slices and there is no location for $v$ that permits it to be connected crossing-free to all $\Gamma_i$ ($0 \leq i \leq 3$).  □

# 7  Conclusions and Open Problems

In this paper we showed that all outerplanar graphs can be drawn in linear volume on a prism – a restriction of the three dimensional integer grid. Proofs

Felsner *et al.*, *Restricted Integer Grids*, JGAA, 7(4) 363–398 (2003)     393

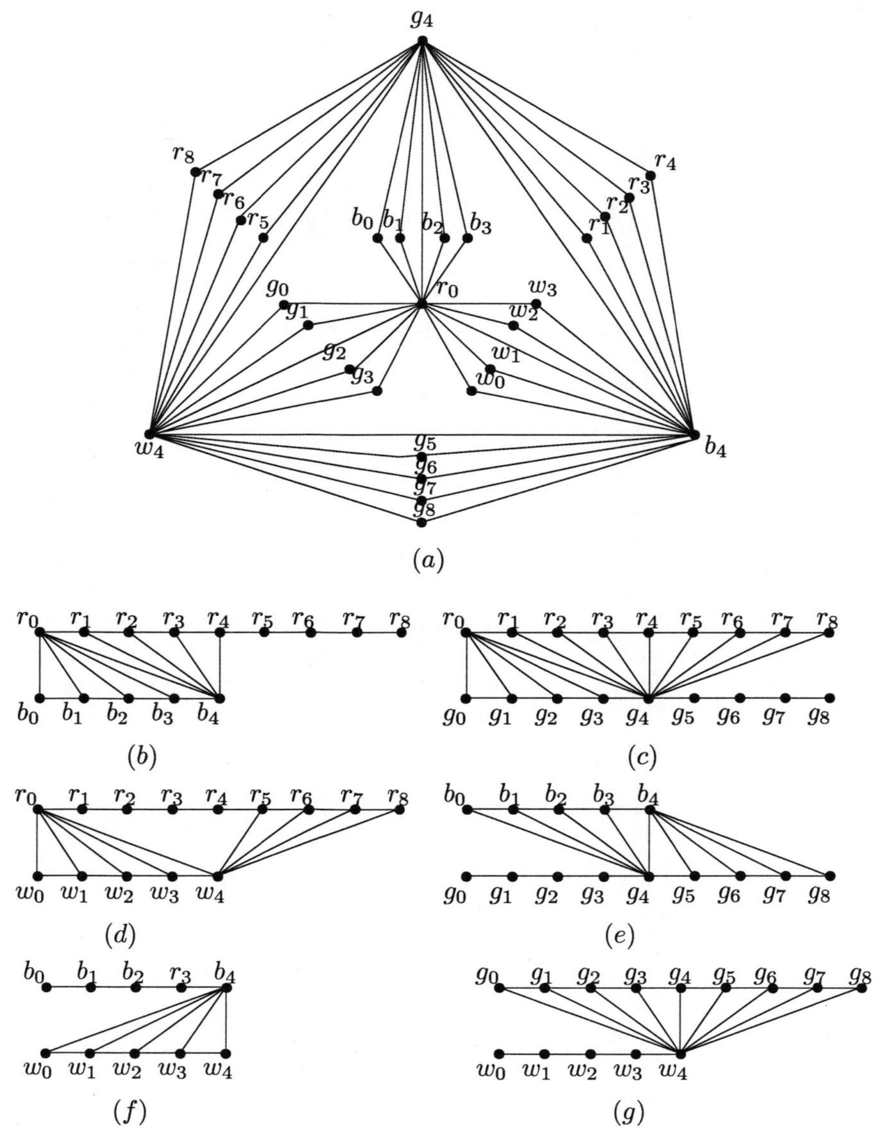

Figure 24: (a) Graph $G$ for Lemma 11. (b)–(e) The subgraphs of $G$ induced by vertices on two different spines are strip-drawable.

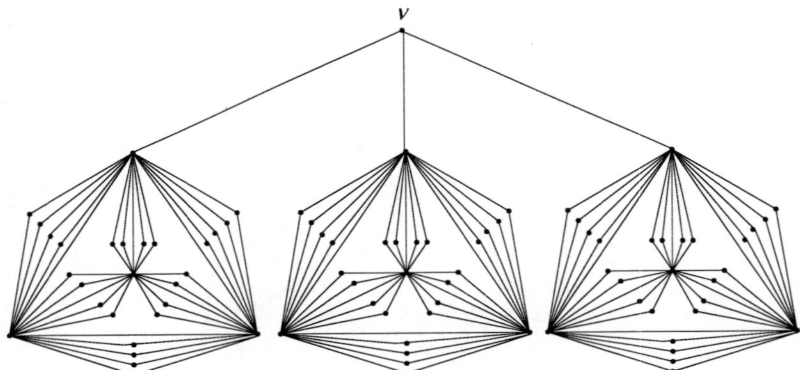

Figure 25: A planar graph that is not box-drawable.

that certain classes of planar graphs are not prism-drawable nor box-drawable were also provided. Although the problem of finding a universal integer 3D grid of linear volume that supports crossing-free straight line drawings of all planar graphs is still far from being solved, the drawing techniques and characterization results of this paper may provide a critical starting point for attacking such an ambitious research programme. We believe that the results on the restricted 2D grid are not only useful preliminary results for the study in three dimensions, but they may also shed some new light on the problem of drawing trees in linear area on the plane. There remain several interesting problems and directions for further research. We conclude the paper by describing some of the more intriguing open problems.

1. Can all outerplanar graphs be drawn in linear area on a 2D integer grid? Does there exist a 2D universal grid set of linear area that supports all outerplanar graphs?

2. Characterize the graphs drawable on an $n \times k$ grid.

3. Can the strong algorithms for recognizing graphs of bounded pathwidth be applied to devise polynomial dynamic programming algorithms to decide $k$-track drawability for fixed $k$? Such an approach has been applied in [18] for the recognition of proper $k$-track drawability for fixed $k$. Also, Schank [36] gave a direct linear time algorithm for the task of recognizing 2-track drawable graphs.

4. Can all planar graphs be drawn in linear volume on a three-dimensional integer grid? Does there exist a 3D universal grid set of linear volume that supports all planar graphs?

5. *The k-lines drawability problem:* A related problem posed by H. de Fraysseix [13] asks if all planar graphs can be drawn on $k$ parallel lines that

Felsner *et al.*, *Restricted Integer Grids*, JGAA, 7(4) 363–398 (2003)     395

lie on the surface of a cylinder, for a fixed value of $k$. Our results on box-drawability imply that $k$ would have to be strictly greater than 4.

6. *The aspect ratio problem:* Our results about linear volume come at the expense of aspect ratio. Is it possible to achieve both linear volume and $o(n)$ aspect ratio for outerplanar graphs? We conjecture that it is in fact not possible in 2D to simultaneously attain linear area and $O(1)$ aspect ratio for some classes of planar graphs.

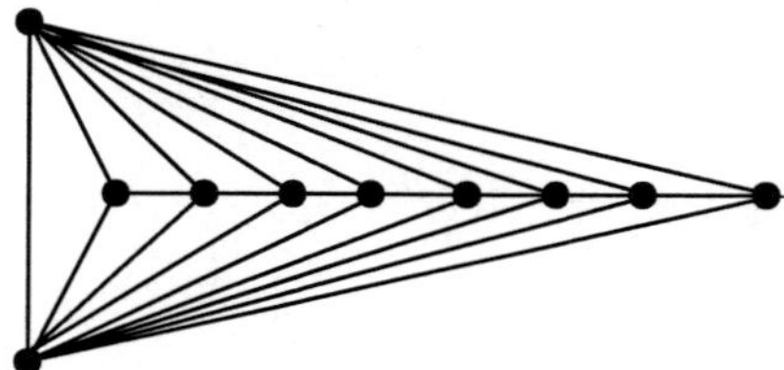

Figure 26: A graph $S_n$ with poor aspect ratio

**Conjecture 1** *There is no fixed constant $k$ for which the family of graphs $S_n$ (in Figure 26) can be drawn in a 2D integer grid of size $k\sqrt{n} \times \sqrt{n}$.*

Note that the graph $S_n$ can be drawn on an $n \times 3$ grid (and hence in linear area but with *linear* aspect ratio). Recently this conjecture was verified and the result is reported in [1].

**Acknowledgments**

We thank Helmut Alt, Pino Di Battista, Hazel Everett, Ashim Garg, Mark Keil, Sylvain Lazard and David Wood for useful discussions related to the results contained in this paper. We are grateful to Matthew Suderman for pointing out an error in earlier versions of this paper and to the referees for their many helpful suggestions.

# References

[1] T. Biedl, T.M. Chan, A. López-Ortiz. Drawing $K_{2,n}$: A Lower Bound. In *Proc. 14th Can. Conf. on Comput. Geom.*, 146–148, 2002

[2] P. Bose, J. Czyzowicz, P. Morin, D. Wood. The Maximum Number of Edges in a Three-Dimensional Grid-Drawing. *Technical Report TR-02-04*, School of Computer Science, Carleton University, 1–4, Oct. 2002.

[3] T. Calamoneri and A. Sterbini. Drawing 2-, 3-, and 4-colorable graphs in $o(n^2)$ volume. In S. North, editor, *Graph Drawing (Proc. GD '96)*, volume 1190 of *Lecture Notes Comput. Sci.*, pages 53–62. Springer-Verlag, 1997.

[4] M. Capobianco and J. Molluzzo. *Examples and Counterexamples in Graph Theory.* North-Holland p.169 1978.

[5] T.M. Chan. A near-linear area bound for drawing binary trees. In *Proc. 10th ACM-SIAM Sympos. on Discrete Algorithms.*, pages 161–168, 1999.

[6] M. Chrobak and G. Kant. Convex grid drawings of 3-connected planar graphs. *Internat. J. Comput. Geom. Appl.*, 7(3):211–223, 1997.

[7] M. Chrobak, M.T. Goodrich, and R. Tamassia. Convex drawings of graphs in two and three dimensions. In *Proc. 12th Annu. ACM Sympos. Comput. Geom.*, pages 319–328, 1996.

[8] M. Chrobak, and S. Nakano. Minimum-width grid drawings of plane graphs. *Comput. Geom. Theory Appl.*, 11:29–54, 1998.

[9] R. F. Cohen, P. Eades, T. Lin, and F. Ruskey. Three-dimensional graph drawing. *Algorithmica*, 17:199–208, 1997.

[10] S. Cornelsen, T. Schank, and D. Wagner. Drawing Graphs on Two and Three Lines. Graph Drawing 2002, *Lecture Notes in Comput. Sci.* 2528, pages 31-41, Springer, 2002.

[11] H. de Fraysseix, J. Pach, and R. Pollack. Small sets supporting Fary embeddings of planar graphs. In *Proc. 20th ACM Sympos. Theory Comput.*, pages 426–433, 1988.

[12] H. de Fraysseix, J. Pach, and R. Pollack. How to draw a planar graph on a grid. *Combinatorica*, 10(1):41–51, 1990.

[13] H. de Fraysseix, personal communication.

[14] G. Di Battista, P. Eades, R. Tamassia, and I.G. Tollis. *Graph Drawing.* Prentice Hall, Upper Saddle River, NJ, 1999.

[15] R. Diestel. *Graph theory.* Graduate Texts in Mathematics. 173. Springer, 2000. Transl. from the German. 2nd ed.

Felsner *et al.*, *Restricted Integer Grids*, JGAA, 7(4) 363–398 (2003)        397

[16] E. Di Giacomo, G. Liotta and S.K. Wismath. Drawing Series-Parallel Graphs on a Box. *14th Canadian Conference On Computational Geometry (CCCG '02)*, pages 149–153, 2002.

[17] V. Dujmović and D.R. Wood. Tree-Partitions of $k$-Trees with Applications in Graph Layout. *29th Workshop on Graph Theoretic Concepts in Computer Science (Proc. WG '03), Lecture Notes in Comput. Sci.*, Springer-Verlag, to appear.

[18] V. Dujmović, M. Fellows, M. Hallett, M. Kitching, G. Liotta, C. McCartin, N. Nishimura, P. Ragde, F. Rosamond, M. Suderman, S. Whitesides, D.R. Wood. A Fixed-Parameter Approach to Two-Layer Planarization. Graph Drawing 2001, *Lecture Notes in Comput. Sci.* 2265, pages 1-15, Springer, 2002.

[19] V. Dujmović, M. Fellows, M. Hallett, M. Kitching, G. Liotta, C. McCartin, N. Nishimura, P. Ragde, F. Rosamond, M. Suderman, S. Whitesides, D.R. Wood. On the Parameterized Complexity of Layered Graph Drawing. *ESA*, 1–12, 2001.

[20] V. Dujmović, P. Morin, and D.R. Wood. Path-Width and Three-Dimensional Straight-Line Grid Drawing of Graphs. Proc. of 10th International Symposium on Graph Drawing (GD '02), *Lecture Notes in Comput. Sci.* 2528, pages 42-53, Springer, 2002.

[21] S. Felsner. Convex drawings of planar graphs and the order dimension of 3-polytopes. *Order*,18, 19–37, 2001.

[22] S. Felsner, G. Liotta and S. Wismath. Straight-Line Drawings on Restricted Integer Grids in Two and Three Dimensions. In M. Jünger and P. Mutzel, editor, *Graph Drawing (Proc. GD '01)*, volume 2265 of *Lecture Notes Comput. Sci.*, pages 328–343. Springer-Verlag, 2002.

[23] A. Garg, M. T. Goodrich, and R. Tamassia. Planar upward tree drawings with optimal area. *Internat. J. Comput. Geom. Appl.*, 6:333–356, 1996.

[24] A. Garg, R. Tamassia, and P. Vocca. Drawing with colors. In *Proc. 4th Annu. European Sympos. Algorithms*, volume 1136 of *Lecture Notes Comput. Sci.*, pages 12–26. Springer-Verlag, 1996.

[25] M. Juenger and S. Leipert. Level planar embedding in linear time. In J. Kratochvil, editor, *Graph Drawing (Proc. GD '99)*, volume 1731 of *Lecture Notes Comput. Sci.*, pages 72–81. Springer-Verlag, 1999.

[26] G. Kant. A new method for planar graph drawings on a grid. In *Proc. 33rd Annu. IEEE Sympos. Found. Comput. Sci.*, pages 101–110, 1992.

[27] G. Kant. Drawing planar graphs using the canonical ordering. *Algorithmica*, 16:4–32, 1996.

[28] L.M. Kirousis and C.H. Papadimitriou. Searching and pebbling. Theor. Comput. Sci. 47, 205–218 (1986).

[29] N. Meggido, S. Hakimi, M. Garey, D. Johnson, and C. Papadimitriou. The complexity of searching a graph. *JACM* 35, 18–44, 1988.

[30] R.H. Möhring, Graph problems related to gate matrix layout and PLA folding. Computational graph theory, Comput. Suppl. 7, 17–51 (1990).

[31] J. Pach, T. Thiele, and G. Tóth. Three-dimensional grid drawings of graphs. In G. Di Battista, editor, *Graph Drawing (Proc. GD '97), Lecture Notes Comput. Sci.* 1353, pages 47–51. Springer-Verlag, 1997.

[32] T. Parsons. Pursuit-evasion in a graph. In Alavi and Lick editors, *Theory and Applications of Graphs*, Springer-Verlag, 426–441, 1976.

[33] F. P. Preparata and M. I. Shamos. *Computational Geometry: An Introduction.* Springer-Verlag, 3rd edition, October 1990.

[34] N. Robertson and P.D. Seymour. Graph Minors I. Excluding a Forest. J. of Comb. Th (B) 35, 39–61 (1983).

[35] N. Robertson and P.D. Seymour. Graph minors - a survey. in Surveys in combinatorics 1985, (Pap. 10th Br. Combin. Conf. 1985), Lond. Math. Soc. Lect. Note Ser. 103, 153-171 (1985)

[36] T. Schank. Algorithmen zur Visualisierung planarer, partitionierter Graphen. Master Thesis, Universität Konstanz, 2001. http://scampi.physik.uni-konstanz.de/ thomas/fmi/download/zula.ps

[37] P. Scheffler. Optimal embedding of a tree into an interval graph in linear time. Combinatorics, graphs and complexity, Proc. 4th Czech. Symp., Prachatice/ Czech. 1990, Ann. Discrete Math. 51, 287–291 (1992).

[38] W. Schnyder. Embedding planar graphs on the grid. In *Proc. 1st ACM-SIAM Sympos. Discrete Algorithms*, pages 138–148, 1990.

[39] W. Schnyder and W.T. Trotter. Convex embeddings of 3-connected plane graphs. *Abstracts of the AMS*, 13(5):502, 1992.

[40] M. Suderman. Pathwidth and Layered Drawings of Trees. *Technical Report, School of Comp. Sci., McGill Univ.*, SOCS-02.8, 2002.

[41] K. Wagner. Über eine Eigenschaft der ebenen Komplexe. Math. Ann. 114, 570-590 (1937).

[42] D.R. Wood. Queue Layouts, Tree-Width, and Three-Dimensional Graph Drawing. 22nd Foundations of Software Technology and Theoretical Computer Science (FSTTCS '02), *Lecture Notes in Computer Science* 2556, Springer-Verlag, pages 348–359, 2002.

[43] G. M. Ziegler. Lectures on Poytopes. *Springer-Verlag*, 1995.

# Journal of Graph Algorithms and Applications

http://jgaa.info/

vol. 7, no. 4, pp. 399–409 (2003)

# Low-Distortion Embeddings of Trees

*Robert Babilon*    *Jiří Matoušek*    *Jana Maxová*    *Pavel Valtr*

Department of Applied Mathematics and
Institute for Theoretical Computer Science (ITI),
Charles University, Malostranské nám. 25,
118 00 Prague, Czech Republic
{babilon,matousek,jana,valtr}@kam.mff.cuni.cz

### Abstract

We prove that every tree $T = (V, E)$ on $n$ vertices with edges of unit length can be embedded in the plane with distortion $O(\sqrt{n})$; that is, we construct a mapping $f: V \rightarrow \mathbf{R}^2$ such that $\rho(u,v) \leq \|f(u) - f(v)\| \leq O(\sqrt{n}) \cdot \rho(u,v)$ for every $u, v \in V$, where $\rho(u,v)$ denotes the length of the path from $u$ to $v$ in $T$. The embedding is described by a simple and easily computable formula. This is asymptotically optimal in the worst case. We also construct interesting optimal embeddings for a special class of trees (fans consisting of paths of the same length glued together at a common vertex).

Communicated by: P. Mutzel and M. Jünger;
submitted May 2002; revised April 2003.

The research was supported by project LN00A056 of the Ministry of Education of the Czech Republic and by Charles University grants No. 158/99 and 159/99.

R. Babilon *et al.*, *Embeddings of Trees*, JGAA, 7(4) 399–409 (2003)     400

# 1   Introduction

Embeddings of finite metric spaces into Euclidean spaces or other normed spaces that approximately preserve the metric received considerable attention in recent years. Numerous significant results have been obtained in the 1980s, mainly in connection with the local theory of Banach spaces. Later on, surprising algorithmic applications and further theoretical results were found in theoretical computer science; see, for example, Indyk [2] or Chapter 15 in [4] for overviews.

Here we consider only embeddings into the spaces $\mathbf{R}^d$ with the Euclidean metric. The quality of an embedding is measured by the *distortion*. Let $(V, \rho)$ be a finite metric space. We say that a mapping $f: V \to \mathbf{R}^d$ has distortion at most $D$ if there exists a real number $\alpha > 0$ such that for all $u, v \in V$,

$$\alpha \cdot \rho(u, v) \leq \|f(u) - f(v)\| \leq D \cdot \alpha \cdot \rho(u, v).$$

This definition permits scaling of all distances in the same ratio $\alpha$, in addition to the distortion of the individual distances by factors between 1 and $D$. Since the image in $\mathbf{R}^d$ can always be re-scaled as needed, we can choose the factor $\alpha$ at our convenience.

We study embeddings with the dimension $d$ fixed, and mainly the case $d = 2$, i.e. embeddings into the plane. A straightforward volume argument shows that the distortion required to embed into $\mathbf{R}^d$ the $n$-point metric space with all distances equal to 1 is at least $\Omega(n^{1/d})$. In [3], it was proved that general metric spaces may require even significantly bigger distortions; namely, for every fixed $d$, there are $n$-point metric spaces that need $\Omega(n^{1/\lfloor (d+1)/2 \rfloor})$ distortion for embedding into $\mathbf{R}^d$.

Better upper bounds can be obtained for special classes of metric spaces. A metric $\rho$ on a finite set $V$ is called a *tree metric* if there is a tree $T = (V', E)$ (in the graph-theoretic sense) with $V' \supseteq V$ and a weight function $w: E \to (0, \infty)$ such that for all $u, v \in V$, $\rho(u, v)$ equals the length of the path connecting the vertices $u$ and $v$ in $T$, where the length of an edge $e \in E$ is $w(e)$. It was conjectured in [3] that, for fixed $d$, all $n$-point tree metrics can be embedded into $\mathbf{R}^d$ with distortion $O(n^{1/d})$. If true, this is best possible, as the example of a star with $n - 1$ leaves and unit-length edges shows (a volume argument applies). Gupta [1] proved the somewhat weaker upper bound $O(n^{1/(d-1)})$.

Here we make a step towards establishing the conjecture. We deal with the planar case, where the gap between the lower bound of $\sqrt{n}$ and Gupta's $O(n)$ upper bound is the largest (in fact, the $O(n)$ upper bound holds even for general metric spaces and embeddings into $\mathbf{R}^1$ [3]). So far we can only handle the case of unit-length edges.

**Theorem 1** *Every $n$-vertex tree with unit-length edges, considered as a metric space, can be embedded into $\mathbf{R}^2$ with distortion $O(\sqrt{n})$. The embedding is described by a simple explicit formula and can be computed efficiently.*

Gupta's result actually states that if the considered tree has at most $\ell$ *leaves*, then an embedding into $\mathbf{R}^d$ with distortion $O(\ell^{1/(d-1)})$ is possible. We show

R. Babilon *et al.*, *Embeddings of Trees*, JGAA, 7(4) 399–409 (2003)      401

that here the dependence on $\ell$ *cannot* be improved. Let $F_{\ell,m}$ (the fan with $\ell$ leaves and path length $m$) denote the tree consisting of $\ell$ paths of length $m$ each glued together at a common vertex (root):

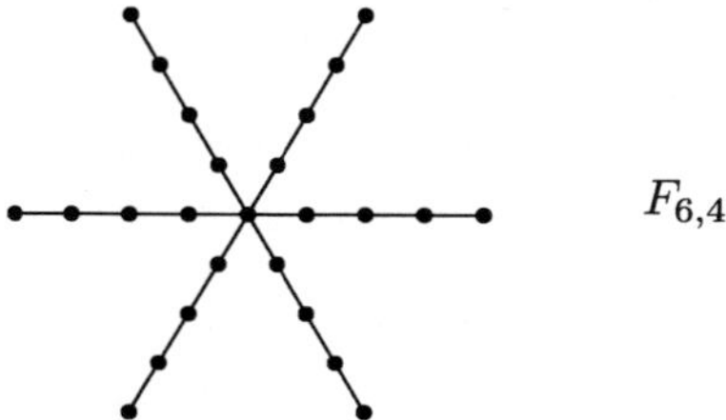

$$F_{6,4}$$

**Proposition 1** *For every fixed $d \geq 2$ and every $m, \ell \geq 2$ every embedding of $F_{\ell,m}$ into $\mathbf{R}^d$ requires distortion $\Omega(\ell^{1/(d-1)})$ if $\ell \leq m^{\frac{d(d-1)}{d+1}}$, and distortion $\Omega(\ell^{1/d} m^{1/(d+1)})$ if $\ell \geq m^{\frac{d(d-1)}{d+1}}$.*

Theorem 1 provides embeddings which are optimal in the worst case, i.e. there are trees for which the distortion cannot be asymptotically improved. But optimal or near-optimal embeddings of special trees seem to present interesting challenges, and sometimes low-distortion embeddings are aesthetically pleasing and offer a good way of drawing the particular trees. We present one interesting example concerning the fan $F_{\sqrt{n},\sqrt{n}}$. Here the "obvious" embedding as in the above picture, as well as the embedding from Theorem 1, yield distortions $O(\sqrt{n})$. In Section 4, we describe a somewhat surprising better embedding, with distortion only $O(n^{5/12})$. This is already optimal according to Proposition 1.

Several interesting problems remain open. The obvious ones are to extend Theorem 1 to higher dimensions and/or to trees with weighted edges. Another, perhaps more difficult, question is to extend the class of the considered metric spaces. Most significantly, we do not know an example of an $n$-vertex planar graph (with weighted edges) whose embedding into $\mathbf{R}^2$ would require distortion larger than about $\sqrt{n}$. If it were possible to show that all planar-graph metrics can be embedded into the plane with $o(n)$ distortion, it would be a neat metric condition separating planar graphs from non-planar ones, since suitable $n$-vertex subdivisions of any fixed non-planar graph require $\Omega(n)$ distortion [3].

## 2   Proof of Theorem 1

**Notation.**

Let $T$ be a tree (the edges have unit lengths) and let $\rho$ be the shortest-path metric on $V = V(T)$. One vertex is chosen as a root. The *height* $h(v)$ of a vertex $v \in V$ is its distance to the root. Let $\pi_v$ denote the path from a vertex $v$ to the root.

For every vertex we fix a linear (left-to-right) ordering of its children. This defines a partial ordering $\preceq$ on $V$: we have $u \prec v$ iff $u \notin \pi_v$, $v \notin \pi_u$, and $\pi_v$

R. Babilon *et al.*, *Embeddings of Trees*, JGAA, 7(4) 399–409 (2003)     402

goes right of $\pi_u$ at the vertex where $\pi_u$ and $\pi_v$ branch. We define

$$\mathrm{sgn}_v(u) = \begin{cases} 0 & \text{if } u \in \pi_v \text{ or } v \in \pi_u \\ +1 & \text{if } u \prec v \\ -1 & \text{if } v \prec u. \end{cases}$$

Furthermore we define $a_v(u)$ as the distance of $v$ to the nearest common ancestor of $u$ and $v$, and $\ell(v) = |\{u \in V(T) : u \prec v\}|$.

## Construction.

Our construction resembles Gupta's construction [1] to some extent, but we needed several new ideas to obtain a significantly better distortion.

First we describe the idea of the construction.

The embedded tree is placed in a suitable acute angle. The apex of the angle is at the origin of coordinates and the angle opens upwards. The root of the tree is placed to the origin.

In the first step we put vertices of height $h$ on the horizontal line $y = 2\sqrt{n}h$. Our angle is divided into subangles that correspond to subtrees rooted at the sons of the root; the sizes of these subangles are proportional to the number of vertices in the subtrees. The vertices of height 1 are placed on the first horizontal line in the middle of the corresponding subangles. We continue in the same way for every subtree, with each subangle translated so that its vertex lies at the root of the subtree.

In the second (and last) step we raise the $y$-coordinates of the vertices by appropriate amounts between 0 and $\sqrt{n}$.

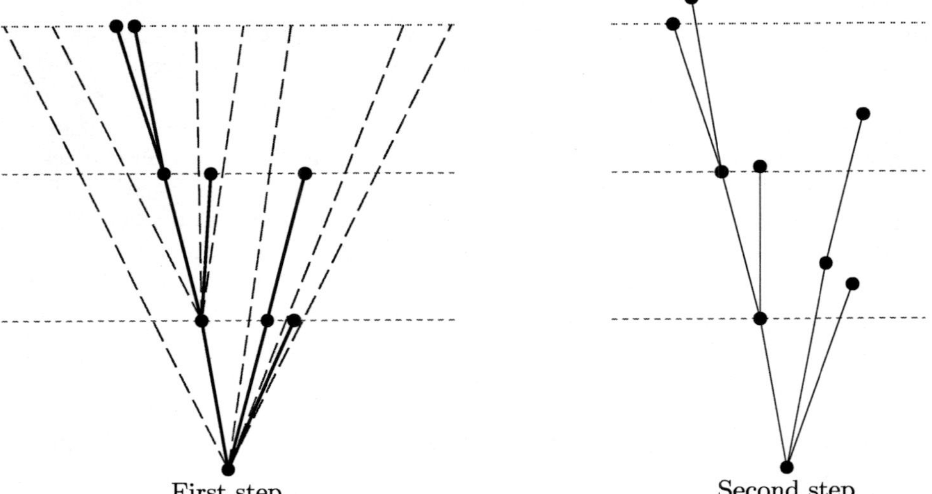

First step          Second step

Now we describe the embedding more formally; actually we use a slightly modified embedding for which the formulas are quite simple.

R. Babilon *et al.*, *Embeddings of Trees*, JGAA, 7(4) 399–409 (2003)    403

First we define an auxiliary function $k(v)$:

$$k(v) = \sum_{w \in V} \mathrm{sgn}_v(w) a_v(w).$$

The embedding $f: V \to \mathbf{R}^2$ is now defined by $f(v) = (x(v), y(v))$, where

$$
\begin{aligned}
x(v) &= n^{-1/2} k(v) \\
y(v) &= 2\sqrt{n}\, h(v) + (\ell(v) \bmod \sqrt{n}).
\end{aligned}
$$

## Distortion.

We now estimate the distortion of the embedding defined above. First we show that the maximum expansion of the tree distance of any two vertices is at most $O(\sqrt{n})$. By the triangle inequality, the maximal expansion is always attained on an edge. Hence, it suffices to consider two vertices connected by an edge. So let $u$ be the father of $v$ in $T$; it suffices to verify that $\|f(u) - f(v)\| = O(\sqrt{n})$. We clearly have $|y(u) - y(v)| \leq 3\sqrt{n}$, and so it suffices to prove that $|x(u) - x(v)| = O(\sqrt{n})$, which will follow from $|k(u) - k(v)| \leq n$. We have

$$|k(v) - k(u)| \leq \sum_{w \in V(T)} \left| \mathrm{sgn}_u(w) a_u(w) - \mathrm{sgn}_v(w) a_v(w) \right|.$$

We claim that every $w$ contributes at most 1 to this sum. First we note that $|a_u(w) - a_v(w)| \leq 1$. If $\mathrm{sgn}_u(w) \neq 0$, we have $\mathrm{sgn}_v(w) = \mathrm{sgn}_u(w)$, and the contribution of $w$ is at most $|a_u(w) - a_v(w)| \leq 1$. The only remaining possibility is $\mathrm{sgn}_u(w) = 0$ and $\mathrm{sgn}_v(w) \neq 0$. But then $u$ is the nearest common ancestor of $v$ and $w$, we have $a_v(w) = 1$, $a_u(w) = 0$, and the contribution of $w$ is at most 1 in this case as well. Therefore $|k(u) - k(v)| \leq n$ as claimed.

Next, we are going to prove that $\|f(u) - f(v)\| = \Omega(\rho(u, v))$ for every $u, v \in V$; this will finish the proof of Theorem 1.

First we consider the situation when $h(u) = h(v)$; this is the main part of the proof, and the case where $u$ and $v$ have different levels will be an easy extension of this. So let $h(u) = h(v)$ and $a = a_v(u) = a_u(v)$, and assume that $u \prec v$.

**Lemma 1** $\ell(v) - \ell(u) \geq a$.

This is because all the vertices on $\pi_u$ not lying on $\pi_v$ are counted in $\ell(v)$ but not in $\ell(u)$. $\qquad\square$

**Corollary 1** If $\ell(v) - \ell(u) < \frac{1}{2}\sqrt{n}$, then $|y(u) - y(v)| \geq \frac{1}{2}\rho(u, v)$.

Indeed, we have $|y(u) - y(v)| = |(\ell(u) \bmod \sqrt{n}) - (\ell(v) \bmod \sqrt{n})| \geq |\ell(u) - \ell(v)| \geq a = \frac{1}{2}\rho(u, v)$. $\qquad\square$

Thus, the difference in the $y$-coordinate takes care of $u$ and $v$ whenever $\ell(v) - \ell(u) < \frac{1}{2}\sqrt{n}$. Next, we need to show that if this is not the case, then the $x$-coordinate takes care of $u$ and $v$.

R. Babilon *et al.*, *Embeddings of Trees*, JGAA, 7(4) 399–409 (2003) 404

**Lemma 2** $k(v) - k(u) \geq [\ell(v) - \ell(u)] \cdot a$.

**Corollary 2** *If* $\ell(v) - \ell(u) \geq \frac{1}{2}\sqrt{n}$, *then* $x(v) - x(u) \geq n^{-1/2}[\ell(v) - \ell(u)] \cdot a \geq \frac{1}{2}a \geq \frac{1}{4}\rho(u, v)$. $\qquad\square$

Combining this corollary with Corollary 1 yields $\|f(u) - f(v)\| = \Omega(\rho(u, v))$ for all $u, v \in V$ with $h(u) = h(v)$.

**Proof of Lemma 2:** We have $k(v) - k(u) = \sum_{w \in V} t_{u,v}(w)$, where $t_{u,v}(w) = \mathrm{sgn}_v(w)a_v(w) - \mathrm{sgn}_u(w)a_u(w)$. First we check that $t_{u,v} \geq 0$ for all $w$. We always have $\mathrm{sgn}_v(w) \geq \mathrm{sgn}_u(w)$, and so the only cases that might cause trouble are $\mathrm{sgn}_v(w) = \mathrm{sgn}_u(w) = 1$ and $\mathrm{sgn}_v(w) = \mathrm{sgn}_u(w) = -1$. We will analyze only the case when $\mathrm{sgn}_v(w) = \mathrm{sgn}_u(w) = 1$, the other one is symmetric. In this case $w \prec u \prec v$, and it is easy to check that then $a_v(w) \geq a_u(w)$, which shows $t_{u,v}(w) \geq 0$ in all cases.

Next, we verify that if $w$ contributes 1 to $\ell(v) - \ell(u)$, which means $w \prec v$ and $w \not\prec u$, then $t_{u,v}(w) \geq a$. We distinguish two cases for such $w$: either $u \prec w \prec v$, or $w$ and $u$ lie on a common path to the root, i.e. $w \in \pi_u$ or $u \in \pi_w$.

In the first case, $u \prec w \prec v$, $\mathrm{sgn}_u(w) = -1$ and $\mathrm{sgn}_v(w) = 1$, and so $t_{u,v}(w) = a_u(w) + a_v(w)$. It is easy to see that if $u$ and $v$ are vertices with $a_u(v) = a_v(u) = a$, then $\max(a_u(w), a_v(w)) \geq a$ for all $w \in V$ and so $t_{u,v}(w) \geq a$ in this case.

In the second case, we have $\mathrm{sgn}_v(w) = 1$ and $\mathrm{sgn}_u(w) = 0$, and so $t_{u,v}(w) = a_v(w) = a$, since the nearest common ancestor of $u$ and $v$ is the same as the nearest common ancestor of $w$ and $v$. The proof of Lemma 2 is complete. $\qquad\square$

**Claim 1** *If two vertices $u, v$ are on different levels, then*

$$\|f(v) - f(u)\| \geq \frac{1}{12\sqrt{2}}\,\rho(u, v).$$

**Proof:** Without loss of generality, suppose that $h(u) < h(v)$ and $u \prec v$. Set $a := a_u(v)$, and let $w$ be the vertex on the path from $u$ to $v$ with $h(w) = h(u)$, $\rho(u, w) = 2a$.

If $\rho(u, v) > 3a$, then $\|f(v) - f(u)\| \geq y(v) - y(u) \geq (\rho(u, v) - 2a)\sqrt{n} \geq \frac{\sqrt{n}}{3}\rho(u, v) \geq \frac{1}{3}\rho(u, v)$.

If $\rho(u, v) \leq 3a$ and $a < 4\sqrt{n}$, then $\|f(v) - f(u)\| \geq y(v) - y(u) \geq \sqrt{n} > a/4 \geq \frac{1}{12}\rho(u, v)$.

Finally, suppose that $\rho(u, v) \leq 3a$ and $a \geq 4\sqrt{n}$. Using the definition of the embedding $f$ and with the observation that $k(v) \leq nh(v)$, it is easy to check that $|x(v)| \leq y(v)$ for any vertex $v$. Thus the whole drawing of the tree lies in the cone bounded by the two rays emanating from the root diagonally upwards and spanning the angle of $45°$ with the $y$-axis. This also holds for every subtree. So the distance of the point $f(u)$ to $f(v)$ is larger than the distance to the left arm of the cone rooted at $f(w)$, which is $\frac{1}{\sqrt{2}}(x(w) - x(u) + y(w) - y(u)) > \frac{1}{\sqrt{2}}(x(w) - x(u) - \sqrt{n}) \geq \frac{1}{\sqrt{2}}(a/2 - \sqrt{n}) \geq \frac{1}{\sqrt{2}}\frac{a}{4} \geq \frac{1}{12\sqrt{2}}\rho(u, v)$ (in the second inequality we used Corollary 2). Thus, $\|f(v) - f(u)\| > \frac{1}{12\sqrt{2}}\rho(u, v)$ in this case. $\qquad\square$

R. Babilon *et al.*, *Embeddings of Trees*, JGAA, 7(4) 399–409 (2003)    405

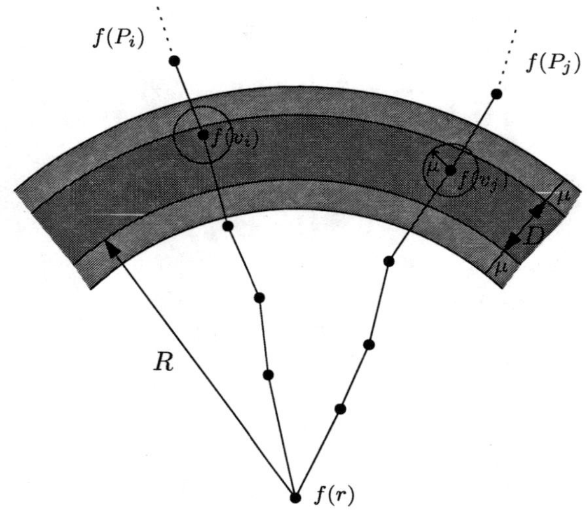

Figure 1: Illustration to the proof of Proposition 1.

# 3    Proof of Proposition 1

Let $F_{\ell,m}$ be defined as in Section 1; that is, $F_{\ell,m}$ denotes the tree consisting of $\ell$ paths of length $m$ glued at the root $r$. We show that every embedding of $F_{\ell,m}$ into $\mathbf{R}^d$ requires distortion $\Omega(\ell^{1/(d-1)})$ if $\ell \leq m^{\frac{d(d-1)}{d+1}}$, and distortion $\Omega(\ell^{1/d}m^{1/(d+1)})$ if $\ell \geq m^{\frac{d(d-1)}{d+1}}$.

Let $f\colon V(F_{\ell,m}) \to \mathbf{R}^d$ be any non-contracting embedding with distortion $D$ (i.e. for all $u, v \in V$, $\rho(u,v) \leq \|f(u) - f(v)\| \leq D \cdot \rho(u,v)$). We choose a real number $R$ such that there are $\ell/2$ of the images of endvertices of $F_{\ell,m}$ contained inside the ball $B$ of radius $R$ around $f(r)$ or on its boundary, and there are $\ell/2$ of them outside the ball or on its boundary.

We consider a ball of radius $m$ around each of the $\ell/2$ images of endvertices contained in $B$. These balls are disjoint and contained in $m$-neighbourhood of $B$. Hence, by a volume argument, $R = \Omega(m\ell^{1/d})$.

Let $S$ denote the sphere of radius $R$ around $f(r)$ (i.e. the boundary of $B$). Let $P_1, \ldots, P_{\ell/2}$ denote the $\ell/2$ paths for which the images of their endvertices are not contained inside $B$. Each $f(P_i)$ intersects $S$ at least once (by $f(P_i)$ we mean $f(V(P_i))$ together with the straight lines between any $f(u)$, $f(v)$ such that $\{u,v\} \in E(P_i)$). Let $m_i$ denote the number of vertices on $f(P_i)$ inside $B$ before the first intersection with $S$. We put $\mu = \min\{m_i \mid i = 1, \ldots, \ell/2\}$. From the choice of $\mu$ it follows that $R \leq D\mu$.

Let $P$ be a spherical shell of width $D$ around $B$; that is,

$$P = B(f(r), R + D) \setminus \text{int } B(f(r), R),$$

where $\text{int} X$ stands for the interior of $X$. At least $\ell/2$ vertices (one from each path

R. Babilon *et al.*, *Embeddings of Trees*, JGAA, 7(4) 399–409 (2003)      406

$P_i$, $i = 1, \ldots, \ell/2$) are contained in $P$. We consider a ball of radius $\mu$ around each of these vertices. These balls are disjoint (because the tree-distance between any two of the vertices is at least $2\mu$) and they are contained in $\mu$-neighbourhood of $P$ (see the picture). Therefore we have, for a suitable constant $c > 0$,

$$R^{d-1}(D + 2\mu) \geq c\ell\mu^d.$$

Thus there are two possibilities: $R^{d-1}D = \Omega(\ell\mu^d)$ or $R^{d-1}\mu = \Omega(\ell\mu^d)$.

If $R^{d-1}D = \Omega(\ell \cdot \mu^d)$, then $D^{d+1} = \Omega(\ell \cdot R)$, since $R \leq D\mu$. And since $R = \Omega(m\ell^{1/d})$ we have $D^{d+1} = \Omega(\ell \cdot m \cdot \ell^{1/d})$. Thus $D = \Omega(\ell^{1/d} \cdot m^{1/(d+1)})$. If $R^{d-1}\mu = \Omega(\ell \cdot \mu^d)$, then since $R \leq D\mu$, we immediately have $D = \Omega(\ell^{1/(d-1)})$. Thus we know that $D \geq \min\{\Omega(\ell^{1/d} \cdot m^{1/(d+1)}), \Omega(\ell^{1/(d-1)})\}$. It means that $D = \Omega(\ell^{1/d} \cdot m^{1/(d+1)})$ if $\ell \geq m^{\frac{d(d-1)}{d+1}}$, and $D = \Omega(\ell^{1/(d-1)})$ if $\ell \leq m^{\frac{d(d-1)}{d+1}}$.

## 4    A low-distortion embedding of a subdivided star

In this section we describe an asymptotically optimal embedding of $F_{\sqrt{n},\sqrt{n}}$ into $\mathbf{R}^2$, with distortion $O(n^{5/12})$. The other $F_{\ell,m}$ can be embedded similarly, with distortion as in Proposition 1; we omit a proof since it is technical and contains no new ideas.

The embedding is sketched in Fig 2. The tree $F_{\sqrt{n},\sqrt{n}}$ is embedded into the shaded trapezoids (there are $n^{1/4} \cdot 2n^{1/3}$ of them, each of height $n^{5/12}$) and into the shaded discs (there are $n^{1/4} \cdot n^{1/4}$ of them). The root of $F_{\sqrt{n},\sqrt{n}}$ is embedded to the bottom vertex of the triangle.

In the following, we describe the embedding in more detail. It will be done in two steps. In the first step, we embed vertices up to the level $h = 2n^{1/3}$ into the dashed trapezoids on the corresponding level. In the second step, we embed all the other vertices on diameters of the discs and change the positions of some vertices on levels between $n^{1/3}$ and $2n^{1/3}$.

**Step 1.**

We start with the vertices on level $h = 2n^{1/3}$. There are $n^{1/2}$ vertices on this level. We divide them into $n^{1/4}$ groups, each of size $n^{1/4}$. Each of these groups is packed into one of the dashed trapezoids (on the last level). This is possible, since the area of each trapezoid on this level is $\Theta(n^{11/12})$, which is exactly the required area (we have $n^{1/4}$ vertices, and any two of them must be at distance at least $\Omega(n^{1/3})$). It follows that no two vertices on level $h = 2n^{1/3}$ are embedded too close to each other.

When all the vertices on the level $h = 2n^{1/3}$ are embedded, we embed the vertices on lower levels. We connect every vertex $x$ on the last level by a straight line to the root. On this line we embed all the vertices on the path from $x$ to the root. We do this in such a way that the distance $\Delta$ (measured on the connecting

R. Babilon *et al.*, *Embeddings of Trees*, JGAA, 7(4) 399–409 (2003)          407

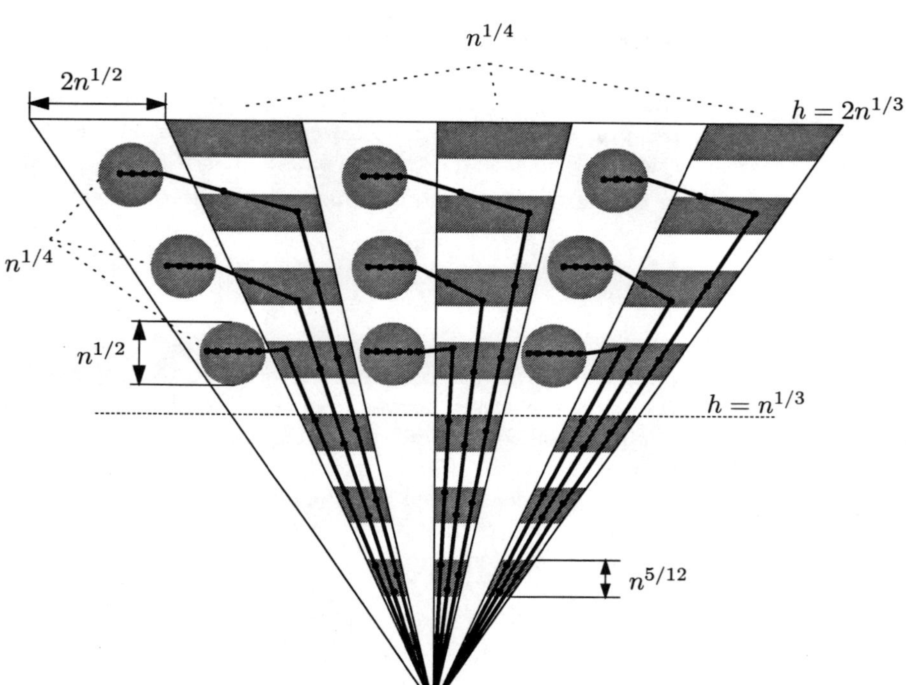



Figure 2: The embedding of $F_{\sqrt{n},\sqrt{n}}$.

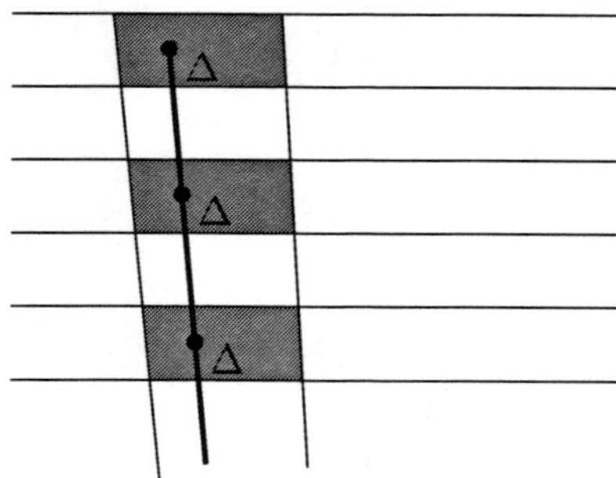

Figure 3: The placement of vertices in lower levels.

line) from a vertex to the bottom of its dashed trapezoid is the same for all the vertices on the path from $x$ to the root (see Fig. 3). It is easy to check that the distance between the vertices on the last level guarantees that the vertices in the lower levels are also far enough from each other. Indeed, if we consider two vertices $x_1$ and $y_1$ of level $h_1$ (i.e., with tree distance $2h_1$) that have distance $d$

in the embedding, then for any $x_2$ on the path from $x_1$ to the root and any $y_2$ on the path from $y_1$ to the root, both at a level $h_2$, their distance is at least $dh_2/h_1$ (while their tree-distance is $2h_2$). Hence, if every two vertices in a higher level are far enough then also every two vertices in a lower level are far enough.

Thus, the construction described so far gives an embedding of $F_{\sqrt{n},2n^{1/3}}$ with distortion $O(n^{5/12})$. In the next step we modify this embedding into an embedding of $F_{\sqrt{n},\sqrt{n}}$.

**Step 2.**

In this step we embed all vertices on the remaining levels. We also change the positions of some vertices on the levels between $n^{1/3}$ and $2n^{1/3}$. There is a column of discs on the left of each column of trapezoids; there are $n^{1/4}$ columns of discs with $n^{1/4}$ discs in each column. Each path of $F_{\sqrt{n},\sqrt{n}}$ starts at the root, goes through some number of dashed trapezoids, having one vertex in each of them, reaches a suitable level, and then it turns into one of the discs, where the rest of it is embedded with vertices packed as tightly as possible. A more precise description follows.

In each column, we consider the discs one by one from the bottom to the top. For each disc, we consider the highest shaded trapezoid lying below the equator of the disc. We take the leftmost vertex embedded in that trapezoid, and we change the embedding of the successors of this vertex: The successors are embedded on the equator of the considered disc and on a straight line connecting the leftmost vertex to the equator of the considered disc. The distance between any two neighbours on the connecting line is $n^{5/12}$ and the distance between any two neighbours on the equator is 1 in the embedding.

We continue with the next higher disc (with a smaller number of paths). In the last level, there is only one path to end. The resulting embedding is indicated in Fig 2.

It is straightforward to check that the resulting embedding of $F_{\sqrt{n},\sqrt{n}}$ yields distortion $O(n^{5/12})$ as claimed. $\qquad\qquad\square$

# Acknowledgments

We would like to thank Helena Nyklová, Petra Smolíková, Ondřej Pangrác, Robert Šámal, and Tomáš Chudlarský for useful discussions on the problems considered in this paper. We also thank two anonymous referees for careful reading and useful comments.

R. Babilon *et al.*, *Embeddings of Trees*, JGAA, 7(4) 399–409 (2003) 409

# References


[1] A. Gupta. Embedding tree metrics into low dimensional Euclidean spaces. *Discrete Comput. Geom.*, 24:105–116, 2000.

[2] P. Indyk. Algorithmic applications of low-distortion embeddings. In *Proc. 42nd IEEE Symposium on Foundations of Computer Science*, 2001.

[3] J. Matoušek. Bi-Lipschitz embeddings into low-dimensional Euclidean spaces. *Comment. Math. Univ. Carolinae*, 31:589–600, 1990.

[4] J. Matoušek. *Lectures on Discrete Geometry*. Springer, New York, 2002.

# Journal of Graph Algorithms and Applications

http://jgaa.info/

vol. 7, no. 4, pp. 411–427 (2003)

# On Cotree-Critical and DFS Cotree-Critical Graphs

*Hubert de Fraysseix*     *Patrice Ossona de Mendez*

Centre d'Analyse et de Mathématiques Sociales (CNRS UMR 8557)
Ecole des Hautes Études en Sciences Sociales
75006 Paris, France
http://www.ehess.fr/centres/cams/
hf@ehess.fr     pom@ehess.fr

### Abstract

We give a characterization of DFS cotree-critical graphs which is central to the linear time Kuratowski finding algorithm implemented in PI-GALE (Public Implementation of a Graph Algorithm Library and Editor [2]) by the authors, and deduce a justification of a very simple algorithm for finding a Kuratowski subdivision in a DFS cotree-critical graph.

Communicated by: P. Mutzel and M. Jünger;
submitted May 2002; revised August 2003.

de Fraysseix & Ossona de Mendez, *Critical Graphs*, JGAA, 7(4) 411–427 (2003)    412

# 1    Introduction

The present paper is a part of the theoretical study underlying a linear time algorithm for finding a Kuratowski subdivision in a non-planar graph ([1]; see also [7] and [9] for other algorithms). Other linear time planarity algorithms don't exhibit a Kuratowski configuration in non planar graphs, but may be used to extract one in quadratic time.

It relies on the concept of DFS cotree-critical graphs, which is a by-product of DFS based planarity testing algorithms (such as [5] and [4]). Roughly speaking, a DFS cotree-critical graph is a simple graph of minimum degree 3 having a DFS tree, such that any non-tree (i.e. cotree) edge is *critical*, in the sense that its deletion would lead to a planar graph. A first study of DFS cotree-critical graphs appeared in [3], in which it is proved that a DFS cotree-critical graph either is isomorphic to $K_5$ or includes a subdivision of $K_{3,3}$ and no subdivision of $K_5$.

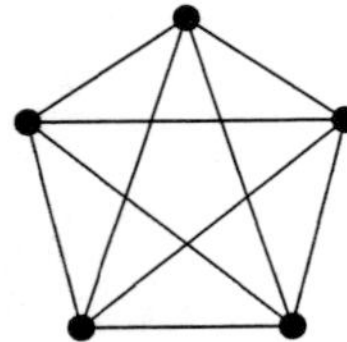
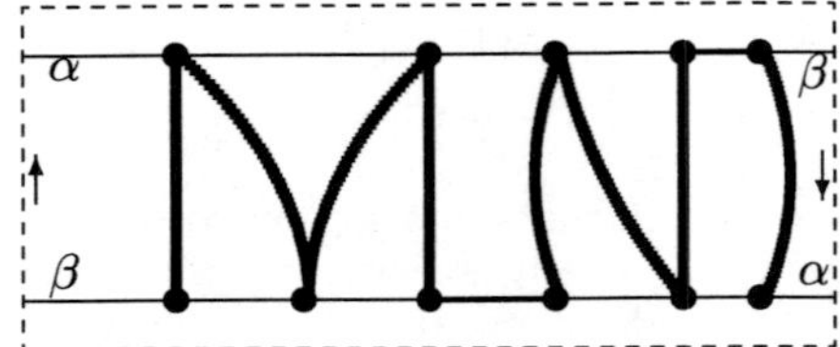


Figure 1: The DFS cotree-critical graphs are either $K_5$ or Möbius pseudo-ladders having all their non-critical edges (thickest) included in a single path.

The linear time Kuratowski subdivision extraction algorithm, which has been both conceived and implemented in [2] by the authors, consists in two steps: the first one correspond to the extraction of a DFS cotree-critical subgraph by a case analysis algorithm; the second one extracts a Kuratowski subdivision from the DFS cotree-critical subgraph by a very simple algorithm (see Algorithm 1), but which theoretical justification is quite complex and relies on the full characterization of DFS cotree-critical graphs that we prove in this paper: a simple graph is DFS cotree-critical if and only if it is either $K_5$ or a Möbius pseudo-ladder having a simple path including all the non-critical edges (see Figure 1).

The algorithm roughly works as follows: it first computes the set Crit of the critical edges of $G$, using the property that a tree edge is critical if and only if it belongs to a fundamental cycle of length 4 of some cotree edge to which it is not adjacent. Then, three pairwise non-adjacent non-critical edges are found to complete a Kuratowski subdivision of $G$ isomorphic to $K_{3,3}$.

The space and time linearity of the algorithm are obvious.

de Fraysseix & Ossona de Mendez, *Critical Graphs*, JGAA, 7(4) 411–427 (2003)    413

---

**Require:** $G$ is a DFS cotree-critical graph, with DFS tree $Y$.
**Ensure:** $K$ is a Kuratowski subdivision in $G$.
  **if** $G$ has 5 vertices **then**
    $K = G$ {$G$ is isomorphic to $K_5$}
  **else if** $G$ has less than 9 vertices **then**
    Extract $K$ with any suitable method.
  **else** {$G$ is a Möbius pseudo-ladder and the DFS tree is a path}
    Crit $\leftarrow E(G) \setminus Y$ {will be the set of critical edges}
    Find a vertex $r$ incident to a single tree edge
    Compute a numbering $\lambda$ of the vertices according to a traversal of the path $Y$ starting at $r$, from 1 to $n$.
    Let $e_i$ denote the tree edge from vertex numbered $i$ to vertex numbered $i+1$.
    **for all** cotree edge $e = (u, v)$ (with $\lambda(u) < \lambda(v)$) **do**
      **if** $\lambda(v) - \lambda(u) = 3$ **then**
        Crit $\leftarrow$ Crit $\cup \{e_{\lambda(u)+1}\}$
      **end if**
    **end for**
    Find a tree edge $f = e_i$ with $2 < i < n - 3$ which is not in Crit.
    $K$ has vertex set $V(G)$ and edge set Crit $\cup \{e_1, e_{n-1}, f\}$.
  **end if**

---

**Algorithm 1:** extracts a Kuratowski subdivision from a DFS cotree-critical graph [2].

# 2 Definitions and Preliminaries

For classical definitions (subgraph, induced subgraph, attachment vertices), we refer the reader to [8].

## 2.1 Möbius Pseudo-Ladder

A Möbius pseudo-ladder is a natural extension of Möbius ladders allowing triangles. This may be formalized by the following definition.

**Definition 2.1** *Let $\gamma$ be a polygon $(v_1, \ldots, v_n)$ and let $\{v_i, v_j\}$ and $\{v_k, v_l\}$ be non adjacent chords of $\gamma$. These chords are* interlaced *with respect to $\gamma$ if, in circular order, one finds exactly one of $\{v_k, v_l\}$ between $v_i$ and $v_j$. They are* non-interlaced, *otherwise.*

Thus, two chords of a polygon are either adjacent, or interlaced or non-interlaced.

**Definition 2.2** *A* Möbius pseudo-ladder *is a non-planar simple graph, which is the union of a polygon $(v_1, \ldots, v_n)$ and chords of the polygon, such that any two non-adjacent bars are interlaced.*
    *With respect to such a decomposition, the chords are called* bars.

de Fraysseix & Ossona de Mendez, *Critical Graphs*, JGAA, 7(4) 411–427 (2003)    414

A Möbius band is obtained from the projective plane by removing an open disk. Definition 2.2 means that a Möbius pseudo-ladder may be drawn in the plane as a polygon and internal chords such that any two non adjacent chords cross: consider a closed disk $\bar{\Delta}$ of the projective plane, which intersects any projective line at most twice (for instance, the disk bounded by a circle of the plane obtained by removing the line at infinity). Embed the polygon on the boundary of $\bar{\Delta}$. Then, any two projective lines determined by pairs of adjacent points intersect in $\bar{\Delta}$. Removing the interior $\Delta$ of $\bar{\Delta}$, we obtain an embedding of the Möbius pseudo ladder in a Möbius band having the polygon as its boundary (see Figure 2).

Notice that $K_{3,3}$ and $K_5$ are both Möbius pseudo-ladders.

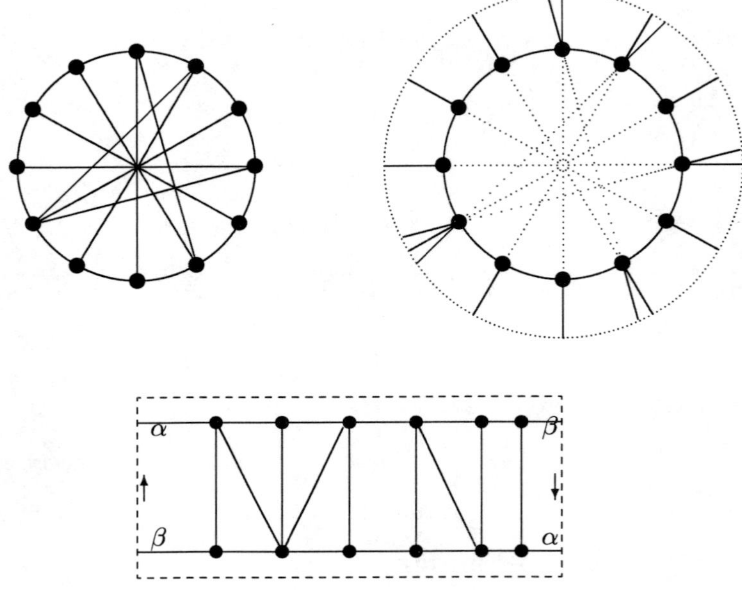

Figure 2: A Möbius pseudo-ladder on the plane, on the projective plane and on the Möbius band

## 2.2   Critical Edges and Cotree-Critical Graphs

**Definition 2.3** *Let $G$ be a graph. An edge $e \in E(G)$ is* critical *for $G$ if $G - e$ is planar.*

**Remark 2.1** *Let $H$ be a subgraph of $G$, then any edge which is critical for $G$ is critical for $H$ (as $G - e$ planar implies $H - e$ planar).*

*Thus, proving that an edge is non-critical for a particular subgraph of $G$ is sufficient to prove that it is non-critical for $G$.*

*Moreover, if $H$ is a non-planar subgraph of $G$, any edge in $E(G) \setminus E(H)$ is obviously non-critical for $G$.*

**Definition 2.4** *A* cotree-critical graph *is a non-planar graph $G$, with minimum degree 3, such that the set of non-critical edges of $G$ is acyclic.*

**Definition 2.5** *A* hut *is a graph obtained from a cycle $(v_1, \ldots, v_p, \ldots, v_n)$ by adding two adjacent vertices $x$ and $y$, such that $x$ is incident to $v_n, v_1, \ldots, v_p$, and $y$ is incident to $v_p, \ldots, v_n, v_1$.*

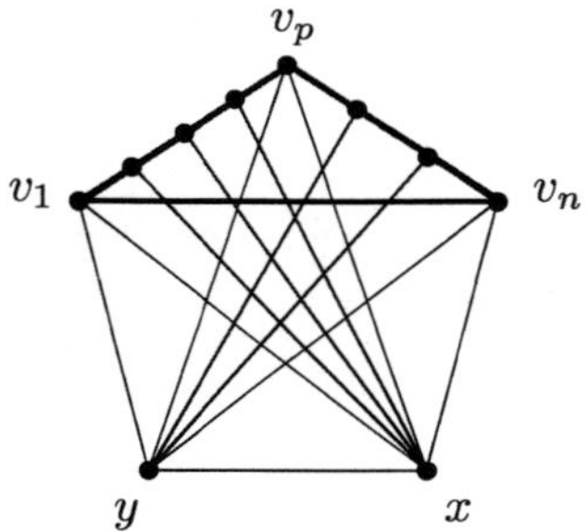


Figure 3: A hut drawn as a Möbius pseudo-ladder

We shall use the following result on cotree-critical graphs (expressed here with our terminology) later on:

**Theorem 2.2 (Fraysseix, Rosenstiehl [3])** *A cotree-critical graph is either a hut or includes a subdivision of $K_{3,3}$ but no subdivision of $K_5$.*

## 2.3  Kuratowksi Subdivisions

A *Kuratowski subdivision* in a graph $G$ is a minimal non-planar subgraph of $G$, that is: a non-planar subgraph $K$ of $G$, such that all the edges of $K$ are critical for $K$. Kuratowski proved in [6] that such minimal graphs are either subdivisions of $K_5$ or subdivisions of $K_{3,3}$.

If $G$ is non-planar and if $K$ is a Kuratowski subdivision in $G$, it is clear that any critical edge for $G$ belongs to $E(K)$. This justifies a special denomination of the vertices and branches of a Kuratowski subdivision:

**Definition 2.6** *Let $G$ be a non-planar graph and let $K$ be a Kuratowski subdivision of $G$. Then, a vertex is said to be a $K$-vertex (resp. a $K$-subvertex, resp. a $K$-exterior vertex) if it is a vertex of degree at least 3 in $K$ (resp. a vertex of degree 2 in $K$, resp. a vertex not in $K$). A $K$-branch is the subdivided path of $K$ between two $K$-vertices. Two $K$-vertices are $K$-adjacent if they are the endpoints of a $K$-branch. A $K$-branch with endpoints $x$ and $y$ is said to link $x$ and $y$, and is denoted $[x, y]$. We further denote $]x, y[$ the subpath of $[x, y]$ obtained by deleting $x$ and $y$.*

*A $K$-branch is* critical *for $G$ if it includes at least one edge which is critical for $G$.*

de Fraysseix & Ossona de Mendez, *Critical Graphs*, JGAA, 7(4) 411–427 (2003)    416

## 2.4   Depth-First Search (DFS) Tree

**Definition 2.7** *A DFS tree of a connected graph $G$, rooted at $v_0 \in V(G)$, may be recursively defined as follows: If $G$ has no edges, the empty set is a DFS tree of $G$. Otherwise, let $G_1, \ldots, G_k$ the connected components of $G - v_0$. Then, a DFS tree of $G$ is the union of the DFS trees $Y_1, \ldots, Y_k$ of $G_1, \ldots, G_k$ rooted at $v_1, \ldots, v_k$ (where $v_1, \ldots, v_k$ are the neighbors of $v_0$ in $G$), and the edges $\{v_0, v_1\}, \ldots, \{v_0, v_k\}$.*

*Vertices of degree 1 in the tree are the terminals of the tree.*

**Definition 2.8** *A DFS cotree-critical graph $G$ is a cotree-critical graph, whose non-critical edge set is a subset of a DFS tree of $G$.*

**Lemma 2.3** *If $G$ is $k$-connected ($k \geq 1$) and $Y$ is a DFS tree of $G$ rooted at $v_0$, then there exists a unique path in $Y$ of length $k - 1$ having $v_0$ as one of its endpoints.*

**Proof:** The lemma is satisfied for $k = 1$. Assume that $k > 1$ and that the lemma is true for all $k' < k$. Let $v_0$ be a vertex of a $k$-connected graph $G$. Then $G - v_0$ has a unique connected component $H$, which is $k - 1$-connected. A DFS tree $Y_G$ of $G$ will be the union of a DFS tree $Y_H$ of $H$ rooted at a neighbor $v_1$ of $v_0$ and the edge $\{v_0, v_1\}$. As there exists, by induction, a unique path in $Y_H$ of length $k - 2$ having $v_1$ as one of its endpoints, there will exist a unique path in $Y_G$ of length $k - 1$ having $v_0$ as one of its endpoints.   $\square$

**Corollary 2.4** *If $G$ is 3-connected and $Y$ is a DFS tree of $G$ rooted at $v_0$, then $v_0$ has a unique son, and this son also has a unique son.*

**Proof:** As $G$ is 3-connected, it is also 2-connected. Hence, there exists a unique tree path of length 1 and a unique tree path of length 2 having $v_0$ as one of its endpoints.   $\square$

Consider the orientation of a DFS tree $Y$ of a connected graph $G$ from its root (notice that each vertex has indegree at most 1 in $Y$). This orientation induces a partial order on the vertices of $G$, having the root of $Y$ as a minimum. In this partial order, any two vertices which are adjacent in $G$ are comparable (this is the usual characterization of DFS trees).

This orientation and partial order are the key to the proofs of the following two easy lemmas:

**Lemma 2.5** *Let $Y$ be a DFS tree of a graph $G$. Let $x, y, z$ be three vertices of $G$, not belonging to the same monotone tree path. If $x$ is a terminal of $Y$ and $x$ is adjacent to both $y$ and $z$, then $x$ is the root of $Y$.*

**Proof:** Assume $x$ is not the root of $Y$. As $y$ and $z$ are adjacent to $x$, they are comparable with $x$. As $x$ is a terminal different from the root $v_0$, $y$ and $y$ belong to the monotone tree path from $v_0$ to $x$, a contradiction.   $\square$

de Fraysseix & Ossona de Mendez, *Critical Graphs*, JGAA, 7(4) 411–427 (2003)    417

**Lemma 2.6** *Let $Y$ be a DFS tree of a graph $G$. Let $x, y, z, t$ be four vertices of $G$, no three of which belong to the same tree path, and such that the tree paths from $x$ to $y$ and $z$ to $t$ intersect. Then, $\{x, y\}$ and $\{z, t\}$ cannot both be edges of $G$.*

**Proof:** Assume both $\{x, y\}$ and $\{z, t\}$ are edges of $G$. If $x$ and $y$ are adjacent, they are comparable and thus, the tree path linking them is a monotone path. Similarly, the same holds for the tree path linking $z$ and $t$. As these two monotone tree paths intersect and as neither $x$ and $z$ belong to both paths, there exists a vertex having indegree at least 2 in the tree, a contradiction.    $\square$

## 3  Cotree-Critical Graphs

**Lemma 3.1** *Let $G$ be a graph and let $H$ be the graph obtained from $G$ by recursively deleting all the vertices of degree 1 and contracting all paths which internal vertices have degree 2 in $G$ to single edges. Then, $G$ is non-planar and has an acyclic set of non-critical edges if and only if $H$ is cotree-critical.*

**Proof:** First notice that $H$ is non-planar if and only if $G$ is non-planar.

The critical edges of $G$ that remain in $H$ are critical edges for $H$, according to the commutativity of deletion, contraction of edges and deletion of isolated vertices (for $e \in E(H)$, if $G - e$ is planar so is $H - e$).

For any induced path $P$ of $G$, either all the edges of $P$ are critical for $G$ or they are all non-critical for $G$. Thus, the edge of $P$ that remains in $H$ is critical for $H$ if and only if at least one edge of $P$ is critical for $G$. Hence, if $H$ had a cycle of non critical edges for $H$, they would define a cycle of non-critical edges for $G$, because each (non-critical) edge for $H$ represents a simple path of (non-critical) edges for $G$ . Since $G$ does not have a cycle of non-critical edges, $H$ cannot have such a cycle either. Thus, as $H$ has minimum degree 3, $H$ is cotree-critical.

Conversely, assume $H$ is cotree-critical. Adding a vertex of degree 1 does not change the status (critical/non-critical) of the other edges and cannot create a cycle of non-critical edges. Similarly, subdividing an edge creates two edges with the same status without changing the status of the other edges and hence cannot create a cycle of non-critical edges. Thus, the set of the non-critical edges of $G$ is acyclic.    $\square$

**Lemma 3.2** *Let $G$ be a cotree-critical graph and let $K$ be a Kuratowski subdivision of $G$ isomorphic to $K_{3,3}$. Then, there exists in $E(G) \setminus E(K)$ no path between:*

- *two vertices ($K$-vertices or $K$-subvertices) of a same $K$-branch of $K$,*

- *two $K$-subvertices of $K$-adjacent $K$-branches of $K$.*

**Proof:** The two cases are shown Fig 4.

de Fraysseix & Ossona de Mendez, *Critical Graphs*, JGAA, 7(4) 411–427 (2003)    418

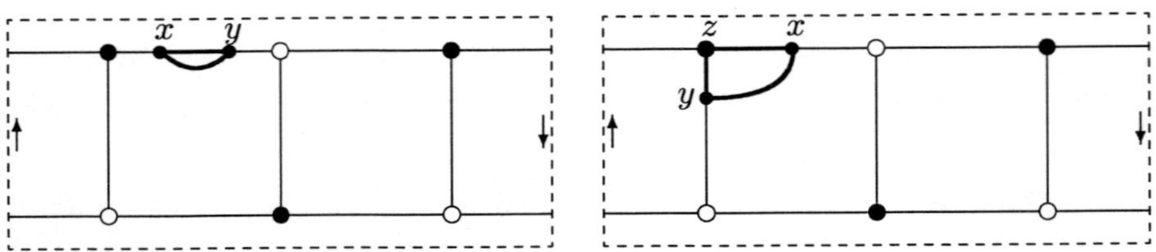

Figure 4: Forbidden paths in cotree-critical graphs (see Lemma 3.2)

If two vertices $x$ and $y$ ($K$-vertices or $K$-subvertices) of a same $K$-branch of $K$ are joined by a path in $E(G)\backslash E(K)$, both this path and the one linking $x$ and $y$ in $K$ are non-critical for $G$. Hence $G$ is not cotree-critical, a contradiction.

If two $K$-subvertices $x$ and $y$ of $K$-adjacent $K$-branches of $K$ are linked by a path in $E(G) \setminus E(K)$, this path is non-critical for $G$. Moreover, if $z$ is the $K$-vertex adjacent to the branches including $x$ and $y$, both paths from $z$ to $x$ and $x$ to $y$ are non-critical for $G$. Hence, $G$ includes a non-critical cycle, a contradiction. □

We need the following definition in the proof of the next lemma:

**Definition 3.1** *Let $H$ be an induced subgraph of a graph $G$. The* attachment vertices *of $H$ in $G$ is the subset of vertices of $H$ having a neighbor in $V(G) \setminus V(H)$.*

**Lemma 3.3** *Every cotree-critical graph is 3-connected.*

**Proof:** Let $G$ be a cotree-critical graph. Assume $G$ has a cut-vertex $v$. Let $H_1, H_2$ be two induced subgraphs of $G$ having $v$ as their attachment vertex and such that $H_1$ is non-planar. As $G$ has no degree 1 vertex, $H_2$ includes a cycle. All the edges of this cycle are non critical for $G$, a contradiction. Hence, $G$ is 2-connected.

Assume $G$ has an articulation pair $\{v, w\}$ such that there exists at least two induced subgraphs $H_1, H_2$ of $G$, different from a path, having $v, w$ as attachment vertices. As $G$ is non planar, we may choose $H_1$ in such a way that $H_1 + \{v, w\}$ is a non-planar graph (see [8], for instance). As there exists in $H_2$ two disjoints paths from $v$ to $w$, no edge of these paths may be critical for $G$ and $H_2$ hence include a cycle of non-critical edges for $G$, a contradiction. □

**Lemma 3.4** *Let $G$ be a cotree-critical graph and let $K$ be a Kuratowski subdivision of $G$. Then, $G$ has no $K$-exterior vertices, that is: $V(G) = V(K)$.*

**Proof:** According to Theorem 2.2, if $K$ is a subdivision of $K_5$, then either $G = K$, or $G$ is a hut, having $K$ has a spanning subgraph. Thus, $G$ has no $K$-exterior vertex in this case, and we shall assume that $K$ is a subdivision of $K_{3,3}$.

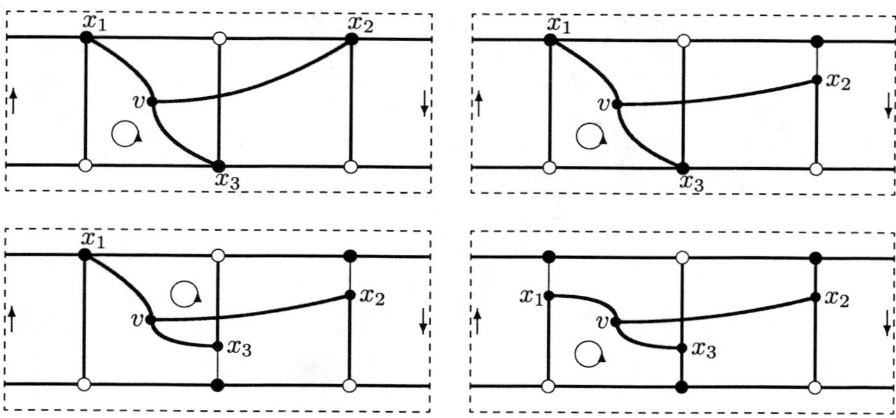

Figure 5: A cotree-critical graphs has no $K$-exterior vertex (see Lemma 3.4)

Assume $V(G) \setminus V(K)$ is not empty and let $v$ be a vertex of $G$ not in $K$. According to Lemma 3.3, $G$ is 3-connected. Hence, there exists 3 disjoint paths $P_1, P_2, P_3$ from $v$ to $K$. As $K + P_1 + P_2 + P_3$ is a non-planar subgraph of $G$ free of vertices of degree 1, it is a subdivision of a 3-connected graph, according to Lemma 3.1 and Lemma 3.3. Thus, the vertices of attachment $x_1, x_2, x_3$ of $P_1, P_2, P_3$ in $K$ are all different. As $K_{3,3}$ is bipartite, we may color the $K$-vertices of $K$ black and white, in such a way that $K$-adjacent $K$-vertices have different colors. According to Lemma 3.2, no path in $E(G) \setminus E(H)$ may link $K$-vertices with different colors. Thus, we may assume no white $K$-vertex belong to $\{x_1, x_2, x_3\}$ and four cases may occur as shown Fig 5. All the four cases show a cycle of non-critical edges, a contradiction. $\qquad\square$

**Corollary 3.5** *If $G$ is cotree-critical, no non-critical $K$-branch may be subdivided, that is: every non-critical $K$-branch is reduced to an edge.*

**Proof:** If a branch of $K$ is non-critical for $G$, there exists a $K_{3,3}$ subdivision avoiding it. Hence, the branch just consists of a single edge, according to Lemma 3.3. $\qquad\square$

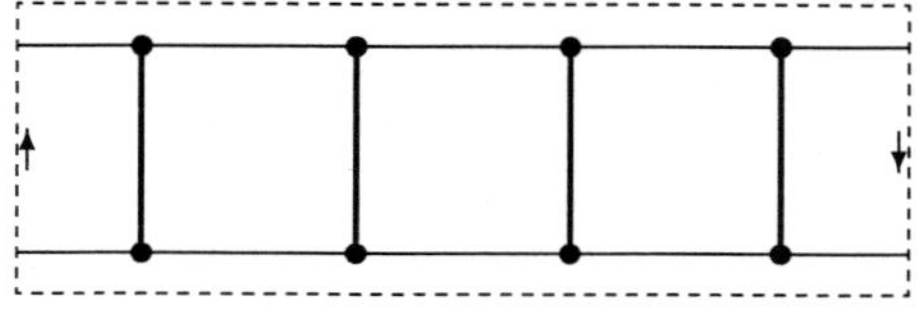

Figure 6: The 4-bars Möbius ladder $M_4$ (all bars are non-critical edges)

de Fraysseix & Ossona de Mendez, *Critical Graphs*, JGAA, 7(4) 411–427 (2003)    420

Let $G$ be a cotree-critical graph obtained by adding an edge linking two subdivision vertices of non-adjacent edges of a subdivision of a $K_{3,3}$. This graph is unique up to isomorphism and is the Möbius ladder with 4 non-critical bars shown Figure 6.

Figure 7 shows a graph having a subdivision of a Möbius ladder with 3 bars as a subgraph, where two of the bars are not single edges.

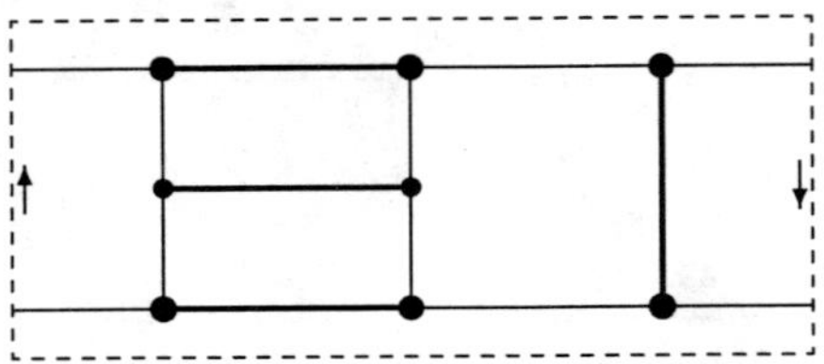

Figure 7: A graph having a subdivision of a 3-bars Möbius ladder as a subgraph (some bars are paths of critical edges)

The same way we have introduced $K$-vertices, $K$-subvertices and $K$-branches relative to a Kuratowski subdivision, we define $M$-vertices, $M$-subvertices and $M$-branches relative to a Möbius ladder subdivision.

**Lemma 3.6** *Let $K$ be a $K_{3,3}$ subdivision in a cotree-critical graph $G$. Not $K$-adjacent $K$-vertices of $K$ form two classes, $\{x, y, z\}$ and $\{x', y', z'\}$, as $K_{3,3}$ is bipartite.*

*If $[x, z']$ or $[x', z]$ is a critical $K$-branch for $G$, then all the edges from $]x, z[=\ ]x, y'] \cup [y', z[$ to $]x', z'[=]x', y] \cup [y, z'[$ and the $K$-branch $[y, y']$ are pairwise adjacent or interlaced, with respect to the cycle $(x, y', z, x', y, z')$.*

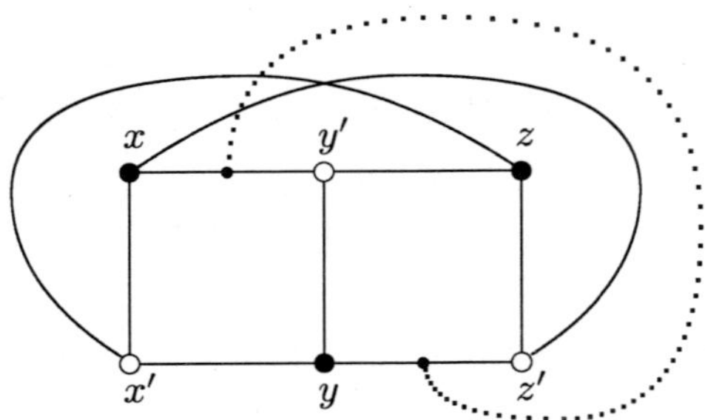

Figure 8: No edges is allowed from $]x, z[$ to $]x', z'[$ by the "outside" (see Lemma 3.6)

**Proof:** The union of the $K_{3,3}$ subdivision and all the edges of $G$ incident to a vertex in $]x, z[$ and a vertex in $]x', z'[$ becomes uniquely embeddable in the plane

de Fraysseix & Ossona de Mendez, *Critical Graphs*, JGAA, 7(4) 411–427 (2003)    421

after removal of the $K$-branches $[x, z']$ or $[x', z]$. Figure 8 displays the outline of a normal drawing of $G$ in the plane which becomes plane when removing any of the $K$-branch $[x, z']$ or $[x', z]$. In such a drawing, given that an edge from $]x, z[$ to $]x', z'[$, if drawn outside, crosses both $[x, z']$ and $[x', z]$, all the edges from $]x, z[$ to $]x', z'[$ and the $K$-branch $[y, y']$ are drawn inside the cycle $(x, y', z, z', y, x')$ without crossing and thus are adjacent or interlaced with respect to the cycle $(x, y', z, x', y, z')$. The result follows. $\qquad\square$

**Lemma 3.7** *If $G$ is a cotree-critical graph having a subdivision of Möbius ladder $M$ with 4 bars as a subgraph, then it is the union of a polygon $\gamma$ and chords which are non-critical for $G$. Moreover the 4 bars $b_1, b_2, b_3, b_4$ of $M$ are chords and any other chord is adjacent or interlaced with all of $b_1, b_2, b_3, b_4$ with respect to $\gamma$.*

**Proof:** Let $G$ be a cotree-critical graph having a subdivision of Möbius $M$ ladder with 4 bars $b_1, b_2, b_3, b_4$ as a subgraph. First notice that all the bars of the Möbius ladder are non-critical for $G$ and that, according to Corollary 3.5, they are hence reduced to edges. According to Lemma 3.4, $M$ covers all the vertices of $G$ as it includes a $K_{3,3}$ and hence the polygon $\gamma$ of the ladder is Hamiltonian. Thus, the remaining edges of $G$ are non-critical chords of $\gamma$.

Let $e$ be a chord different from $b_1, b_2, b_3, b_4$.

- Assume $e$ is adjacent to none of $b_1, b_2, b_3, b_4$.

  Then it cannot be interlaced with less than 3 bars, according to Lemma 3.2, considering the $K_{3,3}$ induced by at least two non-interlaced bars. It cannot also be interlaced with 3 bars, according to Lemma 3.6, considering the $K_{3,3}$ induced by the 2 interlaced bars (as $\{x, x'\}, \{z, z'\}$) and one non-interlaced bar (as $\{y, y'\}$).

- Assume $e$ is adjacent to $b_1$ only.

  Then it is interlaced with the 3 other bars, according to Lemma 3.2, considering the $K_{3,3}$ induced by $b_1$ and two non-interlaced bars.

- Assume $e$ is adjacent to $b_1$ and another bar $b_i$.

  Assume $e$ is not interlaced with some bar $b_j \notin \{b_1, b_i\}$ then, considering the $K_{3,3}$ induced by $b_1, b_i, b_j$ we are led to a contradiction, according to Lemma 3.2. Thus, $e$ is interlaced with the 2 bars to which it is not adjacent.

$\qquad\square$

**Theorem 3.8** *If $G$ is a cotree-critical graph having a subdivision of Möbius ladder $M$ with 4 bars as a subgraph, then it is a Möbius pseudo-ladder whose polygon $\gamma$ is the set of the critical edges of $G$.*

**Proof:** According to Lemma 3.7, $G$ is the union of a polygon $\gamma$ and chords including the 4 bars of $M$. In order to prove that $G$ is a Möbius pseudo-ladder,

de Fraysseix & Ossona de Mendez, *Critical Graphs*, JGAA, 7(4) 411–427 (2003)    422

it is sufficient to prove that any two non-adjacent chords are interlaced with respect to that cycle. We choose to label the 4 bars $b_1, b_2, b_3, b_4$ of $M$ according to an arbitrary traversal orientation of $\gamma$. According to Lemma 3.7, any chord $e$ is adjacent or interlaced with all of $b_1, b_2, b_3, b_4$ and hence its endpoints are traversed between these of two consecutive bars $b_{\alpha(e)}, b_{\beta(e)}$ (with $\beta(e) \equiv \alpha(e) + 1 \pmod 4$), which defines functions $\alpha$ and $\beta$ from the chords different from $b_1, b_2, b_3, b_4$ to $\{1, 2, 3, 4\}$.

As all the bars are interlaced pairwise and as any chord is adjacent or interlaced with all of them, we only have to consider two non-adjacent chords $e, f$ not in $\{b_1, b_2, b_3, b_4\}$.

- Assume $\alpha(e)$ is different from $\alpha(f)$.

  Then, the edges $e$ and $f$ are interlaced, as the endpoints of $e$ and $f$ appear alternatively in a traversal of $\gamma$.

- Assume $\alpha(e)$ is equal to $\alpha(f)$.

  Let $b_i, b_j$ be the bars such that $j \equiv \beta(e) + 1 \equiv \alpha(e) + 2 \equiv i + 3 \pmod 4$. Then, consider the $K_{3,3}$ induced by $\gamma$ and the bars $b_i, e, b_j$. As $b_i$ and $b_j$ are non critical, one of the branches adjacent to both of them is critical, for otherwise a non critical cycle would exist. Hence; it follows from Lemma 3.6 that $e$ and $f$ are interlaced.

$$\square$$

# 4    DFS Cotree-Critical Graphs

An interesting special case of cotree-critical graphs, the DFS cotree-critical graphs, arise when the tree may be obtained using a Depth-First Search, as it happens when computing a cotree-critical subgraph using a planarity testing algorithm. Then, the structure of the so obtained DFS cotree-critical graphs appears to be quite simple and efficient to exhibit a Kuratowski subdivision (leading to a linear time algorithm).

In this section, we first prove that any DFS cotree graph with sufficiently many vertices includes a Möbius ladder with 4 bars as a subgraph and hence are Möbius pseudo-ladders, according to Theorem 3.8. We then prove that these Möbius pseudo-ladders may be fully characterized.

**Lemma 4.1** *Let $G$ be a cotree-critical graph and let $K$ be a Kuratowski subdivision of $G$ isomorphic to $K_{3,3}$. Then, two $K$-vertices $a, b$ which are not $K$-adjacent cannot be adjacent to $K$-subvertices on a same $K$-branch.*

**Proof:** The three possible cases are shown Figure 9; in all cases, a cycle of non-critical edges exists. $\square$

**Lemma 4.2** *Let $G$ be a cotree-critical graph and let $K$ be a Kuratowski subdivision of $G$ isomorphic to $K_{3,3}$. If $G$ has two edges interlaced as shown Figure 10, then $G$ is not DFS cotree-critical.*

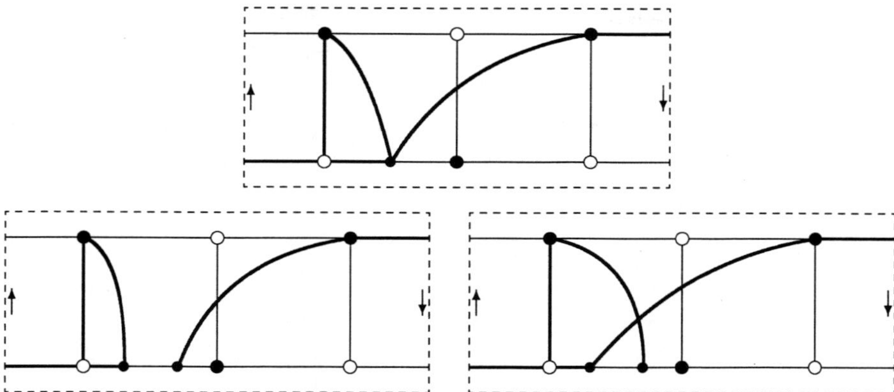

Figure 9: No two non-adjacent $K$-vertices may be adjacent to $K$-subvertices on the same $K$-branch (see Lemma 4.1)

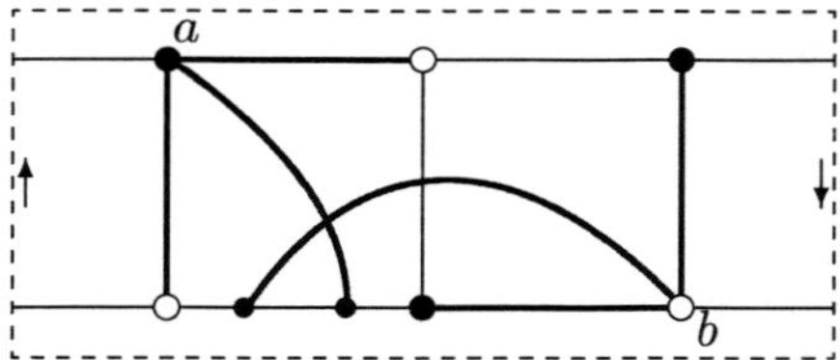

Figure 10: Case of two adjacent $K$-vertices adjacent to $K$-subvertices on the same $K$-branch by two interlaced edges (see Lemma 4.2)

**Proof:** Assume $G$ is cotree-critical. By case analysis, one easily checks that any edge of $G$ outside $E(K)$ is either incident to $a$ or $b$. Hence, all the vertices of $G$ incident to at most one non-critical edge is adjacent to a vertex incident with at least 3 non-critical edges ($a$ or $b$). According to Corollary 2.4, the set of non-critical edges is not a subset of a DFS tree of $G$, so $G$ is not DFS cotree-critical.

$\square$

**Lemma 4.3** *Let $G$ be a DFS cotree-critical graph and let $K$ be a $K_{3,3}$ subdivision in $G$. Then, no two edges in $E(G) \setminus E(K)$ may be incident to the same $K$-vertex.*

**Proof:** Assume $G$ has a subgraph formed by $K$ and two edges $e$ and $f$ incident to the same $K$-vertex $a$. According to Lemma 3.4 and Lemma 3.2, $K$ is a spanning subgraph of $G$ and only four cases may occur, depending on the position of the endpoints of $e$ and $f$ different from $a$, as none of these may belong to a $K$-branch including $a$:

de Fraysseix & Ossona de Mendez, *Critical Graphs*, JGAA, 7(4) 411–427 (2003)    424

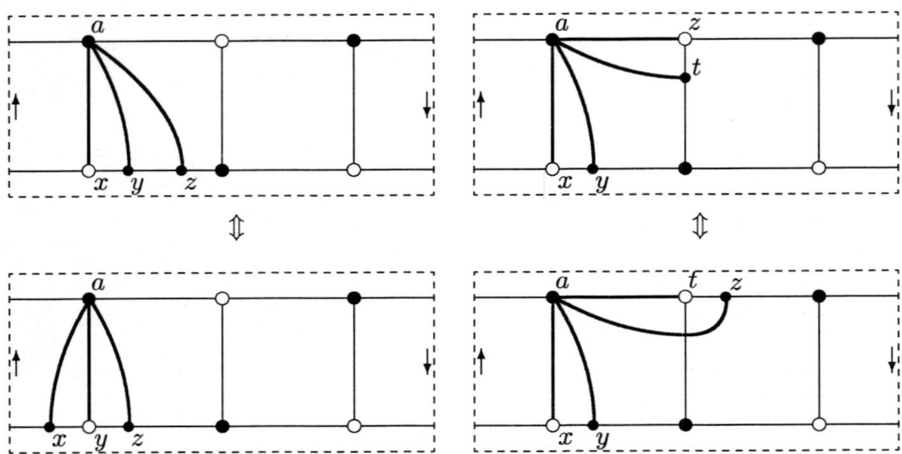

Figure 11: Cases of Lemma 4.3

- either they belong to the same $K$-branch,

- or they belong to two $K$-branches having in common a $K$-vertex which is not $K$-adjacent to $a$,

- or they belong to two $K$-branches having in common a $K$-vertex which is $K$-adjacent to $a$,

- or they belong to two disjoint $K$-branches.

By a suitable choice of the Kuratowski subdivision, the last two cases are easily reduced to the first two ones (see Fig 11).

- Consider the first case.

  Assume there exists a $K$-subvertex $v$ between $x$ and $y$. Then, $v$ is not adjacent to a $K$-vertex different from $a$, according to Lemma 4.1 and Lemma 4.2. If $v$ were adjacent to another $K$-subvertex $w$, the graph would include a Möbius ladder with 4 bars as a subgraph and, according to Theorem 3.8, would be a Möbius pseudo-ladder in which $\{a, y\}$ and $\{v, w\}$ would be non adjacent non interlaced chords, a contradiction. Thus, $v$ may not be adjacent to a vertex different from $a$ and we shall assume, without loss of generality, that $x$ and $y$ are adjacent. Similarly, we may also assume that $y$ and $z$ are adjacent.

  Therefore, if $G$ is DFS cotree-critical with tree $Y$, $y$ is a terminal of $Y$ and, according to Lemma 2.5, is the root of $Y$, which leads to a contradiction, according to Corollary 2.4.

de Fraysseix & Ossona de Mendez, *Critical Graphs*, JGAA, 7(4) 411–427 (2003)   425

- Consider the second case.

  As previously, we may assume that both $x, y$ and $z, t$ are adjacent. $G$ cannot be DFS cotree-critical, according to Lemma 2.6.

  $\square$

**Lemma 4.4** *If $G$ is DFS cotree-critical, includes a subdivision of $K_{3,3}$, and has at least 10 vertices, then $G$ includes a 4-bars Möbius ladder as a subgraph.*

**Proof:** Let $K$ be a $K_{3,3}$ subdivision in $G$.

Assume $K$ has two $K$-subvertices $u$ and $v$ adjacent in $G$. According to Lemma 3.2, $u$ and $v$ neither belong to a same $K$-branch, nor to adjacent $K$-branches. Let $[a, a']$ (resp. $[b, b']$) be the $K$-branch including $u$ (resp. $v$), where $a$ is not $K$-adjacent to $b$. Let $c$ (resp. $c'$) be the $K$-vertex $K$-adjacent to $a'$ and $b'$ (resp. $a$ and $b$). Then, the polygon $(c', a, u, a', c, b', v, b)$ and the chords $\{c, c'\}, \{a, b'\}, \{u, v\}$ and $\{a', b\}$ define a 4-bars Möbius ladder.

Thus, to prove the Lemma, it is sufficient to prove that if no two $K$-subvertices are adjacent in $G$, there exists another $K_{3,3}$ subdivision $K'$ in $G$ having two $K'$-subvertices adjacent in $G$.

As $G$ has at least 10 vertices, there exists at least 4 $K$-subvertices adjacent in $G$ to $K$-vertices. Let $S$ be the set of the pairs $(x, y)$ of $K$-vertices, such that there exists a $K$-subvertex $v$ adjacent to $x$ belonging to a $K$-branch having $y$ as one of its endpoints. Notice that $K + \{x, v\} - \{x, y\}$ is a subdivision of $K_{3,3}$ and thus that $[x, y]$ is non-critical for $G$.

Assume there exists two pairs $(x, y)$ and $(y, z)$ in $S$. Let $u$ be the vertex adjacent to $x$ in the $K$-branch incident to $y$ and let $v$ be the vertex adjacent to $y$ in the $K$-branch incident to $z$. Then, $K + \{x, u\} - \{x, y\}$ is a subdivision $K'$ of $K_{3,3}$ for which $\{v, y\}$ is an edge incident to two $K'$-subvertices. Hence, we are done in this case.

We prove by reductio ad absurdum that the other case (no two pairs $(x, y)$ and $(y, z)$ belong to $S$) may not occur: according to Lemma 4.3, no two edges in $E(G) \backslash E(K)$ may be incident to a same $K$-vertex. Thus, no two pairs $(x, y)$ and $(x, z)$ may belong to $S$. Moreover, assume two pairs $(x, y)$ and $(z, y)$ belong to $S$. Then, $[x, y]$ and $[z, y]$ are non critical for $G$ and thus not subdivided. Hence, $x$ and $z$ have to be adjacent to $K$-subvertices in the same $K$-branch incident to $y$, which contradicts Lemma 4.1. Thus, no two pairs $(x, y)$ and $(z, y)$ may belong to $S$. Then, the set $\{\{x, y\} : (x, y) \in S \text{ or } (y, x) \in S\}$ is a matching of $K_{3,3}$. As $S$ includes at least 4 pairs and as $K_{3,3}$ has no matching of size greater than 3, we are led to a contradiction. $\square$

**Theorem 4.5 (Fraysseix, Rosenstiehl [3])** *A DFS cotree-critical graph is either isomorphic to $K_5$ or includes a subdivision of $K_{3,3}$ but no subdivision of $K_5$.*

**Theorem 4.6** *Any DFS cotree-critical graph is a Möbius pseudo-ladder.*

de Fraysseix & Ossona de Mendez, *Critical Graphs*, JGAA, 7(4) 411–427 (2003)    426

**Proof:** If $G$ is isomorphic to $K_5$, the result holds. Otherwise $G$ includes a subdivision of $K_{3,3}$, according to Theorem 4.5. Then, the result is easily checked for graphs having up to 9 vertices, according to the restrictions given by Lemma 4.1 and Lemma 4.3 and, if $G$ as at least 10 vertices, the result is a consequence of Lemma 4.4 and Theorem 3.8. $\qquad\square$

**Theorem 4.7** *A simple graph $G$ is DFS cotree-critical if and only if it is a Möbius pseudo-ladder which non-critical edges belong to some Hamiltonian path.*

*Moreover, if $G$ is DFS cotree-critical according to a DFS tree $Y$ and $G$ has at least 9 vertices, then $Y$ is a path and $G$ is the union of a cycle of critical edges and pairwise adjacent or interlaced non critical chords.*

**Proof:** If all the non-critical graphs belong to some simple path, the set of the non-critical edges is acyclic and the graph is cotree critical. Furthermore, as we may choose the tree including the non-critical edges as the Hamiltonian path, the graph is DFS cotree-critical.

Conversely, assume $G$ is DFS cotree-critical. The existence of an Hamiltonian including all the non-critical edges is easily checked for graph having up to 9 vertices. Hence, assume $G$ has at least 10 vertices. According to Theorem 4.7, $G$ is a Möbius pseudo ladder. By a suitable choice of a Kuratowski subdivision of $K_{3,3}$, it follows from Lemma 4.3 that no vertex of $G$ may be adjacent to more than 2 non-critical edges. Let $Y$ be a DFS tree including all the non-critical edges. Assume $Y$ has a vertex $v$ of degree at least 3. Then, one of the cases shown Figure 12 occurs (as $v$ is incident to at most 2 non-critical edges) and hence $v$ is adjacent to a terminal $w$ of $T$. According to Lemma 2.5 and Corollary 2.4, we are led to a contradiction. $\qquad\square$

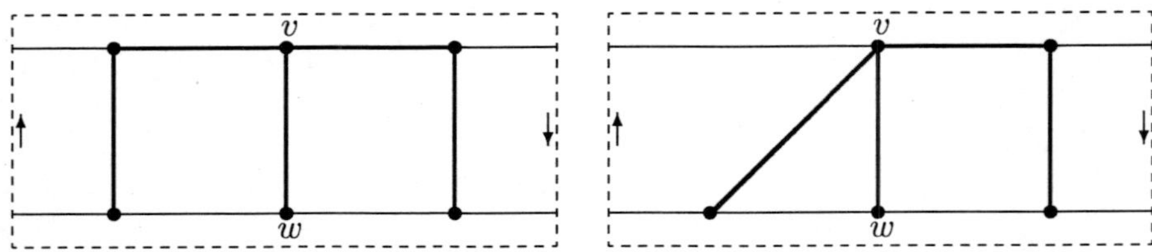


Figure 12: A vertex of degree at least 3 in the tree is adjacent to a terminal of the tree (see Theorem 4.7)

# References

[1] H. de Fraysseix and P. Ossona de Mendez. An algorithm to find a Kuratowski subdivision in DFS cotree critical graphs. In E. T. Baskoro, editor, *Proceedings of the Twelfth Australasian Workshop on Combinatorial Algorithms (AWOCA 2000)*, pages 98–105, Indonesia, 2001. Institut Teknologi Bandung.

[2] H. de Fraysseix and P. Ossona de Mendez. PIGALE: Public Implementation of a Graph Algorithm Library and Editor. Logiciel libre (license GPL), 2002. `http://pigale.sourceforge.net`.

[3] H. de Fraysseix and P. Rosenstiehl. A discriminatory theorem of Kuratowski subgraphs. In J.W. Kennedy M. Borowiecki and M.M. Sysło, editors, *Graph Theory, Łagów 1981*, volume 1018 of *Lecture Notes in Mathematics*, pages 214–222. Springer-Verlag, 1983. Conference dedicated to the memory of Kazimierz Kuratowski.

[4] H. de Fraysseix and P. Rosenstiehl. A characterization of planar graphs by Trémaux orders. *Combinatorica*, 5(2):127–135, 1985.

[5] J. Hopcroft and R. Tarjan. Efficient planarity testing. *J. Assoc. Comput. Math.*, 21:549–568, 1974.

[6] K. Kuratowski. Sur le problème des courbes gauches en topologie. *Fund. Math.*, 15:271–293, 1930.

[7] R. Thomas. Planarity in linear time (class notes).

[8] W. Tutte. *Graph Theory*, volume 21 of *Encyclopedia of Mathematics and its applications*. Addison-Wesley, 1984.

[9] S. Williamson. Depth-first search and Kuratowski subgraphs. *Journal of the ACM*, 31(4):681–693, October 1984.